AN HISTORICAL GEOGRAPHY OF EUROPE
450 B.C.–A.D. 1330

An historical geography of Europe

450 B.C.–A.D. 1330

NORMAN J. G. POUNDS

PROFESSOR OF GEOGRAPHY AND HISTORY, INDIANA UNIVERSITY

CAMBRIDGE UNIVERSITY PRESS

CAMBRIDGE

LONDON NEW YORK NEW ROCHELLE

MELBOURNE SYDNEY

Published by the Press Syndicate of the University of Cambridge
The Pitt Building, Trumpington Street, Cambridge CB2 1RP
32 East 57th Street, New York, NY 10022, USA
296 Beaconsfield Parade, Middle Park, Melbourne 3206, Australia

Library of Congress catalogue card number: 72-75299

ISBN 0 521 08563 2 hard covers
ISBN 0 521 29126 7 paperback

First published 1973
First paperback edition 1976
Reprinted 1980

First printed in Great Britain by Western Printing Services Ltd, Bristol
Reprinted in Great Britain at the
University Press, Cambridge

Contents

Maps and diagrams

Abbreviations

A.A.G. Bijd.	*Afdeling Agrarische Geschiedenis Bijdragen* (Wageningen, Netherlands)
Acad. Inscr. C.R.	*Comptes Rendus, Académie des Inscriptions et Belles Lettres* (Paris)
Acta Arch. (Dk)	*Acta Archaeologica* (Copenhagen)
Acta Pol. Hist.	*Acta Poloniae Historica* (Warsaw)
Actes Coll. Int. Dém. Hist.	*Actes du Colloque International de Démographie Historique*, ed. P. Harsin and E. Hélin (Liège, 1964)
Am. Hist. Rev.	*American Historical Review* (Lancaster, Pennsylvania)
Am. Jnl. Arch.	*American Journal of Archaeology* (New York)
Am. Jnl. Phil.	*American Journal of Philology* (John Hopkins University, Baltimore)
Anc. Pays Ass. Etats	*Anciens Pays et Assemblées d'Etats* (Namur)
Ann. A.A.G.	*Annals of the Association of American Geographers* (Washington)
Ann. Bourg.	*Annales de Bourgogne* (Dijon)
Ann. Brit. Sch. Ath.	*Annual of the British School at Athens* (London)
Ann. Dém. Hist.	*Annales de Démographie Historique* (Paris)
Ann. Géog.	*Annales de Géographie* (Paris)
Ann. d'Hist. Soc.	*Annales d'Histoire Sociale* (Paris)
Ann. Midi	*Annales du Midi* (Toulouse)
Ann. ESC	*Annales: Economies-Sociétiés-Civilisations* (Paris)
Ann. Hist Econ. Soc.	*Annales d'Histoire Economique et Sociale* (Paris)
Ann. Rept. Am. Hist. Assoc.	*Annual Reports of the American Historical Association* (Washington, D.C.)
Ant.	*Antiquity* (Cambridge)
Ant. Class.	*L'Antiquité Classique* (Louvain)
Arch.	*Archaeology* (Cambridge, Mass.)
Arch. Gasc.	*Archives Historiques de la Gascogne* (Paris)
Arch. Jnl.	*Archaeological Journal* (London)
Arch. Pol.	*Archaeologia Polonia* (Warsaw)
Arch. Stor. Sic.	*Archivio Storico Siciliano* (Palermo)
Bibl. Ec. Chartes	*Bibliothèque de l'Ecole des Chartes* (Paris)
Bibl. Ec. franc.	*Bibliothèque de l'Ecole française d'Athènes et de Rome* (Paris)
Bibl. Ec. Ht. Et.	*Bibliothèque de l'Ecole des Hautes Etudes* (Paris)
Bibl. Mérid.	*Bibliothèque Méridionale* (Toulouse)
Bonn. Jahrb.	*Jahrbücher des Vereins von Altertumsfreunden in Rheinlande (Bonner Jahrbücher*, Bonn)
Bull. Com. Roy. Hist.	*Bulletin de la Commission Royale d'Histoire* (Brussels)
Bull. Phil. Hist.	*Bulletin Philologique et Historique de la Comité des Travaux Historiques et Scientifiques* (Paris)
Byz.	*Byzantion* (Paris)
Cah. Ann.	*Cahiers des Annales* (Paris)
Cah. Hist.	*Cahiers d'Histoire* (Grenoble)
Cah. Hist. Mond.	*Cahiers d'Histoire Mondiale* (Paris)
Camb. Anc. Hist.	*Cambridge Ancient History* (London)
Camb. Econ. Hist.	*Cambridge Economic History of Europe* (London)
Camb. Med. Hist.	*Cambridge Medieval History* (London)
Cib. Rev.	*Ciba Review* (Basel)
Class. Jnl.	*Classical Journal* (Chicago)
Class. Philol.	*Classical Philology* (Chicago)
Class. Quart.	*The Classical Quarterly* (London)

Com. Roy. Hist.	*Commission Royale d'Histoire* (Brussels)
Corp. Inscr. Lat.	*Corpus Inscriptionum Latinarum* (Berlin)
Daed.	*Daedalus* (Boston, Mass.)
Dumb. Oaks Papers	*Dumbarton Oaks Papers* (Washington)
Ec. Prat. Ht. Et.	*Ecole Pratique des Hautes Etudes* (Paris)
Econ. Hist.	*Economic History* (London)
Econ. Hist. Rev.	*Economic History Review* (London)
Econ. Jnl.	*The Economic Journal* (London)
E.E.T.S.	*Early English Text Society* (London)
Encycl. Brit.	*Encyclopedia Britannica* (Chicago)
Eng. Hist. Rev.	*English Historical Review* (London)
E.P.N.S.	English Place-Name Society (Cambridge)
Et. Hist.	*Etudes Historiques*, Hungarian Academy of Sciences (Budapest)
Et. Rur.	*Etudes Rurales* (Paris)
Font. Jur. Rom. Antejust.	*Fontes Juris Romani Antejustiniani* (Florence)
Gall.	*Gallia* (Paris)
Gr. & R.	*Greece and Rome* (Oxford)
Germ.	*Germania* (Frankfurt am Main)
Geog.	*Geography* (London)
Geog. Jnl.	*The Geographical Journal* (London)
Hans. Geschbl.	*Hansische Geschichtsblätter* (Leipzig)
Herm.	*Hermes* (London)
Hesp.	*Hesperia* (Cambridge, Mass.)
Hist.	*History* (London)
Hist. (Prague)	*Historia* (Prague)
Inst. Brit. Geog.	*Institute of British Geographers: Transactions and Papers* (London)
Inst. Class. Stud. Bull.	*Institute of Classical Studies, Bulletin* (London)
Ist. Stor. Ital.	*Istituto Storico Italiano* (Rome)
Jahrb. Nat. Stat.	*Jahrbücher für Nationalokönomie und Statistik* (Jena, Stuttgart)
Jahrb. Ver. meckl. Gesch.	*Jahrbuch des Vereins für mecklenburgische Geschichte* (Schwerin)
Jnl. Econ. Bus. Hist.	*Journal of Economic and Business History* (Cambridge, Mass.)
Jnl. Hell. St.	*Journal of Hellenic Studies* (London)
Jnl. Rom. St.	*Journal of Roman Studies* (London)
Kw. Hist. Kult. Mat.	*Kwartalnik Historii Kultury Materialnej* (Warsaw)
Lat.	*Latomus* (Brussels)
Man	*Man* (London)
Mon. Germ. Hist.	*Monumenta Germaniae Historica* (Hannover)
Mon. Pol. Hist.	*Monumenta Polonial Historica* (Lwów and Kraków)
Moy. Age	*Le Moyen Age* (Paris)
Nouv. Et. Hist.	*Nouvelles Etudes Historiques* (Budapest)
Pal. Pilgr. Text Soc.	*Palestine Pilgrims Text Society* (London)
P. & P.	*Past and Present* (London)
Papers Brit. Sch. Rome	*Papers of the British School at Rome* (Rome)
Pat. Lat.	*Patrologia Latina*, ed. J. P. Migne (Paris)
Paul. Wiss.	*Paulys Realencyclopädie der Classischen Altertumswissenschaft* (Stuttgart)
Pet. Mitt.	*Petermanns Mitteilungen* (Gotha)
P.G.	*The Professional Geographer* (Washington)
Proc. Am. Phil. Soc.	*Proceedings of the American Philosophical Society* (Philadelphia)
Proc. Prehist. Soc.	*Proceedings of the Prehistoric Society* (Cambridge)
Quart. Jnl. Econ.	*The Quarterly Journal of Economics* (Cambridge, Mass.)
Rec. Hist. Gaules France	*Recueil des Historiens des Gaules et de la France* (Paris)
Rec. Soc. Jean Bodin	*Recueils de la Société Jean Bodin* (Brussels)

Rel. X Cong. Int.	*Relazione del X Cong. Internationale di Sci Stor* (Florence)
Rev. Belge Phil. Hist.	*Revue Belge de Philologie et d'Histoire* (Brussels)
Rev. Hist.	*Revue Historique* (Paris)
Rev. Hist. Droit Franç Etr.	*Revue Historique du Droit Français et Etranger* (Paris)
Rev. Nord	*Revue du Nord* (Paris)
Rev. Syn.	*Revue de Synthèse* (Paris)
Rhein. Mus. Philol.	*Rheinisches Museum für Philologie* (Bonn, Frankfurt am Main)
Rocz. Dziej. Społ. Gosp.	*Roczniki Dziejów Społecznych i Gospodarczych* (Poznań)
Rom.	*Romania* (Paris)
Script. Rer. Hung.	*Scriptores Rerum Hungaricarum* (Budapest)
Sett. Spol.	*Settimane di Studio del Centro Italiano di Studi sull'alto Medioevo* (Spoleto)
Spec.	*Speculum* (Cambridge, Mass.)
Stud. Mon.	*Studia Monastica* (Montserrat, Spain)
Trans. Am. Phil. Soc.	*Transactions of the American Philosophical Society* (Philadelphia)
Trans. Roy Hist. Soc.	*Transactions of the Royal Historical Society* (London)
Vj. S. Wg.	*Vierteljahrschrift für Sozial- und Wirtschaftsgeschichte* (Leipzig)
Zeitschr. Agrargesch.	*Zeitschrift für Agrargeschichte und Agrarsoziologie* (Frankfurt am Main)
Zeitschr. Soz.	*Zeitschrift für Sozialwissenschaft* (Berlin and Leipzig)

Preface

This book was conceived more than twenty years ago as a series of pictures of Europe at a sequence of periods in European history, each to be linked with the one which follows by an historical narrative. It grew in size until first, it had to be divided into two parts: classical-medieval, and modern, and then the linking narratives had to be omitted. As it stands now, this volume attempts to reconstruct the physical scene at five widely separated periods of time. The choice of periods has necessarily been influenced by the availability of sources. In general, however, sources are most abundant for those periods which have most continuously interested historians: the fifth century B.C., the age of the Flavian and Antonine emperors, the Carolingian period and the late thirteenth and early fourteenth centuries. In a sense these mark climaxes in the flow of European history. It is the scene at these periods, together with – since it is a long haul from about A.D. 800 to 1300 – the early twelfth century, that this book seeks to reconstruct.

The principal elements which made up the geographical scene at any of these periods of time were, in addition to the physical environment itself, the people, the forms and distribution of their settlements, their agriculture, their crafts and industries, and their trade. The purpose has been to present the distribution of each of these elements at each period, with a backward glance over the previous centuries to see how they had come to assume the pattern described.

A closely similar organisation has been adopted for each of the five main chapters, so that it should be possible to read in sequence the sections from each on, for example, agriculture or population, and thus have a survey of these for the classical period and Middle Ages.

In this representation of Europe two significant areas have been omitted: the British Isles and Russia. Any adequate discussion of the former would have made the book a very great deal longer than it already is. Russia has been omitted for somewhat similar reasons; its inclusion would almost have doubled the geographical area covered by the book. Reference is nevertheless made to conditions in both the British Isles and Russia wherever this seemed desirable for an understanding of Europe in the narrower sense. It might have seemed appropriate, in examining the classical world, to have included North Africa and the eastern Mediterranean which were linked so closely with it. This has not been done, partly from considerations of length, but partly also because it was thought desirable to maintain the same geographical coverage for each period.

It has proved impracticable to locate by means of maps every place and area mentioned in the text, and it is assumed that the reader will have at hand a good atlas. Particularly useful are the *Grosser Historischer Weltatlas*, parts 1 and 2 (Munich, 1954, 1970) or *Westermanns Atlas zur Weltgeschichte*, parts 1 and 2 (Brunswick, 1956), but any good desk atlas will serve.

The author wishes to thank those who read and commented upon parts of the manuscript, and who gave him on many occasions the benefit of their specialised knowledge. Foremost among them are his friends and colleagues of the Department of History of Indiana University, in particular Professors Glanville Downey, Arthur R. Hogue, Maureen F. Mazzaoui and Denis Sinor, and also Professor Lowell Bennion, a former colleague.

All illustrations have been drawn by the author, who alone is responsible for any imperfections and inaccuracies. Some have been published previously in journal articles, and he is grateful to the editors of the *Journal of Social History*, *Revue Belge de Philologie et d'Histoire* and the *Annals of the Association of American Geographers* for their kindness in allowing him to republish maps which appeared originally in these journals. All other maps are based on sources which have in general been published. Where relevant these have been acknowledged in the list of maps and diagrams.

August 1972

Department of History,
Indiana University,
Bloomington.

N. J. G. Pounds

1
Prolegomena

More than a century ago Macaulay reminded his readers that the England which he described in his *History* was a very different land from that with which they were familiar. The events of a bare hundred and fifty years had transformed the landscape. But the changes which had taken place between the 'Glorious Revolution' and the mid-nineteenth century were small compared with those which transformed the ancient world into the modern. If Macaulay had difficulty in reconstructing the landscape of the seventeenth century, the task facing the historian who would describe the European landscape when prehistory merged into history is incomparably greater. For the latter era the documentation is scanty; the evidence of geology and botany, slight, and that of archaeology all too often inconclusive.

At a time when the first histories were being written in the Aegean region man was still practising a Stone Age culture in parts of Scandinavia. In the former he could describe, though not always accurately or with understanding, the environment in which he lived; for the latter our evidence derives from the pollen trapped in the peatbogs, from varved clays in the riverine deposits, and from the scanty finds in human burials.

The reconstruction of the face of Europe at the beginning of the historical period must therefore call, not only on sources which are historical in the narrow sense, but also on the researches of the geologist, botanist, meteorologist and archaeologist. This is a task of immense difficulty, and makes demands on too many disciplines ever to be satisfying or complete. This chapter attempts in very general terms to describe the continent of Europe at the beginning of the historical period, and to indicate, though in only the broadest outline, what physical changes occurred during the classical and medieval periods.

'Everything has been changed,' wrote Macaulay, 'but the great features of nature and a few massive and durable works of human art.' The main lineaments of the continent were very much the same about the year 500 B.C. as we see them today. Then, as now, the continent consisted of a northern plain which extended eastwards from the shores of the Atlantic Ocean until it merged into the wider plains of Russia. Across it flowed rivers whose courses time and man have modified only in detail. Towards the north the plain merged gradually, almost imperceptibly beneath its forest cover, into the highlands of Scandinavia and of the north-western regions of the British Isles. To the south the plain was also bounded by hills which lay scattered in compact but irregular masses all the way from central France to southern Poland. Between these hills lay gaps and basins, formed by the movements of the earth's crust and the action of rivers. Through them ran the chief routes of human migration and the avenues of what little distant trade there was at this time. South of this belt of hills and plateaus, mainly forested over their steep slopes and flattened summits, lay the Alps, the most formidable barrier to human movement existing at this time.

These were then, as they remain today, the principal physiographic provinces of Europe. In each of them the terrain, together with soil and climate, imposed conditions and limitations on human settlement and development. It would be going too far to say that to each of these provinces there belonged at each period of human history a particular type of economy and way of life. There were, nevertheless, to be found in each of them certain features of cultural and economic development that could not possibly have occurred in the others.

Scandinavia. The most northerly province was in some respects the simplest. Its geological history is complex in the extreme, and its physical conditions were more restrictive than those of any other region. Norway, all of Sweden except the south, and Finland are made up of rocks of great age and hardness. They had been eroded through a long period of geological time, to an undulating plateau. Recently the whole region had been covered by the ice sheets of the Quaternary Ice Age, the soil removed, and the hills worn to the rounded shapes which we see today. When history began in southern Europe this region had not long emerged from the Ice Age. Coniferous forest had colonised its lower surfaces. Human beings had penetrated the region, but it is doubtful whether they had brought the practice of agriculture with them. This region of short, cool summers and long winters; of poor soil, leached by the rain and without the humus that comes from the autumn leaf-fall, called for a sophisticated, not a primitive agriculture. It was well into the historical period before the farmer made any impression on it. The hunter and the herder reigned supreme, and for long they had no rivals.

On the other hand, this region was well endowed with metalliferous minerals. Iron-ores occurred in abundance and of high quality, and some of Europe's more extensive reserves of copper were to be found in its ancient rocks. It was long, however, before man was able to exploit and use any of the metals, and they did not acquire any great significance until the later Middle Ages.

The European plain. The great plain of Europe begins at the foot of the Pyrenees and extends northwards and eastwards, alternately widening and narrowing, through France, the Low Countries, Germany, Denmark, Poland and into Russia. Its gentle relief, which nowhere exceeds 1,000 feet, is its dominant feature. In other respects it is less uniform. Part was covered by the ice-sheets, which left over much of it deposits of clay, sand and gravel. The glaciation modified the drainage pattern and gave rise to the sandy heaths of Brandenburg as well as the broad, marshy valleys of Poland and the damp pastureland of Denmark.

Beneath the glacial deposits, as well as in parts of the plain which were never covered by the ice-sheets, are rocks younger and softer than those of the northern plateaus. In general they are clays, alternating with limestone, chalk and sandstone. The *relative* hardness of the latter has led to the formation of low ridges, separated by clay vales. The 'scarp-and-vale' of lowland

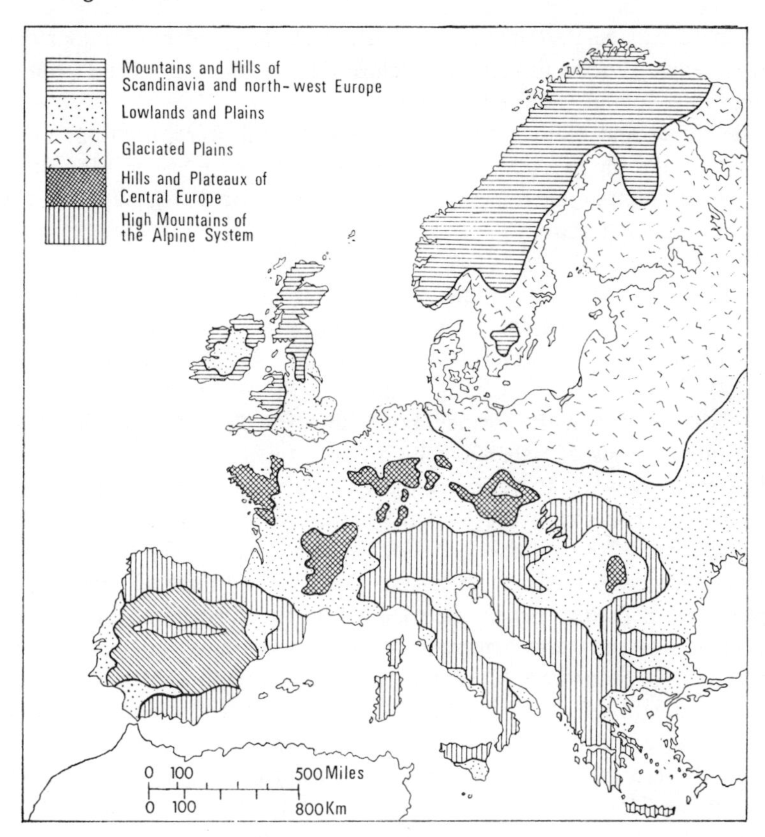

Fig.1.1 Landforms of Europe

England finds its parallel over much of northern France and parts of West Germany. The concentric pattern of ridge and lowland which surrounds Paris extends eastwards into the more rugged *côtes* of Lorraine. The same geological structures and patterns of relief are repeated in south-western Germany and again in southern Poland.

The plain is crossed by rivers which rise in the hills to the south and mostly discharge into the North and Baltic Seas. In Germany and Poland their courses had been greatly modified by the ice-sheets, and there is today a peculiar parallelism in their valleys as they extend from south-east to north-west.

Soil qualities varied greatly. Good, light soils on the limestone outcrops alternated with heavy clays in the intervening vales. Along the rivers was alluvium, fertile but requiring a heavy investment in drainage. Over the more north-easterly parts of the plain lay extensive areas of infertile sands, and along the more southerly parts of the plain and generally beyond the reach of the later extensions of the ice, lay deposits of fertile, wind-blown loess soil. The latter were amongst the best and most easily cleared and cultivated parts of the plain. They never bore a dense forest-cover, such as had colonised the claylands, and they were naturally dry and easily ploughed. Furthermore, movement was easy across them, and the earliest and most important route-ways of early Europe linked the patches of loess which stretched from northern France into Poland and the Ukraine.

The ridges and plateaus of limestone and chalk were often open and easily traversed, like the loess-belt, but their light soils were less fertile. They helped to channel human movement, but never, like the loess, supported a dense agricultural population.

Central European hills. Along the southern margin of the plain, the relief becomes stronger, and the ridges rise like waves about to break against the steep flanks of the central European hills. The latter are in many ways intermediate between the northern plain and the Scandinavian plateau. Their rocks are older than the former, but younger and softer than the latter. They constitute a series of 'islands' which rise from the plain, whose younger and softer rocks flow, as it were, around and between them. Each of the isolated hilly masses rises from the lowlands to a level or undulating surface, which often reaches 3,000 feet and occasionally more than 5,000. On the other hand, the plateaus are sometimes very low – less than a thousand feet – and no perceptible relief feature marks their transition from the plain. Brittany is barely distinguishable in terms of relief from the plain of northern France; the Ardennes rise imperceptibly from the lowlands to the north, west and south, and the so-called Holy Cross Mountains of southern Poland really amount to nothing more than a gently swelling upland amid the Polish plain. By contrast, many of these regions rise steeply and abruptly from the lowland. The steep front of the Harz Mountains looms above the rolling, loess-covered plain of Saxony; the steep slopes of the Vosges and Black Forest face one another across the Rhine valley, and some of the hilly masses which enclose the plain of Bohemia are formidably steep.

This system of hills and plateaus includes the Central Massif of France, Brittany, the Vosges and Black Forest and the series of plateaus, including the Eifel, Siegerland, Sauerland, Hunsrück and Taunus, which extends across Germany towards Bohemia. Bohemia is itself a complex unit in this system, the centre of which has become a low-lying and generally fertile plain.

The rocks of which this system of hills and plateaus is composed yield soils of little value. The higher slopes are covered with conifers, which produce no leaf mould, and everywhere the precipitation is higher than that of the plain, and soils are in consequence heavily leached. These regions have always tended to be pastoral and lumbering regions, and such they remain today. Minerals and mineral fuels are, however, associated with these old rocks. Iron-ore and non-ferrous metals, including copper, tin and lead, occur within them and along their flanks, and here and there, in small, basin-like depressions within them, are deposits of coal. Europe's richest and most extensive coal-field lies along the northern flanks of these hills, from northern France, through Belgium and the Ruhr region, into southern Poland. Though the coal was of little significance during the periods dealt with in this volume, the ferrous and non-ferrous metals were of great importance in the location and development of European manufacturing.

These hills and plateaus have always presented a barrier to movement; it was not that the terrain was difficult to traverse, but the weather was bad more often than not; there were few villages and resting places, and food was not readily available. Early routes followed the gaps between them. From the prehistoric amber traders to the medieval cloth merchants, the travellers kept to the lowlands whenever they could.

The Alps. The distinction between the hills of central Europe and the alpine system is difficult to draw in practice. The Alps are much younger, and were created by the intense folding of rocks which formed in a shallow sea to the south of the area of central European hills. They were folded *against* the latter. Some areas of the older rock were included within the alpine system as the folds of the Alps were piled on top of them. The Bihor mountains of Romania were enveloped by the alpine folds, and in many parts of the system, the older and harder rocks, similar to those of the Harz Mountains and Bohemia, constitute the core of the younger mountains. This is especially the case in the Balkan peninsula, where pre-alpine and alpine structures are superimposed in a pattern of the utmost complexity. In Spain a tabular mass of older rock was caught *between* two lines of folding, and survives today as the Meseta plateau, caught as in a vice by the twin jaws of the Sierra Nevada and the Pyrenees–Cantabrian Mountains.

The alpine system in Europe is part of a much more extensive system of mountain folding which extends almost around the globe. It is, however, broken into segments, around the ends of which movement is possible and even easy. A wide gap separates the Pyrenees from the Alps of France and Italy. The latter extend eastwards without break to the margin of the Pannonian plain. Here the valley of the Danube separates them from their continuation, the Carpathian Mountains of Slovakia and Romania. The Carpathians are continued again beyond the Danube, in the Balkan Mountains of Bulgaria. The Dinaric Mountains of Yugoslavia extend south-eastwards from the Austrian Alps, but, near the head of the Adriatic Sea, they are so low and narrow that they present no serious obstacle to movement.

For most of their great distance, however, the Alps, Pyrenees and Carpathians presented a significant barrier. The Pyrenees separated Spain from France, and forced much of the movement between the two areas to pass around their extremities. The Alps from Provence to the border of the Pannonian plain, a distance of six hundred miles, were unbroken. Their passes were numerous but high. Many lay at more than 6,000 feet above sea level; access to them was difficult, and their summits were clear of snow for only short periods during the year.

It is highly improbable that the higher passes across the Alps were known to and used by man during the earlier phases of his history; only gradually during classical and medieval times were they opened up. Few of them, apart from the Brenner, the Reschen Scheideck and the several low passes between north-eastern Italy and Carinthia, were in fact much used. This fact gave enhanced importance to the routeways – transverse valleys and low cols – by which the continuity of the range was broken at infrequent intervals. The Rhône valley, which in the north is linked with the European plain and in the south discharges into the Mediterranean Sea, was of primary importance both in prehistoric and in historic times. Scarcely less important was the broad, low saddle by which the Dinaric Mountains of Dalmatia were linked with the Alps of Austria. A bundle of routeways crossed it from the head of the Adriatic Sea on the one side to the Danubian plains on the other.

Enclosed by the Carpathian and Dinaric branches of the alpine system lay the Pannonian or Hungarian plain. It was crossed by the Danube, which entered it from the north-west, after cutting across the last, low spurs of the Austrian Alps, and left it in the south-east by the more formidable Iron Gate gorges. Three other routeways led into and out of the plain. To the north, the Moravian Gate formed a low-level route between the Danubian plains and those of northern Europe, and served in some degree to canalise the prehistoric amber trade between the Baltic Sea and the Mediterranean.

To the south-west the saddle already mentioned interposed between the Pannonian plain and that of northern Italy. Southwards lay the Morava–Vardar valleys, which together constituted the easiest and the most important link between central Europe and the Aegean heartland of European civilisation. Branching from this routeway at Niš, the classical *Nissa*, another route crossed the Balkan Mountains by a low pass and followed the Marica valley into Thrace and thence ran to the shores of the Bosporus.

This simple nexus of routes took the only easy ways through the most rugged and mountainous region of Europe. Inland from the Dalmatian coast the Dinaric Mountains constituted a wide belt of extraordinarily difficult terrain It was nowhere particularly high, but formed a plateau dissected by gorges and narrow valleys, none of which provided a route across the region. The opposite sides of the Dinaric Mountains tended to form different worlds, between which there was only a feeble commercial movement during most of classical and medieval times. It was no accident that the divide between the Western and Eastern Roman Empires fell across this region, nor that it separated the jurisdiction of the Orthodox from that of the Roman Church, and formed the westward limit of Turkish conquests in Europe.

The Dinaric system is extended southwards through western Macedonia and the classical Epirus into Greece. Its continuity is broken in northern Epirus – the modern Albania – and the Romans succeeded in building a road – the Via Egnatia – through the mountains from the shore of the Adriatic to that of the Aegean. This tenuous connection proved to be of immense importance in linking Italy and the west with the eastern Balkan and Aegian region.

The Carpathian Mountains curve through Romania, enclosing the fertile basin of Transylvania, and reach the Danube at the gorges of the Iron Gate. South of the river, the mountains again curve to the east as the Balkan Mountains or Stara Planina (the classical *Haemus*). Between the Balkan range, which is relatively narrow and easily crossed, and the Dinaric system lies the Rhodope. This is a compact massif of old, hard rock, resembling in most respects the hilly masses of central Europe. Its steep sides rise to a high, dissected plateau, above which rise the highest summits in south-eastern Europe. At the time when the Alpine system was formed, the Rhodope was a stable mass against which the newly folded ranges were pressed. Throughout historical times its severe climate and poor soil have been hostile to human settlement. It formed an effective northern limit to the Greek world of the Aegean.

The barrier nature of the Rhodope has emphasised the importance of the valleys, which border it: on the west, the Vardar, or *Axios*, and, on the north-east, the Maritza, or *Hebrus*. These have constituted the dominant, at times almost the only, routeways through the Balkans.

The Mediterranean basin. The alpine system comes close to the shores of the Mediterranean Sea, and its branches constitute the backbone of the Italian and Greek peninsulas. Greece, as Plato observed, was but a skeleton of mountain from which the soil had been washed away. The mountains of Italy were little better, but here the soil eroded from their flanks combined with ash from numerous volcanoes to form extensive and generally fertile lowlands. Spain consisted essentially of a high plateau, against which had been folded the ranges of the alpine system. Around the shores of all three peninsulas were coastal plains, some of them fertile; most of them mere embayments between the rugged headlands which reached out into the sea from the mountains of the interior.

The Mediterranean basin differed from the regions lying to the north primarily in its climate. During the summer months, the high pressure system of northern Africa extends northwards. Fair, hot weather, with only rare storms, is the rule over most of the basin. In winter, storms move eastwards from the Atlantic. Rain comes to all parts of the basin, and is heavy and prolonged in the mountains. But temperatures remain mild, unlike those of the rest of Europe, and plant growth continues. In fact, it is during the hot drought of summer that vegetation withers or becomes dormant. It is this reversal of the general European pattern that distinguishes the Mediterranean and has conditioned its cultures. Unless irrigation is possible, the winter and spring, not the summer, are the seasons of growth. Pasture withers away during the hot months; there is little for animals to eat, and much of the farm stock has traditionally moved to the mountains where, amid the melting snow, the grass remained green. Nor are the soils (see below, p. 20) suited to heavy cropping. They break down rapidly in the warm climate; the natural vegetation yields little leaf-mould, and erosion had already stripped the hillsides bare in many parts of the region by classical times. The plains were fertile, especially those which bordered the Italian peninsula. But nowhere were they large, and the most extensive of them needed to be drained during the winter and irrigated in summer, if their crops were to be commensurate with their natural fertility.

These major features of nature have not greatly changed during the span of historical time. Most of the changes in the physical environment that have been of significance to man have been on a narrow and local scale. But there have been more widespread modifications in certain aspects of the environment, notably in the climate, the rivers and coastlines, the level of the sea and the vegetation, and these require a more detailed consideration.

The Ice Age. The whole of human history has been lived under the shadow of the Ice Age. Ice-sheets spread outwards from Scandinavia and from the Alps. Much of the northern plain was covered by ice for prolonged periods, and glaciers moved down the alpine valleys and covered the nearby lowlands. The ice advanced and retreated at least four times in a period of about a million years. Each time the water, locked up in the form of ice, brought about a lowering of the sea level. Each time the immense weight of the ice depressed the land which it covered. With the retreat which followed each glacial advance, water was returned to the sea and its level rose. As the ice-sheets melted away, they left behind them large areas of boulder clay, while the escaping melt-waters scoured valleys and formed gravel terraces along their sides.

Between the advances of the ice and after each retreat from its previous maximum, dry and in all probability cold conditions prevailed over at least the northern half of Europe. Dust, whipped up from the drying surface of the boulder clay, was carried by the wind and redeposited as beds of loess, or limon. Each glacial advance tended to destroy the evidence for the previous glaciation, so that only the last of the series is really well documented by its geological deposits.

The sequence of changes, initiated when the last glaciation reached its maximum and the ice began to melt away around its margin, has continued until the present. Oscillations of sea level and of climate, accompanied by the slow and irregular re-colonisation of the wasted continent by plants and animals, have continued without pause throughout recent geological time. The climatic fluctuations of the historical period and the small changes in marine levels are but the most recent phase in the recovery from the last glacial advance.

The sea level. The historical period has, in general, been one in which sea levels have risen in relation to the land in much of Europe. To some extent this has been due to a continued diminution of the size of ice-caps and glaciers and the return of their melt-water to the oceans. To this extent the rise of sea level must have been worldwide, but the amount of melting that has taken place during the historical period has not been great, and the change in sea level that can be attributed to it is measurable but very small.

Much of the change, which locally at least has been very considerable, is to be attributed to a continued warping or distortion of the earth's crust. Scandinavia had been depressed beneath the weight of ice during the glacial period, and this in turn led to a very slow, deep-seated movement outwards of material beneath the earth's crust. This material forced the crust upwards beyond the limits of the ice-sheets. With the disappearance of the latter, a very slow return flow of this deep-seated material took place, so that, superimposed upon the rise in sea level which resulted from the melting of the ice, there was an upward movement of the crust in Scandinavia and an opposite movement elsewhere.

The return of water to the oceans at first brought about far more momentous changes than the slow and delayed recovery of the crust. The coasts of Scandinavia were inundated, and the Baltic was transformed into a sea of perhaps twice its present size. The size and shape of this proto-Baltic Sea fluctuated with the uneven rise of sea level and the delayed recovery of the land. During the historical period the latter process has been dominant. The land is today rising, at a maximum rate in northern Sweden of about 10 millimetres a year. This is sufficiently rapid – about a metre per century – to have brought about fundamental changes. The Baltic Sea is becoming shallower and less saline; its area is growing smaller and its surrounding coastal plain wider. Ports and harbours are getting shallower, and places where the Viking and even the Hanseatic ships tied up are now high and dry and quite inaccessible to shipping. The contraction of the water area must also have had some effect on the climate, probably making it less temperate.

Elsewhere in Europe the changes in sea level have been less regular and are less susceptible of measurement. Iceland and the north-western parts of the British Isles have long been subjected to an uplift of similar origin to that of Scandinavia, but of smaller magnitude. The rest of western and all of Mediterranean Europe is in general sinking very slowly in relation to the sea. But other factors make this movement less easily measurable and less regular than the opposite movement in Scandinavia. In the first place the mountains of the alpine system are somewhat unstable, with a slight tendency to rise in relation to the sea. Parts of this area are liable to earthquakes, and in southern Italy, Sicily and the Aegean basin volcanic activity has further disturbed the pattern of post-glacial change.

Subsidence is most marked along the flat coasts such as those which surround the North Sea. There periodic inundations, notably in the ninth century and again in the fifteenth, destroyed the lands which man had reclaimed from the coastal and riverine marshes, and such losses might have been even greater if defences against the sea had not been built. The changes were slow and continuous but they generally showed themselves when a combination of physical circumstances brought about exceptionally high tides. Then the sea defences were breached and large areas flooded. During the Middle Ages it took a long time to repair such damage. The floods of February 1953 in the Netherlands and along the shores of the Thames estuary are a reminder that the modification of the coastline by changes in the relative level of sea and land is a continuing process.

The sinking of the land in relation to the sea has been almost as rapid along the alluvial coastline of the Mediterranean, notably on the shore of Languedoc and at the head of the Adriatic. Disastrous flooding has here been less significant, partly because the Mediterranean Sea is tideless and less stormy than the North Atlantic, partly because the rapid silting and progradation of the coastline have given it a degree of natural protection from the sea.

Much of the southern coastline of Europe is high and rocky. Changes in the relationship of sea and land would be difficult to detect and could have had little human significance. Only where harbour works and other coastal buildings were constructed in classical times is it now possible to form some estimate of the amount and direction of change. The submarine exploration of Kenchreai, the port of Corinth lying on the Saronic Gulf, has demonstrated the submergence not only of harbour works, but also of buildings which must originally have been close to sea level.[1] Nevertheless, a rise in sea level of no more than 5 feet is suggested. It does not follow that Dhílos, 130 miles away, experienced a similar submergence, but it would be surprising if levels there had remained constant. Elsewhere in the Aegean region there is evidence of a considerable, though indeterminate, rise in sea level.

Pausanias records the disappearance of the town of Helica, on the Corinthian Gulf, as the result of an earthquake and the submergence of the site on which it had been built. Such drastic though local changes are known in other parts of the ancient world. The pillars of the Temple of Serapis at Pozzuoli, near Naples, appear to have undergone submergence to a depth of at least 20 feet, followed by re-emergence since their construction during the later Roman Empire. This was, however, an area of strong volcanic action, and local factors far outweigh the general in bringing about changes in the relationship of sea and land. It is probable that the Mediterranean region as a whole stood a few feet higher in relation to the sea at the beginning of the historical period, but that the general tendency for the land to sink was locally offset by contrary forces, tectonic or volcanic, which have brought about a small uplift of the land.

Around the shores of the North Sea, where the sinking of the land in relation to the sea can be measured most accurately, the changes in level probably had greater consequences for human history than elsewhere. The site of London has been shown to be subsiding, and remains have been found of Romano-British settlements along the Thames estuary up to 13 feet below present high water. The destruction of coastal settlements in the East Riding of Yorkshire and on the coast of East Anglia, as well as the drowning of Old Winchelsea in 1287, is part of this general picture of subsidence.

The coastline of north-western Europe evolved as the resultant of two dominant forces. The first was the tendency, already discussed, for the coast to sink in relation to the sea. The other was an opposite tendency for the islands to grow larger and the sea to become more shallow, as silt brought down by the rivers was laid over the sea floor. The latter process was, through much of the historical period, aided by man, who, by warping, embanking and draining, has aided the process of natural reclamation.

The coast of Flanders illustrates these opposing trends. It was in process of continuous modification through natural causes until recent years. In early classical times it was made up of a discontinuous line of islands backed by shallow lagoons. The classical period appears to have been one of gradual sedimentation, but late in the Roman period there was a rise in the water level; some low-lying settlements were abandoned; others were raised on artificial mounds, or *terpen*. This marine transgression, known as Dunkerque II, spread marine sediments over part of the plain; widened and deepened the estuaries, and probably produced a great number of small offshore islands all the way from Calais to the Danish peninsula. It also opened the *Lacus Flevo*, ancestor of the Zuider Zee, to the ocean.

A period of sedimentation followed until, some time in the tenth or eleventh century, a renewed advance of the sea – Dunkerque III – inundated low-lying land and opened up the historically important estuaries of the Zwin, Yser and Aa. The cycle of sedimentation and transgression was repeated in the later

Middle Ages, with the extensive flooding of the early fifteenth century. By this date, however, the coastline and lowlands of Flanders were no longer shaped only by natural forces. The landscape was moulded increasingly by human agency. Dykes were built to restrain the sea, and the protected lowlands were kept dry by pumping, so that the risk of extensive flooding was greatly reduced, though never entirely removed.

The coastline of Europe was thus, about 500 B.C., radically different in many places from that of today. Nowhere, in fact, has it gone entirely unchanged. Cliffed coasts have everywhere receded under the impact of the waves; flat coasts have in general prograded with the addition of silt, sand and shingle. On straight coasts the longshore travel of beach material has closed or constricted estuaries and created spits and bars. With dominant onshore winds sand-dunes formed, and tended, with the increased storminess of the later Middle Ages, to move inland and to encroach on farmland.

The west-facing coast of northern France, from Boulogne to Le Tréport, has prograded in some areas as much as four miles. At the same time, the sinking of the coastline has tended to 'drown' the valleys of the rivers – Canche, Authie and Somme – and to form long arms of the sea. These, in turn, have filled with riverine deposits, and gradually turned into damp alluvial plains. The continued sinking of the land, however, has left this region exposed to the danger of inundation by a combination of storm and tide. It was not until modern times that was made safe by embanking and draining.

Another area of marked progradation is the flat coast of southern Brittany and La Vendée. While the estuaries of the larger rivers, the Loire and Gironde, have been kept clear by the scour of the tide, the mouths of others have been silted. During the Middle Ages salt-pans were constructed here for the production of 'Bay' salt, and in modern times they have been drained and reclaimed.

The straight coast of Aquitaine is, like that of Flanders, a product of the smoothing action of the waves as they break obliquely against the beach. The drying sun and the onshore winds have piled up sand-dunes which have throughout the historical period moved inland, overwhelming cropland and producing the sterile *Landes* which today extend over 4,500 square miles.

In southern Spain the slow submergence of the coast produced a deep embayment, while the waves built up a sandy spit, partially cutting it off from the sea. This, the *Lacus Ligustinus* of the ancients, slowly silted to form the blighted and malarial *las Marismas* which have only recently been drained and reclaimed. At countless points around the coast of southern Europe this process has been repeated: submergence, sedimentation, reclamation, so that nowhere do the flat coasts even remotely resemble those at the dawn of history.

Changes have been greatest near the mouths of the largest rivers: Ebro, Rhône, Po and Vardar. The channels of the Rhône delta, wrote R. J. Russell, 'are all modern',[2] and have throughout the historical period been in process of slow but continual change. None of its present channels can with certainty be ascribed to the prehistoric period, and the town and port of Aigues Mortes, established by Louis IX in the thirteenth century, was effectively silted within a few generations.

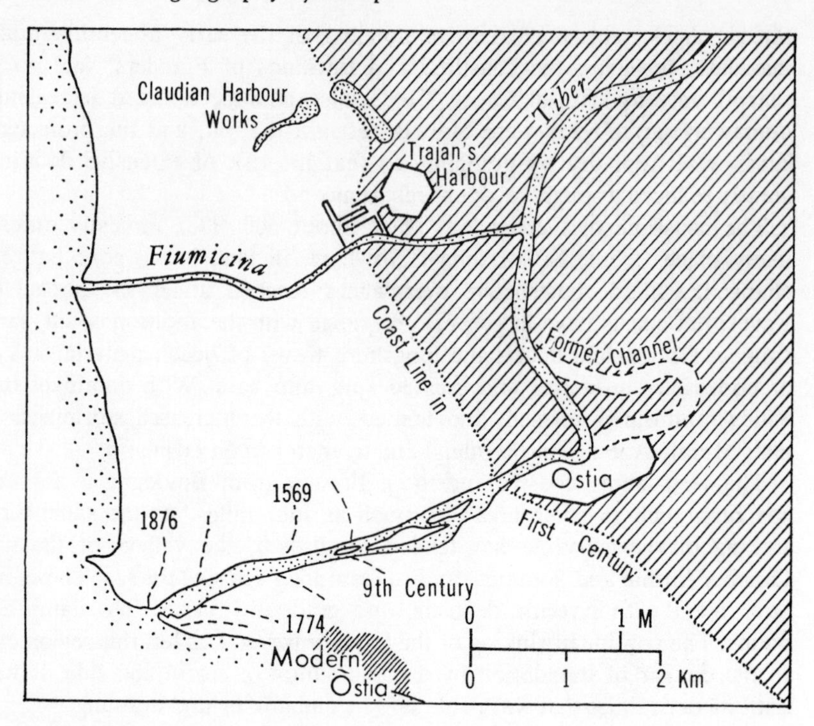

Fig.1.2 Coastal change at the mouth of the Tiber

The advance of the Po delta into the Adriatic Sea has been more rapid during the historical period than that of the Rhône. Furthermore, it can be more easily measured because it has left behind it a trail of abandoned ports. Spina, a port which served to link the Etruscans with Greek traders of the sixth century B.C., lay at that time on or very close to the coast. The Pseudo-Scylax of the late fourth century B.C. infers that it lay 3.5 kilometres from the shore, while Strabo placed it 15 kilometres inland. By this time it was derelict, and its role had passed to Adria, which in its turn had been silted by the later years of the Roman Empire. For a time Ravenna served as port for the Po valley until the rise of Venice in the early Middle Ages. An advance, locally at least, of up to 20 miles can in this way be documented since the beginning of historical times.

Changes of smaller amplitude have taken place at the mouths of other rivers and along all the flat coasts where sedimentation occurred. These have been reconstructed for the Tiber (fig. 1.2) and doubtless could be demonstrated for many other Mediterranean rivers.

The river systems. Europe's river system had assumed its present shape long before its history began, and over the southern half of the continent this system antedated even the Quaternary Ice Age. Yet in detail the rivers have changed greatly since about 500 B.C. Physical changes in rivers and their flow are bound up, in the first place, with variations in sea level and climate, and secondly with the process of erosion by which rivers transport sediment from their upper reaches to their lower and to the sea.

Uplift of the land, relative to the sea, leads the river to incise its bed more deeply. If the movement has been of sufficient magnitude, as it was in Scandinavia, a narrow valley was formed within the limits of the older valley, which remained as a terrace. In this way the waterway is narrowed and navigation restricted, as has happened at Stockholm. Elsewhere in Europe the rise of sea level led to the drowning of valleys and the creation of conditions favourable to navigation. At the dawn of history much of coastal Europe was marked by long, branching inlets of the sea which might have carried seaborne shipping far into the interior of the continent.

Within the historical period, however, this tendency along the submerging coasts, to form wider and deeper waterways, was more than offset by the silting of river channels. There were few rivers whose navigable tracts were not shortened in this way. The increasing size of ships has also helped to bring the limit of navigation closer to the sea, though in a few instances modern engineering has been able to re-open waterways that had been slowly silting for many centuries.

Above their maritime tracts most rivers meandered across a floodplain of varying width. The bordering lowlands were fertile, but damp and often flooded in winter. In much of the European plain, these riverine floodplains were formerly marginal, in the sense that a little more moisture could exclude them as an area of settlement and agriculture, while a slight diminution in rainfall could bring them within the limits of profitable farming. The alternate use and abandonment of such lands was clearly dependent upon the local water level, which was thus a function of precipitation and evaporation.

Climatic change. The years about 500 B.C. were marked, at least in the northern two-thirds of Europe, by a changing climate, and it was probably not until more than a thousand years later that settlement on a large scale began to penetrate the damp woodland which had spread over the floors of the larger river valleys. This movement towards the valleys may have been due to increasing population; it was assisted by the appearance of drier conditions, which made agriculture possible on the alluvial valley plains. The onset of moister conditions in the fourteenth and fifteenth centuries led to the abandonment of some of these lands, which became in all probability too heavy to plough and perhaps too wet for regular use as meadow. Such changes can be documented in Brandenburg and the Low Countries.[3] It cannot be claimed that the movement away from the valley floors was on a very large scale; but in conditions of declining population, these were marginal lands that could be sacrificed without loss.

Changes in both sea level and the human occupation of the valley soils were, in large measure, responses to fluctuations of climate. From the time, about 25,000 B.C., when the last (Würm) advance of the ice reached its maximum, there has been a gradual amelioration of the climate, interrupted by periodic returns to cooler and wetter conditions. These oscillations have continued, with diminishing amplitude, into modern times, and are with us still.

The last centuries before the beginning of historic times had been relatively dry over most of Europe. During the sixth century B.C. the climate became perceptibly more moist; summers were probably cooler and more cloudy, and winters wetter. This change initiated the climatic period commonly known as the Sub-Atlantic.

The change, as measured in millimetres of average rainfall, average monthly temperatures, or average duration of sunshine, was undoubtedly so small that it would make little, if any, difference to human settlement on the coastal plains of Greece or Italy, or on the loess soils (see below, p. 19) of central Europe. The human body would in all probability not have been sensitive to such changes, but they were great enough to influence the competition between plant species. There were marked changes in those marginal areas where certain plants and animals had only just been able to subsist and reproduce. There one species yielded to another better able to endure the changed physical conditions.

The moister conditions led to the spread of peat – the so-called 'blanket-bog' – over many upland surfaces. Prehistoric trackways and even agricultural tools were buried and preserved under a renewed growth. The moorlands of south-western England, which had been colonised during the Bronze Age or even earlier, were abandoned at the beginning of the Iron Age. The reason for this was unquestionably the deterioration of climate, especially as these moorlands were not again settled until the warmer and drier period of the early Middle Ages. At the same time the frontier of human settlement retreated in Scandinavia, as shown by archaeological finds, from about 68° N during the Bronze Age to 60° N in the ensuing Iron Age. In much of Europe north of the Alps the lighter and drier soils began to be settled and cultivated. The moisture content of the heavier soils increased, and poorly drained and low-lying land was often given over to grazing or entirely abandoned.

The period of damper and cooler climate in northern Europe appears to have lasted through classical times, but in the Mediterranean region the climate was probably not perceptibly different from that of today. Arguments regarding climatic change during the classical period are mainly based in the subjective judgments of ancient writers. These have little scientific value, and are indicative rather of the extremes experienced than of changes in long-term averages. The suggestion that the decay of Roman civilisation was in any way due to climatic change does not, at least on the basis of literary evidence, bear serious examination.

On the other hand, there is good evidence for an improvement of climate during the early Middle Ages, though this may also have been a phenomenon of northern and north-western Europe rather than of southern. There is evidence for the abandonment of some settlements, because presumably of their growing shortage of water, and highland areas, abandoned a thousand years earlier, were again colonised. Folklore and legend, particularly Celtic, are strongly suggestive of drier conditions, if not actually of a water shortage. At the same time, in the seas of north-western Europe, diminished storminess facilitated the Atlantic voyages of the Norsemen. The snowline in the Alps crept a little higher, and the glaciers began to retreat. The alpine passes could be used for longer periods in summer, and the traffic across them increased. New alpine routes were opened up, and it was no accident that, towards the end of this drier and warmer period, the St Gotthard Pass came into use.

The chronology of the early medieval warm period is far from clear. It was punctuated by short periods of relatively damp climate, and conditions became cooler and moister perhaps in the twelfth century, though several writers date the change from about the year 1000 or 1050. If the change to wetter and cooler conditions did begin as early as this it must have been extremely slow, since it was not reflected in any contraction in the pattern of settlement or the extent of cultivated land for another two centuries.

Whatever may have happened during the previous century, there can be no question of a sharp change of climate over much of Europe early in the fourteenth century. Literary sources indicate a succession of very wet years. The years 1315–17 were a period of unusually severe weather, poor harvests and famine, and similar conditions recurred not infrequently during the succeeding centuries. There is reason to suppose that the wet and stormy conditions of the fourteenth century led some time in the fifteenth to conditions that were colder but somewhat drier.

This climatic fluctuation initiated the so-called 'Little Ice Age', which lasted into the nineteenth century. There is reason to suppose that this change was not merely local, but resulted from diminished insolation, and was experienced in the Mediterranean region as well as in northern Europe. The evidence for these changes is unmistakable. Settlements at higher altitudes and on wetter soil were abandoned; the tree-line crept lower in the mountains, and glaciers began again to advance. The North Atlantic sea routes became hazardous; communication with Iceland was endangered, and that with Greenland terminated. Increased storminess combined with the small rise in sea level to breach the dykes of the Low Countries, and to inundate coastal settlements elsewhere in northern Europe (see above, p. 11).

The climatic deterioration with which the Middle Ages ended represented only a small change in average temperatures. The overall change from the twelfth century to the fifteenth may have been no more than a drop in average monthly temperatures of 1.2° to 1.4° C. An average change of this amplitude, accompanied as it was by some increase in rainfall, was nevertheless highly significant under marginal conditions, though probably of no great importance in such areas as central Italy. The change, however, took the form of an increased number of bad years, when a long, cold and wet winter was followed by a late spring and a short, cool summer. A single bad year caused hardship and suffering, but it was not disastrous if it was followed by a year of above average or even of normal weather conditions. In the later Middle Ages it became increasingly likely that one bad year would be followed by another, with no opportunity to replenish the granaries and rebuild farm stocks. The *cerche des feux* of fifteenth-century Burgundy noted of one village after another that it had been 'tempestez du temps en ceste présente année, tant qu'ils n'ont recueillis ne blefs ne vins'.[4] Under these conditions famine led to epidemics, and these in turn to high mortality and the abandonment of settlements. Thus did the Middle Ages end in some parts of Europe, not in the

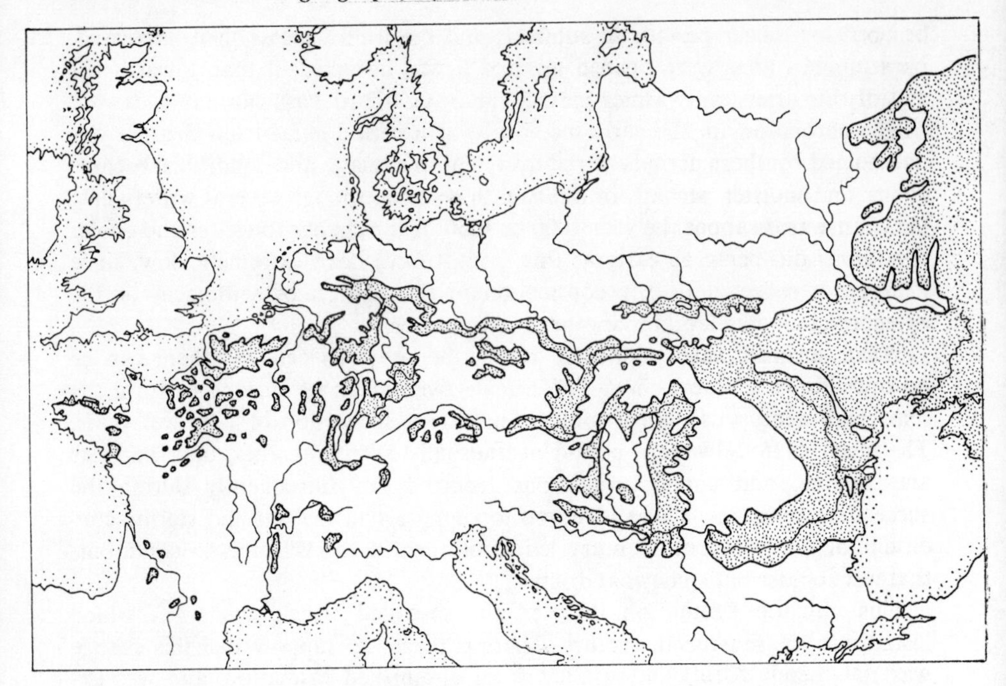

Fig.1.3 The loess soils of Europe

bright dawn of the Renaissance, but under cloudy skies, in poverty, misery and starvation.

Vegetation of Europe. As the climate improved after the retreat of the ice-sheet, coniferous forest, followed by broad-leaved, spread northwards. The boundary between the two fluctuated with post-glacial climatic oscillations, moving north with the warmer and south with the cooler periods. A similar movement took place on the mountain sides, where the deciduous woodland gave place upwards to coniferous, and the latter to grass, tundra and rock. Here too the boundary between these broad vegetation types moved up and down with climatic changes.

The generally accepted view is that forest had spread across most of the continent by the time of the so-called climatic optimum, or Atlantic phase (*c.* 5000–*c.* 2500 B.C.). At this time coniferous forest was restricted to sub-arctic regions and higher altitudes. Elsewhere broad-leaved trees predominated. The following period, sometimes called the Sub-Boreal, was cooler and drier. It was formerly held that the reduction in rainfall was sufficient to replace the forest by a 'steppe-heath' (*Steppenheide*) vegetation on the drier and lighter soils – primarily the loess. This more open vegetation, it was argued, permitted the Neolithic farmers to spread through the region. This view, associated primarily with the German geographer, Robert Gradmann, is no longer tenable.[5] Primitive man was quite capable of destroying the forest trees that stood in the way of his agriculture, and he chose to cultivate the loess soils, not on account of their presumed open character, but because they were suited to cultivation with simple tools.

By the beginning of the fifth century B.C. the continuity of Europe's forest cover was broken in many places. Much of the loess-belt had been cleared by man; in southern Europe inroads had been made in the forests to obtain timber for housing, shipbuilding and charcoal-burning. In northern Europe, the forest was disappearing in marginal areas under the spreading blanket-bog. Otto Schlüter prepared, on the basis of place-names, soil-types and literary evidence, a map showing the presumed expansion of agricultural land and the corresponding contraction of the forest in central Europe.[6] He has been criticised for representing too large an area as cleared of forest in early historical times.[7] Nevertheless, large areas, especially on the higher and more easily cultivated soils, were under at least intermittent cultivation. Elsewhere the forest was not always dense, and was unquestionably interrupted by breaks resulting from the activities of man or the forces of nature. The idea of the primeval forest, untouched by man and shaped only by climatic changes and conditions of soil, is a myth. Everywhere, except in high latitudes and at considerable altitudes, the vegetation of Europe bore the stamp of human activity even before the beginning of written history.

It has been shown by pollen analysis that beechwoods predominated in western and southern Germany, and an association consisting largely of oak, elm and lime in north-eastern Germany and in Poland. These broad-leaved trees extended far into Scandinavia before they were entirely replaced by conifers. In central Europe conifers formed a belt high up on the flanks of the mountain ranges, and formed stands on some of the plateau lands of central Germany. Firs were to be found in the Black Forest, and conifers had not been entirely supplanted on the sandy soils of the northern plain. In western Europe the forests were predominantly of broad-leaved trees, their species varying somewhat with the soil. Beechwoods tended to prevail on the lighter and more calcareous soils; oakwoods on the heavier, while everywhere the willow, aspen and poplar grew along the rivers and over the damp valley floors.

This forest cover had established itself during the favourable climate of late prehistoric times. It reacted only very slowly to the climatic change of about 500 B.C., except on the climatic margin of broad-leaved species and in those areas where rising ground-water led to the replacement of forest with bog. The forest changed its character rather through the destruction of trees and the failure of the former species to regenerate. In this way the conifers gained ground at the expense of the deciduous trees in the later Middle Ages and probably also during the recession of classical times. It is to be presumed that the major advance of coniferous trees – in so far as this has not been due to modern plantation – was in the later Middle Ages, when so much of the forest was cut over for charcoal-burning and lumber-supply.

The Mediterranean region presents a special case. Its climate was, in the main, one of hot and generally dry summers and of cool and moderately rainy winters. A climatic deterioration such as took place in northern Europe would not have had any drastic consequences here. Much of the Mediterranean region had been colonised by a dry forest, consisting mainly of evergreen oak, and of certain varieties of deciduous oak, poplar and pine. This was, as Pierre George has emphasised, a highly unstable plant association.[8] Once disturbed, it did not readily regenerate. Rather, low-growing xerophytic plants took over and reduced the high forest to those typically Mediterranean vegetation forms, the *maquis* and the *garrigue*.

Classical Greece experienced a shortage of timber, and even charcoal was scarce. The destruction of the forests had begun much earlier. A layer of wood ash with datable finds shows that deforestation had already made great progress in Achaia by the seventh century B.C. By the fifth century, the urbanised regions of Greece were largely treeless. Elsewhere in the Mediterranean world, deforestation was later in beginning and probably slower. It was not until the Middle Ages that the Dalmatian forests were depleted to supply shipbuilding timber. In moister regions of the Apennines, the Píndhos and the mountains of Spain the forests were able to regenerate and in general to survive their cutting and burning at the hand of man.

In most hilly environments soil erosion is a natural accompaniment of the destruction of the forest. However quick the revival of tree growth, some loss of topsoil is to be expected, and in a Mediterranean environment the loss could be drastic. No doubt the well-known passage in the *Critias* exaggerates:

> The whole of the land lies like a promontory jutting out from the rest of the continent far into the sea . . . the soil which has kept breaking away from the high lands during these ages and these disasters, forms no pile of sediment worth mentioning, as in other regions, but keeps sliding away ceaselessly and disappearing in the deep. And . . . what now remains compared with what then existed is like the skeleton of a sick man, all the fat and soft earth having wasted away, and only the bare framework of the land being left.

But by the beginning of classical times the hillsides were in some degree scoured by erosion, and the rivers choked with the silt.

Sedimentation is the complement of erosion. Rapid soil-erosion leads to silting of river channels and enhances the dangers of flooding in winter and spring. In this way marshes were formed along parts of the Aegean coast, especially in Macedonia and Thrace. On the west coast of Italy marshes due primarily to excessive sediment brought down from the Apennines, grew during classical and medieval times, until they formed a fetid and malarial swamp which extended intermittently from Tuscany to Campania.

The soils of Europe. The variety and distribution of soils have been a major factor in economic development within the historical period, as they had been in guiding human settlement in the prehistoric. It was not the intrinsic qualities of the soils that mattered, but rather the ease with which they could be ploughed. It mattered little that the land might have to be abandoned after a few years, provided simple wooden ploughs, or even hoes, were effective in tilling it. Thus the earliest soils to be cultivated north of the Alps were not the clays, intrinsically rich in soil nutrients, nor the valley alluvium, which defied

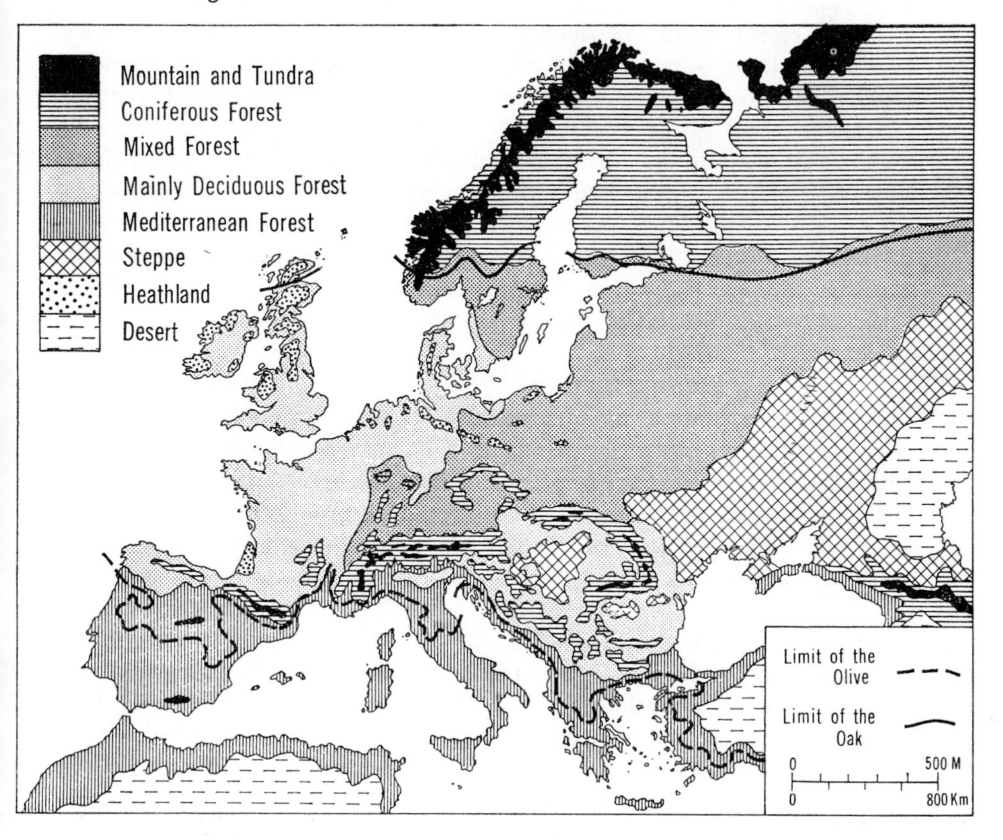

Fig.1.4 Vegetation belts of Europe during historical times

man's early attempts to drain it, but the loam soils, the gravel terraces, the loess, and even the sandy heath-lands. It was not until the dry conditions of the early Middle Ages replaced the damper climate of classical times that significant advances were made on the forested clays.

The soil map of Europe is highly complex, and in itself provides no certain guide to conditions two and a half millennia ago. Man has modified soils as much as he has the vegetation. One can, however, draw a broad distinction between the relatively light and dry, though by no means fertile, soils which had developed on areas of limestone and chalk, and the heavier clays which tended to form narrow belts of land between the outcrops of limestone. These contrasted soils prevailed over much of western Europe, and human settlement, which began on the one, spread in the course of its history to the other.

Locally the rocks were covered by wind-blown loess. This combined the chemical qualities of a good soil with a lightness of texture which made it easy to work. In many parts of Europe the earliest agriculture was practised on the loess, and by the beginning of the historical period its extent – an irregular belt lying along the margin of the European plain and extending to the lowlands along the valley of the Danube – was roughly known.

Across northern Europe, from Flanders to Poland, the surface rocks were mainly of recent origin, and derived from the advance and retreat of the ice. As the ice-sheets retreated they laid down an extensive, uneven spread of boulder clay. Its soil quality varied greatly. In some places it could approach a loam; in others, it was so intractably heavy that little attempt has ever been made to cultivate it. Its surface was uneven; lakes formed in the hollows, some of them to fill slowly with peat-bog; it was poorly drained and commonly supported a complex net of small rivers. The boulder-clay plains of Germany, Poland and southern Scandinavia were able to support a not inconsiderable population, but the cultivation of much of this area had to await the coming of a more sophisticated agricultural technology than was available to prehistoric man.

Associated with the final retreat of the ice and with the vast floods which discharged from it were the large deposits of sand and gravel which extend in elongated strips across the plain, from the Campine of Belgium and the Netherlands (see p. 281) to the heaths of Germany and the sands of Poland. Today they are amongst the least fertile soils of Europe. In the first millennium B.C. their very lightness gave them some value, though they were quickly abandoned as soon as it became technically possible to cultivate the heavier soils.

The soils of Mediterranean Europe, lastly, were generally light and friable. The long summer drought, which made cultivation precarious, had at least served to prevent the leaching away of soil nutrients. With irrigation, the lowland soils of Greece and Italy could be made very productive, and even without, gave every other year a moderate return on the wheat or barley sown. But in much of this region tree crops – the olive, vine and fig – deep-rooting and more tolerant of drought, were the most profitable form of land use. It was with good reason that the Roman agronomists urged upon their wealthy clients the advantages of orchard and vineyard cultivation in physical conditions like those of the Italian peninsula.

Conditions of agriculture. Throughout the eighteen centuries spanned by this volume, agriculture remained the dominant occupation in almost all of Europe. Except in a few areas of very restricted extent, over 80 per cent of the population lived by farming; in much of Europe and for most of the time this proportion was considerably above 90 per cent. Europe never at any time formed an entirely closed economic system. In classical times grain was imported from Africa to supply its largest cities, and there was an export of wine and olive oil. But throughout the Middle Ages foodstuffs did not enter into the trade between Europe and neighbouring continents in any significant amounts, and even the volume of internal trade in foodstuffs became important only in the later Middle Ages. The knightly classes, the churchmen and the burgesses were supported by the peasantry, and their numbers were limited by the ability of the peasant to feed them.

Very little progress was made in the science of agriculture during this period. Crop yields were probably no greater and the labour of the peasant no less hard when it ended than it had been when it began. There were technical inventions; a heavier plough was invented, fitted with coulter and mould-board, and drawn by a large team, and this, in turn, allowed the clay soils to be cultivated. The alternation of grain-crop and fallow, which had been practised in classical Europe, was continued with little change. Elsewhere, a three-field system was evolved, but this only meant that the desirable, autumn-sown bread-crops could be grown one year in three, instead of every other year. It brought about a greater *total* grain production, but meant that many people lived on the less palatable spring-sown oats and barley (see below, p. 371).

There was, over this span of eighteen centuries, a very considerable expansion of the area under crops, but it is doubtful whether this could anywhere have done more than keep pace with the growth in population. In some areas of Europe it did not even do this; the average size of farm units became smaller and the level of welfare of those who cultivated them, lower. Nor was there any significant change in the pattern of cropping. Almost all crops cultivated at the end of the Middle Ages had been known to the Romans. There had, however, been changes in the ratios in which they were cultivated. The adoption of a three-field system over large areas of northern Europe had led to the growing of far more oats and barley than previously. The cultivation of poorer soils in northern Europe resulted in the predominance of rye over other bread-crops, and oats, which had in late prehistoric times been little more than a weed, became a crop of very great importance for man and beast. The increased cultivation of coarse grains contributed to the appearance of some form of mixed farming in many areas.

But most other crops, the legumes and roots and fodder crops, such as lucerne, were as familiar to Columella as to Walter of Henley.

Animal husbandry was probably a good deal less important in the later Middle Ages than it had been in classical times. It remained important – even dominant – in most mountainous areas, and transhumant sheep were increasing in number on the Spanish Meseta during the later Middle Ages. But over much of the continent a growing population necessitated a more intensive use of the land. Fewer animals meant less manure, though the medieval peasant had learned to make the most of what his small stock of animals provided.

There is no conclusive evidence of soil deterioration on a wide or general scale during classical and medieval times. Columella repeated the old complaint that 'the soil was worn out and exhausted by over-production of earlier days', but went on to deny its validity.[9] On the contrary the surviving manorial account rolls and other evidences of crop yields indicate rather an improvement towards the end of the Middle Ages. But if the returns from the land did not decline significantly, this is probably because the farmer, both classical and medieval, never demanded a great deal of it.

Seed was sown thinly and the return of nature was low. The yield-ratio is *one* measure of the efficiency of farming. A low ratio means that the farming community barely satisfies its own needs; a high ratio suggests the possibility of a surplus, to be sold into the market and used to feed non-agricultural communities. In fact, however, a low yield-ratio from a thinly peopled area, where each farm family cultivates an extensive area, *may* yield a sizeable surplus. This was almost certainly the case in eastern Europe, which in the later Middle Ages provided large quantities of rye and wheat for export to western Europe. On the other hand, a high yield-ratio from a densely populated rural area, with small farm units, might give only a very small surplus. The size of the agricultural surplus, on which, during classical and medieval times, all forms of human progress were dependent, was thus a function of land quality, the density of rural population and the expected yield-ratios.

The anonymous *Hosebonderie*[10] of the thirteenth century considered that barley should yield eightfold; rye, sevenfold; wheat, fivefold, and oats, four. Thierry of Hireçon expected larger yields from his land in northern France.[11] But yields of this magnitude were unusual, and could have been obtained regularly only from soils of high quality, with a considerable use of manure. Walter of Henley thought that a threefold yield was an absolute minimum, but there is good evidence that far more lands approximated this lower limit than achieved the high production of the anonymous *Hosebonderie*.

Table 1.1 *Ratio of harvest to amount sown*

	Before 1250	1250–1500	1500–1750	After 1750
England	3.7	4.7	7.0	10.6
France	3.0	4.3	6.3	
Germany		4.2	6.4	

Slicher van Bath has tabulated and analysed an immense volume of data on crop yields during the Middle Ages and modern times. There was often only a twofold return on oats, and a fourfold return on wheat and barley. But yields not only varied with the soil, but showed a tendency to rise during the later Middle Ages. Yields seem to have been on average consistently higher in England than in France or Germany. Comparable figures for the classical period are not available.

If we confine our attention to the bread-crops, it seems that, on average, a quarter of every crop had to be set aside for the next year's sowing, but that in bad years this proportion might rise to a half. There are even instances of the total loss of a crop through the chances of weather or the depredations of war.

The seed was sown more thinly than is usual today, but this varied greatly with the soil and, doubtless, the availability of seed. 'A *iugerum* of rich land,' wrote Columella,[12] 'usually requires four *modii* of wheat (*triticum*); land of medium quality, five; it calls for nine *modii* of spelt if the soil is fertile, and ten if it is ordinary.' Others, he added, sow 8 *modii* of both wheat and spelt on both good and medium quality land. Translated into modern terms this might range for wheat from 1.5 to 4 bushels to the acre.

The classical writers reported a very great range in both the intensity of sowing and the yields. Some of the ratios which they quoted are unreasonably high, and probably do not reflect any careful measurements. Columella's more sober statement: 'we can scarcely remember a time when, over the greatest part of Italy, corn returned a fourfold yield',[13] was probably nearer the truth, and suggests yields that were not much different from those of the earlier Middle Ages.

A great deal is known about crop yields, but very little about yields per unit area. Manorial book-keeping kept good account of the former, but rarely concerned itself with the latter. Yet it was the area yield which became the more important as population grew during the Middle Ages and farm size diminished. The intensity of sowing clearly varied greatly, but, without heavy manuring, heavy seeding was likely to be wasteful. Columella indicates that there was a considerable range in the intensity with which cereal crops were sown. No doubt some of the better soils were made to crop heavily, but over much of Europe, both classical and medieval, seed was sown relatively thinly. Large areas were cropped with but a small return.

The average English wheat production is said to have been from five to six quintals to the hectare (8 to 10 bushels per acre). In Dombes (Franche Comté) at the end of the Middle Ages, it was eight quintals to the hectare, while rye yielded nine.[14]

These figures express an order of magnitude that may have been expected on medium to good soil towards the end of the Middle Ages. A 15-acre arable holding, cultivated on a three-field system, might have yielded:

	Winter grain	*Spring grain*
(a)	45 bushels of wheat	35 bushels of oats
or (b)	50 bushels of rye	35 bushels of oats

Farms would have been larger a century or two earlier, but the land would have been cultivated less intensively and the yield-ratios might have been lower.

The bread-grains, supplemented by legumes and very small quantities of meat and dairy produce in northern Europe and oil in southern, provided the diet for the rural population. How well a farm family could subsist on the produce of about ten cultivated acres depended, of course, on the size of the household, the quality of the soil and the amount of payments in money and in kind that it was called upon to make. To none of these variables is it possible to give a precise value. A family of five would not find in 45 bushels of wheat or 50 of rye an abundant diet. This amount would, furthermore, be subject to considerable fluctuations with the level of the harvest, and would have been reduced by the obligation to pay tithe and make payments to the lord. Even under normal conditions the family would have been obliged to dip into the coarse grains which in turn would have restricted the amount available for the stock, thus hampering ploughing and cutting down on the manure supply.

The rural surplus which supported urban and all other non-agricultural society derived in the main from the peasant's services and payments to his lord. These provided the marketable produce which fed the cities and maintained the administrative, ecclesiastical and feudal superstructures. How large was it? No data exists for any form of national accounting during the span of classical and medieval history. It is sometimes said that the non-rural population during the later Middle Ages amounted to 10 per cent of the whole. The proportion was, locally at least, a great deal higher (see p. 358), but it was also lower in some areas. During the earlier Middle Ages non-rural population made up an appreciably smaller proportion of the total, but during the earlier Roman Empire it may have been as large as in the thirteenth and fourteenth centuries.

It is very probable that the non-agricultural sector lived better than the agricultural; its diet was larger and more varied; it wore better clothes and its houses were in general larger and better built. If it amounted overall to 10 per cent of the population, it nevertheless consumed perhaps from 15 to 25 per cent of the total production.[15] This was as large a superstructure as the broad backs of the peasantry could support. Until there was a revolution in agriculture, and farm productivity could be greatly increased, a narrow limit was set to the size of the urban and non-agricultural segment of the population.

How big, in fact, was the population? To this question, David Hume replied two centuries ago that 'the facts delivered by ancient authors are either so uncertain or so imperfect as to afford us nothing positive in this matter'.[16] In the following chapters an attempt is made to assess the limits between which the population of Europe lay at five sequential periods. For each of these periods one can estimate with only a relatively small margin of error the population of small and restricted areas. In extrapolating these data for a large part of Europe one assumes immense risks, and the probability of error rises sharply.

Seen through the perspective of historical time, the population increased with the improvements in agricultural methods. Within any short period of time there were fluctuations. To some extent these demonstrate the tendency of population to expand until the Malthusian checks operate to reduce it. In some degree they reflect the secular changes in the physical environment. The data at present appear to be too slight for analysis to be possible.

Human society and the environment in which it lives have always been in process of change. To some extent these changes result from the influence of each on the other, as successive generations, each with small modifications in their social order and technical equipment, tried to create a place for themselves. The method adopted in the following chapters has been, not to trace through time the several elements in the man–land relationship, but to attempt to synthesise for five discrete periods the whole complex of conditions that has resulted.

The choice of periods for this study has been conditioned on the one hand by the availability of sources; on the other by the intrinsic interest and importance of the periods themselves. It so happens that periods which in retrospect appear to have been creative and significant in the formation of western society were well documented. Perhaps they appear significant because they have left relatively abundant sources. In any case it can hardly be questioned that the periods of Pericles and of the Antonine emperors were, in the traditional judgment of historians, periods when it was good to be alive. Whether this judgment is a good one may be questioned (see below, p. 439). The age of Charlemagne again is more brightly illuminated than the ages which both preceded and followed, even if it shines more dimly than other periods that have been chosen.

Another traditional judgment regards the thirteenth century as the crown and glory of the Middle Ages, the 'greatest of centuries'. It would have been logical to have surveyed Europe at about the time when the Crusades ended and the Hohenstaufen suffered their final eclipse. Instead, a date has been chosen about three-quarters of a century later. This was done because economic growth, according to the uncertain indicators available, continued into the fourteenth century, before the late medieval economy took its downward trend. The final chapter thus presents a picture of man and the land towards the end of the first third of the fourteenth century.

From the early ninth to the early fourteenth century was a period of growth in the European economy. About 1100 this expansion was at the flood-tide, and changes in the relationship of man and the land were taking place at a faster pace in all probability than at any other period before the nineteenth century. Chapter 5 attempts, then, to cut a section through this time of rapid change.

This method raises a final question: how long is the 'period' which is considered in each chapter? Though it is possible to conjecture what Europe was like in 1100 or 1300, it is impossible to be certain or precise, because the body of strictly contemporary archival material is relatively small. One must draw upon sources which originated both before and after the date in question, but they must cluster around it. If their point of origin lies too far from the central date, their evidence will prove to be anachronistic. Clearly the time-span from which sources may be taken is greater in times when change was slow than in those of rapid development. This is a matter of historical judgment. If at times the presentation appears static, this may be because there was little or no change in some particular aspect of the relationship between man and his environment. At others, the relationship may have been in flux, and if it was, an attempt is made to indicate the direction at that time.

In the basic issues discussed in this book the rate of change was slow in classical and medieval Europe: In some respects – the crop yields discussed above, for example – there was in all probability no change at all. It is thus possible to space these views of Europe fairly widely – five over a span of some eighteen centuries. After the sixteenth century the speed of change is greater, or at least appears so in the light of a vastly more abundant documentation. What in earlier times was accomplished in a few centuries could now be done in a few decades. The method of taking a few discrete periods breaks down, and a 'vertical' study of population, land, resources, development and so on must be substituted. The second volume will begin with the Europe of the early sixteenth century, and will pursue the same themes until the middle of the nineteenth.

2

Europe in the mid-fifth century B.C.

In the middle years of the fifth century B.C. the long war between the Greeks and the Persians was drawing to a close. Hostility was deepening between Athens and Sparta, the foremost protagonists on the Greek side, but the Delian League, over which Athens had presided since its foundation in 478 B.C., was still intact, though a couple of its members had rebelled against the growing control which Athens was exercising over it. In 454 its treasury had been transferred from the island of Dhílos to the city of Athens, an event which marked the final transformation of a voluntary league of Greek states into an instrument of Athenian imperialism. Athens had at the same time become, by the standards of classical Greece, an exceedingly wealthy city, and its riches were matched by its pride and its self-assurance. Work was beginning on the cluster of buildings on the Acropolis – the Parthenon was begun in 447 – which was to make Athens the most beautiful city of antiquity. The last of the plays of Aeschylus had just been presented; Herodotus, Sophocles and Euripides were in middle life, and Thucydides and Aristophanes were children. The quarter of a century which began in 450 saw the climax of Athenian civilisation.

At this time Rome was a small town spread over the low hills that rose above the Tiber marshes. The Etruscan kings had been driven from south of the river, but their threat to the city was far more real than that presented at this time by the Persians to Athens. The Etruscans were the dominant people of central Italy, and their authority still extended from the Greek colonies in Campania to the foothills of the Alps.

Beyond the Alps, the Hallstatt, or Early Iron Age culture had been replaced by the more refined and sophisticated La Tène civilisation. Iron-working had been developed to a high level of artistry, and the Celts, armed with iron weapons and shields, were spreading outwards from south Germany and the basin of the Middle Danube. They were the bearers not only of more ferocious methods of warfare, but also of a more developed agriculture. They spread their 'Celtic' fields over the areas of lighter and more easily cleared soil in north-western Europe and southern Britain. But in northern Britain the earlier nomadic and pastoral way of life lingered on, and its weapons and tools continued to be made of bronze, if not also of polished stone. Nor had the use of iron yet penetrated more than the southern borders of Scandinavia. The Bronze Age still covered much of Sweden, and the few inhabitants of northernmost Europe still hunted with the weapons of the late Stone Age.

Never, perhaps, in the span of European history was the cultural gradient between the south and the north, between Attica and Finland, as steep as it was in the middle years of the fifth century B.C. At no time was the contrast in cultural levels as great. It is this cultural and social scene which this chapter seeks to reconstruct.

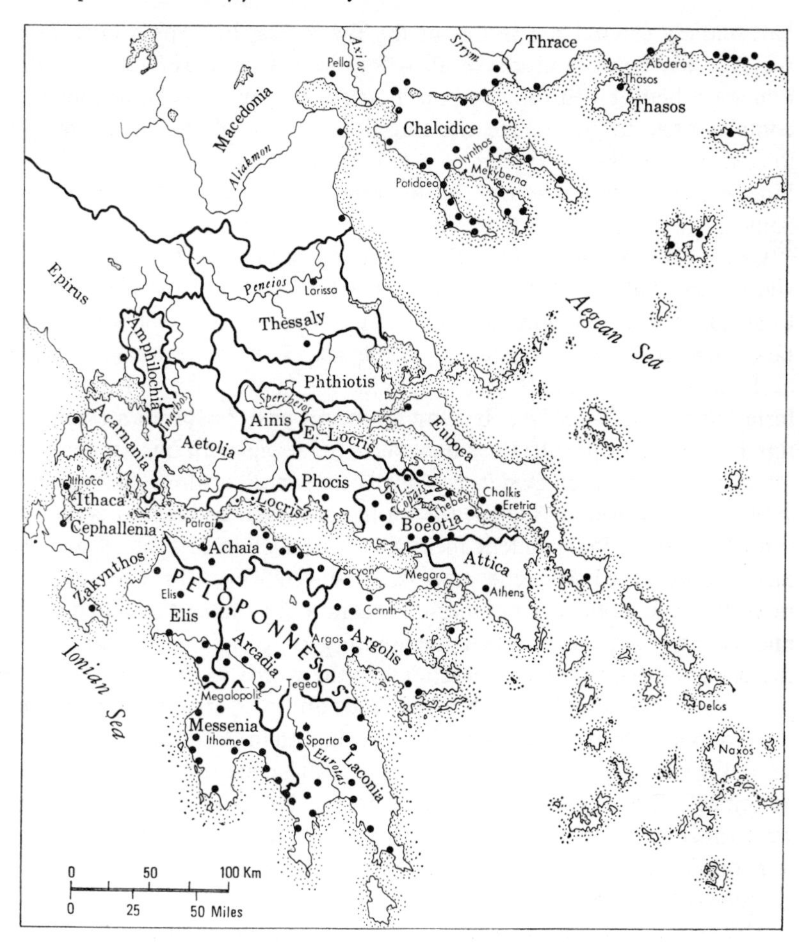

Fig.2.1 Political map of Greece

POLITICAL GEOGRAPHY

It is difficult to speak in any formal sense of a political geography of Europe, that is, of a geography of politically organised areas and peoples. Over much of the continent political organisation was of a very rudimentary order, and over some parts it did not exist at all. In such regions we can speak only of tribal areas, at best ill-defined and always in process of change. There is some evidence that the tribes in central and western Europe recognised boundaries separating them from one another, and that they observed them as long as it was convenient to do so. But only in the Greek and Italian peninsulas were there states, in the sense of areas of land, however indefinite, organised politically with governments in effective control. There, however, they were too numerous to count and too fluid for convenient and easy description and analysis.

Greece. Hellenic civilisation in the fifth century enclosed the Aegean Sea, and had outliers around the shores of the Black Sea, in Cyprus and Egypt, along the Adriatic coast, and above all in southern Italy and Sicily. It spanned the boundary between Europe and Asia, and it becomes necessary, for the sake of completeness, to go outside the traditional limits of the continent. Byzantion looked across the Bosporus to Chalcedon; the Crimean cities lay opposite Sinope, and eastward across the Aegean from Athens, Megara and Sparta lay Samos, Miletus and Halicarnassos.

Greek civilisation did not extend over all of Greece. It developed around its shores, and had but little impact on the mountains of the interior. It is usual to speak of it as thalassic, though many of the city-states of which it was composed had little or no contact with the sea. The Greek world was predominantly made up of small political units which we have come to call by the term 'city states' – πόλεις. It is by no means easy to form an estimate of their number owing in part to their fluctuating relations with one another, but above all to the absence of evidence. Athens had, not too long before our period, subjected Eleusis and Aegina; the former she absorbed, whereas the latter continued to be an independent member of the Delian League. Sparta had similarly incorporated Messene, and had reduced Tegea to a status of political satellite. In Boeotia the number of separate city-states fluctuated between ten and twenty. The *polis* was a community of people who acted together in the management of their local affairs. In the words of the pseudo-Aristotle, the *polis* was 'an assemblage of houses, lands and property sufficient to enable the inhabitants to lead a civilised life'.[1] No less than 343 such *poleis* belonged at some time or other to the Delian League, and if we add the indefinite number of 'cities' which existed at this time in the Peloponnesos and in central and western Greece, we probably have a total of a thousand units to which contemporaries might have accorded the title of πόλις. The total may have been considerably more.

The Greek city-state is generally thought of as a small area of lowland, surrounded and protected by hills. If this view is too simple, it nevertheless embodies an element of truth.[2] The *polis* centred in an area of lowland, which provided its agricultural land, and its frontier generally ran through the incultivable hills, where the shepherds and goatherds of rival communities were accustomed to meet and dispute their mutual boundary. But there were many *poleis*, like Athens itself, whose territory was divided by mountain ranges, and there were plains, which seemed, like Boeotia, destined for political unity, that remained divided throughout the Hellenic period. The *polis* was not a geographical concept; it was a community united in the management of its territory.

To the Greeks themselves the city-state was the product of the synoecism of numerous village settlements. 'When several villages,' wrote Aristotle, 'are united in a single community, large enough to be nearly or quite self-sufficing, the state comes into existence.'[3] Local traditions often ascribed the process to a single person: Theseus in Athens, Cadmus in Thebes; and we have the clear record of the foundation of Megalopolis in Arcadia by Epaminondas in 370 B.C. to replace a number of villages.[4] But in most instances the process was slower and less dramatic. The city became the focus of an ill-defined region,

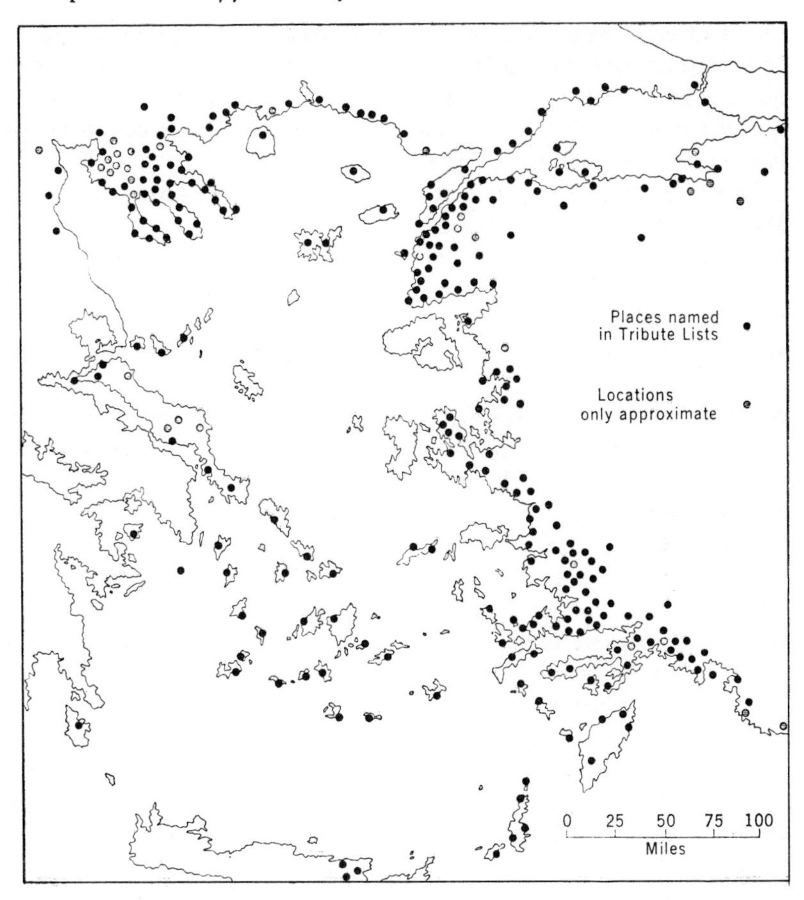

Fig.2.2 The Delian League, total membership

within which it was usually the only settlement of size and importance. It was the seat of the government of the *polis*. It was usually walled, and though only a comparatively small proportion of the citizen body may have lived there, its beautification was a matter of pride to all citizens. The city that was poorly built and lacking in public buildings was an object of scorn. Thus Pausanias wrote of Panopaeus, 'a city of Phocis, if city it can be called that has no government offices, no gymnasium, no theatre, no market place, no water conducted to a fountain . . .'[5] Sparta was an exception in this, as it was also in many other respects. It was 'not built continuously, and has no splendid temples or other edifices; it rather resembles a group of villages like the ancient towns of Hellas'. But Thucydides warned his readers not to be deceived; 'the greatness of cities should be estimated by their real power and not by appearances'.

Synoecism did not always lead to a large-scale transfer of population from the villages to the newly created city. 'The ancient Athenians,' according to Thucydides, 'enjoyed a country life in self-governing communities; and although they are now united in a single city, they and their descendants . . . from old habit generally resided with their households in the country where they had been born.'[6] Athens was the largest city in Greece, but it probably contained no more than half the population of its city-state. The basis of every *polis*, even of Athens itself, was its agriculture, and Greeks continued after synoecism to live close to the land. The central-places of the smaller *poleis* can have been little more than large and pretentious villages.

The areas of the *poleis* varied greatly. Sparta was the largest, with an area of about 3,000 square miles. Athens at its most extensive had a little more than 1,000 square miles. But apart from these 'there was hardly a *polis* of the homeland or the islands with more than 400 square miles. Some, and some of importance, had less than 40.'[7] Corinth covered about 340 square miles; Sicyon, 140, and Aegina, only 33. It is impossible to estimate with any pretension to accuracy the area of the *poleis* of Boeotia, or of Phocis, where the 22 city-states could each have had on average no more than 30 square miles.

Most *poleis* were separated from one another by frontiers rather than by boundaries. The shepherds of Athens and Plataea met in summer on the upper slopes of Cithaeron, above the limits of cultivation, and their quarrels led not infrequently to disputes between their respective states, and ultimately to the demarcation of a precise boundary. The territorial limits of Attica gradually acquired precision, and were in part defended by forts and a wall. Elsewhere rivers often became the agreed boundaries between city-states, but over much of Greece the process of delimitation had not been completed, if, indeed, it had really even begun.

The distribution of *poleis* was in some measure determined by geography. They clustered round the Aegean Sea and were spread over its islands, whereever agricultural land was adequate. Yet the conditions of physical geography do not wholly account for this distribution. Thessaly, despite its agricultural wealth, had not yet developed a *polis*-based society. *Poleis* were few in the mountainous interior of the Greek peninsula and along its western shores. They were not numerous along the northern shore of the Aegean Sea, in Macedonia and Thrace, but in the coastal region of Asia Minor, between the Anatolian plateau and the sea, they were to be found in scores (fig. 2.2).

It is not easy to define Greece – Ελλας – as that term was understood in the fifth century. That it was the area within which Greeks lived goes without saying, but the Greek people themselves spoke several dialects, and near the borders of the Greek world these passed into the distinct languages of 'barbarian' peoples like the Illyrians and Thracians. Epirus formed no part of Greece, and in the fifth century Greek commerce and culture had made little impression upon its tribes. It is doubtful whether the tribes of Aetolia and Acarnania should be considered Greek,[8] and even Homer's 'wooded Zacynthus' (Zante), 'rugged Ithaca', and 'sandy Pylos' lay on the margin of the Greek world in the age of Pericles. These regions were all rugged and forested; coastal lowlands were few and restricted in area, and communications within them were difficult. The Aegean cities had established trading posts along the coast of western Greece, but in the interior the primitive pattern of village settlements still prevailed. Euripides described the Aetolian Tydeies, though a Greek, as 'half-barbarian'. Despite the differences in dialect, in religious cult and in political organisation, the Greeks thought of themselves as united against all other peoples and superior to most of them. It does not follow that they had any clear concept of a Greek land, Hellas. A political unity to match their cultural distinctiveness never once entered their minds. They were first and foremost members of their *polis*-community, and their mental horizon rarely extended far beyond its boundaries.

The Peloponnesos was politically more advanced than the north-west, but even here the *polis*-organisation had not yet come to the whole region. Arcadia, the mountainous interior of the Peloponnese, was a refuge into which the earlier peoples of Greece had fled in the face of invaders from the north. The 'autochthonous people of holy Arcadia'[9] still lived in villages. When, nearly a century later, Epaminondas founded the city of Megalopolis amid the mountains of Arcadia, he gathered for the purpose the population of 36 villages as well as of three 'cities' which can themselves have been no more than large villages.[10]

Between Arcadia and the Gulf of Corinth lay Achaea, made up of small areas of lowland separated by the limestone hills which dropped from the mountains to the sea. The cities of Achaea, Messene and Elis were built for security on the tops of rocky spurs. They were very small, and in function were self-governing village communities, and the structure of their society was still tribal and almost Homeric. Messene had been under Spartan rule for several generations. Its people, reduced to the status of helots (see below), revolted in the mid-fifth century, but were crushed, and the mountain fortress of Ithome, which they had occupied, was captured by the Spartiates.

Across the Taygetus Mountains from Messene lay the main area of Spartan Laconia. Its city of Sparta consisted of four villages, and it did not become a city in the architectural sense until Hellenistic times. The Spartiates were few in number – perhaps not more than 5,000 adult males at this time – in relation to the total population of the Spartan *polis*. The greater part was made up of helots (εἵλωτες), agriculturalists and craftsmen, whose status was but little better than that of slaves. They seem to have lived in simmering discontent under Spartiate control, but nevertheless sometimes served in the army and were even given their freedom on occasion, Scattered through the territory of Laconia were the *perioikoi* (περίοικοι), who lived in self-governing communities, fought in the Spartan armies, and enjoyed some of the rights of citizenship. It was probably they who were responsible for what manufacturing and trade were carried on in Sparta. The Spartiates may have been outnumbered by as much as eight to one, and it speaks well for the efficiency of their police state that they were able to control so many resentful helots and unreliable allies.

The north-eastern corner of the Peloponnese had long been politically precocious. It had been the focus of the Mycenean civilisation more than seven centuries earlier, and was now distinguished by the *poleis* of Corinth, the most developed commercially after Athens itself; Sicyon, Corinth's neighbour on the Corinthian Gulf to the west, and itself an industrial and commercial state, and Argos, victim of a recent aggression by Sparta, which had already absorbed Tiryns and other towns of the Argive plain. In the interior were the little *poleis* of Tegea, Phlius, and Cleonae, and around the coast of Argolis, the small trading towns of Epidauros, Troezen and Hermione.

The narrow isthmus of Corinth linked the Peloponnese with the Greek mainland. At its narrowest it was less than 5 miles across, and a short portage, regularly used in classical times, linked the westward-looking Corinthian Gulf with the Saronic Gulf which opened eastward towards the Cyclades and Ionia. The isthmus became 'a natural crossroads of the Greek world'.[11] There was talk of cutting a canal to join the two seas, but in the fifth century small ships made use of the *diolkos,* a road over which they could be pushed. It appears, however, that merchandise was generally transshipped and portaged across the isthmus, and that only warships were actually manhandled along the *diolkos.*

The isthmus provided also a land route southward into the Peloponnese. The gulfs to west and east were narrow and consituted no great obstacle, but Greece's invaders from the north were not usually provided with ships. The isthmus thus provided a short line of defence against them. A Mycenean wall, of cyclopean masonry, was built here, probably in the thirteenth century B.C. to guard against invaders. Another wall was hastily built early in the fifth century B.C. to hold back the Persians, and Herodotus records the debate among the Peloponnesian allies of Athens who wished to withdraw from Attica and hold the isthmian line. The line of this wall can be traced, though it was in large part destroyed or obscured in the building of later lines of defence.

It was one of the great advantages of Corinth that the city possessed not only its own port of Lechaeum on the Corinthian Gulf, but also that of Kenkreai at the head of the Saronic Gulf. At the north-eastern end of the isthmus was Megara, neighbour and rival of Athens. East of Megara lay Attica, and to the north Boeotia. Attica had been united politically at a relatively early date, and had expanded to include both Eleusis to the west and Oropus beyond the Parnes range to the north. Boeotia, with its neighbouring provinces, Phocis and Locris, was politically more fragmented than almost any other part of ancient Greece. The region is mountainous; ranges extend eastward from Aetolia, and between them lie a series of basins, well protected by the terrain but often damp and ill-drained. In these hill-girt depressions arose a large number of communities, many of which ripened into *poleis.* Most were very small; all were predominantly agricultural, and many probably resembled the Phocian city of Panopaeus, for which Pausanias showed so much contempt. Others must have resembled the fishing port of Anthedon, on the shore of the Euripus, which had almost no agricultural land, but lived by the sale of its fish.

The ruggedness of the terrain provides an explanation for the political fragmentation of Locris, Phocis and the territory around the head of the Malian Gulf. It does not, however, explain the division of Boeotia. Here a large alluvial plain – one of the largest in Greece – lay between the Helicon and Cithaeron ranges on the south and the more northerly Parnassos and Oeta ranges. Its soil was rich, at least by the standards of Hellas, and Athenians never tired of ridiculing the foolish, well-fed Boeotians. Yet within its 1,200 square miles, Boeotia seems never to have had less than about ten small, quarrelsome *poleis.*

Faced with the paradox of Boeotia, one is almost tempted to revert to the

Toynbee model; to assume that ease bred indolence, and to agree that others 'reacted, in their degree, to the prick of Necessity's spur, while comfortable Boeotia cared for none of these things'.[12] The answer lies more reasonably, as Jardé has emphasised,[13] in the contingencies of Boeotian politics. Boeotia did not experience the rule of tyrants in that pre-classical period when unity was forced on so many of the Greek *poleis*, and it must be remembered that, in spite of their endemic inter-city strife, the *poleis* of Boeotia were in fact united in the Boeotian League during much of the fifth century.

Thessaly, beyond the Othrys range and ringed by mountains of its own, was scarcely more Greek than Aetolia and Acarnania. It was 'a border province, a march between Greece properly so-called and the barbarian world'.[14] Its climate, more continental than that of most of peninsular Greece, was unsuited to the olive, and the grape-vine did not grow well. It was a ranching province; its plains were extensive and fertile, but ill-drained; its mountains cut it off from cultural and commercial contacts. Its cities were few and small and its society consisted primarily of a land-owning aristocracy which controlled a servile population, descended in all probability from earlier and possibly pre-Hellenic inhabitants. In mid-fifth century Thessaly was a 'feudal' state, without even the nominal unity that comes from a common overlord.

Off the coast of Locris, Boeotia and Attica lay the long, narrow and mountainous island of Euboea. Its outer or north-eastern coastline was rockbound and harbourless. Where it looked across the narrow Euboean Channel to the mainland were small coastal plains, and on the largest of these Chalcis, Eretria and other smaller *poleis* had grown up.

The direction of the mountains of Euboea and of Attica is continued south-eastward towards the shores of Caria by the rocky islands of the Cyclades. Though differing greatly in size, these were alike in their general configuration: rugged, mountainous, lacking in good harbours, and almost destitute of level cropland. Yet each had its city-state, some rich and powerful like Naxos and Paros; others deriving prestige from their religious associations, and yet others little more than fishing communities.

The northern shore of the Aegean lay far from the hearths of Hellenic civilisation, in an environment that was in many respects un-Mediterranean. Winters were cold and the rivers froze; while in summer the thundery heat was reminiscent of central Europe rather than of its Mediterranean peninsulas. Inland were the plains of Macedonia and Thrace, un-Hellenic in both their extent and the limitations and the opportunities which they offered to agriculture. Yet along this coast, from the Thermaic Gulf to the Propontis (Sea of Marmara), lay a succession of Greek colonies, most of them small, strongly protected, and living under continuous threat from the Thracian tribes of the interior. Most were colonies founded from the city-states of southern Greece; those of Chalcidice were, as their name suggests, in part settled by emigrants from Euboean Chalcis, and Potidaea was exceptional in being a Corinthian colony. Others derived their citizenry from Eretrea and Andros. What, one may ask, drove them to this foreign coast, where they lay skin-deep between the steppe and the sea? Thucydides was probably correct in attributing these settlements to overpopulation and land-hunger. Though they subsequently played an important role in Greek commerce, they were almost certainly not established for this purpose. The Greek settlers later opened up mines, particularly for silver, in the Thracian hinterland, and the metals were exported through the small coastal towns. Some of these northern *poleis* were well placed to control the trade in grain and timber, which was so important for Athens, and Athenian politicians were very sensitive to political movements among them. But these were not the original reasons for their foundation. The Macedonian and Thracian towns remained small and predominantly agricultural. One of them, Antisara, was said to contain 'in a microcosm . . . all the essential elements of Greek city-state without being in itself more than a mere diminutive settlement'.[15]

Although only two states, Athens and Sparta, covered areas that by the standards of classical Greece could be called large, there were several instances in the fifth century of groups of states forming alliances or leagues for political or even religious purposes. The Achaeans, for example, managed whatever affairs they had in common through a democratic league, which met in the sanctuary of Zeus at Aegium. The Chalcidic League, made up of the numerous cities of Chalcidice, differed from the Achaean League in being under constant pressure from the peoples of the interior, and its close association arose in part from the need for mutual defence.[16] Olynthus became in the fifth century the focus of the League's activities, and a common coinage, bearing the word ΧΑΛΚΙΔΕΩΝ, was minted.

Other leagues gave some semblance of unity to the quarrelsome states of Acarnania, Aetolia, Arcadia, Boeotia and Phocis at this time. Most had their origin 'in a common religious festival and organization to protect and supervise a special cult',[17] which itself derived from an earlier tribal unity. Some never developed beyond this stage; others became rudimentary confederations with specifically political functions. Of those leagues whose functions were primarily religious the Delphic Amphictyony was the most significant and influential, and it can be said that its oracle of Apollo was an important channel of communication and bond of union between Greek peoples.

All these regional associations became insignificant beside the Delian League, founded in 478 with the ostensible motive of resisting the Persians and of driving them from Greek soil. Athens took the initiative in its formation and remained its dominant and eventually its controlling member. The membership was called upon to pay a tribute, assessed in the first instance by Aristides, according to their resources. The treasury of the League was established on the island of Dhílos, in the Cyclades, but was removed to Athens in 454 on the pretext that a muster of Persian ships presented a threat to its safety.

The extent of the Delian League is difficult to discover. Corinth and Sparta, traditionally unfriendly to Athens, played no role. Megara and the cities of Boeotia and Phocis did not join. Many small city-states traditionally friendly to Athens do not appear in the tribute lists, and it is probable that they made their tribute payment through another city-state. Others appear only intermittently in the tribute lists, and in some instances payment of tribute was dependent upon Athens' ability to collect it. As late as 454, fourteen cities were

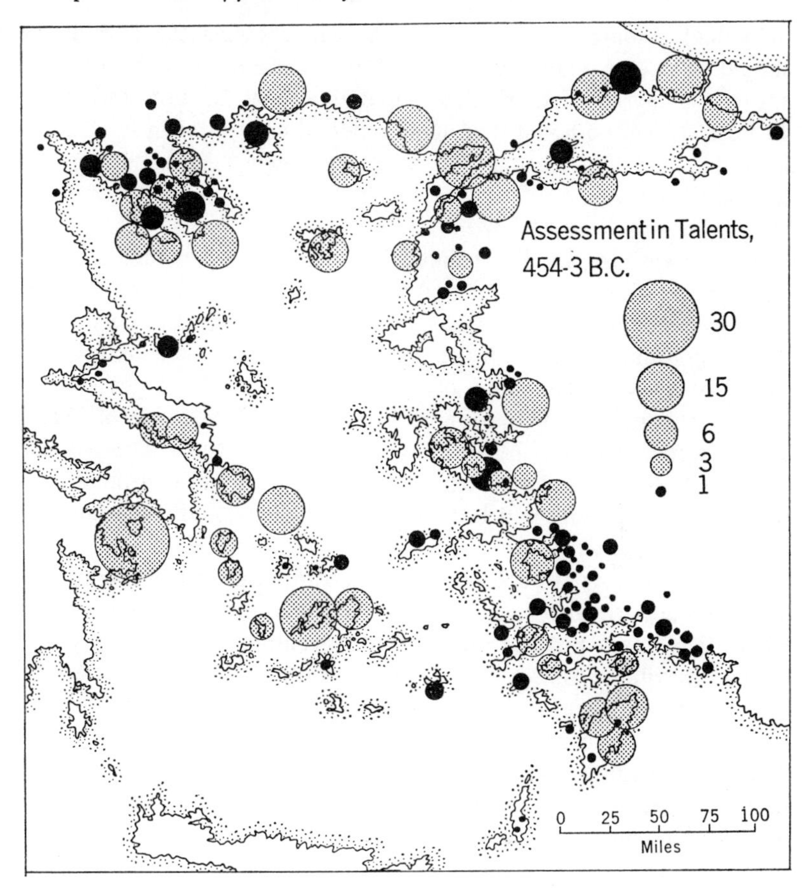

Fig.2.3 The Delian League, tribute paid in 452 B.C.

probably still paying their tribute by the service of their ships, and in the early days of the Confederacy the number that discharged their obligations in this way was far greater. Figure 2.2 shows the distribution of known members of the Confederacy. Including those whose identification is only tentative, 208 cities are marked on the map.[18] A further nine names appearing in the tribute lists have not yet been identified.

The map shows tight clusters of cities in Chalcidice, around the Hellespont and in Caria. The great majority of these cities were very small; many must have numbered their population in hundreds rather than in thousands, and some were extinguished a century later when Philip of Macedonia devastated Chalcidice, or decayed and disappeared when the great Hellenistic cities arose in Asia Minor. A rank–size graph of their monetary contributions[19] suggests that the great majority of the League's members were very small, and were overshadowed by the few large *poleis*.

The amount of tribute had been assessed in the first instance by Aristides, and for the original members it does not appear to have changed significantly before the mid-fifth century. Ostensibly, the tribute varied with the ability of the cities to pay, and thus may be said to bear some relationship to their individual resources and incomes. A map showing the size of the tribute should be one of relative wealth, and thus, since most of them were primarily engaged in agriculture, of their population. Fig. 2.3 uses the assessment of 452 B.C. It will be noted that many cities which appear in fig. 2.2 are omitted from this map. For reasons that we do not know they failed to pay tribute in this particular year.

The Overseas Greeks. The tribute lists give a very incomplete picture of the number and diversity of Greek city-states. Very few *poleis* of the Greek mainland were members of the Delian League, and Greek cities outside the Aegean world stood aloof from both the struggle with the Persians and the Confederation.[20] At an early date many of the Greek cities of the Aegean had sent away their surplus population to found colonies. Most of the towns of Chalcidice and Thrace were colonies of *poleis* lying to the south of them. The greatest colonising activity had been, however, outside the Aegean basin: around the shores of the Black Sea, in southern Italy and eastern Sicily, and, even farther afield, in the western Mediterranean.

The chief motive for colonisation was overpopulation and land-hunger. The colonies were primarily agricultural settlements, and the colonists sought homes where they could practise the form of agriculture to which they were accustomed. Inevitably most were founded within the region of Mediterranean climate, if not always within the area where the olive flourishes. In most cases they were established on unfriendly shores, with Sicels, Illyrians or Samnites watching them from the hills of the interior. The foundation of such colonies was fraught with danger, and they had necessarily to be able to defend themselves. Thus the colonists sought a hill, as at Massilia, or an offshore island, as at Syracuse, on which to base their operations. For the rest, Homer described the ideal site of a colony when he depicted the Island of the Cyclops:

> Along the shore of the grey sea there are soft water-meadows where the vine would never wither; and there is plenty of land level enough for the plough, where they could count on cutting a deep crop at every harvest-time, for the soil below the surface is exceedingly rich. Also it has a safe harbour, in which there is no occasion to tie up at all . . . All your crews have to do is to beach their boats . . . Finally, at the head of the harbour there is a stream of fresh water, running out of a cave in a grove of poplar trees.[21]

The colony was generally bound by ties of loyalty and sentiment to its mother city, but only exceptionally was there any political connection. Corinth, for example, continued to appoint the chief magistrate in its colony of Potidaea, to the chagrin of Athens, but this was unusual. The hostility between Corinth and its daughter-city of Corcyra, on the other hand, was as unusual as its friendship with Syracuse, which it had also founded, was normal. It is almost as difficult to compile a list of colonies as it is to prepare one of the *poleis* of Aegean Greece. Fig. 2.4 attempts to show those that are known. Almost without exception they were coastal in their location. A few grew to be quite large, or at least very much larger than most of the diminutive cities of the Aegean.[22]

The Black Sea coast was lined with small Greek settlements. Its physical environment – flat coastline backed by interminable plains and a climate of bitter winters and summer rains, was unfamiliar to them; the dangers from the Scythian and Sarmatian hordes were more than many of them could withstand, and the difficulties of navigation through the Straits deterred some.

The Adriatic Sea was also, though in different ways, unsuited to Greek colonisation. The rugged Dalmatian coast offered little space between the mountains and the sea, and on the opposite or Italian coast, the plain was narrow and harbourless and its hinterland was occupied by tribes more numerous and warlike than those encountered on most other Mediterranean shores. Few important colonies, therefore, grew up along the Adriatic shore. Spina, near the Po mouth (see below, p. 42), may have canalised Greek trade with the Etruscans, but Epidamnos in Epirus, and Epidaurus, near the site of the future Dubrovnik, can have had little opportunity for trade and, despite their rough terrain, must have been mainly agricultural.[23]

Southern Italy and Sicily were the sphere of Greek colonisation *par excellence*. This was a kinder land than the coastal regions of the Black Sea and Adriatic. It was, indeed, a richer land than Greece itself, and some of the cities of Magna Graecia had a reputation for opulence and luxury unequalled in the austere Aegean region. The alluvial plains, while no more extensive than in Greece, were more manageable and stood less in need of drainage and irrigation. The soil, much of it enriched for centuries by volcanic dust, was fertile, and the rainfall was greater and more reliable.

These western colonists were sent out mainly from the smaller and poorer among the Greek city-states, those of Achaea, Euboea and the Corinthian isthmus, where there was little space for expanding agriculture. The new western cities in many instances exceeded by far their parent states in size and wealth. They occupied likely urban sites along the south and west coasts of the Italian mainland from Apulia to the vicinity of Naples, and along the shores of all except the western third of Sicily. They began in the north-west with Cumae, the farthermost of the Greek settlements in Campania, established upon a steep-sided hill, and suggesting by 'the precarious nature of their foothold that the colonists [wanted] the fullest protection against the aborigines on the landward side'.[24] Nearby were Naples and Poseidonia, the Paestum of Roman times. South of Paestum, the Cilento and Lucanian Mountains came down to the coast, leaving no space for a large city-state and, indeed, little enough for a road between the hills and the sea. On this coast were only two small Greek settlements, the hill-top town of Velia, and Laos. The latter was a colony of Sybaris, established on the northern shore of Calabria, and was probably conceived as a strategic and commercial dependency of Sybaris itself. Calabria, the 'toe' of Italy, is largely occupied by the rugged mountain masses of Sila and Aspromonte. It offered in itself little inducement to settle, but the voyage around it was sometimes more dangerous and difficult than the journey across. Just as Sybaris established its north coast station of Laos, so Croton,

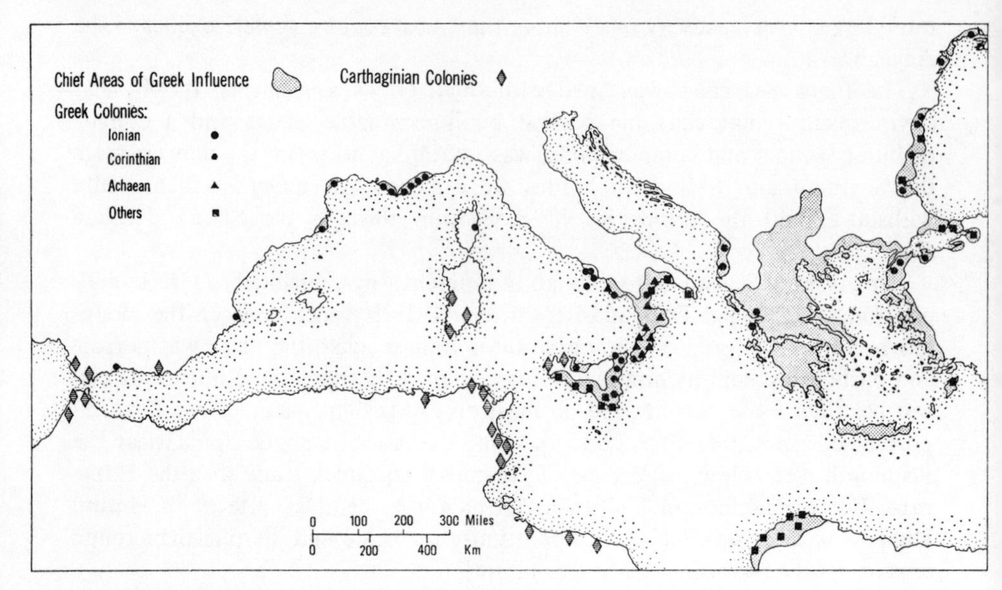

Fig.2.4 Greek Settlement in the Mediterranean

perhaps in conjunction with Scylletion, founded Terina on the opposite coast; Locris, the settlements of Medina and Hippinion, and Siris, that of Pyxus.

The cities of Tarantum, Metapontum, Siris, Croton and Thourioi, the last founded by the Athenians about 443 B.C. near the site of the earlier Sybaris, lay around the Gulf of Taranto, where the coastal plain was wider and more fertile than elsewhere south of Poseidonia. The southern shore of Calabria was more hospitable than the north, and its hills sloped towards the sun. Here, on the narrow ledge between mountain and sea, lay the colonies of Skylletion, Caulonia, Locris and, at the tip of the Calabrian peninsula, Rhegion.

The eastern coast of Sicily was not only a natural landfall for Greeks sailing west, but also the part of the island with the greatest potentialities for settlement. The mountains of the interior of Sicily sloped gently towards the sea. The hills in which they terminated were mostly cultivable, and the wide coastal plains were among the most productive in the classical world. Naxos was the first city to be founded on this coast, below the hill on which Taormina now lies. To the south of the huge, smoking mass of Etna were founded Catania, Leontinoi, Megara and, above all, Syracuse. These *poleis* extended their influence into the interior, and Syracuse even established colonies, not all of which have been located, amid the hills of south-eastern Sicily. It grew to be the largest and most powerful of these city-states, the only one which could rival Athens in size and wealth and was able to defeat the Athenian armies in the field.

The south coast of Sicily offered less opportunity than the eastern. The mountains came closer to the coast, and the area of lowland was small. Nevertheless, Acragas, in the middle of this coast, was in the mid-fifth century a wealthy and powerful *polis*. To the east lay Gela; to the west, Selinunte and its dependent colonies. The north coast of Sicily was as inhospitable as that of

Calabria. There also the mountains came down steeply to the sea, and here the Greeks founded only two colonies, Himera and, within the shelter of the north-eastern promontory of Sicily, Zancle – now the city of Messina.

Throughout Magna Graecia the Greek colonists could do no more than cling to the coast. The interior of both Italy and Sicily was occupied by numerous and generally hostile tribes, and in each of these areas the Greeks came into contact and conflict with states as highly organised and as powerful as their own: the Etruscan and the Carthaginian. The Sicels of Sicily were undergoing slow hellenisation, but in the process they learned also the advantage of organisation. The Greeks taught them how to resist Greek penetration, and in the middle years of the fifth century the Sicel tribes formed a loose federation, with its focus at Palice, in the hills of the eastern interior. They even succeeded in temporarily seizing the city of Catania. On the other hand, commercial relations developed between the Greek cities and the Sicel tribes, and there was a flow of manufactured goods from such cities as Acragas and Gela into the interior. In the west of Sicily the Sicels were less exposed to Greek influence, and their closest relations were with the Carthaginian settlements of the extreme west of the island.

Greek colonisation was feebler and less enduring in the western basin of the Mediterranean Sea. In fact, only one Greek city showed any deep and lasting interest in these western lands, little Phocaea, on the distant coast of Ionia. The chief Phocaean colony was Massilia, well placed to the east of the Rhône delta for both commerce and defence. The Phocaeans may not have had any great interest in commerce when they founded it, but it became in the fifth century the most important avenue of Greek trade with the Celts. From Massilia small colonies were established at Alalia, in Corsica, and at several sites along the coast of Spain, the farthermost of which was near Velez Malaga, within 80 miles of the Strait of Gibraltar. The Phocaeans had even dared to interest themselves in the forbidden lands of Tartessos, beyond the Pillars of Hercules. Inevitably they aroused the hostility of both Carthaginians and Etruscans. They survived an attack made jointly by these two peoples, but found it desirable to abandon their settlement in Corsica, and to use its population to reinforce their holdings along the Spanish coast. During the middle years of the fifth century, the Carthaginians, smarting from their recent defeat by the Sicilian Greeks, tolerated the Phocaean colonies, while awaiting the opportunity to destroy them. Indeed, the Spanish coast appears to have been divided informally into a Greek and a Carthaginian sphere of influence, with a dividing line somewhere in the vicinity of Cape Gata.

Greek settlements along the eastern shore of the Aegean Sea were at least as numerous as those in the Greek peninsula itself, and were far more conspicuous in the columns of the tribute lists. They include the *poleis* of the Troad and of the north-eastern Aegean; the large and powerful cities of Ionia, and the almost countless settlements of Caria, in south-western Asia Minor. They were as Greek as those of the Greek mainland; they had, according to their own traditions, been colonised from Greece proper, and, in their turn, they had played a prominent role in Greek colonisation. Their cultural and political achievement had reached its highest point in the sixth century, before

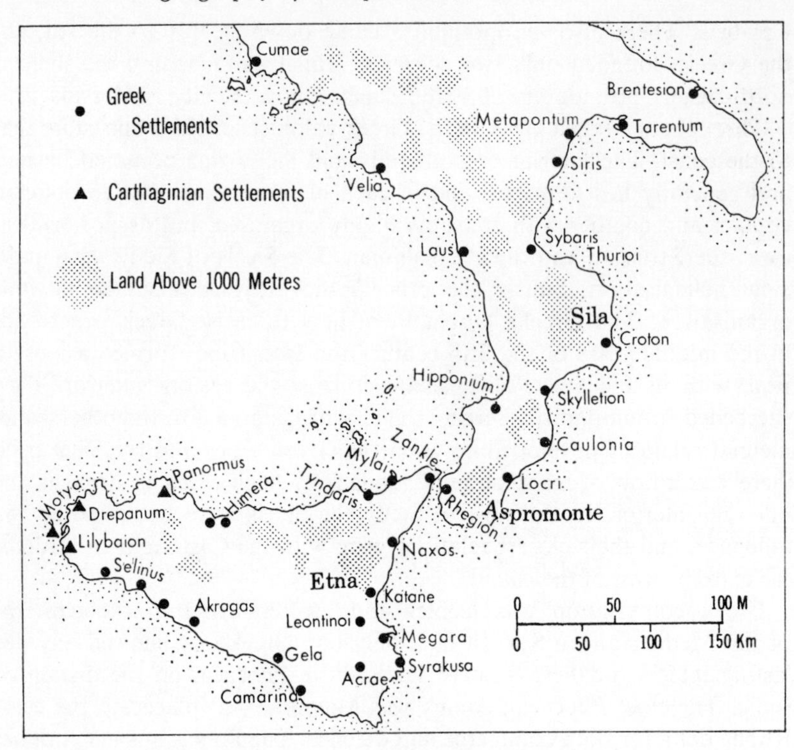

Fig.2.5 Greek and Carthaginian settlement in southern Italy and Sicily (*Magna Graecia*)

the Persian Wars, when they could fairly claim to have been the foremost *poleis* of Greece.

They were then subjected first to the slightly hellenised kingdoms of Lydia, then to the tyranny of the Persian 'barbarians'. At the very beginning of the fifth century, the Ionian cities, with some aid from mainland Greece, revolted against Persian rule. They were crushed; Miletus was destroyed, and it was only after the defeat of the Persians by Athens and its allies that the city-states of the eastern Aegean were able to throw off Persian rule. They had, however, only exchanged an oriental tyrant for an Athenian master. They mostly became members of the Delian League, which fell increasingly under the control of Athens. The League passed into the Athenian Empire, and Athens compelled the obedience of the city-states of Ionia, Caria and the Troad.

Carthaginian settlements. The city of Carthage had been founded by Phoenicians from Tyre in Syria during the ninth century B.C. During the following century Phoenicians, either from Carthage or direct from the shores of the Levant, settled in Sicily, in Malta and in Sardinia, and later still in the Balearic Islands and along the Spanish coast. The priority of Greeks or Carthaginians in the western basin of the Mediterranean has been disputed but is not significant in our discussion of their distribution in the fifth century. The Greeks were able to exclude the Carthaginians from Sicily except its western extremity, and there the latter established strongly fortified settlements. In the mid-fifth century Carthaginian power was restricted to the two cities of Motya and Panormus – the modern Palermo – and their surrounding districts. It had probably been more extensive earlier, but a decisive military defeat at the hands of the Greeks of Syracuse and Acragas in 480 had set them back for a considerable period. Farther west, however, the Carthaginians controlled at least southern Sardinia and maintained control of the coast of southern Spain and the Strait of Gibraltar. Beyond the Strait the Carthaginians had founded Gades (Cadiz) and, a half-century earlier, had overthrown the Iberian state of Tartessos, which in all probability had spread across the Sierra Morena. The Carthaginians thus came to control the Guadalquivir valley and the rich mining region which lay to the north. The Massiliotes seem to have tried, though unsuccessfully, to reach this mineral-bearing region by advancing across the Meseta from those parts of the Spanish coast which they already controlled.

The Carthaginian settlements of Sicily, Spain and the islands never enjoyed the degree of independence that the Greek cities possessed. Distance and the slowness of travel prevented Carthage from exercising more than a loose control over them, but the colonies did supply agricultural and mineral products to their capital city in Africa as well as manpower for its armies. This was indeed a far-flung empire, the only one of its kind in Europe of the fifth century.

The Etruscans. In addition to the organised opposition of the Carthaginians the Greeks in the western basin of the Mediterranean faced that of the Etruscans. This enigmatic people occupied much of central and northern Italy, where they constituted a highly organised though far from unitary state. They have traditionally been regarded as invaders who had come by sea from south-western Asia Minor, and the difficulties of their language were attributed to its non-Indo-European origin. This view has now been challenged. 'We can speak of a homogeneous (Etruscan) culture,' wrote von Vacano, 'only with reserve. In particular, any attempt to draw a line between them and their Umbrian and Latin neighbours meets with great difficulties,'[25] and Pallottino claimed that 'the formative process of the nation can only have taken place on the territory of Etruria proper'.[26] Even the Asiatic origin of their language is now suspect, and it is beginning to be held that Etruscan embodied important elements of the pre-Indo-European Mediterranean speech. Perhaps it is the Etruscan culture which is autochthonous, while all the other Italian languages have been brought into the peninsula by immigrants.

In the mid-fifth century Etruscan rule extended from the alpine foot-hills in the north to the neighbourhood of the Greek settlements of Cumae and Naples in the south. The focus of their civilisation, however, lay in Etruria. where their cities, generally said to have been twelve in number, though actually more numerous, formed a confederation or league. The league of Etruscan cities resembled the associations of Hellenic cities, in that the local patriotism of each city was intense and always unwilling to sacrifice local interests to those of the league as a whole.

The area controlled by the Etruscan cities was well endowed. Over the south, the volcanic tuff yielded a good soil; the coastal marshes were fertile when drained, and the valleys and basins of the interior and the north provided agricultural possibilities unknown to ancient Greece. There was an abundance of timber on the hills, and above all there was the mineral wealth of the region. The wealth and prosperity of Etruria were founded in part on its iron workings, and iron goods were traded from its ports over the western Mediterranean.

Etruscan political control extended southward between the Apennines and the Tyrrhenian Sea as far as Campania, where the river Sele seems to have formed its boundary. There, however, the Etruscans met not only with the opposition of the Greeks, but also with that of the Latin and Oscan tribes. Furthermore, this region did not possess mineral wealth to attract them, and their rule was relatively short-lived. They were expelled from Rome a half century before our period. They retained their more southerly settlements, including their capital city of Capua, for another century. But in the mid-fifth century, Etruscan power in Campania was declining and a couple of decades later they were to be defeated by the Greeks of Cumae. Yet the heirs to the Etruscan power in Campania were not the Greeks, but the Italian tribes. Excavations have shown that the first city of Pompeii was Etruscan; the second, established in the late fifth century, was Samnite.[27] Almost certainly, at the period with which we are concerned, the Etruscans who remained in Campania were cut off from Etruria by the Italian tribes, which had invaded the plains, and by the revolt of Rome.

Towards the north, Etruscan power extended beyond the Apennines and even beyond the Po. Evidence for the political organisation of these more northerly Etruscans is scanty, but they appear at our date to have constituted another confederacy, not unlike that of Etruria itself. About a dozen cities are said to have grown up in the Po valley, including Felsina (Bologna), the chief northern seat of Etruscan power; Spina, their port in the Po delta, through which they traded not only with the Hellenic world, but also with the Illyrians and other peoples who lay around the head of the Adriatic Sea, and Marzabotto, near Bologna, whose excavations have thrown so much light on the planning of Etruscan towns.

The Etruscans occupied also the eastern shore of Corsica and the island of Elba, and may also have held, if only lightly and intermittently, parts of north-eastern Sardinia. Here their alliance of convenience with the Carthaginians had led to the withdrawal of the Greeks from this area. The Etruscan power in the western Mediterranean has been called, like that of the Phocaeans, a 'thalassocracy'. It has even been said that their 'only genuine connecting link [was] the coast'.[28] This appears to be an overstatement. They possessed a fleet and carried on a large and important overseas trade, but few of their cities were coastal, and their territorial expansion was almost entirely overland.

Rome. There was nothing in the site and appearance of Rome about 450 B.C. to suggest its future greatness. In size and function it was a city resembling in all probability any one of the federated cities of the Etruscans. In language and culture it resembled the Latin tribes of central Italy, from whom the

Roman people had sprung. Rome was the focus of a small city-state, lying mainly to the south-east of the Tiber. It had passed through stages of economic development broadly similar to those postulated for the Hellenic cities. Small village settlements, located on the summits of the Roman hills, had during the seventh century merged to form the *Septimontium.* During the sixth century the city passed under Etruscan control. According to Roman tradition a wall of great size was built around the whole by Servius Tullius. Though excavation has revealed something less pretentious (see below, p. 62), it is likely that the site was given some form of protection at this time. By the mid-fifth century we may be confident that the Etruscans had gone, and that Rome was again a small Latin town, ringed by enemies more powerful than itself.

At this time Rome could easily have succumbed to its neighbours. The Etruscans tried to regain the city, and succeeded in destroying the Servian wall, so that the city was unprotected when in 390 B.C. it was captured by the Gauls. The Etruscans continued to hold the city of Veii, which dominated the Tiber valley from its hilltop, only 10 miles to the north of Rome. The Romans had been able, however, to maintain their control over the Campagna at least as far up the Tiber valley as Fidenae, and up the Anio to Collatio. To the east the low, volcanic Alban Hills, with their ancient settlements of Tusculum and Alba Longa, formed the Roman frontier. Tradition ascribes to the Romans the control of the Tiber mouth at this date, and attributes to them the foundation of the port of Ostia. Excavation, however, has revealed no evidence of a settlement here earlier in date than the fourth century, and we must presume that in the mid-fifth century the Romans had no seaborne commerce and, in fact, no association with the sea.

The Italian Tribes. Beyond the limits of Rome lived the Italian tribes, familiar through the pages of Livy. They belonged to two related language-groups, the Latin and the Osco-Umbrian, both of which had numerous and distinctive dialects. It was from one of these, the Latin group, that the language of Rome and the Campagna was derived. But the Osco-Umbrian peoples were more numerous and more powerful. They lived in villages and were organised in tribes which at intervals federated into larger units to make war on their neighbours. The Sabellians had driven the Iapyges, probably an Illyrian people, into the Apulian peninsula, which derived from them its name; the Samnites pressed continuously against the Greek settlements of southern Italy, and later in the century overwhelmed the more southerly Etruscan cities. Nearer Rome, the Sabines advanced down the Tiber and Anio valleys, occupying the plain of Fidenae. To their left lay the Aequi and Hernici, who inhabited the hill country about Avezzano and the Fucine Lake. Between the Hernici and the coast, and spanning the more southerly part of the plain of Latium, lived the Volsci. In the strategy of the fifth century these tribes had cut the land connection between Etruria and the Etruscan settlements in Campania and were now closing in on Rome itself.

The cause of this drive of the mountain peoples towards the sea 'was beyond doubt the progressive overpopulation of the mountain country'.[29] The danger from the middle Tiber and Anio valleys, that is from the Sabines, was the smallest, perhaps because the political organisation of these peoples was the least effective. Indeed the Roman population appears at this time to have absorbed groups of Sabines, and the Sabine threat may well have been much reduced by the mid-fifth century. The danger from the Aequi, Hernici and Volsci, however, remained undiminished, and beyond them to the south-east lay the yet greater strength of the Samnites, against whom the Romans were to pit their strength a century later.

Iron Age Europe. Europe, beyond the sphere of Greek, Etruscan and Carthaginian settlement, had no territorial states and only the most rudimentary political organisation. Its population lived in small and almost wholly self-sufficing communities. Trade with distant areas; though important, was small in volume and restricted to metals, salt, and luxury goods. Yet it is on the basis of this narrow range of commodities that the most widely used and most fully authenticated classification of these peoples and their cultures is based. The Bronze Age, which had begun in Europe almost two thousand years earlier, was now nearing its close. Iron was still a rare luxury in many parts of northern Europe, and was unknown in some, but over most of Europe south of the Baltic and North Seas, there had spread the Iron Age culture, known from its most famous type-site in Upper Austria as the Hallstatt. Indeed, at the time with which we are concerned, a more refined and sophisticated Iron Age culture, known, from its type-site near Neufchâtel, as La Tène, had appeared and was now found over parts of south-west Germany and eastern France.

These material cultures were being carried outward from their central European source in part by the physical movement of loosely organised tribes; in part also by the less predictable movement of individual technicians, most of whom would have been metal-workers and potters, and by the slow creep of ideas communicated from one community to the next, from one individual to another, in the course of their possibly infrequent contacts. The chronology of these movements is extremely difficult to reconstruct, and it is even more difficult to assign precise geographical limits to the cultures during the fifth century.

In northern Italy, the Villanovans were still to be found wherever the Etruscans had not imposed their urban and commercial civilisation upon them. These were an Iron Age people, politically and technically less advanced than the Etruscans. The Villanovan culture, recognisable by its pottery and burials, still existed in much of northern Italy. In north-east Italy the Atestine culture, so called from its type-site at Este, spanned much of the area between the Po and the Alps. Its people appear to have been related closely both in their material culture and also in race and language to the Villanovans. We know this culture from its burials, rather than from its settlements; it possessed iron tools and weapons, but was particularly noteworthy for the excellence of its beaten bronze work. In fact, much of the little that is known of its social and economic organisation is derived from the embossed designs on its bronze *situlae*. These portray an agricultural society without any trace of urban development, though there was clearly a well-developed metal industry and some trade in metal goods. The Atestines belonged to the Venetic people, whose language was unquestionably Indo-European, related to Illyrian, and 'in many ways intermediate to Greek and Latin'.[30] The linguistic pattern is even less clear than the pattern of material cultures. These north Italian peoples used an alphabet at this time, but they have left no literature and only a few inscriptions.

West of the Atestines, and also between the Po and the Alps, lived the Comacines. They were akin to both Atestines and Villanovans, but 'their material culture was sufficiently independent and original to justify us in regarding them as a separate nation'.[31] It is usual to regard them as part of the Ligurian people, who inhabited at this time a very extensive area, reaching from the valley of the Po westwards at least to that of the Rhône. Ligurian, like Venetic, was an Indo-European language, but, in the opinion of Whatmough, was 'neither Italic nor Keltic, but . . . linguistically, as also geographically, intermediate between them'.[32] North of the Veneti and within the ranges of the eastern Alps lived the Raeti. Culturally these people fell under the influence both of the Atestines, whose decorated bronze ware has been found as far north as the Danube, and of the Iron Age Hallstatt cultures. They, too, were Indo-European, speaking dialects intermediate between Celtic and Illyrian.

Beyond the range of these north Italian peoples and cultures there appears, superficially at least, to have been a far greater degree of homogeneity. This springs only from the greater paucity of evidence and from the complete absence of inscriptions which might throw light on the spoken languages. In this extensive area, from Spain to the Baltic and the Balkans, there were large and important peoples or 'nations'. In Spain lived the Iberians. From northern Spain to the upper Danube valley were the Celts; in the middle Danube valley and extending from there southward to Epirus were the Illyrians, and to the east of them the Thracians. The Slavs already in all probability inhabited the great plain of northern Europe from roughly the Oder river eastwards to the Pripyat region.

Our knowledge of the peoples of northern Europe twenty-four centuries ago is slight. Archaeological evidence provides no certain guide either to their social and political relationships or to their linguistic affinities. It is presumed that southern Scandinavia and perhaps parts of the north German plain were at this time inhabited by proto-Germanic peoples; it is presumed also that in the east Baltic provinces, beyond the Slavs, lived the Baltic peoples who survived into the historic period as Prussians and Lithuanians.[33] Beyond these lived nameless and unknown peoples, probably neolithic in their culture and Finno-Ugric in their linguistic affiliation. The Russian steppe, already at this date an open grassland, was the home of the Scythians and Sarmatians, who are very much better known to us than most other peoples of this age, in part, from their art, and in part also from their frequent contacts with the Greek settlers of the Black Sea coastline. But of north-western Europe and the British Isles, beyond the areas inhabited by the Celts, almost nothing – linguistically speaking – is known. Although one can speak of tribes and, with reservations, of 'nations', or groups of tribes tied together by bonds of language

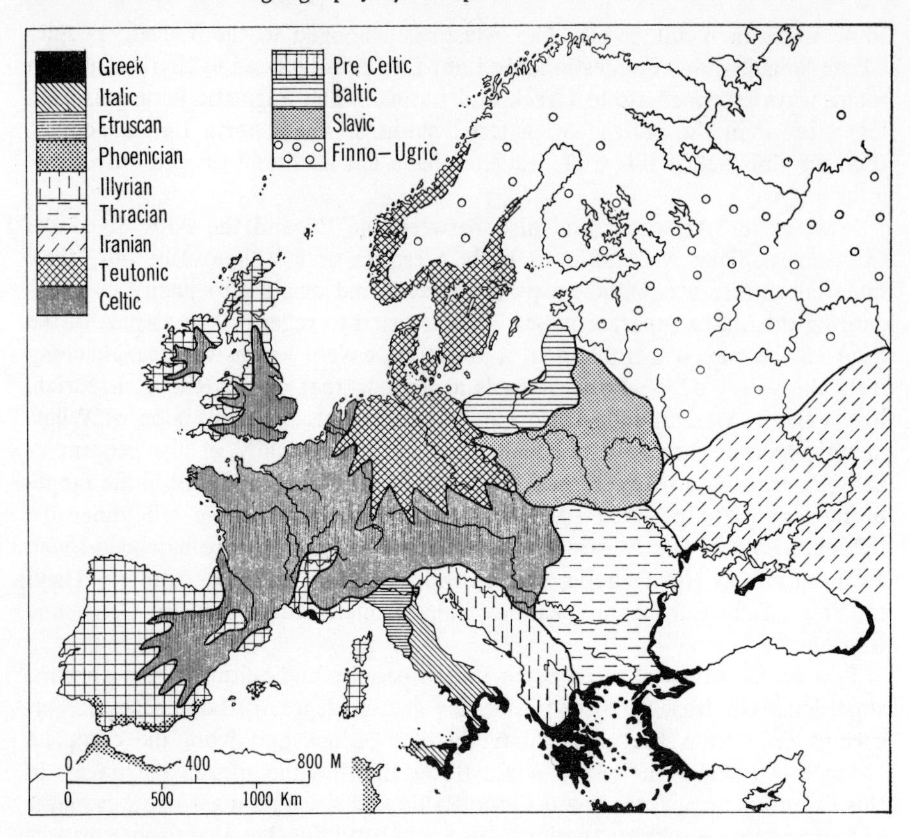

Fig.2.6 Ethnic-linguistic map of Europe in the fifth century B.C.

and material culture, one cannot speak of 'states'. Peoples were not bound closely and indissolubly to a specific area of land. A concept of territorial sovereignty had not emerged.

The Celts. Most important and possibly also most numerous of these peoples were the Celts. Their language appears to have arisen in south Germany, and to have been carried thence westward into Gaul and south-westward into Spain. In the fifth century they inhabited the Spanish Meseta, and were pressing against the Ligurians in northern Italy. From northern France they had crossed the English Channel and had by now spread over south-eastern England. To the east, they had advanced down the Danube valley; settled Bohemia to which they gave its name, and reached as far as Beograd. Probably by the mid-fifth century Celtic bands had begun to raid the Venetic and Ligurian settlements of the north Italian plain, though their more extensive and destructive attacks lay a half century in the future.

What, one may ask, was the nature of this outward expansion of the Celtic peoples? Was it a mass movement involving large numbers, or was it a movement of relatively small war bands, rather like the Viking raids of later centuries, or, lastly, did the Celts merely supply leadership to earlier and non-Celtic peoples? It was probably all of these. There is evidence at this time of Celtic withdrawal from the plains of north Germany, in the face, presumably, of the pressure of Germanic peoples. In Britain and probably also in much of Gaul and Spain, the Celts provided an *élite*, which conquered, ruled and eventually celticised the local, pre-Celtic population. Celtic inroads into peninsular Italy and the Balkans were probably made by small bands, which were in the end defeated and either eliminated or assimilated. The simple matter of logistics and food supply makes it probable that the Celtic expansion proceeded generally by the movement and settlement of small communities and their gradual assimilation of those among whom they came to live. There is good reason to believe, also, that some groups of Ligurians and Veneti sought refuge in the Alps and northern Apennines, where they continued, perhaps for centuries, to constitute islands of refuge, culturally distinct from the peoples of surrounding areas.[34]

That the Celts were enabled to expand so far and so fast was due primarily to their military equipment. Though retaining bronze in much of their artwork, they used iron swords; their war bands were well organised, and classical writers speak with some considerable dread of their warlike qualities. Our uncertainty regarding the precise nature of their expansion, and our ignorance of whether all who took part in this movement actually spoke a language that was recognisably part of the group which came later to be known as Celtic, make it difficult to show on a map the extent of Celtic settlement in the fifth century. It is usual to regard the Hallstatt culture as approximately coterminous with the Celts, on the grounds that without the iron culture of Hallstatt the Celts could never have achieved the expansion which in fact they did. The revolution brought about by the introduction of iron influenced other branches of activity than warfare. Iron tools allowed the forest to be cut more easily and the iron coulter permitted the plough to break up for cultivation the heavy clays of northern Europe. The revolution effected by the coming of iron was more significant in northern than in Mediterranean Europe, because the forests were denser and the soils heavier in the former and less amenable to the tools of the Bronze Age peasant.

By the mid-fifth century the smelting and use of iron had spread from its central European hearth over most of Europe south of the Baltic and North Seas, and while simple iron tools were still in use on the outer margin of this cultural province, refinements were being made nearer its centre which culminated in the mid-fifth century in the appearance of the La Tène culture in southern Germany and eastern France. Knowledge of iron-working had been brought to Italy from the Middle East, probably from Anatolia, and its penetration of the Alps and central Europe 'can be most satisfactorily attributed . . . to the arrival of expert miners and prospectors in the several metalliferous regions; the new method of mining, as illustrated in the Alps, is so sophisticated and so novel, that one feels it must have been initiated at least by experts trained in more civilized regions'.[35] We may be reasonably sure that the knowledge of iron-working was not brought to its Hallstatt centres of dispersion by way of the Balkan peninsula and the Danube valley. On the contrary, it reached Hungary, possibly with the Celtic invasion, from Austria, and there was at this time no developed iron culture north of the Greek settlements of Macedonia and Thrace.

The increasing use of iron brought about a political no less than a social and economic revolution. In the late Bronze and even in the early Iron Age society was still fragmented. 'Politically . . . we must envisage,' in the words of Childe, 'a multiplicity of distinct societies closely juxtaposed and often competing with one another for arable and pasture. None of these were yet demonstrably united into petty kingdoms'.[36] One gets the impression of democratic peasant societies. Such societies survived the advent of iron and lasted through the fifth century in many of the marginal areas of the Hallstatt culture, possibly in Spain and probably in southern Britain. But elsewhere within the Celtic area society was becoming more stratified and political units larger. The iron sword and the war-chariot were developed. The horse began to be used in warfare. A political *élite* appeared, and Celtic society began to acquire characteristics which, for lack of a more suitable term, are sometimes described as 'feudal'. In terms of political geography this showed itself in the appearance of 'states' – the Romans called them *civitates*, and the Celts themselves, *tuath*. They were small in size and population, 'and normally conformed to an area with natural topographical boundaries'.[37] Very many of these political groups survived in the 'tribes' of Roman Gaul and Britain. They began during the fifth century to acquire some centralised direction; a threefold social structure – king, nobles and free commoners – appeared; there may even have been a fixed political capital, but the Celts did not at this time create an urban civilisation or an extensive and stable state-system.

The Iberians. By the fifth century the expansion of the Celts had reached more than halfway across the Spanish peninsula, bringing with it an iron-using, late Hallstatt culture. In Spain they found an Iberian people, whose cultures were 'as diverse as are the corresponding geographical divisions of the land'. In the south they were related to those of north Africa; in north-eastern Spain, the peoples appear to have been Ligurian. The complexity of peoples and cultures was intensified by the arrival of Greeks and Phoenicians. The former, as we have seen, settled along the Mediterranean coast, and were probably interested primarily in the tin and copper to be had in the interior of the Iberian peninsula and in the lands to the north beyond the seas. But their settlements seem to have been essentially trading posts. They may not have been agriculturally self-sufficing, and they certainly did not establish their authority over their hinterlands. Although a great deal of Attic pottery has been found near the sites of the Greek colonies, native settlements within 30 miles showed 'only a very low percentage of Greek wares at the end of the fifth century',[38] and beyond this the Greeks had virtually no cultural impact. There was, however, one exception; strong Hellenic influences seen today in coastal settlements and cemeteries, extend inland from such towns as Hemeroskopeion and Akra Leuke towards Tartessos and the metalliferous regions of the Sierra Morena. In general, however, the Greeks seem to have maintained friendly – or at least commercial relations with the Phoenicians, and metals from the west and north probably passed from one to the other of the Greek cities of the Iberian coast, whither they were brought by the Phoenicians themselves.

The Phoenician sphere of influence was southern Spain. Here they had overthrown the state of Tartessos, the only city-state of the pre-Roman west (which lay on the sandy spit which from the north had almost closed the mouth of the Guadalquivir), and had transferred its commercial functions to Gades, situated Phoenician-fashion on a small island to the south of the mouth of the Guadalquivir. Tartessos had commanded an extensive hinterland with a rich agriculture. The Phoenicians may not have inherited all this territory, but Gades nonetheless ruled a territorial state of far greater extent than that of any of the Greek colonies.

Northern Europe. Despite the apparent precision of the map (fig. 2.6), there were no clear-cut boundaries to the Celtic world. Much of the Low Countries 'appear to have been a species of no-man's-land during most of the La Tène period',[39] and, farther to the east, much of Saxony was inhabited by mixed Celtic-Nordic peoples. In the north-west of Gaul the Celts were mixing with the Bronze Age peasants, of whose linguistic associations we know nothing for certain.

At what date the Germanic peoples began to impinge on the Celts is far from clear. We may be certain, however, that in the fifth century their pressure was considerable, and that the Celts, despite their superior armament, were, locally at least, retreating before them. The Germans inhabited southern Norway and Sweden, the Danish peninsula and islands and the north German coast to the east and west. Though not falling within the limits of the Hallstatt culture, they had gained some familiarity with iron. By 450 B.C. iron was in use in Denmark, but does not appear to have become important in Sweden until a half century later.

Social and political organisation in Scandinavia was one of small peasant communities; there is no evidence at this date of tribal groupings even on the small scale demonstrated by the Celts. This was a difficult time for the settlers in the central lowland and around the coast of southern Sweden. The climate was becoming wetter: peat-bog was encroaching on lowland areas, and forest was gradually yielding to moorland on many of the uplands. The area of cropland and pasture must have been growing smaller, and it is difficult not to see in this physical change at least a partial reason for the pressure of the proto-German peoples on the Celts and Slavs.

Eastern Europe. By the fifth century the proto-Slavs had probably appeared as a distinct people. Their homeland was almost certainly in an area lying between the Carpathian Mountains and the marshes of the Pripet and bounded on the west by the Vistula and on the east by the Dniepr. This corresponds roughly with south-eastern Poland and the north-western Ukraine. To the north lived the ancestors of the later Balts, and beyond them lay Finno-Ugric peoples. The economy of the latter was based mainly on hunting and fishing, though agriculture was gradually pressing into this region from the south. The population was certainly small and partly nomadic, and social groups were possibly no larger than the family.

The steppes of southern Russia, from the Carpathians eastwards into Asia, were at this time occupied by the Scythians. It is now generally agreed that this powerful people was substantially of Iranian origin, though not without a possibly large admixture of Tatar blood. Some of their social practices are said to suggest 'a Tatar clan in the last stage of degeneracy'.[40] That they were Indo-Europeans, speaking a Persian tongue, seems now to be well established. Herodotus gives us a long and detailed account of the 'Scythians', but unfortunately the Greeks tended to use the term for any 'northern barbarian from the east of Europe',[41] and part of his description is, therefore, applicable to the northern forest dwellers – Balts or Slavs – rather than to the Scythians of the steppe.

The Scythians appear to have had a tighter political organisation, and were more capable of concerted action than other barbarian peoples. Their tribes were united under a king, and in time of war joined their forces to make up what was probably the most formidable military machine in Europe at this time. They 'managed one thing, and that the most important in human affairs, better than anyone else on the face of the earth: I mean their own preservation,' wrote Herodotus.[42] They appear, like the Celts, to have had a 'feudal' social structure, with a horse-riding, pastoral and semi-nomadic aristocracy, which dominated the settled and agricultural masses. This division is likely to reflect the initial difference in the ethnic composition of the people, the aristocracy deriving mainly from the invaders – perhaps Tatar – from the east, while the peasantry derived from the settled Tripolye people of the later Neolithic.

The extent of Scythian settlement is roughly indicated by the area of their elaborate burials. According to Herodotus, they had no cities; this is, however, doubtful in view of the hill forts that have been excavated at several places in Scythia.[43] The nomadic pastoralism of the Royal Scyths seems to have prevailed over most of the steppe eastward from the river Bug to the Kuban. To the west, however, the wooded steppe had a thin population of settled peasant farmers, who may have been the forward edge of the slowly advancing wave of Slavs. To the north, where the steppe merges into the broad-leaved forest, there were also settled, agricultural communities, which may have come only temporarily under Scythian domination. They are known, from a type-site near Kiev, as the Zarubintsi culture. These may have been the settlements of the 'farming Scyths', mentioned by Herodotus;[44] they may, on the other hand, have been Finno-Ugric peoples, or Balts, or even Slavs. From their base in the south Russian steppe the Scythians raided westward. They *may* have destroyed the Lusatian culture; they must have disrupted peasant life in Transylvania and the middle and lower Danube basins.

The Balkan peninsula. The classical writers leave us in no doubt that the Balkan peninsula, north at least of the Greek settlements, was divided between the Illyrians and the Thracians. The cultural influence of the Greeks on the Balkan peninsula was slight. The Vardar and Morava routes, so important in Roman times and later, were little used, and Greek influence on the territory lying to the north of Hellas came largely from the Adriatic and Black Sea coasts. A number of settlements on the west coast of the Balkan peninsula, from Acarnania northward to Trogir in present-day Croatia, provided bases from which the Greeks carried on a very small trade with the Illyrian tribes. The Greeks found the overland route from the Aegean into Illyria extremely difficult, and appear to have used it only on the rarest occasions.

The Illyrians occupied the area lying to the west of the Vardar and Morava valleys and extending to the Adriatic Sea and the eastern Alps. The Veneti were Illyrians, and the Messapians of Apulia were almost certainly immigrants from the Illyrian coast. Illyria received the Hallstatt iron culture from the north-west, but in the fifth century seems to have been almost impervious to Greek civilisation diffused from the south-east. It was, in Casson's words, 'a self-contained and conservative region... of all barbaric races the Illyrians remained until Roman times the most vigorous and most exclusive; their presence was always a menace to cultured Hellenic centres'.[45]

The Thracians who lived east of the Illyrians and stretched to the Black and Aegean Seas, are said to have 'differed only dialectically from the Illyrians'.[46] Thrace and the Thracians were more accessible than the Illyrians. In the opinion of Casson the mountains of Thrace constituted no insuperable barrier, and much of the area was overrun by the armies of Xerxes. Nevertheless we have it on the authority of Herodotus that at least one Thracian tribe, the Satrae, owed its independence to the rugged nature of its environment. Thrace was well peopled, but its population was made up of small and semi-independent tribal units, whose divisions alone, in the opinion of Herodotus, prevented them from becoming 'the most powerful nation on earth'.[47]

POPULATION

For the estimation of early populations, wrote David Hume, 'the facts delivered by ancient authors are either so uncertain or so imperfect as to afford us nothing positive'. There was no census, and the only statistical material available consists of estimates of the size of armies, the number of ships, or the size of tribute paid to the Athenian League. For the rest, we have only what we can infer from the size of the grain trade, the area of cities, or the extent of agricultural land. It is at least possible that some classical Greek writers considered Greece to have been overpopulated. Both Plato and Aristotle[48] condemned large cities, and Plato suggested[49] as the ideal population of a city-state a total of 5,040 men, with their dependants. The long history of Greek colonisation is witness to a continuing population pressure. Athens, it is true, established no true colonies, and her few overseas settlements were established more for military and strategic reasons than demographic. Attica, indeed, seems not to have been densely peopled in the Archaic period, and internal colonisation appears to have been able to absorb the increasing population. The same is broadly true of Boeotia and Thessaly, where not even the *polis*-basis of society had appeared. But in much of the Aegean world the pressure of population against the meagre agricultural resources could reach a dangerous level all too easily.

The size of a population is clearly related to the level of its technological advancement and to its ability to produce food. The Greeks made very little advance in agricultural science; one can point to no significant land reclamation, and they appear even to have allowed Lake Copais to revert to swamp and lake. The pressure of a growing population was thus felt all the more acutely until the total began to decline late in the fifth century. The scanty evidence points to a population that rose until the period of the Peloponnesian War, and then declined through the fourth century.

We start with Attica, for it is to Attica that most of the sources relate. A comparison of the figures given by Herodotus and Thucydides for the Athenian army indicates that the population must have grown 'at a prodigious rate'[50] during the intervening years. The estimates given by Thucydides of the number of hoplites, or second-line troops, and of ships have suggested totals ranging from 35,000 to 60,000 for the citizen population, or, with their families, from 110,000 to 180,000. But the estimation of a total population from the number of those of military age calls for a knowledge of the age structure of the people, and this can only be inferred from known populations which are considered to be at a similar stage of social, technical and economic development. To the citizen body of Athens must be added the number of *metoikoi* or resident aliens, and of slaves, and for these the data are even less satisfactory. Estimates can be little more than guesses, and if there appears to be some degree of unanimity among historians, this is probably only because the giant figure of Julius Beloch still casts his shadow over the work of his successors. Table 2.1 collates such estimates as appear to have been based on an adequate knowledge and use of the available statistics. Estimates of the total population range from 200,000[51] to 334,000[52] and the reality probably lies between these extremes. A population of this order of size is suggested also by the scanty evidence regarding food production and import. These data, however, are fragmentary and relate mainly to the following century. In 329 B.C. the grain crop of Attica, according to the Eleusis table, was 28,500 *medimni* (about 40,470 bushels) of wheat and 340,350 (483,300 bushels) of barley; 368,850 *medimni* in all.[53] After allowing for seed, about 295,000 *medimni* would have been available for human consumption. We do not know the volume of grain import for this year, but 355 Demosthenes claimed in a political speech[54] that Athens derived 400,000 *medimni* from the ports of the Bosporus, and as much again from all other foreign sources. The total amount of grain available may thus have been of the order of 1,200,000 *medimni* or a little less. At the accepted rate of 7.5 *medimni* for a male adult and 5 for women and children, this suggests a population of about 180,000; we may be sure that the Eleusis tables tended to underestimate the production within Attica, and it is possible that Demosthenes for political reasons may have exaggerated the import from the Bosporus. Thus the total grain available may have been sufficient for perhaps as many as 200,000, a figure which is in line with one of near 300,000 at the beginning of the Peloponnesian War.

Beloch based his discussion of the population of the classical world on the estimated food-producing capacity of the land.[55] Parts of Greece were, however, fed on imported grain, but only for Athens is there any significant data on the grain trade.[56] Beloch's estimates were cautious and, in general, low.

Others have argued for a high population density. The arguments and conclusions of Beloch have, in fact, stood up to the criticisms of the last 80 years. We may agree, then, with Beloch's claim that in the fifth century Attica was one of the most densely peopled areas in the world, because here was to be found not only a well-populated countryside, but also a large city, fed in large part with imported grain.

Table 2.1 *Population of Athens (in thousands)*

	Citizens		Metoiki		Slaves	Total
		with families		with families		
Beloch	35	135	(5)		100	285
Ehrenberg and Kahrstedt	35–45	110–150	10–15	25–40	80–110	215–300
Gomme	43	172	9–5	28–5	115	315–5
Tod	—	150–170	—	35–40	80–100	265–310

Sources
Julius Beloch, *Die Bevölkerung der Griechisch-Römischen Welt* (Leipzig, 1888), pp. 367–74.
Victor Ehrenberg, *The Greek State* (Oxford, 1960), pp. 32–9; Ulrich Kahrstedt, 'Die Bevölkerung des Altertums', *Handwörterbuch der Staatswissenschaften*, II, 655–70.
A. W. Gomme, *The Population of Athens in the Fifth and Fourth Centuries B.C.* (Oxford, 1933); 'The Slave Population of Athens', *Jnl. Hell. St.*, 66 (1948), 127–9; 'The Population of Athens Again', *ibid.*, 79 (1959), 61–8.
Marcus N. Tod, 'The Economic Background of the Fifth Century', *Camb. Anc. Hist.*, V (1953), 1–32.

It is difficult, if not impossible, to arrive at any estimate of the population of other *poleis* of the Aegean world. For none of them is it possible even to guess the proportion of their population that lived respectively in the nucleated, urban central-place, and in the surrounding villages and hamlets. Perhaps some 80 to 90 per cent may have lived in the dominant settlement, but for this there is no real evidence.

The area of the central settlement may perhaps provide a very rough measure of *its* population. But it is difficult to establish the ground plan of most of them, and even more so to discover how densely their area was built over. The evidence suggests that, apart from Athens, most of the central-places of the Greek *poleis* were small and the majority little larger than a modern village.

The only area in Europe that could possibly have equalled the population density of the Aegean lands was Magna Graecia. The cities here were not numerous, but a few of them were large by the standards of the Aegean. Catania was of well above average size and Syracuse was probably the only European city that could rival Athens. Several of the city-states of southern Italy, shut in between the rugged Bruttian Mountains and the sea, must have been very small, some no larger than the petty urban centres of Chalcidice. On the basis of the actual size of their walled central-places and of the extent of cultivable land, the present writer would not put the population of the Greek *poleis* of south Italy above 150,000, and of those of Sicily, above 350,000.

Any estimate of the population of the Etruscan world of central Italy is made very difficult by our ignorance of their settlement pattern. Most of their cities are sufficiently familiar for a not unrealistic estimate of their size to be possible. On the evidence of its cemeteries, the population of the important Etruscan city of *Caere* (Cerveteri) has been put at about 25,000,[57] and Ward Perkins estimated the urban population of Etruria at about 200,000. On the basis of a careful examination of sites within certain restricted areas, the same author concluded that 'the picture of settlement at the height of the Etruscan period was probably very similar to the picture that one would find, for example, in the 12th, 13th or 14th centuries'.[58]

Despite the intensive archaeological work that has been carried out on the site of Rome, even the extent of the city in the fifth century is not clearly known, and any guess at the size of the Roman population must be based on the extent of the cropland which it controlled. It might not be unrealistic to say that the whole Etruscan–Latin settlement area had a population density not less than that of the Peloponnese, and locally considerably higher. The evidence points to a relatively dense population in Latium, and attempts at land reclamation suggest some considerable pressure on agricultural resources.

The few Greek and Carthaginian colonies of the western Mediterranean were individually very small. Massilia, in virtue of its commerce with the Celts, was the most prosperous and probably the largest. The Carthaginian settlements of western Sicily were small, and occupied narrow and restricted sites strongly protected by nature. It is highly improbable that any settlement had more than about 30,000 people all told, and most almost certainly had less than 10,000.

Any estimates of the population of the rest of Europe must be based on the economies practised. The Villanovans and the related Atestines and Comacines of northern Italy and perhaps also the La Tène Celts probably had a density as great as that to be found in the agricultural region of Thessaly. The mountainous areas of Italy and Sicily, with an economy that was still predominantly pastoral, may have resembled Epirus. The plains of northern France and the chalk country of southern England with, in all probability, a fairly thin agricultural population, may have been as densely peopled as marginal areas of the Greek world, such as Macedonia and the interior of Thrace.

Economy and population were, however, very closely related to the quality of the land. Iron tools were known, and could have been used to cut down the forest trees, but it is doubtful whether much progress had been made in this direction. Large areas of clay soil lay waiting to be cleared and cultivated, but without a heavy plough, fitted with a metal coulter, such soils were too intractable. Soils much inferior in fertility to the clays were used in preference to them because they were more easily cleared and tilled. Settlements were, for example, relatively numerous on the sandy heaths of Denmark and north

Germany and on gravel terraces. On the other hand, Iron Age, and also Bronze Age finds are rare from the areas of heavy clay. A similar pattern is visible in France and Germany. Throughout northern and western Europe there were large areas where there was probably little or no settled population.

The lowest density of population was to be found in those areas where agriculture was either impossible or had not yet made its appearance. These would have included much of the alpine mountain system, some of the cool, damp uplands of north-western Europe and much of the northern regions of forest and tundra. In some of these areas pastoralism was practised. It is even possible that refugee groups, driven by invaders into such areas, had reverted from crop-farming to a predominantly pastoral economy. Much of Scandinavia, Finland and northern Russia, however, knew only a hunting-fishing economy, and its population can have been no denser that that of the Canadian north a century ago.

SETTLEMENT

The Europe of the mid-fifth century was in the early phases of what Childe has called the 'urban revolution'. This development had made most progress in the Greek world of the Aegean and Magna Graecia. This world was made up of *poleis*, communities – most of them small, 'face-to-face' communities – which ran their own affairs and acted in almost complete independence of one another. The *polis* was not a city or town, though its area was likely to embrace some urban-type central-place. Its citizens lived both in the central 'city' and over the surrounding rural areas. Most of the population in the majority of *poleis* lived by agriculture. Crafts were probably carried on in the central-place, and possibly also in smaller and more rural communities, and if there were specialists there must also have been some local exchange and trade on however modest a scale. Such activities were focussed in the central-place of the *polis*. There, too, its citizens met to conduct their political affairs and to enjoy the company of one another and such spectacles as may have been provided for them. In most *poleis*, furthermore, it was a journey of only an hour or two on foot from the farthest corner of the *polis* to its focal city. There were, however, a few *poleis* larger in area and population than most, and with a very much more diversified economy. These included Athens and Corinth, where the merchant and the manufacturer constituted an important but nonetheless small segment of the community.

Urban settlement

The Greek city, the focus of all political and economic activities within the city-state, was commonly located near, but not on the coast. Piracy had long been a danger, and on the islands the settlements were in classical times – as they are today – often sheltered by a spur of a hill so that they were invisible from the sea. Thus, to quote only the most familiar instances, Athens, Megara, Corinth, Sicyon were founded a mile or two from the coast, upon which they each developed a port, separate from the city itself.

There is an interesting contrast between those cities established on hilltops

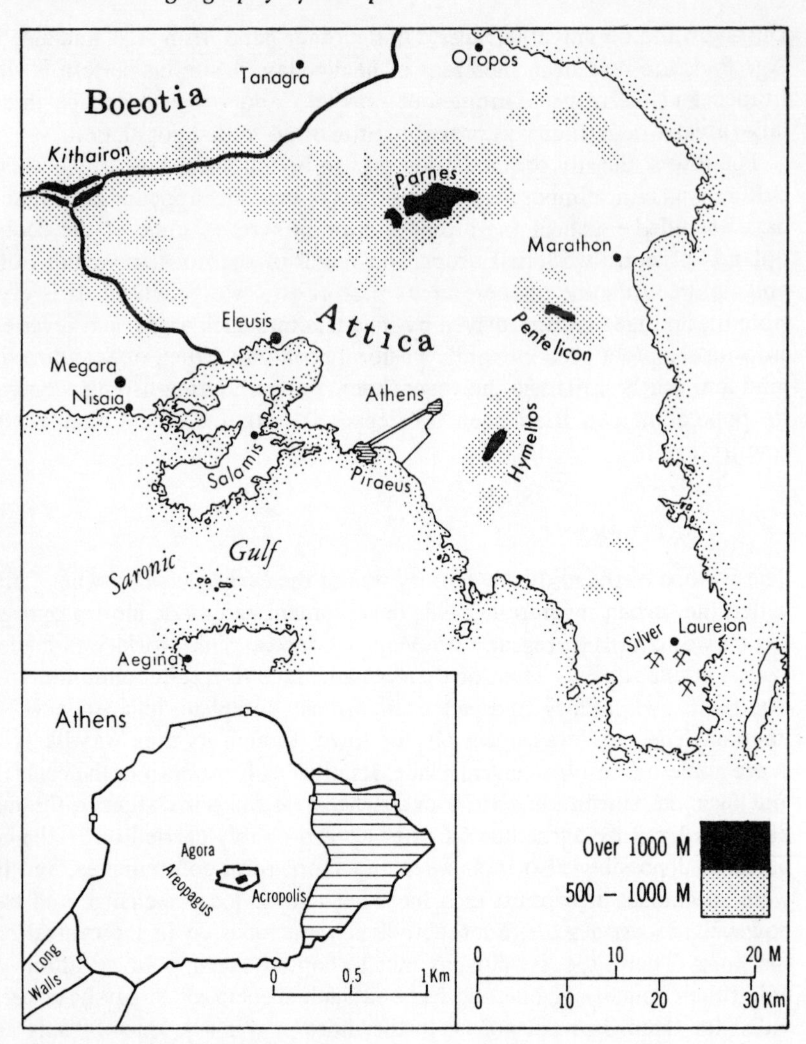

Fig.2.7 Athens and Attica

and those established in the plains, and this may reflect differences in their origin. Apart from colonies which arose, like Athena springing fully armed from the head of Zeus, from the deliberate act of their parent cities, we have two categories of city: those formed by the synoecism of villages, such as Athens, Sparta, Tegea, and those which had grown gradually from smaller village settlements. The latter are most numerous in the west of Greece. They were commonly located on hilltops, like Boeotian Tanagra, which, in the words of Dicaearchus, 'stands on high and rugged ground. Its aspect is white and chalky; but the houses with their porches and encaustic paintings give it a very pretty appearance.'[59] It is not difficult to visualise the little city, so resplendent when seen from a distance, shining out from its hilltop above the olive-groves. Such also must have been the hilltop settlements of Arcadia which were abandoned between 370 and 362 B.C. when Epaminondas of Thebes founded and populated his new valley settlement of Megalopolis. Asea, also in Arcadia, must have been similar, an oval enclosure only about 200 metres by 100; walled and containing a predominantly agricultural community of at most two or three thousands.[60] There can be little doubt that these settlements were built on the hills for security, and that in more peaceful times they tended to be abandoned in favour of valley sites, only to be re-occupied in times of invasion and war. There is no means of knowing how numerous were such upland settlements in the interior of Greece, nor the extent to which each formed the nucleus of a distinct *polis*. We may suspect, however, that during the fifth century there was a tendency to abandon them in favour of lowland settlements.

The lowland settlement was commonly the product of the synoecism of a number of village settlements. It was usually larger than the hilltop settlements which it replaced. Many lowland cities had as a nucleus a fortress, or ακροπόλις, which occupied a low hill rising above the level of the plain. The Athenian Acropolis is the most familiar example, but there were many others, including the Kadmeia at Thebes, the Acrocorinth at Corinth, and the Larissa at Argos. These constituted the archaic city, which expanded over the surrounding lower ground, as a result perhaps of the natural growth of its population, perhaps of an organised synoecism. In most instances the city retained some physical contact with its acropolis. At Plataea, as also at Corinth, the newer city actually moved away from the old. The early-fifth-century city lay high up on a spur of Mount Cithaeron; the later city was built on lower ground, protected to some extent by a break in the surface of the land, but within easy reach of its fields. Both the process of synoecism and also the movement of settlement from the hills to the valley floor are perhaps to be associated with the increasing importance of crop-farming and the decline – relative at least – of pastoralism, which in general had demanded smaller and more widely scattered settlements.

Greek cities differed also in the layout of their streets, houses and public buildings. Beside the older unplanned cities were newer towns laid out upon a regular, grid-iron plan. In the fifth century the typical city was still but a jumble of narrow winding streets and one-storeyed houses amid which lay somewhat incongruously the αγορα, or market place, and the public buildings upon which it had perhaps lavished a large amount of its income. It was, in fact, little more than an overgrown village. Even Athens had its slums. 'The city itself,' wrote the Pseudo-Dicaearchus, 'is dry and ill-supplied with water. The streets are nothing but miserable old lanes, the houses mean, with a few better ones among them.'[61] The majority of the houses consisted of a single room, with floor of beaten clay and walls of sun-dried bricks. In the case of Athens the oldest and poorest housing lay nearest to the Acropolis. The excavation of a street of fifth-century Athens, lying at the foot of the Areopagus, revealed 'a minor street and narrow alleys... built up with lesser houses, commercial establishments, and workshops'.[62] Drainage was always a problem, and in the fifth century 'the rain-water apparently ran down the surface of the roadway'. There was no water supply, and the populace was dependent on a few fountains. The rooms, whether workshops, as R. S. Young claims, or dwellings, or, as was probably the case, serving both functions, were grouped around a series of small courtyards, entered through narrow tunnels. Their walls had footings of rough masonry, but the walls themselves were of adobe, and the roofs were perhaps tiled. The buildings may in some instances have had a second storey, but this is uncertain. 'In private life', said Demosthenes, the Athenians 'were severe and simple';[63] their poor remains show just how simple.

In sharp contrast with the domestic architecture stood the public buildings. Life in classical Greece was public, and the *agora*, usually though not correctly translated as 'market place', was it focus. The Persian Cyrus was reported to have said that he never feared people who stood around talking and cheating in the public *place*; but the *agora* was more than a market place; here people passed much of their leisure time, instead of in their dank and narrow homes. Around it were most likely to be found the *prytaneion*, the nearest approach to a city hall; the temples and public monuments and, in all probability, the chief water supply of the city. At Anthedon in Boeotia the *agora* 'is all planted with trees and flanked by colonnades',[64] and Anthedon was a very small town. The *agora* was as diverse in size and plan as the market places in Mediterranean cities today, but we may assume that it was invariably present and became, in fact, a distinguishing mark of the central-place of the *polis*.

Temples and other religious buildings were more varied in their style and their relationship both to one another and to other structures within the city. Their function was to give a greater degree of individuality to the city-state and to intensify its cohesiveness. On their most ambitious scale, realised at Athens, Corinth, and at several cities of Magna Graecia – notably Akragas, they were among the most impressive buildings erected in the classical or any other period. Many cities also possessed a theatre, and some also a stadium for athletic contests.

In sharp contrast with the unplanned city of the Greeks, was the planned, which first made its appearance in the fifth century. It had been foreshadowed almost two centuries earlier,[65] but was not to achieve its fullest development until Hellenistic times. According to Aristotle,[66] Hippodamus of Miletus introduced 'the method of dividing cities', and applied it to the Piraeus. This is usually taken to mean that he devised the method of laying out a series of straight streets, intersecting at right angles and enclosing rectangular blocks. Piraeus continued to demonstrate this plan, but Hippodamus' ideas received little application within the century in which he lived and worked. Thourioi, founded by the Athenians near the site of Sybaris, in 443 B.C., was allegedly 'divided lengthways by the four streets and crossways by three'.[67] Other examples are too late to concern us here, though we shall encounter a method of city planning in Italy that may in fact antedate that of Hippodamus.

The Greek city, whether planned or unplanned, seems in general to have been fortified. Wherever possible, advantage was taken of the contours of the ground to strengthen its defences. Even when the city was established in the plain, it still made what use it could of natural defences. The fifth-century walls of Athens were begun by Themistocles, but after the Persian Wars only 'a small part was left standing'[68] of the old line of the wall. The Spartans, who had never enclosed their own city, proposed that the Athenians 'join with them in razing the fortifications of other towns outside the Peloponnesos which had them standing'.[69] The whole episode has a peculiarly modern ring. The Athenians rejected the Spartan proposal and went ahead with their own rebuilding. Saflund claims that this wall of Themistocles is 'the oldest city-wall which is not merely an acropolis fortification' on the Greek mainland.[70] The pre-Themistoclean wall of Athens, however, almost certainly enclosed land below the Acropolis, and it cannot be assumed that the many cities that were besieged during the middle years of the century had only recently built their walls.

The little evidence that is available suggests that in the fifth century the walls were roughly built, generally of sun-dried bricks, placed perhaps on a masonry footing. This is borne out by surviving fragments of the Themistoclean wall at Athens and by excavations at Potidaea. City walls of the fourth and later centuries were in general far more elaborately built of squared stones, and it is usually these which have survived or have been excavated.

Not all cities were walled or, in fact, capable of defence. Sparta remained throughout the fifth century an unwalled city, made up of four villages. It is remarkable that the only Homeric state to survive as an important political unit into the fifth century should have retained for so long its archaic character. It was enabled to do so by its remoteness and natural protection, 'deep in the hills', by its military strength, its ruthless egoism, and the support which it derived from its membership of the Peloponnesian League.

In other parts of Greece the process of synoecism was only beginning in the mid-fifth century. Only a year or two earlier the villages of Mantinea, an upland basin in eastern Arcadia, had merged to form a walled city which lasted until its destruction by the Spartans in the next century. In all probability its villages had previously constituted some kind of loose political union; but only now, and perhaps in self-defence, did they merge physically to form a single, urban-type settlement. Only a few years earlier (471 B.C.) Elis, in the western Peloponnese, had been formed in the same way, but appears never to have been walled. In the next century, Megalopolis and Messene were also created by a process of synoecism.

The size of the Greek city. It is difficult to estimate the area of the walled enclosure of a Greek city. Not even at Athens is the line of the walls known with precision. There is reason to suppose that in some instances, especially those of the larger cities, the urban area was far from built-up. In fact, one is constantly amazed at the huge areas enclosed by their walls. This area was increased yet farther by the extension in some instances of the city walls to enclose a port on the coast nearby. The Long Walls of Athens, built 461–456 to enclose the port of Piraeus within the same enclosure as the city itself, are the most familiar example, but both Corinth and Megara also had their own 'long walls'. In the words of Wycherley, 'the wall was loosely flung around the city; it was not the frame into which the rest was fitted, and it was not normally a dominant factor in the plan'.[71] The total area enclosed by the walls of Athens was about 475 acres and, including Piraeus and the area between the Long

Walls, about three times this area. The walls of Corinth, excluding both the Acrocorinth and the Long Walls down to the Gulf, were 10 kilometres in length. There were extensive open spaces within the city, in addition to such public places as the *agora*.

When, in the face of the Spartan invasion, the villagers of Attica took refuge within the city walls of Athens, 'the majority took up their abode in the vacant spaces of the city', and even the vacant area of the Pelargikon[72] was occupied. The open nature of the Greek city raised serious military problems; the length of wall to be defended was out of all proportion to the available defenders. Gomme estimated[73] that the total circuit of the walls of Athens was 148 stades or about 16 miles, which in turn suggests that each soldier had about 12 feet of wall to defend on the assumption that he was on duty for twelve hours each day.

At the opposite extreme were the countless small cities, like the thirty towns of Chalcidice recorded in the tribute lists. Today this area of about 3,280 square kilometres has a population of about 135,000;[74] in the fifth century the population was probably smaller, and Beloch[75] suggested 100,000. Its cultivated area is today about 470 square kilometres, and it is doubtful whether it could ever have greatly exceeded 500. The author has argued[76] that the food-base provided by Chalcidice could not have supported more than some 50,000 persons in classical times. In 454–453 B.C. twenty-eight *poleis* in Chalcidice paid altogether a total of 67 talents and 5,000 drachmae to the treasury of the Delian League. This suggests that one talent of tribute can be taken to represent about 750 people. No one would ever claim that tribute bore any precise and regular relationship to the size of the population of each member *polis*. But the assessment must have been some crude measure of ability to pay, and this latter was in most *poleis* dependent upon their agricultural production. Furthermore, the payment represented the *polis*, not only its central-place, and if one derives a population figure from the scale of assessments for tribute, the central-place is likely to have been smaller, perhaps much smaller than this. More than half the *poleis* recorded in the tribute lists each paid less than two talents. The above argument would suggest that their population did not exceed 1,500. All that we can say with reasonable assurance is that they were very small; that every citizen could have been known to every other, and that it truly was a 'face-to-face' society.

Among the Chalcidic cities, Olynthus, rebuilt at the end of the century, ultimately became a city of perhaps 6,000 to 8,000,[77] but can have done this only by absorbing some of the smaller *poleis* of its region. The amount of tribute paid by the latter suggests that some were communities of well under 1,000. What resources had they, one may ask, for building the city wall, which was needed on this frontier of the Greek world? The answer is possibly supplied in Eden's picture[78] of the fortified village, medieval in date, of Mesta, on the island of Chios. Here the outermost houses of the village formed a continuous wall, without window or other opening, strengthened with towers and capable of defence.[79]

Cities of Magna Graecia. These were in most respects similar to the towns of the Aegean region. They were fewer than were to be found in a comparable area of Greece or Ionia, and it is possible that their average size was somewhat larger. They differed also in having been founded on the edge of a barbaric and often hostile country; they were in consequence more strongly fortified. Syracuse, the largest and wealthiest, and Ischia, one of the earliest, were founded, Phoenician-style, on small islands; only later did Syracuse spread to the mainland. Most others conformed in their geographical location and layout to the common Aegean pattern: near but not on the sea, like Acragas, Locri and Sybaris, whose site was later occupied by the Athenian colony of Thourioi. They were backed by a fertile plain, large enough to attract the settlers from Euboea or Achaia; and endowed with a natural harbour, like Tarantum, or at least with a sandy beach where, Homeric-fashion, ships might be pulled above the lap of the waves.

At least two of these cities, Naples and Thourioi, seem to have had a planned layout even in the mid-fifth century, but little is known of most others. Either they were destroyed by the Romans or by the mountain tribes or, like Naples, *Rhegion* (Reggio) and *Zancle* (Messina), they were continuously inhabited and the ancient site and plan obscured by later generations of building. It is thus difficult to form an estimate of their size. That several of them were larger in general than the Aegean cities is apparent. Syracuse may at its greatest have approached the size of Athens. Nowhere in eastern Sicily or in Calabria do we find the swarms of petty city-states such as occurred in Chalcidice and the Hellespontine region. This rule is more nearly one *polis* to each discrete area of fertile, coastal plain. This contrast between the settlement patterns must go back to the earliest days of the settlement of Magna Graecia by relatively large, organised expeditions, while in Chalcidice, Thrace, and perhaps the Hellespont, small bands of Greeks may have infiltrated the half-hellenised rural communities of Thracians.

Urbanism in central Italy. The origins of city life in central Italy are as obscure as the origin of the Etruscans themselves. The older view was that the city as an institution derived from the Ionian world of Asia Minor. But this is now by no means as certain as it once seemed, and it is at least possible that the Etruscan city, like the Etruscan language, may have derived from local roots. The Etruscans were in the fifth century an urban people, and in political organisation were ahead of the Greeks, in so far as their cities formed a permanent confederation. These were far more numerous than the twelve of the Etrurian League, and they were also large and well defended. Veii and Caere each covered over half a square mile within their fortifications, and Orvieto had an area of 260 acres. In general, the occupied sites naturally strong and easily defended, but within easy reach of lower ground. With a few exceptions, such as Veii, Orvieto and Perugia, they were not hilltop towns, and several were coastal, or at least located within a mile or two of the shore.

About 500 B.C. the Etruscans established a settlement near their northern frontier city of *Bonona* (Bologna). It was destroyed by the Gauls a century later; its name is unknown, and it is called today by that of the nearby village of Marzabotto. Excavations over a period of many years revealed a city planned on rectilinear lines.[80] Greek merchants frequented Spina, and Greek influence in its planning cannot be excluded. On the other hand, both Spina and Marzabotto were probably laid out before the period of Hippodamus, and there is, indeed, no good reason for denying that this style of planning could have originated here where we find it.

Etruscan influence in the foundation and growth of Rome, however, is indisputable. Rome is probably the best known city of antiquity because, for reasons that will be examined in the next chapter, excavation of the ancient city was more practicable here than in Athens, Syracuse, or any other great city of the past. At the site of Rome a number of low hills of tufa rise beside the navigable Tiber. Those which came to be known as the Capitol, Palatine and Aventine were detached and steep. The others, Quirinal, Viminal, Oppian and Caelian, merged eastward, without perceptible break of slope, into the plain of the Campagna. Between them were damp valleys which added to the natural protection of the site.

On these hills there grew up small villages of huts in the eighth century. During the seventh century settlement overflowed from the hills to the depression north of the Palatine, which later became the site of the Forum, and at the same time the villages merged to form the federation known as the Septimontium. The archaic Rome of mud-built huts merged about 575 B.C. into the Rome of stone-built houses and public buildings. This change corresponded approximately both with the synoecism of the villages to form a city and also with the occupation of the site by the Etruscans who remained in possession until perhaps well into the first half of the fifth century.

The Etruscans probably laid out the great drain, the *Cloaca Maxima*, along the depression between the Palatine and Capitoline, and they almost certainly built some form of defence for the city. The fragments of wall, once ascribed to the Etruscan Servius Tullius, are now, however, thought to belong to the fourth century. In the mid-fifth century, the Palatine, Capitoline and Aventine relied for protection upon the natural steepness of their slopes, supplemented by fortifications. It is uncertain whether earthen banks had been built to link up the hills and to enclose the Forum area. The chief threat to the city came from the north-east, where the city's 'hills' merged into the plateau of the Campagna. It is almost certain that here an *agger*, or bank, was built across the base of the Quirinal, Viminal and Oppian hills. For the rest, the city is likely to have relied for protection primarily on the river and the marshy ground which bordered it.

There is nothing inherently improbable in a wall of the length of the alleged Servian wall of Rome, but the archaeological evidence is against it. The fact is that Rome, after the expulsion of the Tarquins, was not a large or important city; only, in Bloch's words, 'a Latin city of medium importance, threatened by the mountain peoples surrounding her'. Nothing in her site, form or function suggested in any way her future importance. The site had been chosen by the early Latins, probably because it could be easily defended, and at the same time had access to crop and grazing land. That it had strategic potentialities, and was well placed to extend its authority over central Italy, probably did not occur to and possibly could not have been comprehended by its founders and early settlers. Rome was, however, built on low-lying ground. A valley or plain site was regarded by Saflund as characteristically Etrusco-Roman.[81] The cities founded by the Romans themselves were invariably on low ground, and their establishment led to the abandonment of many upland settlements. The lowland site had many advantages; it was usually closer to good cropland and to commercial routes; it permitted a planned layout, and also encouraged later growth and expansion. But it could not come into being until it became possible to build strong walls to replace the natural protection afforded by the steep slope of the hills.

The cities of the Italic tribes were in the fifth century mostly hilltop settlements. There were, however, exceptions. The Italiote city of Arpi, in Apulia, lay on low ground; covered an immense area, and was enclosed by a bank and ditch which must have enclosed cultivated fields as well as houses. Perhaps we have here little more than a barrier to separate cropland from pasture.

Western Mediterranean. Two urban civilisations, the Greek and Phoenician, had invaded the western basin of the Mediterranean. The general nature of Phoenician colonisation has been discussed. It was commercial rather than agricultural; it did not demand large areas for cultivation, but rather defensible sites which compensated for the small numbers of the Phoenicians. Their typical sites were coastal: a peninsula like Carthage or Gades, or an offshore island, like Tyre. The only Sicilian site that is at all well known is Motya. In the mid-fifth century Motya occupied a small island, lying about half a mile offshore and protected from the open sea by sandbanks. Its area of about 125 acres was completely enclosed by its strong masonry walls. Even a small harbour had been excavated within the island and was approached by a gap in the enclosing wall. Motya was a commercial city. It is unlikely that the whole island was built-up, but it must necessarily have been dependent on imports for its food supply. Little is at present known of its street pattern.

Other Carthaginian cities appear to have occupied similar sites, always chosen to give a maximum protection against the native peoples of the hinterland. In Sardinia, the city of Sulcis was on a small island; others – Nora, Tharros and Caralis – were built on promontories for greater security. On the Spanish coast the situation was similar. The most famous of Carthaginian settlements in Spain was *Gades* (Cadiz), which lay at the northern extremity of a long, narrow, sandy island, which partially closed the mouth of the Guadalete. Carthaginian Ibiza occupied a small island and the much later foundation of New Carthage (Cartagena) was built on an easily defended promontory.

Greek settlements in the western Mediterranean were less aloof. Since they were *colonies de peuplement* the small defensive site was less useful to them than one with easy access to cultivable land. They appear therefore to have followed the characteristically Greek pattern of a fortified hillock, or acropolis, with a town extending out over the surrounding lowland. Foremost among these western Greek cities was Massilia, which conformed closely to this plan.

An urban civilisation was restricted to the Mediterranean basin, and even here was well developed only in areas of Greek, Etruscan and Carthaginian settlement. Beyond its limits, however, were settlements, urban in scale if not also in function, most commonly associated with the Iberians and the Celts. These usually occupied the rounded summits of hills and were protected by ditches, earthen banks and sometimes by masonry walls. Their permanent inhabitants may have been comparatively few, but they formed refuges for the population of a large area, and were surrounded by small villages of huts, which were abandoned by their inhabitants in time of danger in favour of the hilltop forts.

In Spain these *castros* may have derived certain of their stylistic features from the Greek settlers along the Mediterranean coast, but at least in the north-western half of the peninsula they were built by the Celtic invaders. They were most frequent in Old Castile, where several later formed the nuclei of important modern cities. Some are associated with large enclosures for cattle, and their internal arrangements vary from scattered round huts to rows of stone-built houses. France became in a similar way a country of hill forts, which grew in size and elaboration through later centuries, until the defences of Gergovia, Alesia and Bibracte were to tax to the uttermost the military resources of Julius Caesar. Hill forts were introduced into Britain by the Celtic invaders, who also brought with them the use of iron. Hill forts were fewer in those areas east of the Rhine where the Celts did not settle, or came only in small numbers. They were numerous in southern Sweden, where most were of later date than the fifth century.

On the other hand, settlements which might almost be called urban already existed in the fifth century in the plains of central Poland. The best known of them lay at the site now known as Biskupin. It was founded upon an island, surrounded by the waters of a small lake, which subsequently rose and forced the abandonment of the city in the third century B.C. In this way its remains were preserved until, in 1933, the lake level again sank and permitted them to be excavated. The city was oval in plan, was enclosed by a stout stockade, 463 metres in length, and covered an area of about 4.9 acres. Within the palisade were thirteen rows of wooden houses, lying straight and parallel to one another. There were between 102 and 106 separate dwellings[82] and, on the assumption that each housed a family, the total population may have been between 500 and 1,000. The settlement was primarily agricultural, though not without crafts and commercial activities. In size and function it was clearly not unlike many of the smaller cities of the Greek world. Biskupin was one of several such towns in central Poland at this time; others – much less well preserved and less well known – were at Kruszwica, Kamieniec, Sobiejuchy and Iżdębnik.

Rural settlement

The cities of the classical world have left sufficient evidence of themselves for us to discern the main features of their plan. This cannot be said of villages and hamlets. The rural homes were rarely, and in some parts of Europe, never, built of stone; most often they were of wood which decays or burns. The fields which surround them have in many cases been continuously cultivated since the fifth century, and all evidence of earlier cultivation obliterated by the later. Only rarely and locally have settlements – as distinct from burials – been uncovered, or the pattern of Iron Age fields revealed in aerial photographs.

And from this slight evidence it is difficult and probably unwise to generalise regarding rural settlement. At best we can sample the patterns of settlement outside the cities, wherever archaeological research permits, for the literary evidence is negligible.

Rural settlement in Greece. Even in Attica, the best documented and best known part of Europe at this time, it is difficult to form any picture of the countryside. Gomme estimates[83] that about 160,000, somewhat more than half of the total population of Attica, lived outside the city of Athens-Piraeus. The evidence suggests that they lived in compact villages, whose sites have in many instances been continuously occupied since classical times. The villages seem to have varied greatly in size, and the more important became the foci of the *demes*, the smallest administrative divisions of the Athenian state, corresponding roughly with the English parish. There may have been scattered peasant farms, and there were certainly some large establishments, resembling the *villae* of a later age, such as the farm described by Xenophon.[84] In the mountains, where a pastoral economy predominated, dwellings were in all probability scattered, and some were perhaps only inhabited in summer. The commonest building material seems to have been sun-dried bricks, sometimes placed on a footing of rough masonry. Roofs may have been of shingles or thatch. Wood construction may have been used in the mountains, but timber was by no means abundant in Attica, and was not much used anywhere in Greece except for shipbuilding.

The settlement pattern of the rest of Greece was probably similar. In the plains villages were built small and compact, both for easier protection and to economise the scarce cropland; amid the mountains they were set on hilltop sites which had the advantage of being easily defended. Some of these sites were later enclosed by walls whose remains today are extremely difficult to date, but are more likely to be of the fourth century B.C. than of the fifth. Yet others were probably enclosed by the outer walls of the houses, built as a continuous screen in the manner recommended by Plato for his ideal Cretan city.

Outside the Aegean basin, Greek colonies clung to a narrow coastal plain backed by the mountains and the mountain tribes. In the fifth century B.C. the latter had been little influenced by the Greeks. They lived mainly in villages and in small hilltop towns, such as Livy described among the Samnite hills. In detail they may have resembled the collection of huts uncovered on the Palatine at Rome, but only rarely is it possible to form any clear picture of the geographical pattern and the architectural style of such settlements.

An exception is provided by the *nuraghi* of Sardinia, tall round, masonry structures, sometimes three storeys in height and dominating the circular village huts which were grouped closely around them. That they were built for defence is very probable, but why structures of this degree of elaboration instead of a rampart around the settlement we do not know.

Northern Europe. In non-Mediterranean Europe, the dominant settlement pattern was one of small villages and scattered farmsteads. These varied greatly in location, layout and architectural style. The data, however, are at present quite inadequate to permit the compilation of a map of settlement types. One can, nevertheless, recognise rudimentary street and nucleated villages, and one can only generalise from these regarding the morphology of settlements at this time.

Over much of central and western Europe the economy was predominantly based on crop-farming, though nowhere exclusively so. No settlement is known to have been without its stock, and in the Alps and even in the hills of central Europe, animal husbandry probably replaced crop-farming as the dominant means of livelihood. Towards the north crop-farming entirely gave place to pastoralism, and in Scandinavia, pastoralism was itself replaced by an economy based upon hunting and fishing. The type of settlement was conditioned by the economy. Large villages were probably rare, and could only have occurred where crop-farming was practised relatively intensively. Most settlements in all probability were hamlets, consisting of only two or three farming families. The isolated farmstead was far from uncommon, even in good farming country, and was probably more abundant in the highland regions where pastoralism predominated. Some of these may have been the seasonal homes of transhumant herders, the predecessors of the shielings and saetars of more recent times.

In central and north-western Europe, the fifth century B.C. was a relatively peaceful period. Conditions were favourable to widespread settlement, uninfluenced by military considerations. Cultivable land was probably abundant; migrations were slow and involved relatively small numbers; the period of frantic fortification, represented by the Scottish brochs, the Gallic *oppida* and the late Iron Age forts of southern Britain and Sweden, and associated with the outward movements of the Marnian (La Tène) tribes, still lay in the future. Few settlements of the fifth century B.C. were fortified, and such protection as they had seems to have been designed as much to keep the animals from wandering as to deter human beings from entering. Furthermore there appears often to have been a surprising continuity in the occupation of a particular site. At Ginderup, in Jutland, remains were found of seven superimposed houses. They were built of wood or sod, which decayed quickly or easily caught fire. For a period of maybe several centuries there was apparently no violent disturbance. At numerous other sites the evidence points to long-continued settlement through the last phases of the Bronze Age and the earliest of the Iron.

At Wasserburg Buchau, on the shore of the Federsee, in Württemberg, a village grew up in the late Bronze Age. It was compact, and, with 38 houses, was probably large for its time. These were replaced in the early Iron Age by nine very much larger farmsteads, irregularly scattered within a palisaded enclosure. This represented a prosperous farming community; the village contained granaries, and the farmsteads themselves were built on a courtyard plan, each forming three sides of a square. It was, perhaps, typical of Iron Age settlements in central Europe.

As a general rule houses in continental Europe were rectangular in plan, with a length considerably in excess of their width. Walls were most often of vertical wooden posts, with wattle-and-daub sidings, supporting a ridge roof. In Denmark and Sweden, the roof was commonly supported by two rows of internal wooden posts, outside which was a wall of clay or sod. Such 'long-houses' go back to the practice of the Danubian peasants who colonised the loess areas of Europe in the Neolithic. Remains of such clusters of houses are not uncommon in Germany and Scandinavia; they are mostly later in date than the fifth century, but their architectural style is much older than their physical remains.

At Vestervig, in northern Jutland, was a small street of such houses, of which five on one side and one on the other have been excavated. Here is probably an ancestor of the 'street-village' of central Europe, and Hatt considers that such villages were 'general in North Jutland in the Iron Age'. On the islands of Oland and Gotland quite large villages of such houses developed later in the Roman Iron Age. This was a period of worsening climate (see p. 13) and of rising water levels. Groups of houses were being built on small artificial mounds, or *terpen*, near the Frisian coast and amid the lakes of the Netherlands, and on crannogs in the shallow lakes of the Alps, south Germany and the British Isles.

Less is known of settlement patterns in eastern Europe, because fewer sites have been excavated, but the evidence suggests that eastward through the north German and Polish plain and on through at least the southern margin of the great forest belt of Russia, a similar type of settlement prevailed. The common pattern consisted of small villages, rarely of more than ten or a dozen houses with their barns and sheds. These villages, however, differed from those met with in central and western Europe, in being fortified and located, furthermore, on sites well suited to defence. The reason is to be found in the disturbed condition of eastern Europe, and the defences may have been erected against the Scythians who at this time were raiding deeply into the forest zone. It is probably in this context that we must understand the fortified town of Biskupin. A Baltic village, apparently typical of this kind of settlement, was excavated at Starzykowe Małe, in northern Poland. It consisted of eight houses with their outbuildings, set in a rough circle, enclosed by a double rampart of stones, and located on a peninsula of dry land protected by marsh on three sides and a multiple wooden rampart on the other. The houses were rectangular in plan but smaller and probably simpler than those found at Buchau and Biskupin, and the population of the settlement was probably between forty and sixty. At present some twenty fortified villages of this kind and period are known along the lower Vistula and in Masuria and Samland, all of them strategically placed and as strongly protected as the nature of the site and the technology of the period allowed. No less than seven are found close to Starzykowe Małe, but as not all have been excavated it is uncertain whether they were contemporary. If it can be demonstrated that they were, this might imply a relatively dense population of over 100 per square mile in this restricted area, a figure which seems high for this early date.

Eastward from the Baltic region the same type of settlement extended through the region of broad-leaved forest to the basins of the Oka and Volga. Here were many hundreds of fortified hilltop villages; they lay on the bluffs above the rivers or on islands of dry land amid the marsh. Some sites had been occupied for many centuries, possibly since the time of the early Bronze Age

Fat'janovo culture. Most consisted of a few huts, irregularly placed in general, but sometimes arranged on a street plan, and protected by fence, bank and ditch, as well as by the nature of the site. The village of Svinukhovo, on the Urga river in central Russia, is typical of such settlements, built against both the Scythians from the south and the Finno-Ugric nomads to the north. Some such villages were quite large; that of Starshy Kashir, described by Mongait, had 32 huts; others showed round instead of rectangular huts, showing in this perhaps a Tatar influence.

In Scandinavia the land extended far to the north of the limits of cultivation and of the village settlements. There is evidence for small hamlets of settled fishermen along the Norwegian coast as far as Lofoten and Vesteraalen, where they had lived since the Neolithic. By contrast, in the vast interior lived, hunted and fished a small, scattered and nomadic population of Ural-Altaic peoples, ancestors of the Lapps. Yet even they were not without homes, fixed points to which they returned seasonally in the course of their wanderings. The nomadic hunter-fishermen seem to have had both winter and summer dwellings, the former strongly built and located on the coast or lake-shore, the latter more lightly built hunting camps on the high fjeld. These semi-nomadic communities seem indeed to have been in some instances larger than the villages of Slavs and Balts found in central Europe. In northern Norway there were 16 village communities 'with a total of over 200 houses'.[85] The largest settlement, that at Karlebotn in Finmark, had no less than 72 houses; some of them round or four-sided turf huts; others, oval in shape, dug deep into the beach terraces and approached by tunnels. They were arranged in an irregular line along the outer edge of a fjord terrace.

This was a Stone Age community, with no evidence by which it could be dated precisely. But here the Stone Age lasted into times which farther south are termed historic, and some of its inhabitants, with their tools and weapons of slate and bone, could well have been the contemporaries of Pericles.

AGRICULTURE

The previous pages 55–68 have shown how steep was the cultural gradient at this time from Athens northwards through the farming villages of the Celts to the pastoral settlements of north-western Europe and the Stone Age communities of the north. This transition is reflected in the pattern of agriculture, though the cultivation methods of the Greeks were less advanced than their skills in many other directions.

Attica. The picture of agriculture in the fifth century is clearer for Attica than for any other part of contemporary Europe. This was a small and infertile region and its soil was 'poor and thin'.[86] Of its total area of about 1,000 square miles probably no more than a third could be cultivated and another third was made up of forest and rough grazing.[87] A further 60,000 acres perhaps were planted with tree crops, especially the olive and the vine. The best agricultural land was naturally the small, broken areas of coastal plain which lay between the mountains and the sea. There the soil – in part alluvial – was richer and capable of holding some moisture through the hot, dry months of summer, but over much of Attica the limestone either came to the surface or was hidden only by a thin mantle of alkaline soil. Rainfall was small but often violent, and soil-erosion a danger which the Greek farmer tried to guard against by building low terraces and laying out his fields across the slope of the hills. The extent of cropland, as revealed by crop marks on the land today, may have been more extensive than at any subsequent period. Cultivation terraces are traceable up to almost 1,000 feet on the slopes of Kiapha Drisi, and the limit of grain cultivation may have been much higher.

Forest covered much of the higher ground, and charcoal was burned on Parnes and Pentelikon, where it was customary also to graze the sheep in summer:

> My flock I was a-herding . . .
> . . . Came this fellow – he's a charcoal man.
> . . . Unto this self-same place to saw out tree-stumps there.[88]

But the forest was disappearing fast before the charcoal-burner and the goat, and rough grazing land, scarred perhaps by erosion, was spreading up the mountain sides in the wake of the forest. The herdsman in the *Bacchae* describes himself as coming from Cithaeron, 'where the ground is never free from dazzling shafts of snow', where he 'was pasturing my cattle and working up towards the high ground'.[89]

There were no meadows. There was grazing land along the banks of many rivers, like that which Hephaestus showed upon the shield of Achilles, or 'the banks of Tanaii', on which the shepherd 'tends his flocks',[90] but in winter it was likely to be waterlogged and unsuitable for the autumn-sown crops or even for grazing animals. It is not surprising that in his ideal city Plato allocated to the same person the tasks not only of providing water for the crops but also that of eradicating it from the fields. Drainage was the counterpart of irrigation.[91]

The crops grown in Attica were mainly grain or bread crops. Wheat was the most highly esteemed; barley, the most widespread, because in the dry climate and alkaline soils of Attica it cropped appreciably more heavily than wheat. Oats were known primarily as a weed found in the wheat field, and their cultivation did not seem to be worth developing. Climate and soil militated against rye; maize was unknown, and millet and rice, though known, were not seriously cultivated.[92] Figures have survived in the Eleusis tribute lists of 329–328 B.C. of the wheat and barley production of Attica. Of a total yield of 528,000 bushels about 86 per cent was barley and the remainder wheat. Making allowances for the somewhat heavier yield of barley, this suggests that about a seventh of the grain-growing area was sown with wheat. The figures cannot have been very different in the fifth century.

Cereals provided the bulk of the food consumed in most parts of Greece. The wheat – most of it of a 'naked' *durum* type – was made into bread, but barley, owing to its lack of gluten, does not produce a light bread, and is most palatable when cooked, with seasoning or flavouring, as a kind of porridge. A number of vegetables, including beans, lentils, beet and garlic, were grown and used. Milk, mainly from the goat, was drunk; meat was eaten very sparingly; fish was an important food, but olives were the chief source of fats. Gone from Greece were the heroic feasts described by Homer. These belonged to a more pastoral and less crowded age.

The province of Attica has today an area of about 1,460 square miles, of which about 24 per cent is under cultivation.[93] The current yield of wheat is about 15.2 bushels to the acre, and of barley, about the same. If these conditions had obtained in the fifth century B.C., and the cropland had been devoted exclusively to wheat and barley, food could have been produced for over 200,000 people. On the other hand, allowance must be made for the area under fruit trees and vegetables, for the fact that cropland was cultivated in alternate years, and probably for lower yields. This calculation would appear to suggest that the agricultural production of Attica was sufficient for the rural population, with perhaps a small surplus to help feed the city.

This assumes that both agricultural technology and the area under cultivation have remained unchanged. Farming methods have indeed remained very simple, and it is hard to believe that yields per acre have greatly increased, until very recently, since classical times. It cannot be assumed, however, that the cultivated area has not altered. The field patterns suggest that agriculture covered, in fact, a larger area in the fifth century B.C. than it has done at any time since. Some of the fields lay on land of such poor agricultural quality that one must suppose that over a third of the total area of Attica may, in some way or another, have been under cultivation. Even if we allow for the fact that some of this land yielded well below the average; that part was occupied by buildings and paths, and that vineyards, olive groves and vegetable plots may have covered up to 10 per cent of the cultivated land, it may still have been possible to raise enough grain from the soil of Attica to feed at least half the Athenian population in the mid-fifth century.

It would, indeed, have been surprising if this were not the case, since more or less half the Athenian population lived in the rural areas and worked on the land (see above, p. 65). To say that grain production was smaller than this is to suggest that the farm population was unable to feed even itself. It can, of course, be argued that a high proportion of the farmers of Attica were specialised in viticulture and olive-growing, and that they purchased their bread-crops by the export of oil and wine. That there was such an export is apparent, but, as will be shown later, it could not have been on such a scale as to detract greatly from the self-sufficiency of the Attic farmer.

On the related questions of the area under cultivation and the volume of production we must suspend judgment, suggesting however that it may well have been appreciably higher than has generally been supposed. On the structure of agriculture we are no more adequately informed. That Attica was a region of small peasant farms is apparent. The desires and the problems of the small farmer run through the plays of Aristophanes: he gazes

> . . . fondly country-wards, longing for Peace,
> Loathing the town, sick for my village-home,
> Which never cried, 'Come, buy my charcoal,

My vinegar, my oil, my anything',
But freely gave us all.[94]

A large number of Athenians, if not a majority, appear to have had some rural property, on which many of them lived until the fortunes of war drove them to find refuge within the city walls of Athens. Most of these holdings were small, and Finley was able to find evidence of only five holdings which could possibly be called large.[95] The largest of which we have any clear evidence is the estate of Phainippos, estimated to have been of 700 to 1,000 acres.[96] Such an estate, like the one described by Xenophon[97] or hinted at in the spurious Aristotelean treatise[98] on farming, must have been run by slave labour. But the great many small farms were certainly cultivated by their owners, with only such help as they could get from their families and perhaps from a slave. Evidence of the size of peasant farms is very scanty, but a 14-acre farm seems to have been considered small, while farms of 45 to 70 acres were probably above the average. Glotz thought that a medium-sized farm had 30 to 50 acres of cropland.[99] If we make allowances for the fact that each parcel of land was allowed to lie fallow every other year, we have an area under crops not dissimilar to the 30 acres which made up the standard holding in the early Middle Ages.

Even less is known about the yield per hectare. For barley, Beloch quotes several estimates of between 4 and 8 hectolitres per hectare, and one of 18 to 19; wheat yields would have been somewhat lower. If we disregard the higher estimates for barley as quite unrealistic, we have for a 40-acre farm, each parcel of which was cropped in alternate years, a yield of between 90 and 180 bushels. Allowing a quarter of this as seed, we nevertheless have a production sufficient for 8 to 16 people, and this was certainly well in excess of the size of the average Attic family.

Each farmstead had around it a small croft for vegetables, vines, and a few fruit trees. Aristophanes' elderly peasant planned to have:

First a row of vinelets . . .
Next the little fig-tree shoots beside them growing lustily,
Thirdly, the domestic vine . . .
Round them all shall olives grow to form a pleasant boundary.[100]

The fields were almost certainly small, and it has generally been assumed that they were square or at least rectangular in plan. Field boundaries, where they existed, were usually of loose stones gathered from the fields. Ridgway had claimed that the Greeks used a long, narrow 'strip-acre', like those of the medieval open-fields. In this he has been supported by George Thomson.[101] It seems likely, however, that strips, which undoubtedly did exist in fifth-century Attica, were the result rather of deliberate terracing on hilly land. Field shape has always been adjusted to the plough and conditions of ploughing. The Greeks used a light plough, that could be carried to the fields on the shoulders of the ploughman; could be drawn by a single ox, and turned and manipulated within a small space. Such was the plough described by Hesiod,[102] and a plough of this type has remained in use in Greece until today. It was far better suited to small compact fields than to strips. The division of land between heirs was the general rule, and must have contributed to the fragmentation of farm holdings that undoubtedly existed. Disputes about the boundaries and even the identity of minute parcels of land were not infrequent.

The cropland was allowed to lie fallow in alternate years. The fallow was ploughed in spring to kill the weeds before they could go to seed, and again in summer, so that such weeds as survived might be torn up by the plough and shrivelled by the sun. Manure – such as was available – was scattered on the fields, and the grain was sown after the first autumn rains. The protection of the crop against soil-wash during the rains and against choking weeds during the ensuing period of growth was among the obligations of the farmer. Xenophon's Ischomachus, no less than Jethro Tull, believed that the more the hoe was among the roots the better for the crop. The crop was harvested, usually with short, curved sickles, in early summer.

Greece was not well suited to animal farming, owing to the long summer drought and the lack of green fodder at this season. Nevertheless cattle were important as draught animals for the plough.[103] The animals which could fend for themselves on the mountain pastures were more important as a source of food. It appears that many farms in Attica possessed some rights on the rough hill-grazing,[104] and Aristophanes shows us Strepsiades driving in his goats from the waste.[105] Sheep and goats were grazed on the hills, and were probably returned to the lowlands in winter to pick a living from the fallow. Pigs were also reared and probably fed, as they have been throughout history, on domestic waste. A picture thus emerges of the Attic farmer as a peasant, living in a small village, cultivating his holding with a peasant's loving care, and more at home with his transhumant flocks on the slopes of Parnes than in the streets of Athens. The Balkans have not greatly changed in twenty-five centuries.

The rest of Greece. Attica was not typical of Greece, because it was probably more densely populated and certainly drier and less suited to crop-farming than much of the peninsula. The contrast with neighbouring Boeotia was familiar to every Athenian, who jeered at the comfortable and well-fed man from its leading city of Thebes. The alluvial soils of Boeotia yielded better, and the damper climate favoured animal husbandry. 'Pray, how's cheese selling in Boeotia now?'[106] asked the sausage-seller in the *Knights*. To the west, in Aetolia, Acarnania and Arcadia, and to the north in Thessaly, animal husbandry was even more important. The mountain *poleis* probably depended mainly on their flocks and herds. Thessaly was characterised by its great estates, owned by chieftains and cultivated by an unfree peasantry, and it even had a grain surplus which it sometimes sent to the Athenian market.[107] Cattle were kept on large ranches, and sheep migrated seasonally between their winter grazing on the Thessalian plains and their summer pastures on the surrounding mountain slopes. Thessaly was clearly a land of greater potential wealth than Attica, though this was not reflected in its economic and political development until the century following.

Sparta, 'deep in the hills' of Laconia, resembled Thessaly in its wide extent of cultivable land. 'A land of corn, wine and oil', wrote Jardé,[108] not without some exaggeration (see below, p. 88). 'Laconia was unacquainted with any activity but agriculture', and, as in Thessaly, the land was cultivated by unfree helots. Only Corinth, a *polis* small in area and dependent on commerce, had problems of supply that were at all serious, though here they were far less significant than those which faced Athens.

Agriculture was even less important in the far west of the Greek mainland. The mountains were forested. Along 'the shore of the grey sea' there was 'land level enough for the plough', and alluvial soil lay along the valleys, but this was on the whole 'a rough land', and, as in Ithaca itself, there was 'no room for horses to run about in, nor any meadow at all. It is a pasture land for goats',[109] and such most it has remained.

Italy. Around the coast of southern Italy and the more easterly parts of Sicily lay a fringe of Greek agriculture, between the mountains and the sea. The cities of the Greek west were more generously endowed with land than those of Hellas, and the soil, furthermore, was better. Some – Sybaris, for example, and Acragas – were noted for their agricultural wealth. Metapontion placed an ear of barley on its coins; others made a wreath of wheat a part of their symbolism, and the mythology of the grain goddess was particularly strongly developed in Sicily. One presumes that the western Greeks cultivated the same crops as Greece itself, modified to a degree by the somewhat moister conditions. Barley was the main cereal, though wheat may have been more important relatively than in Greece, especially on the volcanic soils of Sicily and Campania. The vine was grown, and the olive was everywhere the mark of Greek settlement. One presumes also that a similar peasant agriculture prevailed, and that the majority of the citizens possessed small farms even if not all lived on them.

Sicily was far from achieving its later reputation as a source of grain, but the plains of Leontinoi, Syracuse and Acragas had a surplus for export, and in the fifth century Athens certainly derived wheat from this source. In the hinterland of Gela was a scattering of small villages, but of the structure of agriculture we know almost nothing, and can only conjecture that the great estates of a later age had not yet appeared. We do know, however, that animal husbandry was important; that swine were pastured in the woods, and that cheese was amongst the exports of Sicily.[110]

Campania merges into Latium, and Latium into Tuscany. The climate is modified with the higher latitude; wheat begins to rival barley in importance; the vine is less important, and the olive disappears. These changes were imposed by nature. On the other hand, the Etruscans and their pupils, the Romans, displayed an agricultural skill that may have been more advanced than that of the Greeks. Latium was populous, and, possibly under Etruscan supervision, the Latins had constructed drainage works to reclaim the damp plain for cultivation.

The traditional Roman farm was of two iugera – about 1.25 acres, clearly insufficient to support a family. If farms were indeed as small as this, they must have been supplemented by land held in common and also by hill grazing.

It is evident that Romans and Etruscans both practised a mixed husbandry, and it seems probable that they used the mountain pastures for summer grazing. Livy praised the 'rich ploughlands of Etruria', and stated that the Etruscans had an exportable surplus of grain.[111] Latium appears, however, to have suffered from recurring shortages, which were made good from the Greek colonies of southern Italy.

There was a sharp contrast between the agriculture of the Italian plains and that of the hill folk. The latter practised a predominantly pastoral economy, and, according to Livy, 'used to ravage the regions of the plain and coast, despising their cultivators, who were of a softer character'. Roman armies had difficulties in provisioning themselves on their campaigns into these hills, where large numbers of sheep, cattle and probably goats lived a partially transhumant life. Little can be said of the Atestines, Comacines and Ligurians of northern Italy, but in the plain of the Po there were settled agricultural peoples.[112]

Western Mediterranean. Colonies in the western Mediterranean seem to have been established primarily for commerce in metals, and very little is known of their agricultural production and consumption. They were probably able to grow most of their grain, though wine and olive oil may have been imported. The Phoenicians, who had inherited the lands of Tartessos, probably had a broader agricultural base, and may in the fifth century have been growing the olive. But around most of the rim of the western Mediterranean, agriculture was probably simple corn-growing, supplemented by transhumant pastoralism.

Northern Europe. Beyond the Alps, the Pyrenees and the Rhodope there was no sudden lapse into barbarism. Agriculture had been practised by settled cultivators for well over two thousand years. The Neolithic peasants had picked out the soils best adapted to their tools and farming methods. The loess soils which they cultivated were not naturally clear of forest, but they could be broken up by a light and primitive plough, and their natural fertility ensured that they could be cultivated for a prolonged period. The coming of metals effected a twofold revolution. Metal axes, first of bronze and then of iron, allowed trees to be cut, fields to be cleared, and crop-farming to be extended into areas which had hitherto been densely forested. At the same time, the use of metal tools permitted the construction of more strongly made ploughs, of metal-tipped ploughshares, and ultimately of ploughs equipped with coulter and mould board. This revolution allowed early man to crop the clay soils, and thus to tap a vast source of food that had hitherto gone unused, but in the fifth century B.C. it had scarcely begun.

Over much of northern Europe a primitive 'slash-and-burn' method of cultivating the land was in use, and some variety of digging-stick was probably the most used agricultural implement. The course of these early revolutions in agriculture is extremely difficult to trace, because, at least on the better soils, subsequent generations of farmers have completely obliterated the evidence of prehistoric occupance. Hatt has, for example, traced and mapped the evidences of early Iron Age fields in Denmark,[113] and has found that the majority lay on the poor, light and sandy soils of the west and north of the Danish peninsula. 'The farmers of the early Iron Age,' wrote Hatt, 'preferred such light soil to the more heavy loam or clay soils,' which, he suggested, remained uncultivated. To a degree this is true, but Brøndsted has shown that the greatest number of early Iron Age burials and artifacts come, not from the region of sandy heaths, but from the region of damper and heavier glacial soils in eastern Denmark.[114] It is evident that some of the population lived in this latter area from which all traces of agriculture have vanished. It remains true, however, that the major assault on the heavier soils had to await the coming of the mould-board plough with its large plough-team, and as a general rule this was not until the early Middle Ages.

In the Low Counties, traces of Iron Age settlements are rare on the sandy heaths but relatively numerous on the light, easily tilled but also more fertile soils of central Belgium. On the unusually well documented island of Gotland in the Baltic, Iron Age settlements generally avoided both the areas of light, sandy outwash and also the heavier boulder-clay soils.[115]

The climate of northern Europe was becoming wetter and cooler during the fifth century (see above, p. 13). Upland settlements were being abandoned, and their fields were given over to rough grazing. While the frontier of settlement dropped lower in the mountain sides, it also retreated along the sub-arctic margin of settlement.

Agriculture was carried on in most of central and western Europe in small fields, approximately square in plan and roughly grouped around the home-stead. These fields varied in size from a fraction of an acre to several acres; sometimes they were enclosed by banks of earth or stone; sometimes they were aligned in terraces along a hillside. Such fields are commonly described as Celtic, and their area of distribution bears some similarity to the settlement pattern of the Iron Age Celts. They appear also to have been adapted to the use of a light plough, or *ard*, commonly drawn by two oxen. This plough did not penetrate the soil deeply, and was quite incapable of turning the sod. There is good evidence, both from later literary descriptions and from scratch marks left on underlying clay, that the fields were commonly cross-ploughed, with the two sets of furrows intersecting at right angles.[116] Light ploughs, such as could have been used in such fields, have been found, carbonised and preserved, in the Danish bogs.

Some of the soils cultivated in this way were very far from productive. Manure may have been used, but the land nevertheless lost fertility quickly, and frequent fallowing was necessary. In fact, the fields appear to have been cropped in alternate years, just as in Greece. A few fossilised stores of grain, together with the imprints made by grains in the soft clay of unbaked pots, provide the only evidence for the crops cultivated. At Vallhagar,[117] the remains were of small and common spelt and of emmer wheat. Modern bread-wheat was not found, but rye was known, and a hulled barley appears to have been widely cultivated.

The Danish evidence presents a not dissimilar picture during the late Bronze and early Iron Ages. Barley was the dominant bread crop, and about three-quarters of the grain and grain-impressions found were to be of naked or husked barley. Other grains represented were emmer, common wheat and oats, both in its cultivated and in its wild form.

The small spelt, found at Vallhagar, had been the crop of the Neolithic farmers, and had once been widely cultivated in central Europe. Its importance had long been declining in the face of heavier cropping cereals, and the Vallhagar site may well prove to be its last known place of cultivation on a significant scale. Spelt itself had been known in central Europe since the Bronze Age, and appears in the Iron Age to have been cultivated over a broad belt of territory in north-west Europe and to have been introduced into southern Britain. Outliers from its main area of cultivation were found in Denmark and Gotland. It was, unlike small spelt, a man-made cereal, the response to natural selection within the primitive cornfield.

The only other member of the wheat (*Triticum*) genus to be widely cultivated was emmer, which had also been brought westward through central Europe by the Danubian peasants. Common wheat had made its appearance but was certainly not important or widely grown. Oats were cultivated in central Europe as well as in Britain, and were associated especially with the Celtic peoples. Rye was the last of the common cereals to appear. It had been a weed of the cornfield, and as cereal cultivation was extended northward into soils more acid and less fertile, rye was able to establish itself at the expense of its less hardy rivals, until it became a crop in its own right. Its adoption as a cultivated crop belongs to the early Bronze Age. The most extensive of the bread grains was barley. The majority of the finds from sites as widely spaced as Denmark, Gotland, and southern Britain suggest that it may have occupied up to three-quarters of the cropland.

In the natural competition between rival species of cereals man was beginning to play a part. We have already seen that in the Mediterranean region he preferred, when he could afford it, a grain sufficiently rich in gluten to make a light bread, and, for this reason, there was a tendency for wheat to replace the heavier cropping barley. He preferred also a grain which separated itself naturally from its hulls, because this facilitated both milling and transport. In general he was forced to compromise between what his palate preferred, and what his soil could yield most abundantly. It is not surprising then that the history of grain cultivation is one of the slow, uneven progress of the 'naked' wheats at the expense of both the hulled varieties and also of barley, oats, and rye.

Animal rearing. In no part of Europe north of the Alps was animal husbandry unimportant, and in hilly and mountainous regions, as well as in sub-arctic regions, crop-farming yielded place completely to pastoralism, hunting or fishing. The small agricultural communities, which were spread unevenly across Europe from the Spanish Meseta to the Russian Steppe, used animals for pulling the plough; the Celts harnessed horses to their waggons and war chariots; everywhere meat and probably milk and milk-products were an essential part of the food supply. The early Iron Age farmer in non-Mediterranean Europe was essentially a mixed farmer.

The modern farmer would be likely to recognise among the stock of his

Iron Age predecessor most of the denizens of his own farmyard. Both the long-horned (*primigenius*) and the shorthorned (*longifrons*) cattle were known. Sheep were reared; goats were kept, though not frequently, and pigs, descended from the native European wild pig, were bred. The evidence, which consists almost exclusively of their bones, tells little of their relative importance. It does however, permit certain generalisations. In the Neolithic and early Bronze, cattle and pigs had predominated; sheep had been few, and horses rare. This began to change during the Iron Age. The pig, in the fifth century, was of relatively small importance; cattle were still probably the most important farm animals, but sheep had greatly increased in numbers and, locally at least, exceeded cattle.

The reason for the change of emphasis is almost certainly to be found in the drastic modification of the natural vegetation during the later Bronze Age. The natural habitat of pigs was the woodlands, where acorns and beechmast provided their accustomed food. The introduction of metal tools led to a gradual destruction of the woodlands, and the growing use of the plough ensured that the newly cleared land was cultivated with some regularity. As the woodland, the natural habitat of the pig, became more restricted, the area of stubble and fallow, suited to the grazing habits of sheep, became more extensive. One supposes that flocks of sheep grazed the plains of eastern France and the cleared areas of the Rhineland and central and southern Germany. Elsewhere, however, the evidence suggests that cattle were more numerous. At the few sites for which there is sufficient evidence – the Masurian region of northern Poland, Groningen (Netherlands), Denmark, southern Sweden, south-western Germany and much of southern England – cattle predominated.

It is only rarely that we can gain an insight into the structure of a north European settlement of this period, and Applebaum's analysis[118] of the early Iron Age settlement on Figheldean Down in Wiltshire is thus of immense importance. The Figheldean community may have numbered about 275 persons, and cultivated some 370 acres. These occupy 'continuous blocks of downland, each compactly covered by fields, with stretches of pasture land in between'. These correspond, he suggests, with the in- and out-fields of a later day. In summer, the stock was out on the rough grazing which separated the cultivated areas and stretched away to the neighbouring settlements. In winter, much of it was brought in and folded on the enclosed fields. This necessitated the cultivation over at least part of the land of a spring-sown crop, which was most probably barley. In the highland regions of Great Britain the retention of spring-sown cereals was more marked than in the lowlands because of the need to use the croplands for winter grazing. In this way the cropping system was adjusted to the practice of transhumance.

Within the European region of mixed agriculture lay many areas, most of them mountainous and isolated, in which pastoralism predominated even to the exclusion of crop-farming. Agriculture was certainly practised in the Alpine and Pyrenean valleys, and it is probable that the mountain pastures were used in the summer. The plateau surfaces of the Massif Central of France were probably occupied by pastoral peoples who preserved cultures older than those of the surrounding lowlands. The same was probably true of the uplands of central Europe and the Carpathian Mountains.

It is difficult, if not impossible, to determine the northern limits of arable husbandry. Towards its margin it became less important and in every way subsidiary to animal husbandry, but even at the remote settlement of Jarlshof, in the Shetland Islands, primarily a pastoral settlement where cattle and fish provided the bulk of the food, crops were nevertheless grown. It was certainly not a plough agriculture, and the farmers probably used some kind of a cash-rom, or foot-plough, such as is still not unknown in these parts. In central Sweden, agriculture was important; in western Norway it had been recently introduced and was superimposed on a basically pastoral and fishing economy.

East of the Baltic Sea, the predominantly agricultural Baltic culture passed gradually into the hunting-herding cultures which lay around the gulfs of Bothnia and Finland and extended across northern Russia. Almost certainly crop husbandry did not extend beyond 60° N. (see above, p. 68) and in Russia nowhere approached this latitude.

The Steppe. South-east of the Slavs and the Balts, whose settlements in the main occupied clearings in the forests, lay the south Russian Steppe, the settlement area of the Scythians. Herodotus used this term to embrace people of the wooded Steppe and forest margin to the north and west, who were more likely to have been Balts or Slavs. The true Scythians of the Steppe seem to have been in a transitional phase from a predominantly pastoral to a primarily agricultural economy. The importance of the horse in their society is implicit in their art. The horse motif runs through it, and on the highly decorative Chertomlyk jug is extended to a series of horse-training scenes.[119] The true Scythians remained semi-nomadic. Whether they also practised agriculture is uncertain; if they did not, they must have obtained grain from the settled or agricultural Scythians, who are likely to have been Balts or Slavs under nomadic Scythian domination. It is perhaps permissible to think of the central Steppe, from the river Bug eastwards to the Kuban, as still the sphere of the nomadic Scyth, but to regard the people of the lightly wooded margins as abandoning slowly their nomadic ways – if indeed all of them had even been nomads – and forming settled, agricultural communities.

The southern margin of the Steppe was the source of much of the grain imported by Greece. This surely was not produced by the Royal Scyths of the Steppe; the producers are more likely to have been the native and pre-Scythian peoples, possibly related to the Thracians. We can only guess at the extent of commercial wheat farming. Given the problems of overland transport, it is unlikely to have extended many miles inland from the coastal cities through which it was marketed.

INDUSTRY AND MINING

Manufacturing in the fifth century B.C. had not risen above the level of a domestic craft. Some writers claim to see in Athens the beginnings of a factory system, based upon the labour gangs of slaves, herded into large *ergasteria*.[120]

There is evidence from the fourth century, much of it in the speeches of Demosthenes, for large workshops, and in the late fifth century the shield factory of Cephalus employed 120 slaves. It is impossible, however, to say how common such undertaking were or whether they were entirely exceptional. The poor craftsman, whom Lysias defended, was perhaps more typical in his hope that he might some day have a single slave to help him in his old age.[121] The typical Athenian workshop 'probably contained the master and at most two or three hands'.[122] Only in mining was the scale of operation generally larger, and here it sometimes ran to the opposite extreme. In 413 no less than 20,000 slaves, 'most of them *cheirotechnai*', or skilled workers,[123] are reported to have deserted and joined the Spartans who had occupied Decelea. It has generally been assumed that they came from the silver mines near Laureion, in southern Attica, and it is certain that slaves were generally employed in the mining operations of the Greeks.

Greek manufacturing. Only for Athens can a picture be formed and even here the gaps in the literary and epigraphical record are large and numerous. We must visualise a city inhabited in part by craftsmen. They lived and worked in small establishments, which embraced both living quarters and workshop. These surrounded the small courtyards, which opened off the narrow streets of the poorer and topographically the lower quarters of the city. Here lived the masons and bronze workers, tanners and leather workers, the fullers, dyers and potters. The congestion, with its accompanying heat, noise, and smells, are left to our imaginations.

Some craftsmen – possibly those who worked for hire – lived in the surrounding country, and commuted, presumably daily, to their work. The Erechtheon building accounts show that the *deme* affiliations of the masons lay up to 15 miles from the Acropolis on which they were at this time employed.[124] It is even possible that those who lived at a distance from the city were part-time farmers. Within the city itself the various crafts were not in general segregated from one another, though some tended to be concentrated in certain quarters of the city. There was a 'Street of the Marble Workers', and the Kerameikos district was presumably given over mainly to the potters. Masons appear to have predominated in some of the more southerly of the city's *demes*, perhaps because of the proximity of Hymettus and its marble quarries. Wood-working predominated in the village of Acharnai, which provided carpenters for work on the Erechtheon.

Food processing and spinning and weaving do not appear to have been widely practised on a commercial scale. The archaeological evidence suggests that no well-to-do home was without its small, hand-operated corn-mill and its bakery, though the poorer homes, where there was little room for such luxury, relied upon small commercial bakeries. In every home wool and flax were spun and woven. Every housewife was, like Penelope in the Odyssey, employed almost continuously, but doubtless more effectively, at the distaff and loom,[125] which was of the tall, narrow variety, suited to the confined space of the Greek home. Cloth-finishing, however, seems to have been conducted on a commercial basis. The cloth was fulled with the feet, using a fuller's earth or gypsum, and then dyed and finished for the market. But of the commercial organisation of these operations nothing is known.

In a well-known passage in the *Cyropaedia* Xenophon compared the narrow specialisation of craftsmen in large towns with the breadth of their activities in small: 'In small towns the same workman makes chairs and doors and ploughs and tables, and often this same artisan builds houses, and even so he is thankful if he can only find employment enough to support him . . . In large cities, on the other hand, inasmuch as many people have demands to make upon each branch of industry, one trade alone, and very often even less than a whole trade, is enough to support a man: one man, for instance, makes shoes for men, and another for women; and there are places even where one man earns a living by only stitching shoes, another by cutting them out, another by sewing the uppers together, while there is another who performs none of these operations but only assembles the parts.'

The Greeks probably used flax and wove linen cloth, but the matter is obscured by the uncertain meaning of the terms which they used. Though they have provided our own age with much of its scientific terminology, the Greeks showed little precision in their own use of technical terms. Nor is it clear whether they were familiar with silk and cotton; certainly these fibres were not in regular use.

The level of Greek technology was low, and the Greeks did little to raise it. For this Aymard blames the institution of slavery, which removed the incentive to mechanise. In fact, the techniques of the fifth century show little – if any – advance on those described by Homer. Despite the more widespread use of iron, the methods of producing it changed little. It continued to be smelted by a direct process in a small, low furnace, operated by a hand-worked bellows. An impure iron was obtained in blooms and hammered as soon as it had been taken from the furnace, not only to expel as much slag as possible but also to reduce the metal to a merchantable shape. There is nothing to suggest that the Greeks were able to produce cast-iron. Illustrations of iron-making on Greek vases suggest a very small scale of operation in workshops employing only three or four men.

A similar furnace was probably used for making bronze, but in this case the metal was run directly into moulds, and not forged, as was the case with iron. In the metal industries, however, the Greeks operated under serious difficulties. Metalliferous ores were scarce, and the only available fuel, charcoal, was far from abundant in the more densely settled areas of Greece. That iron-smelting was practised in most urban communities is self-evident; without it there would have been neither weapons nor tools. Nevertheless some cities appear to have achieved some distinction for their products; Chalcis made swords; Boeotia, where iron-ore was more abundant than in most other parts of Greece, was known for helmets; Argos, for breast plates, and Laconia for knives.[126] We know, too, that in Athens there were several establishments which turned out swords, knives, and shields.[127]

No branch of manufacture of comparable importance was conducted on so small a scale as the craft of the potter. The pots were turned by hand, and baked in small, domed kilns. The potter could command few technical aids in shaping and decorating his pots; his clay was of indifferent quality and he used low kiln temperatures, and yet Attic pottery of the fifth century is found from southern Gaul and the Lombardy Plain to the shores of the Black Sea. Fine pieces were among the most important of Athenian exports and their fragments are today the best evidence for the extent of Athenian commerce, while a coarser ware was used to ship wine and oil to those parts of the Mediterranean basin not fortunate enough to be able to produce their own.

The Greeks were familiar with glass, but do not appear to have acquired the technique – specifically the high furnace temperature – for its manufacture. Their bricks were sun-dried, though they used kiln-fired roofing tiles. They sometimes used lead dowels in fitting heavy pieces of masonry, but it is doubtful whether lead pipes for water supply were used as early as the fifth century. In Athens, at least, masons, sculptors and carpenters were numerous on account of the ambitious programme of public works. Some of them were probably itinerant like the medieval master-masons; many of them were certainly *metics*, freeborn aliens whom the Athenians welcomed and employed; about half the Erechtheon workers whose status is known to us were *metics*, and about a quarter, slaves. It is a chastening thought that the Caryatides may have been the product of slave labour.

In no other Greek city was manufacturing so varied and extensive as in Athens, but in many others and local crafts did more than supply local needs, Corinth exported its bronzes and pottery; Megara was noted for its manufacture of rough tunics;[128] metal goods were exported by Chalcis and Sicyon.

The Greek cities of southern Italy and Sicily carried on a similar range of industries. Tarantum was noted for its linen, and Syracuse for its woollen cloth. which was distributed over the Greek world. In Italy, however, the foci of industrial activity were Etruria and the Greek and Etruscan cities of Campania. Ceramic and cloth manufactures were important, but both yielded in importance to the iron and bronze industries. The list of commodities contributed by the Etruscans to the expedition of Scipio Africanus included, in addition to corn and timber for ships, the sail-cloth of Tarquinia, the weapons and tools of Arretium, and the iron of Populonia.[129] 'The metallurgical activity of the Etruscans was the most intense in all the central Mediterranean';[130] its focus was Populonia, a city lying on the Etruscan coast opposite the island of Elba. The mines of Elba furnished most of the iron ore, which was carried to the mainland and smelted, evidently with charcoal from the Tuscan hills. The industry appears to have been rural. In the Fucinaia valley mounds of slag have been found, together with remains of a number of small furnaces. Iron workings lay scattered northward to Volterra, south to the hills of the Massa Maritima and inland to Monte Amiata and the Alpi Apuani. The inland sites doubtless used local iron ore as well as charcoal, but some of the mines and smelters were for copper and silver-lead. There was also a well developed metal industry north of the Apennines, in the Atestine territory of the lower Po valley. Here in the fifth century bronze *situlae* and small, decorative articles of the highest artistic quality were being made, some of which were exported northwards to Rhaetia.

Hallstatt-La Tène Europe. By the middle years of the fifth century the use of iron had spread over all Europe except northern Scandinavia and parts of the British Isles. In many areas, the metal must have been acquired by trade and used more for ornament and decoration than for weapons and tools. While this was happening in the outer, expanding periphery of the Iron Age civilisation, a newer and more developed culture was appearing near its centre. By the mid-fifth century the La Tène culture, with its superior armament and its more sophisticated art forms, reflecting Greco-Etruscan influences, was spreading over south-western Germany and eastern France. During the preceding Hallstatt periods, the use of iron had not wholly displaced bronze even in the manufacture of tools and weapons. Bronze continued to be made throughout the La Tène period; it was important chiefly because it could be cast in moulds and hammered into patterns and shapes. But the bronze worker was slowly being displaced by the craftsmen in iron, and those who could not adapt themselves to the new technology had perforce to move on: perhaps the earliest example of structural unemployment. But at its periphery, the bronze culture was expanding into the area of the Neolithic herders and cultivators of the outer margin of Europe. About this time a bronzesmith arrived in the Stone-Age settlement at Jarlshof in the Shetlands. He 'set up his workshop in one of the courtyards, and began to mould swords, knives, pins and other articles that must have impressed the locals as representing the height of luxury and scientific wonder. We can suppose that the smith, on the contrary, felt himself wretchedly benighted, and often cursed an unjust fate which had deprived him of his more comfortable livelihood farther south. For he may well have been one of the victims of the blacksmiths and their new metal, whose arrival in Britain threatened many of the bronze-founders with unemployment.'[131]

Iron was more difficult to smelt than copper. It required higher temperatures and could only very rarely have yielded a fluid metal that could be run into moulds. It is highly probable that iron-smelting was first practised in furnaces built to take copper ore. But once established, it could spread easily because iron-ore is one of the commonest of metalliferous substances in the earth's crust. Two kinds of iron-ore were used, a relatively high-grade haematite or siderite, which had to be mined, and a limonite, or bog ore, which had only to be dug or dredged from the marshes of northern Europe. The latter was widely distributed, easily mined, and quickly exhausted, and its availability helps to explain the rapid spread of iron-working, hard on the heels of bronze, through northern Europe. But it was the ore which occurred in beds and veins that supplied the more important and long-lasting centres of iron-working at this time: the eastern Alps, the Bernese Jura, the Siegerland – very approximately the areas which were to remain pre-eminent through Roman times and the Middle Ages. Iron works were very numerous in north-western Germany. Within an area of only 130 square kilometres near Siegen, considerably over a hundred slag-heaps have been counted, and 63 of them have been assigned with certainty to the prehistoric period. In the Engbachtal, the furnaces – no less than 25 altogether – lay along the valley bottom, close to the stream, but nearly always in a position exposed to the prevailingly west winds. The furnaces could not all, however, have been worked together, and their construction and use probably spanned a long period of time. The ore was obtained from shallow mine workings cut into the valley bottom and sides, and charcoal from the higher slopes.

Mining and quarrying. Classical Greece was poorly mineralised, and the Greeks were heavily dependent on imported metals. A little iron was obtained in Boeotia and Laconia, but was supplemented by imports, certainly from the Black Sea coast, perhaps also from Etruria. Copper was obtained from Cyprus, and tin ore from north-western Europe through some west Mediterranean intermediary. Only in the precious metals was Greece well endowed. Gold, obtained from the alluvium along the streams which descended the slopes of Pangaeus, was used to supply both Athens and the Thracian cities. Gold also occurred along the Axios, Strymon and Hebrus valleys, but the only source known to have been worked in the fifth century, apart from Pangaeus, was the island of Thasos.

The chief source of silver, as also of lead, was the mines of Laureion, at the south-eastern extremity of Attica. The ores, consisting mainly of galena and blende, occurred at the junction of the marble and schist, as these outcropped along the eastern flanks of the Laureion range. Traces of iron ore occurred, and there was some zinc whose usefulness was unknown to the ancient world. The galena was worked on account of the silver it contained. The ore was reached at first by drifts cut into the hillside, and later by shafts sunk from the surface. After it had been raised to the ground level, it was crushed by hand, sorted by gravity, and the ore was then smelted. The object was to extract as much as possible of the silver, and a great deal of the lead remained in the slag. The mines drained naturally, and appear to have been generally dry. Indeed, the supply of water for washing the ore raised an acute problem, partially solved by constructing large storage tanks. Though there is no evidence, it seems likely that the supply of timber for roof supports and for use as fuel must also have proved difficult.

Mining at Laureion was organised in a large number of small enterprises, each of which was leased from the Athenian state. The terms of the leases defined the boundaries of the concessions. Though very few of these can now be recognised, it does seem that a single lease related to only a very small area, probably no more than a single shaft or cutting. The number of mines is not known, nor the manner in which the royalty was paid to the Athenian treasury. Labour in the mines was provided by slaves, and Ardaillon estimated their number as at most 20,000. More recently Lauffer has suggested[132] that their number may have risen to 30,000 during the period of most intensive mine working.

Attica and, indeed, much of Greece were at this time noted for their building stones. Only four miles to the east of the Acropolis rose the steep ridge of Hymettus, its western flanks scarred by the quarries which had furnished marble for the city's buildings. Somewhat farther to the north-east was Pentelicus, which yielded a marble similar to that taken from the lower beds of Hymettus. Marble was brought to Athens, and doubtless also to other cities, where monumental building was in progress, from Aegean islands, such as

Paros. Most other parts of Greece had an abundance of limestone, which was used in public buildings for which marble was either unsuitable or too costly. Nor must one forget the clay, inferior in quality though it was, from which the bricks and pottery were made, and the emery which was already being quarried on the island of Naxos and used for sharpening tools.

After iron (see above, p. 80) the most important of the metals was copper. Bronze was still in widespread use, but, relatively at least, its importance had declined and some of the copper mines which had supplied its principal metal had closed down. On the other hand, bronze was capable of being re-used indefinitely, and the bronze-worker of the Hallstatt and La Tène may well have been the earliest dealer in scrap metal.

Unquestionably the copper-mining regions of Salsburg and Tyrol were still active and provided ore for smelters in the neighbouring valleys. The ore, predominantly pyrites, was first roasted to oxidise it, and then smelted to a matte or concentrate before being exported for further refining elsewhere. Pittioni has identified a number of mines and related smelters in the Kitzbühel Alps and Hohe Tauern, and has estimated, on the basis of the amount of ore removed, that the Mitterberg field alone must have yielded about 20,000 tons of raw copper. 'The total yield of the eastern Alps,' he estimates, 'would be probably . . . about 100,000 tons of raw copper,' and 'the annual production in the Alps might be reckoned at 100 tons.'[133] Doubtless this quantity of copper could at this time have supplied a good deal of the needs of central Europe.

At an earlier period the copper ores of central Germany had been of far greater importance than those of the Alps. The former, which occurred in the Harz Mountains, the Thuringian Basin, and the Erzgebirge of Saxony and Bohemia, contained small quantities of arsenic, nickel, silver and other minerals, and can thus be traced in the artifacts made from them. During the Bronze Age, copper from these deposits was traded widely in central Europe, especially that from the arsenic-bearing deposits near Zwickau, in Saxony. Metal from this source has been traced southwards across Bohemia to the Danube valley and eastwards to the Burgenland. One cannot say how long these ores continued to be exploited, but there was a continuing demand for copper, and one cannot suppose that such well established and abundant sources of ore would be readily abandoned.

Copper was probably also worked at this time in western Hungary, where, it has been claimed, ore production declined in importance during the Bronze Age before the competition of that from central German mines, and in Transylvania, Bulgaria, Serbia and possibly also in the Carpathian Mountains of Slovakia.

Tin in a proportion of up to 8 per cent is an essential ingredient of bronze. In very few places in Europe was tin found in close proximity to copper, but one of these was the Vogtland district of the Erzgebirge; others were north-western Spain and Cornwall. It is presumed that bronze was first cast in central Germany, but the art of bronze-founding and, with it, the trade in the alloy metal, tin, spread over much of the continent. During the fifth century the only significant sources of tin had been reduced to the north-western Spain, Brittany and Cornwall.

Unlike copper, tin does not occur native, but its only important ore – the oxide, cassiterite – is both heavy and inert, and is easily preserved amid the sands and gravels of stream beds. It is presumed that most of the prehistoric mining of tin took the form of the panning of placer deposits. Earlier estimates of both the antiquity and the range of the prehistoric tin trade are probably greatly exaggerated. Archaeological evidence is slight and conflicting, and the literary evidence consists of little more than echoes of early classical authors, repeated, possibly inaccurately, in the writings of the later. Foremost among these latter is the *Ora Maritima* of Avienus,[134] which embodies, in addition to other fragments, part of a periplus, compiled in the sixth century B.C. by a Phocaean sailor from Massilia. At this time Tartessos was the emporium of the tin trade, but whether the metal was obtained from the Iberian hinterland or brought by sea from farther afield, as Avienus supposed, is not clear.

Both Brittany and north-west Spain contained tin, but have yielded no certain evidence of tin mining at this time. On the other hand there is abundant proof of prehistoric working in Cornwall, though it is generally somewhat difficult to date. Countless attempts have been made to identify the Cassiterides of the Classical writers. We must remember, however, that the source of tin was deliberately shrouded in obscurity, and that to many who used the term, the Cassiterides had no more precise a location than the 'Happy Isles'. There *may* have been some contribution from Galicia and Brittany, but it is probable that in the fifth century much of the tin used in the Mediterranean world came from Cornwall. Here it was obtained from the sands and gravels lining the short valleys which drop from the more westerly granite massifs, Land's End and Carnmenellis, to the sea. Diodorus Siculus is quite explicit: 'They ... work the tin, treating the bed which bears it in an ingenious manner. This bed, being like rock, contains earthy seams and in them the workers quarry the ore, which they melt down and cleanse of its impurities ... On the island of Ictis the merchants purchase the tin of the natives and carry it across the Strait of Galatia, or Gaul.'[135] But who the merchants were, whether Carthaginian, or Greek or Celt, we do not know.

TRADE

In a continent in which most communities were largely if not entirely self-sufficing, long-distance trade was limited in volume and restricted in its range and variety. Over much of the continent we must assume an economy of very limited markets, based on the autonomous and self-sufficient family group. Elsewhere, trade was restricted to a few essentials, like salt and metals. Only in parts of the Mediterranean world was there a high degree of agricultural and industrial specialisation, and only here was long distance trade well developed.

Greece

The poverty of classical Greece in naturally occurring resources was a commonplace of both ancient writers and modern historians. Civilisation was dependent upon trade, because without it the necessary variety of commodities could not have been assembled at one place. Though we can, with some pretension to accuracy, describe the variety and direction of Greek trade, its volume is incapable of precise estimate, and its size and importance relative to the economy of the polis have been a matter of controversy. It is evident that even the smallest city-state engaged to some degree in 'foreign' trade. 'It is almost impossible,' wrote Plato, 'to found a state in a place where it will not need imports.'[136] Each must export in order to import as well as make provision for merchants, mariners and shipbuilders.

The Greeks distinguished between the local trader – κάπηλος – who sold the products of the fields and workshops in the local markets, and the ἔμπορος, who dealt in foreign goods. The former did business in the market-place, which was a feature of every Greek city. The comedies of Aristophanes are full of references to the peasants who brought their farm surpluses to the Athenian markets, to their devious tricks, and to the low, even the marginal level of subsistence of such people. Game and charcoal; wool, skins and hides; vegetables and garlic were all exposed for sale. It even seems likely, on the evidence of Aristophanes, that there were particular market-areas or market-places allocated to specific types of commodities.[137]

Most *poleis* must have been able to derive the greater part of their food supply from this local traffic, though many appear to have carried on some kind of long-distance trade. Athens is the best documented of the *poleis*, but even here the evidence is almost entirely qualitative, and it is possible to derive widely opposed hypotheses from it. At one extreme stands Beloch,[138] who, basing his argument on slight and far from unambiguous evidence for Piraeus harbour dues, tried to show Athens as a kind of industrial city, importing foodstuffs and raw materials and exporting manufactured and processed goods. At the opposite end of the spectrum of opinion are Bücher and Hasebroek.[139] They admit the importance of the grain trade, and Bücher claims further some importance for commerce in luxury goods, but both deny the existence of a large or significant trade in industrial products.

They have a certain amount of circumstantial evidence on their side. Though references to the grain trade are numerous, there is little literary evidence for trade in other commodities. The sailing season was a short one. In Hesiod's day it had occupied only the months of August and September. It was longer in the fifth century, but may still have lasted less than half the year. Most ships were small, and had a carrying capacity of not much over 13 tons, though there were also larger vessels of up to 250 tons.[140] Ships were generally owned by the captains who sailed in them. The captain was his own master, and took his cargo wherever the winds or his whim dictated. Only the grain trade was subject to any overall supervision, in so far as the supply of grain was a matter of continual political concern and its sale at Piraeus was overseen by officials, *sitophylakes,* of the city government.

It is probable that the largest vessels were engaged on the Black Sea and Sicilian runs. The small scale and the unorganised nature of the trade in all other commodities, the lack of state supervision and control, and the low status of the merchant all suggest that commerce may have been of little importance in general in classical Greece. Greek society, including that of Athens, was fundamentally not geared to 'international' trade. It did not produce primarily for a foreign market, and its *poleis*, though safeguarding their food supply, did not aim to develop and protect foreign markets and long-distance commerce.

On the other hand, there is some evidence to support the views of Beloch. No one disputes that the grain trade was considerable, and its import must have been compensated by commensurate exports of goods or bullion. Secondly, the archaeological finds of Attic pottery over considerable areas of the western Mediterranean suggest that the manufacture and export of ceramics was important. On this matter, however, the literary sources are almost wholly silent. Lastly, the evidence for states interfering with the trade of their neighbours is abundant; 'upon the whole', wrote Boeckh, 'war was as much carried on by impeding commerce as by force of arms'.[141] But here again the most famous and significant example of interference with commerce, the Spartan occupation of the Hellespont, was aimed only at the grain trade.

The Grain Trade. The statement of Demosthenes that Athens imported 400,000 medimni of grain from the Black Sea region and as much again from other sources has been widely used as a measure of the importance of the grain trade. The political circumstances of the speech suggest that he may have exaggerated. His figures do, however, indicate a very great dependence on imported grain, and it is unlikely that this dependence was any less a century earlier. Athens was not alone in its reliance on imported grain. One presumes that such cities as Corinth, Sicyon and Argos also imported grain, and one learns with some surprise that Teos and even the cities of Elis, Acarnania and Thessaly on occasion bought bread-crops.[142] But Thessaly was normally an exporter of grain through its port of Pagasae,[143] and one is tempted to assume that under normal conditions most *poleis* were self-sufficing, and that imports became necessary only when war or weather had destroyed the crops. It seems likely, however, that all coastal cities at least on occasion imported corn, but that inland cities had of necessity to be self-sufficing. A passage in Xenophon[144] describes the feelings of the Acarnanians 'that inasmuch as *their cities were in the interior* they would be just as truly besieged by people who destroyed their corn as if they were besieged by an army encamped around them'.

Thessaly,[145] Thrace,[146] and possibly Macedonia, Chalcidice and Cyprus may have shipped grain to Athens; so also in all probability did Euboea, whence it was taken across the narrow Euripus channel and carried overland through Oropus to Athens. It is generally assumed, however, that wheat from the Black Sea region formed the larger part of the Athenian import. It seems to have done so in the fourth century, when the Black Sea port of Theodosia was founded to handle the trade, but it is by no means certain that the Pontic region had this degree of importance in the fifth. At the time of the second Persian invasion, Gelon of Syracuse offered 'to provision the entire Greek army for as long as the war may last',[147] thus implying that there was normally a considerable wheat surplus in Sicily. When the Athenians were debating the fatal expedition against Syracuse, Nicias, recounting the advantages of the Sicilian cities, noted that they 'grow their own grain instead of importing it'. He did not suggest, however, that the Athenians might live off the country they were to invade, and advocated the export from Athens of the food supply of the army.

There is, nevertheless, good reason to suppose that Sicily shipped grain to the Peloponnesos, and it is not improbable that the Athenian expeditions of both 427 and 415 aimed to deny this source of food to their enemies if not also to gain possession of it for themselves. Gernet considers,[148] and is probably correct in this, that Magna Graecia was a more important source of grain to Greece in the fifth century than the Black Sea coastlands. On the other hand, the scanty evidence provided by Thucydides suggests that the Athenians were deeply touched by the loss in 411 of control of Euboea, 'on which they were more dependent for supplies than [on] Attica itself'.[149]

Although the Athenians had long been deeply interested in the Hellespont and the Black Sea, the really significant references to the Pontic grain trade are later in date than the Spartan occupation of Decelea and Euboea, and it is not impossible that Athens turned to this source only after others – Sicily and Euboea – had been lost. The history of the Greek settlements around the Black Sea suggests that their commercial apogee may not have been reached until the fourth century. Herodotus' description of Scythia contains no reference either to grain production or to the grain trade, which he might have been expected to make, if it had been of vital importance to the Athens of his day. He did, however, tell the story of how, from the shore at Abydos, Xerxes saw 'boats sailing down the Hellespont with cargoes of food from the Black Sea for Aegina and the Peloponnese'. He inquired their destination, and was told that they were sailing 'to Persia's enemies . . . with a cargo of grain'. To this he replied with typical arrogance 'are we not bound ourselves for the same destination . . . I do not see that the men in those ships are doing us any harm in carrying our grain for us.'

We may perhaps conclude that the grain trade of Athens and of some other Greek cities was highly important in the mid-fifth century, but also that it conformed to no regular pattern. This is perhaps one reason why it was regularly discussed in the meetings of the *ecclesia*, or citizen body. Athens, Michell wrote, 'got her wheat from whatever source was available. Barred from one market, she turned to another . . .'[150] In the words of Xenophon, 'merchants sail to wherever they hear there is most of it; and cross the Aegean, Euxine, and Sicilian Seas'.[151]

Lumber and other commodities. The Greek cities had less freedom of choice among the sources of their other imports. Greece itself had but slender resources in timber. The forests which remained were adequate to provide the little timber that was needed in buildings, which were, however, predominantly of masonry and sun-dried brick. They also yielded charcoal, but ships' timbers had to be imported. The western side of the Greek peninsula, with its heavier rainfall and more extensive forest cover, probably provided most of the material from which the Corinthian galleys were built. But cities around the Aegean Sea were in a more difficult position. Writers of a later age, like Theophrastus and the elder Pliny, give a sketch of the forest resources of the Greek world. How many of these were being tapped in the fifth century B.C. is not known. There is good evidence that Athens relied heavily on timber from Macedonia and Chalcidice, and this may have been supplemented from Thessaly and the southern coastlands of the Black Sea. Certain details of the campaigns of the Peloponnesian War have been explained as attempts by the protagonists to cut their enemies off from their supply of shipbuilding timbers, and Alcibiades even declared that 'the timber which Italy supplies in such abundance' was one of the objectives of the Sicilian expedition.[152] It seems that the timber whose lack was felt most acutely was that used in masts and the long oars of the galleys, and for the tall straight-grained silver fir which provided it the Greeks had to go as far as southern Italy. The import of timber must have been almost as important for Athens as that of grain, and its safety was no less a preoccupation of the Athenian government.

That iron as well as copper and tin were imported is evident, but the sources of these metals are far from clear. Euripides mentioned 'iron from Sicily',[153] which may very probably have been Etruscan iron transmitted by the Greek merchants of Magna Graecia. Iron was also obtained from the south-eastern coastland of the Black Sea, the land of the Chalybes, and perhaps also from Asia Minor, though there is no real evidence of this. Nor is the source of copper and tin apparent. One presumes that Cyprus and central Italy were most important for copper, and that the tin came from the 'Cassiterides', but whether the Greeks made their own alloy or imported bronze in ingot form is quite unknown.

Other imports of the Greek world of the Aegean included wool, primarily from the mountainous regions of western Greece and Asia Minor. Later writers specify the sources and quality of the various kinds of wool produced, but such detail is not available for the fifth century. Flax was grown near the eastern shore of the Black Sea, and in Egypt, Macedonia and Thrace. Hemp, for the manufacture of ropes, came from the Scythian lands north of the Black Sea. Tanning was an important industry in fifth-century Athens, and the import of hides was probably large in a land where cattle rearing was as difficult as it was in Attica. The nearest source lay along the west coast of the Greek peninsula, but in the following century the Black Sea region supplied hides to the Greek cities. In addition to foodstuffs and industrial materials there was an import of slaves. The silence of Greek writers on this subject is notorious; no estimate of the numbers acquired is possible; nor do the sources throw much light on the areas which provided them and the means by which they were obtained. It seems probable that the Black Sea coast and the region of the lower Danube were important sources, though others – Asia Minor and Illyria – probably made some contribution.

The exports of the Greek cities form a miscellaneous group of agricultural and manufactured goods. Pottery, both the artistic figured ware and also the more utilitarian vessels and *amphorae*, were among the exports of Attica. Fragments of these wares are found from Massilia to south Russia and the Nile delta, and are evidence of the great range of Athenian trade. Wine and olive oil were prominent among the goods shipped from the Aegean to areas climatically less well endowed. The olive of Attica, which was naturally well suited to produce it, was perhaps the most important item in the export trade, but other Greek cities, notably those along the eastern coast of the Aegean also exported olive oil in return for grain. The market was primarily in the Black Sea region, but also in all probability in those parts of the western Mediterranean where the cultivation of the olive had not yet been established.

The grape-vine was more widely grown than the olive, but the export of wine was nonetheless important. Many Greek cities illustrated the vine on their coins, and we must assume that most were able to satisfy their local needs. A few produced wines of quality which were widely known and presumably exported. Athens was among the states which exported wine, but there is no evidence for the volume of export.

Fish also entered prominently into Greek trade. The Aegean and Black Sea were intensively fished, and the Pseudo-Dicaearchus described the inhabitants of Anthedon, 'almost all fishermen, living by their hooks, by the purple shell and by sponges, growing old on the beach among the seaweed and in their huts'.[154] Obviously there was a large and unrecorded trade in fish and in other products of the sea from many such fishing ports as Anthedon. Doubtless Athens also received fish from this source. The Greeks ate sea-food of every kind, 'from barnacles to whales'.[155] Much of it was dried or pickled, and shipped out in pots made in Attica. Athenaeus, writing at a very much later date, lists the delicacies that were obtained from the waters of Greece: 'Sicilian lampreys . . . tunnies caught off Phalerum . . . fish from Sicyon, eels from Boeotia . . . from the Hellespont mackerel and all kinds of salt-dried fish.'[156] The tunny, which commonly bred in the northern recesses of the Black Sea, was caught in immense quantities off the Bosporus. Many of the cities along the shores of the straits were engaged in the fisheries. Aristophanes compared Cleon to the man set on the cliffs to look out for the shoals of fish,

> watching from the rocks the tribute,
> tunny-fashion, shoaling in . . .[157]

The Bosporus was 'rich in salt fish', and Byzantion, 'the mother . . . of tunnies, pickled in the right season . . . as well as of deep-sea mackerel and well-fed sword fish'.[158] add to these the sea-perch from Azof and sturgeon from the Danube, and we see how great a volume and variety of fish was available at least to those Greeks who lived within reach of the sea.

The extent to which some ports of Greece served as entrepots is far from clear. The large number of foreigners, many of them merchants, who made Athens their base, suggests that the entrepot trade was important, and according to Thucydides, Pericles regarded their presence as Athens' great good fortune. Xenophon, writing early in the next century, is quite explicit. 'Every traveller,' he wrote, 'who would cross from one to the other end of Greece passes Athens as the centre of a circle, whether he goes by water or by road . . .' And Athens has 'the finest and safest accommodation for shipping . . . at Athens [ships] have the opportunity of exchanging cargo and exporting very many classes of goods that are in demand, or, if they do not want to ship a return cargo of goods, it is sound business to export silver'.[159] Xenophon advocated both the intensification of this entrepot trade and also the increased production of silver as a further encouragement to merchants to use the port of Piraeus.

Athens, in the middle years of the fifth century, possessed a great commercial advantage in its port. The Long Walls from the city were made to enclose the peninsulas of Acte and Munychia, while jetties extended partway across the three small bays which constituted the port itself. In this way the harbours had some protection against both high seas and human enemies. Around the shore of these protected harbours were the 'ship-houses', long, narrow, covered buildings into which ships could be pulled for repairs and laid up during the winter season.

This pattern of sheltered harbour and 'ship-houses' could be repeated around the Aegean. It became a common practice for the walls of the city to be continued into the sea, and thus partially to enclose the harbour, even if this necessitated building 'long walls' of extreme length, as at Athens, Megara and Corinth. These harbour walls consisted usually of rubble masonry dumped into the sea until it formed an embankment high and firm enough to serve as a foundation for a wall of regularly laid masonry.

The jetty, beside which ships might tie up for loading or unloading, was similarly built, and, in the few instances known, projected at right angles from the shore. It is probable, however, that only the larger vessels tied up in this way; the smaller were merely dragged up on the beach, as in Homeric times, beyond the reach of the waves of this almost tideless sea.

The Greek sailor was rarely out of sight of land. He had few navigational aids, and his ship was ill-built for storms. Sailing was easy within the Aegean Sea, and on the longer voyage to Magna Graecia, ships followed 'the coastwise route from Sicily to Peloponnesus'.[160]

The western Mediterranean

Commerce in western Greece and Magna Graecia was smaller in volume and incomparably less well documented than that of the Aegean region. Indeed, it is possible only to sketch the broad outlines of the commercial pattern, and even so knowledge is restricted to imperishable articles of trade – bronze, tin and pottery – that have been recovered in recent excavations.

Only a few years earlier a commercial revolution had taken place in the western Mediterranean. Until the last quarter of the sixth century the Greeks of the Aegean and also, in all probability, of Magna Graecia, had carried on trade with the Etruscans of central Italy and the Phocaean colonies of the western Mediterranean. Excavations at Massilia have revealed a large quantity of Attic pottery, over 90 per cent of which belongs to the period 580–535 B.C., and none to the fifth century. A similar date has been given for the Attic pottery found in Etruria and southern Italy. The finds indicate a considerable Athenian trade in the sixth century, diminishing in volume and importance towards its end and becoming relatively unimportant in the fifth century.

This decline of Hellenic trade in the western Mediterranean, amounting at Massilia to its virtual extinction, is clearly to be associated with the domination of the western seas which Carthaginian and Etruscan forces achieved at the naval battle of Alalia, about 539 B.C. Before this Greek trade had been considerable in these waters. At one time, it had penetrated beyond the Pillars of Hercules to Tartessos and had perhaps extended even farther. This trade was 'first and foremost a quest by Mediterranean folk for tin'.[161] Sometime before 509 the Carthaginians seized Tartessos, and excluded the Greeks from the western trade. It was probably at this time that the Greeks began a probing movement from their bases on the east coast of Spain towards the Meseta, in an attempt to outflank the Carthaginians in the south. This failed, and the Greeks were left with the route across Gaul to the English Channel.

Hatt has demonstated[162] that at this time Greek commercial and cultural contacts with central Europe were intensified. They were shown in the adoption, especially at Heuneberg, on the upper Danube, of a Greek pattern of fortification, and above all in the grave finds at Vix in Burgundy. The evidence points to the use of the river route from the north coast of Gaul by way of the Seine, Saône and Rhône, with a portage from near Vix to the vicinity of Dijon. Tin had come by this route and had been carried by way of the Rhône valley to Massilia or across the Alpine passes to northern Italy.

The movement of tin towards the Mediterranean Sea was probably requited by the export of wine and of other Mediterranean products, in addition to art works like the great bronze crater found at Vix. This trade would account for the Attic pottery found in Gaul and for Greek influences in La Tène art. Most of the goods imported by the Celts, 'whether pottery or bronze, were ultimately connected with the carrying, storing, mixing and drinking of wine . . . La Tène art may largely have owned its existence to Celtic thirst'.[163] The Greeks, however, were far from monopolising this trade, and Etruscan bronzes have been found in central Europe, suggesting that the Etruscans were also in some way involved.

The actual source of the tin is far from clear. During the Spanish Bronze Age tin had been obtained from north-western Spain, but the sources incorporated into Avienus' poem suggest that by the sixth century B.C. the metal was being sought much farther to the north. Dion is probably correct[164] in his identification of the Oestrymnides of the poem with the islands off the south coast of Brittany, and of the Breton ores as a source of Carthaginian tin. This does not, however, exclude Cornwall as a secondary source at this time. Hencken claims that mining had begun here as early as 2000 B.C.,[165] and that during the fifth century the tin was collected by the commercially minded Veneti of Brittany and despatched across Gaul. Diodorus Siculus gave a very circumstantial account of tin-mining in Cornwall, which he may have derived ultimately from the lost narrative of Pytheas (see above, p. 85). If this was indeed the case, it suggests that by the fourth century B.C. Cornwall had become the only significant source of tin in western Europe.

Diodorus described the Cornish tin-miners as a mild and gentle people, their manners softened by commerce.[166] After they had mined and smelted the ore, they took it to the island of Ictis (St Michael's Mount), where merchants purchased the tin and conveyed it across the Channel to Gaul. Cornish folklore has it that the Phoenicians came to Cornwall for tin. Though there is no incontrovertible evidence that they did, the legend may well have a basis in fact.

St Michael's Mount, an island at high tide and today linked with the shore at low, is just such a site as the Phoenicians commonly chose. A famous painting in the Royal Exchange, London, shows the Phoenicians trading purple cloth to the natives; this also is folklore, without, on this occasion, even a remote possibility of approximating the truth. If they came, we do not know what they brought, and it is far more likely that the merchants who came to the coast of Cornwall for tin were in fact the Veneti of the opposite shores of Brittany.

Eastward of the probable routes of the transcontinental tin trade, commerce was practised on only a very small scale and was limited to a narrow range of articles. Without an urban society, such as we meet with in the Mediterranean region, there was little trade in foodstuffs. Presumably there may have been some traffic in skins and furs, but the objects of commerce consisted overwhelmingly of metals and minerals, localised in their occurrence, but in wide demand. One of the oldest of these was the amber dug from the coastal regions of the Baltic, and despatched southward to the Mediterranean. de Navarro has demonstrated[167] that the amber took a progressively more easterly route, until in the fifth century it followed the Vistula and upper Oder, and then passed through the Moravian Gate and around the eastern end of the Alps to the head of the Adriatic. The amber may have been paid for by the export of Etruscan art-work. Whether amber was also carried down through the Balkans to Greece is unclear, but extensive finds have been made in Bosnia.

Salt was an important article of trade, and the salt deposits at Hallstatt and Hallein in Austria were worked and the salt distributed. But iron, copper and bronze perhaps made up the larger part of the commerce of central Europe at this time. Metal-working, especially in bronze, was to be found in places far from the natural occurrences of the metal. Kujawy, in north-east Poland, was in the fifth century a centre of bronze working, and we must presume that the copper was obtained from the eastern Alps and the tin perhaps from Bohemia. The study of trace elements in bronze shows that the deposits of the eastern Harz Mountains, and of the Thuringian Basin and Erzgebirge probably provided the copper in bronze articles as far afield as Hungary, Yugoslavia and Italy.

A final problem in tracing the pattern of trade in the fifth century is the alleged commercial connection between the Greek world of the Aegean and the Balkan peninsula and Danube valley. The evidence in favour of commerce between Macedonia-Thrace and the Hungarian plain is very slight. The Greek writers make no mention of it, and the argument[168] that the Greeks obtained iron by this route from the eastern Alps finds very little support.

CONCLUSION

In 431 war broke out between Athens and Sparta, and soon engulfed the classical Greek world. When it ended, twenty-seven years later, Sparta and her allies had been weakened; the Athenian League had ceased to exist, and Athens was prostrate and her hegemony in the Greek world ended. The role of Athens passed to other regions, to Thebes, to Thessaly and lastly to Macedonia. The age of the *poleis*, small in size, jealous of their independence, and incurably quarrelsome and bellicose, was effectively over. Power passed to larger territorial states, which could support bigger battalions, and tyranny replaced the open society of the *polis*.

The Macedonian state became the nucleus of the empire of Alexander, which embraced the whole Aegean world. After his death it broke up, and its Macedonian core passed, together with parts of peninsular Greece to the family of Antigonus. The Greek *poleis* of Asia Minor mostly fell under the rule of the similar territorial states of Pergamon and the Seleucids. The rest of Greece again broke up into *poleis*, grouped into the two rival leagues, the Aetolian in central Greece and the Achaian in the Peloponnesos, while to the north a new state appeared, the Kingdom of Epirus.

In Italy also a change was taking place from *poleis* to territorial states. Most significant historically was the growth of the city-state of Rome. It defeated its earliest rivals, the similar states of southern Etruria; extended its authority over the Campagna, and then came into conflict first with the Italic tribes of the mountainous interior of the Italian peninsula and then with the Greek city-states of the south. Each victory, each conquest, raised another danger and presented the prospect of another war of conquest in a seemingly endless progression.

In this way the expanding power of the Roman *polis* was brought into conflict with the sea-based power of Carthage. The defeat of the latter, most significantly in the Second Punic War (218–202 B.C.), gave Rome control over much of the western Mediterranean littoral, and created the problem of her relations with the peoples of the hinterlands. But even before the conquest of the western basin of the Mediterranean had been completed, Rome found herself involved in the affairs of the Aegean.

Contact between the two spheres was inevitable. The Greeks belonged to both, and those of Magna Graecia were always ready to appeal for help against Rome to the inhabitants of Old Greece. The Kingdom of Epirus lay nearest, and it was King Pyrrhus who first answered the call and invaded southern Italy. Then Philip V of Macedonia came to the assistance of Hannibal, ineffectively as it proved, but the Romans in southern Italy remained mindful of the danger which threatened from the populous, warlike and well-trained tribesmen of northern Greece.

The result was the Macedonia Wars, which led to the war with the Seleucids and the Conquest of Greece. By the end of the second century B.C. the Romans had brought under their control the coastlands of the Mediterranean from western Asia Minor to southern Spain, as well as the hinterland of Carthage in North Africa.

There was a certain logic in the expansion of the Roman Empire, each advance being conditioned by the circumstances of the previous conquest. The Empire reached its greatest extent in the second century A.D. Over six centuries of territorial growth came to an end, the basic problems of Roman imperialism still unresolved. These were the question of defence against the barbarian world beyond its borders, and that of reconciling imperial rule with a socio-political structure which in its essentials derived from the *polis*. The Romans never really reconciled *polis* with empire.

3

Europe in the age of the Antonines

The reigns of Hadrian and the two Antonines, which together spanned much of the second century A.D., constitute one of the few periods in human history to which subsequent generations have looked back with reverence and longing.[1] It fell short in artistic achievement of the level set by Periclean Athens or even Augustan Rome, but more than made up for this by its attention to human welfare and happiness. 'Their united reigns,' wrote Gibbon, not without a certain exaggeration, 'are possibly the only period of history in which the happiness of a great people was the sole object of government.' Aristeides, who lived under the benevolent rule of Antoninus Pius, claimed in a rhetorical flourish that he could not remember when this peace was broken, and relegated stories of war to the category of legend, so profound was the peace amid which he lived.[2] While he wrote, however, there was conflict along the northern frontier of the Empire and revolt in Britain and in some of the African and Asian provinces.

A long period of economic growth had preceded the age of the Antonines, The population had increased; agriculture had penetrated the forest and waste, and villas and scattered farms had, at least in the more peaceful provinces of the Empire, replaced the nucleated and defensible villages. 'Cultivated fields have overcome the forests,' wrote Tertullian, 'the sands are being planted, the rocks hewn, the swamp drained; there are as many cities today as there were formerly huts.'[3] Tertullian's powers of exaggeration equalled those of Aristeides. This was no golden age, but it did mark the culmination of a period of growth and was the prelude to a long decline.

This period of relative peace followed centuries of intermittent, even of continuous war. The Emperor Augustus, more than a century and a half earlier, had closed the doors of the Temple of Janus, yet sporadic fighting continued both within and along the borders of the Empire. Even the peace of the Antonines was not unbroken, and it was but the lull before fighting became more intense both along the frontiers and even at the heart of the Empire.

The face of Europe must have seemed strangely similar in the second century A.D. to what it had been six centuries earlier. The political scene had undergone changes during these centuries, but the ways in which people lived were modified only in detail. The techniques of production were little altered. This was not because classical peoples were uninventive; there are many instances of their ingenuity, but their inventions and discoveries were diffused very slowly or not at all. The areas under cultivation had been expanded to supply a growing population, but there was little in the practice of agriculture under the Antonines that would have seemed strange to people of the fifth century B.C. No serious attempt was made to economise on labour, partly because the aristocratic cast of mind of the ruling classes did not readily concern itself with invention; partly because, throughout most of antiquity, there was an abundance of labour in the form of slaves.

The Graeco-Etruscan-Carthaginian world of the fifth century B.C. had been a mosaic of small, quasi-independent, quarrelsome and even hostile political units. They were in process of destroying themselves by their feuds while at the same time they laid the theoretical foundations of an ideal society which in practice they were unable to achieve. Within Greece itself political units of increasing size and economic resources in turn achieved an hegemony: Sparta, Thebes (Boeotia) and Thessaly, until all were subjected to the power of a state with an even richer territorial base, Macedonia. In the Italian peninsula also the territorial units best equipped socially and economically gradually overcame their weaker rivals, gained in strength by doing so, and thus prepared themselves for the next round in this elimination contest.

The Romans first overcame the tribes of the Campagna, then defeated in turn the Etruscans, the Samnites and the other tribes of central and south Italy. This brought them into contact with the Hellenistic settlements of Magna Graecia and thus into conflict with Carthage. And so the stage of Roman political action grew wider. Inspired both by a search for military security and a desire for lands to exploit and cities to loot, and led by generals who longed for fresh fields to conquer and the political rewards of victory, the Roman armies spread into the Celtic and Iberian lands to the west, and the Illyrian and Greek to the east, until, in the closing years of the Republic, they had cast a circle around the whole Mediterranean basin.

Yet, despite a supreme administrative and engineering talent, the Romans ran their Empire as if it was merely an overgrown city-state. Its affairs were inseparable from the politics of the city, and the graft and peculation, the rivalries and feuds of small town politics were reflected in the military councils of Roman armies in Egypt and Spain, Numidia and Gaul. It ended with the savage internecine wars of the first century B.C., and from the ruins of this bankrupt political system emerged the Principate.

Augustus, for all the republican legacy in his style and title, was the first of a line of emperors, of whom the Antonines, Antoninus Pius (138–161) and Marcus Aurelius (161–180), were respectively the fifteenth and sixteenth. There was no formal rule of succession, beyond a rather vague principle that the Principate belonged to the family of Augustus. Emperors from beyond its ranks had in some sense to be made legitimate; right of conquest was never quite enough.

If one of the foremost purposes of the Empire was to bring peace to a troubled continent, it succeeded by and large. Augustus rationalised and stabilised the European frontier, so that his successors found it necessary to make only minor territorial adjustments. With some hesitation the 'forward' policy of conquering ever broader areas as a means of protecting the Empire was given up, all the more readily, perhaps, because central and northern Europe offered little economic advantage to the Romans. The frontier was established along the Rhine and Danube, and this line was transcended only by the Claudian invasion of Britain, the incorporation of the *Agri Decumates*, between the upper Danube and Rhine, and Trajan's conquest of Dacia.

It is true that administration was centralised in Rome; that the Emperor controlled the movement and stationing of the legions, and Emperor and Senate between them nominated senior provincial officials and thus dictated policy. Yet the functions of government were essentially decentralised, and most were carried out at the local level, in the *poleis* and *civitates.* Administrative questions, even those of trifling importance, were sometimes referred from these local units to the provincial and even to the imperial levels of administration. But the Empire was, nonetheless, in essence an informal federation of city-states, around which, as it were, the Empire had cast a net and established a kind of protectorate (see p. 101), 'a vast experiment in local self-government'.[4]

The establishment in this pragmatic fashion of the Roman Empire and the extension to most parts of it for most of the time of the pax romana, had two consequences of great importance. In the first place, the diffusion throughout the Empire of a way of life that was essentially classical was facilitated. 'Let the wild beasts,' wrote Cassiodorus,[5] 'live in fields and woods; men ought to draw together into cities.' Despite continual literary expressions of a love of the simple life of the countryside, it was urban living which attracted the Romans. Urban life distinguished the civilised from the uncivilised, and the Romans, extending the practice of the Greeks and Etruscans, made the city the focus of urban living.

The spread of urban living had brought about changes in the countryside. The city populations had to be supplied and fed, and agriculture and transport were geared in some measure to meet this need. In Italy itself the primitive peasant society had been destroyed during the Second Punic War and its aftermath. Ownership of the land had passed to richer men, and land use had changed, in part at least, from the cultivation of wheat, spelt and barley for the peasant family to that of wine and oil for the city. It was for the new class of rural landowners that Cato and later Varro and Columella wrote their agricultural treatises.

Peace and unity within the Mediterranean basin also facilitated movement of peoples, diffusion of ideas and the transport of goods on a scale not known hitherto. Even without the conscious policy of the Emperors, localisms would have gradually broken down and local and tribal dialects would have given way to Latin. It was the Roman policy to encourage these developments. It also allowed Middle Eastern mystical beliefs to spread throughout the Empire, carried by soldiers like a talisman or borne white-hot by preachers and fanatics. The acculturation process which had been going on within the Empire was of necessity in two directions: the diffusion of the culture of Rome and also that of non-Roman peoples. If the *forum* in every small town of the Empire represented the cultural legacy of Rome, the temple of Mithra and the Christian church reflected the external influences which Rome was gradually absorbing and assimilating. From persecuted minorities, their followers often came in time to be part of the institutional framework of the Empire.

The cultural borrowings were not all in one direction. During the period of imperial expansion, as Toynbee has observed, trade anticipated the flag, and Roman artifacts were spreading far out into barbarian Europe. Roman commerce penetrated the plain of Poland, and objects of Mediterranean origin were to be found as far afield as northern Norway. Germans and Slavs were acquiring the technology of the Romans, and their craftsmen were imitating, and not without some success, the arts of the Empire.

POLITICAL GEOGRAPHY

Augustus had left as his testament to his successors the advice to 'keep the bounds of empire within the existing limits'.[6] The disastrous defeat of A.D. 9 had led to the abandonment of adventure into central Europe, and the boundary was for a period stabilised along the Rhine and Danube. It was not until a generation after Augustus' death that Moesia Inferior became a province, and the boundary was established along the lowermost course of the Danube. The frontier system was not, however, regularised until the second century A.D., and then primarily by the efforts of the Emperor Hadrian. Pelham spoke of 'its marvellous completeness and uniformity ... inspired by one policy and executed on a single plan'.[7] Hadrian knit together the remains of earlier defensive works, establishing a line of walls, ditches and forts, as if to say to the tribes beyond: 'thus far, and no farther'. It was throughout its length a static boundary against which barbarian waves would beat in vain. How ineffective in the end Hadrian's superb engineering proved to be is implicit in the 'fall' of the Empire which it was intended to guard.

The Roman frontier in Britain, a province with which this chapter is not primarily concerned, was the simplest. At this time it consisted of a simple fortified line drawn across the north of England, advanced temporarily under Antoninus Pius to the Scottish lowlands.

Rhine-Danube frontier. The boundary of the Empire between the North Sea and the Black Sea was more complex. From the time of Julius Caesar himself, the boundary of the Empire had been established along the Rhine, and under Augustus it was continued along the Danube towards the Black Sea. The attempt to advance its limits into Germany had failed, and for much of the first century A.D. the Romans had been content to hold the line of the river and to maintain an uninhibited screen to the east. After about A.D. 70 the Romans allowed this zone to become settled by Germans, who in the age of the Antonines maintained commercial and generally peaceful relations with the Romans. Many of the tribes east of the Rhine constituted tributary states, providing auxiliary troops for the Roman army, and provisions for the frontier garrisons. Such relationships were, however, only temporary, and appear to have terminated by the mid-second century.

Weaknesses soon became apparent in the boundary as it was drawn along the two rivers. The first showed itself near the Rhine mouth, where the river broke up into a number of distributaries, of which the most northerly flowed directly into the *Lacus Flevo*, ancestor of the Zuider Zee. The river did not present a clear and unambiguous boundary, and the Romans had no alternative but to take under their supervision – it hardly amounted to control – the whole area up to the farthest branch of the river. Forts were established, but little attempt was made to settle the marshy Netherlands. The eastern bound-

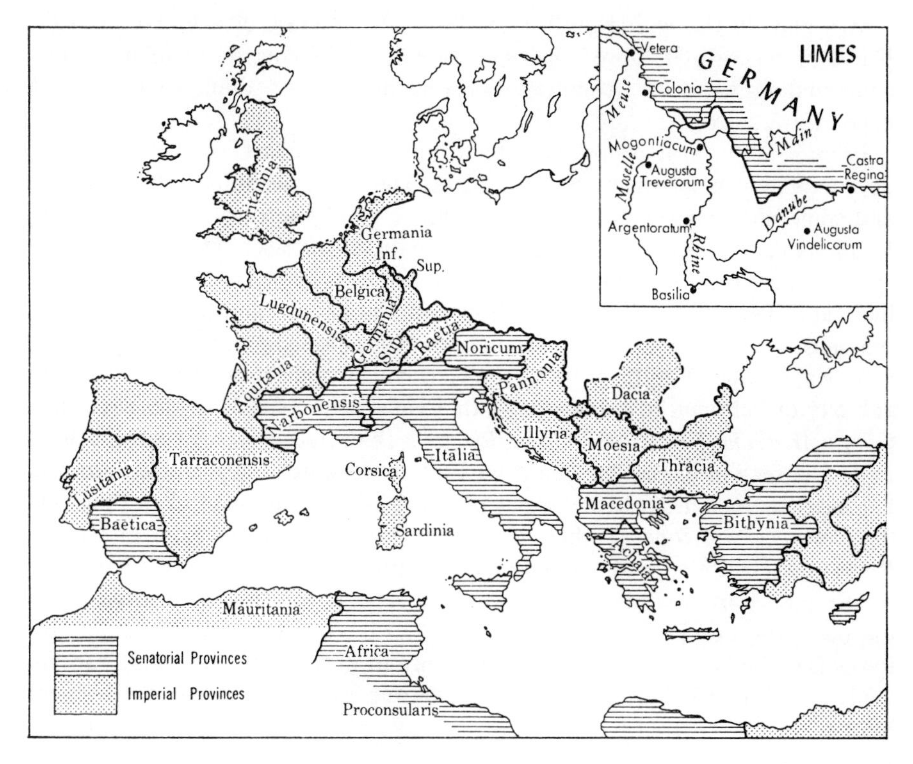

Fig.3.1 Political map of the Roman Empire, with inset of the German *limes*

ary of the Gallic provinces was made to follow the river Rhine up to Lake Constance.

The second weakness in the line of the rivers lay between the upper Rhine and upper Danube, where a salient of German territory extended deep into the Empire. Between the time of Vespasian and that of Hadrian much of this was brought within the Roman boundary. The motives for the annexation of the so-called *Agri Decumates* were mixed. It shortened the boundary and eased communications between the Rhineland provinces and Raetia and Noricum. The Chatti, who lived in the present territory of Hesse, were constantly threatening the Wetterau plain, near Frankfurt, and thus the Empire itself, and its incorporation seemed the surest way to remove this danger. At the same time the area incorporated was one of relatively high agricultural value and the line finally chosen proved to be more easily defended. Hadrian finally organised the piecemeal conquests of his predecessors in this area into a unitary defensive line, protected in part by a masonry wall, but more often by a palisade and ditch, strengthened by masonry towers, and reinforced at less frequent intervals by forts. The defensive line departed from the Rhine to the north of Coblenz, crossed the Taunus, enclosed the Wettarau, and ran down to the Main. From a point near Miltenberg it struck southwards and then eastwards, through the infertile sandstone and limestone uplands of Swabia, to meet the Danube 20 miles above Regensburg. Tactically it was carefully drawn and, in the words of Brogan, bore 'the impress of a grand strategic plan'. In general it followed the outer or eastward-facing slopes of hills, thus commanding Germanic country, and was never itself dominated by land beyond its control.[8]

From near Regensburg the Danube formed the boundary as far downstream approximately as the junction of the Tisza. The river, generally wide and swift, and sometimes fringed with marshes, made a good boundary. Fortified cities, legionary camps, and forts lined the south bank of the river. They were closely spaced where the river could be easily crossed; less frequent where the Danube had cut a deep valley between the Bohemian mountains and the Alps. One of the most exposed sectors of this boundary was that opposite the plain of Moravia, inhabited at this time by the Germanic Quadi. Here were the fortress and city of Petronell (*Carnuntum*), with Vienna (*Vindobona*) to the west, and Györ (*Arrabona*), Komárno (*Brigetio*), and Buda (*Aquincum*) to the south-east. The section of the Danube from Buda southward to Osijek (*Mursa*) was bordered by marshes, and defended by a line of forts. Roads linked the frontier defences with one another, facilitating rapid movement from one point to another, and also with the cities and legionary stations of the interior of the Empire. Forts were built even along the crags of the Iron Gate gorges until the occupation of Dacia rendered them superfluous. The boundary along the lower Danube had been subjected to pressure from two directions, the Transylvanian region and the Russian Steppe. During the first century the Romans built a series of defensive walls across Dobrudja and southern Bessarabia, which, with fortresses along the Danube itself, were intended to safeguard the boundary of Moesia. They were not enough, and between 101 and 106 Trajan twice led expeditions across the Danube into Dacian territory. The Dacians had in recent years become more closely united and belligerent, and only by the conquest of their homeland within the curving line of the Carpathian Mountains could the Roman boundary, it was thought, be made safe. This was Trajan's achievement. The new frontier was a great deal less definite than it is represented in historical atlases.[9] Roman authority faded out in the northern Carpathians, just as, at the base of Trajan's commemorative column in the Roman Forum, where the scroll tapers to a point, the masonry castles give way to wooden watchtowers, and these to small sentry posts and warning beacons along the frontiers of Empire.

The Romans never incorporated the Hungarian Plain between Dacia and Pannonia. For a time the friendly Sarmatian tribe of Jazyges, who inhabited it, formed a dependent state, and the Romans were generally able to dominate the plain and to cross it at will. Nevertheless, the southern part of this plain must have given the Romans cause for alarm, as they constructed a number of forts along the Danube and defensive dykes from the Danube to the Tisza.

The river Danube, eastwards from the Iron Gate, with its steep bluffs along the south bank and marshes fringing its northern, provided an effective barrier. But where the river swings to the north to enclose the low platform of Dobrudja the danger was greater, and artificial barriers were erected.

During much of the second century there was peace along the northern boundary of the Empire, and not all the frontier cities were even walled. The

Fig.3.2 The Roman frontier (relief from Trajan's column)

Germanic and Sarmatian tribes were generally quiet, and Roman culture and commerce were slowly spreading among them. Only towards the end of the period of the Antonines was this peace broken by the attacks of the Chatti, Marcomanni and Quadi, and by the desperate campaigns of Marcus Aurelius.

The Roman provinces. The system of provincial government which had evolved under the Roman Republic was modified and adapted, but in its main features continued through the Empire. There were over twenty European provinces in the age of the Antonines, but the number varied with the exigences of local administration. The chief administrative officer in each was nominated by and responsible to either the Emperor or the Senate. The distinction between imperial and senatorial provinces, made in most historical atlases and texts, was in fact less clear than appears, and the view that the imperial provinces were those in which legions were regularly stationed has been challenged.[10]

Rome was careful to interfere relatively little with the native institutions of the conquered peoples, and wherever possible administrative boundaries respected the unity of the tribe. Within each province the tribal area, if the tribe was still a tangible unity, became a *civitas*, and its chief settlement was often made the local capital and administrative centre, usually adding to its name that of the tribe which it served. But the tribal centre became the focus of the Roman administration only where there was no recognisable urban system already in existence. The Romans took over the pre-existing inhabited site for their administrative convenience. Sometimes they moved it a short distance to a more convenient site or one less easily defended (see below, p. 119).

The province of *Gallia Lugdunensis* thus had 26 such tribal cities. Tacitus found 64 *civitates* in Gaul, while the *Notitia Galliarum* enumerated 114.[11] In Britain there were over a dozen *civitates* organised around tribes, to which should be added five cities of Roman creation. In all, there are said to have been no less than 1,000 units of local government in the eastern provinces of the Empire, and over 900 of them were *civitates*. In the words of Rostovtzeff, 'the Empire in the second century presented more than ever the appearance of a vast federation of city-states',[12] though it was a federation of which membership was strictly involuntary.

The *civitas* as a general rule consisted, like the *polis*, which in many ways it resembled, of a nucleated or urban settlement, surrounded by a *territorium*. The city itself became not only the chief expression of the local pride of the *civitas*, but also the medium for the romanisation of the *territorium*.

The *civitates* varied greatly in size. Those of Britain were relatively large; in southern France, more densely peopled and urbanised over a longer period of time, they were small and closely spaced. This mosaic of city-states was the Roman ideal, but it was fully realised only in those parts of the Empire which had been most thoroughly romanised. The pattern of *civitates* was, however, far from permanent, and there are wide discrepancies in their estimated numbers. Pliny mentioned 189 cities in *Hispania Tarraconensis*,[13] whereas Ptolemy, using perhaps later source material, listed only 248 self-governing cities for the whole of Spain.[14] By contrast with the western provinces of the Empire, the Danubian and Balkan provinces contained few such city-regions, and most were of fairly recent origin. In mountainous areas, which covered much of these provinces, the native Thracian and Illyrian peoples still lived in small villages and were members of no *civitas*. If, indeed, they were self-governing, their autonomy was unrecognised by the Empire whose object was to urbanise the population and to regularise the local administration. Greece too provided a partial exception to the general pattern of tribally based *civitates*, but for different reasons. Here the legacy of classical Greece lay heavy on the pattern of Roman administration. The federalism, which had characterised the Hellenistic age, survived still in the leagues – Achaean, Arcadian, Boeotian, Thessalian – of city-states, intermediate in the administrative hierarchy between the provincial governor and the *poleis*. The latter, unlike the *civitates* of Gaul and Britain, in no sense represented tribal units. In all provinces there was a council, *concilium provinciae*, analogous in some ways to the Greek leagues, but its duties were largely honorific, and were concerned mainly with the imperial cult. It did, however, have the power to report directly to the Emperor regarding the conduct of the governor and the administration of the province.

There were areas, especially near the remote fringes of the Empire, where there was no tribal unity to serve as a vehicle of Roman administration. In the *Agri Decumates*, a notional *civitas* might be created, with a *caput* which developed during the course of the Empire from imperial estate to *municipium*. Elsewhere the tribe and its tribal area might be fairly clear, without, however, having any form of quasi-urban centre. The Celtic *Dumnonii* of south-western Britain and the Illyrian *Scordisci* of Pannonia both retained much of their previous custom and organisation, subject to a general overview by Rome. Neither developed a clearly urban centre, for *Isca Dumnoniorum*, on the eastern fringe of the Dumnonian territory, scarcely qualified. In the Balkan peninsula, not all the tribes were sufficiently settled to have had distinct tribal territories, and the Bessoi remained nomadic in some degree into the imperial period. They had villages, but probably never an urbanised *caput* as in Gaul.

Italian regions. The influence of its past was no less important in the administrative structure of Italy itself. The peninsula was 'a conglomerate of several hundreds of fairly independent town territories', directly dependent upon the Emperor, the Roman Senate and the prefects and commissioners of the city.[15] Not infrequently an imperial official (*curator civitatis*) was appointed to supervise their finances, and roads were the responsibility of the *praefecti vehiculorum.* The reform of the Italian administration effected by Augustus consisted in the division of Italy into eleven *regiones.* They conformed very roughly with the historic divisions: Etruria, Samnium, Picenum, but their boundaries were simplified and made to correspond, perhaps as an administrative convenience, with rivers as much as possible. This remained substantially the geographical framework of Italian administration through the age of the Antonines, though the 'regions' never achieved any great importance in the administrative structure, and remained, in fact, somewhat theoretical.

PEOPLES OF THE ROMAN EMPIRE

The movement of peoples during the long peace of the Roman Empire probably exceeded in the aggregate that due to the more sensational *Völkerwanderung.* The culture of much of Europe was changed during the imperial period, and many millions of people became 'romanised', learned to frequent cities and to enjoy urban entertainment, to accept the gods of Rome and to speak, however imperfectly, the Latin language.

Ethnic relations. The Romans displayed little antipathy towards those alien peoples whom they had little reason to fear. There was an element of race prejudice among some of the Roman writers; 'I cannot', wrote Juvenal, 'endure a Rome that is full of Greeks',[16] and there is good reason to suppose that many Romans feared those sharp business qualities which have always made the Greeks formidable competitors. There was also some dislike of the Jews, based apparently on their exclusiveness and their alienation from Hellenistic and Roman society. On the other hand, dislike of the non-Roman peoples was tempered by the tendency in some writers to see in them the simple virtues which Romans had lost. Lucian contrasted the vulgarity of Rome with the supposed simplicity of Athens,[17] and Tacitus used some of the Germanic virtues to point a Roman moral.

What the Romans expected of their non-Roman subjects was the acceptance of a code of values. Strabo's account of the Celto-Iberians of southern Spain – they 'have completely changed to the Roman mode of life ... And in the present *synoecised* cities ... manifest the change to these civil modes of life'[18] – showed at least part of what was hoped for them.

Romanisation, however, defies definition and it cannot therefore be said how 'romanised' the Empire became. The building of towns and the practice of urban living were aspects of romanisation. So also was the use of the Latin language and the erection of inscriptions.

'There were always barbarians within the Roman Empire,' wrote Ramsay MacMullen,[19] 'left to themselves in remote mountains and deserts, or recently incorporated by conquest.' Some of these were Roman citizens, and after A.D. 212, all of them. Yet Rome had very little impact on them, and its most valued contribution to their welfare was peace in which to live their traditional mode of life. In all parts of Roman Europe we would find such peoples, with the possible exception of Italy, and even here there is some evidence that languages other than Latin continue to be spoken in parts of the north and even in the south.[20] In Spain, such communities were numerous, most of them speaking a form of Celtic or Iberian. One of them, the Basques, with a language all of its own, has retained its identity to the present. In Wales, the prevailing Brythonic Celtic language acquired a few loan-words, but was otherwise untouched by Latin, while in the Roman *civitas* of the *Dumnonii* it is very doubtful whether romanisation amounted to very much more than the acquisition of a few – doubtless treasured – bits of hardware. The same is probably true of northern Britain and southern Scotland. In lowland Britain, even where the imprint of Roman civilisation was most marked, the native peasantry probably had little or no Latin. Their villages showed little of Roman sophistication, and their language was probably Brythonic. Indeed, the Roman development of town life, as Collingwood has shown, had the effect of creating a society which might almost be described as stratified at two levels, the romanised urban and the Celtic pagan, between which the Roman villa came in time to constitute some sort of a bridge.[21]

The situation was broadly similar in Gaul; Iberian and Basque survived in the south; there must have been islands of Celtic speech in other parts of the country, and it may even have been dominant in north-western Gaul.[22] It has, indeed, been argued that there was in fact a Celtic revival in the latter years of the Empire.[23]

Why, it may be asked, are most inscriptions in Latin, if other tongues were so widely spoken? The usual, and probably the correct answer is that most of those who could read were literate in Latin; that Latin was the prestige language, and that only the wealthier – and presumably romanised – segments of society ever put up monuments and inscribed tablets. As evidence that other languages than Latin were accepted in the Latin west there is the third-century acceptance of the validity of legal enactments in the local Celtic idiom as well as in other non-Roman languages.[24] There is lastly the well known evidence from St Jerome that the *Treveri* in his day still spoke a Celtic language.[25] It has been shown that, from the epigraphic evidence around Trier, the land of the *Treveri*, that personal names with a Celtic root make up no more than 15 per cent of the total in and near the city, but 40 to 50 per cent in outlying districts. The study could not be extended deeper into the countryside for lack of inscriptions. Presumably, the more 'Celtic' the area became, the fewer were evidences of literacy. It might be added that the writing of inscriptions was a specifically Graeco-Roman practice, and that other peoples would probably not have done this even if they had been literate in their own languages. In fact, there were

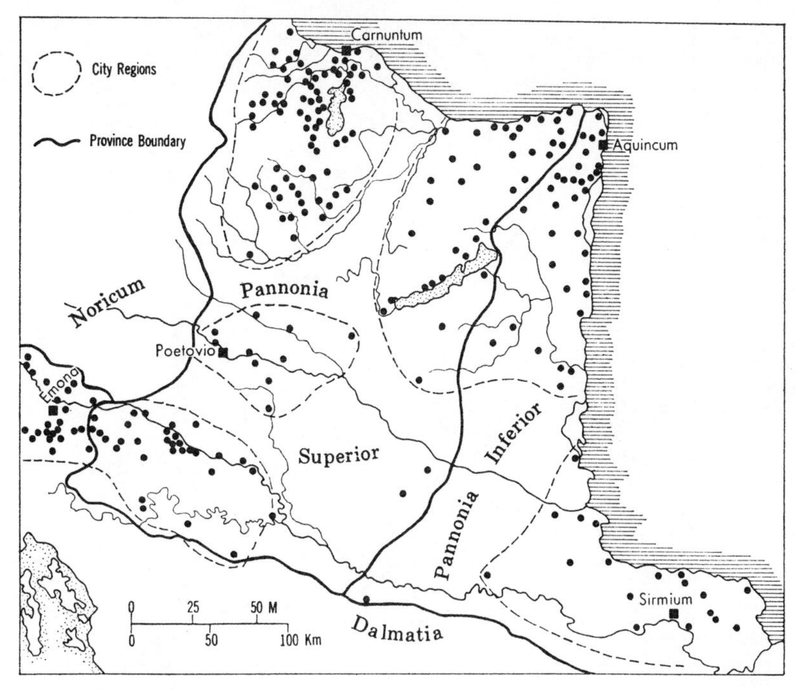

Fig.3.3 Inscriptions in Roman Pannonia

alphabets other than Greek and Latin, such as the Iberian, Punic and Etruscan, which may still have been in use.

In the Alpine, Danubian and Balkan provinces of the Empire, there was an even lesser degree of assimilation of the conquered peoples. The rewards of empire were smaller than in the west. Agriculture was less developed and, though there were important mining operations, conducted by Romans, the native peoples were in general less exposed than in the west to Mediterranean civilisation. They were mainly Illyrian and Thracian, with Celtic peoples settled among them. Towns were few compared with those of the western provinces of the Empire, and were possibly considerably smaller. The road net was less dense, but over much of the region the terrain made roads difficult to construct and probably rendered them unnecessary.

The distribution of both inscriptions and villas, at least in Pannonia, shows a clustering around the towns (figs. 3.3 and 3.4) and it is reasonable to suppose that these aspects of romanisation were diffused outward from the few urban centres. Between the clusters were neither inscriptions nor the remains of villas, and one must suppose that the impact of Rome was little felt in these areas. Indeed, there is evidence of continuing banditry in the remoter areas of the lower Danubian and Balkan provinces.

Greece was perhaps even less pervious to Roman culture. Its civilisation was older; its literature richer, and many Romans thought of themselves as assimilated to it; as Horace wrote:

> *Graecia capta ferum victorem cepit et artis*
> *intulit agresti Latio.*[26]

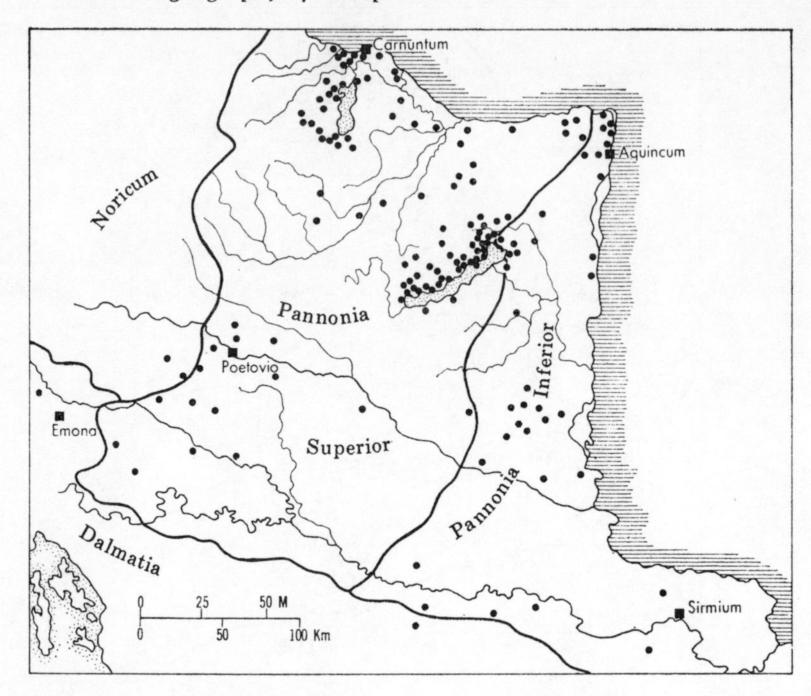

Fig.3.4 Villas in Roman Pannonia

If little was added during the imperial period to the Greek heritage, the Greeks, nonetheless, lived quietly through this long twilight of their classical greatness. Greeks poured into Italy, and many Romans for their part were attracted by the culture and the quiet dignity of Athens. They added to the buildings of the city – notably the Emperor Hadrian – and they were patrons of a Greek literary revival. Even within Italy itself there survived cities which were predominantly Greek in language and culture. Strabo singled out *Taras* (Taranto), *Rhegium* (Reggio di Calabria), and *Neapolis* (Naples) as cities that were still Greek in his day.[27] How many expatriate Romans enjoyed the Greek delights of *otiosa Neapolis*?

Roman writers of the first century A.D. described with varying degrees of emotion the influx of foreigners, mostly from the Middle East, into Rome. Seneca described the crowds of footloose people who flocked to the city, some 'brought by ambition, some by the call of public duty, or by reason of some mission, others by luxury which seeks a harbour rich and commodious for vices . . .'[28] 'The Syrian Orontes,' wrote Juvenal, 'has long since poured into the Tiber.'[29] Tenney Frank does not consider the picture at all overdrawn. A study of the inscriptions indicated that up to 90 per cent of those commemorated were of foreign origin.[30] Some were transients, with no intention of making Rome their permanent home; many were involuntary immigrants, swelling the ranks of the slave population. Most bore Greek names, which suggested only that they came from the Levant. Names, however, are an uncertain guide to their places of origin. Rome had become a great cosmopolitan city, like Alexandria, in which both the ancient Roman strain and, if one accepts the strictures of the satirists, the classical Roman virtues had both been overwhelmed by foreign peoples and strange mores.

The influx of Middle Eastern peoples was most marked at Rome, but was present in some degree in most provinces. Syrians were to be found in Spain and Gaul, where they may have been engaged in trade. In southern Spain there was already in the second century a community of Jews. The recruitment of the legions also contributed to the diffusion of races within the Empire. At first the legions formed 'enclaves of latinity and of Roman civilization generally'.[31] Until the time of Hadrian they had been recruited only from the ranks of citizens. Thereafter, non-citizens were enlisted: the legions rarely moved outside the provinces in which they were stationed, and their ranks tended to be filled from their local regions.

In the frontier provinces of Europe, lastly, there had been a continual immigration from barbarian lands beyond. Not only did Germans, Sarmatians, Dacians and others come across to sell their products in the frontier towns, but some also took service under the Romans. In addition to this migration of individuals into the Empire, there was the movement of whole tribes, Getae into Moesia, Germans into the Rhineland provinces, Quadi into Pannonia. Marcus Aurelius himself settled some 3,000 Navistae within the Empire. During the imperial period the immigrants 'included most tribes bordering on the empire, in a total certainly amounting to the millions'.[32]

Beyond the Roman frontier

The line of Roman defences which extended from the river Clyde to the Black Sea was no ordinary boundary. It separated two worlds. Truces and treaties might be negotiated, but there could be no lasting peace. The dialogue which takes place between states was lacking across this boundary, and in its place was only mutual distrust and hatred. Trade crossed it, and goods of Roman origin found their way deep into barbarian territory, but standards of living and of conduct differed sharply on each side. Occasionally the Romans would make an alliance of convenience with a Germanic tribe, or even allow it to migrate and settle within the Empire. But this was unusual; the Romans preferred to pit one tribe against another, and few things delighted them more than to see the barbarians weaken one another by their mutual strife.

The Roman impact, even on tribes living close to the frontier, was commonly very small. The Celtic peoples of northern Britain and Ireland were little influenced by either the armies or the civilisation of Rome. Even the Votadini of the Tweed valley, only a few miles in advance of Hadrian's Wall, gave the impression of being 'squalid barbarians living little above subsistence level'.[33]

Beyond the Rhine and Danube peoples lived under the simpler organisation of tribe and clan, and one is tempted to assume that their ethnic composition was also simple, clear-cut and capable of being shown clearly and unambiguously on a map. This probably was not the case. The degree of 'racial' mixing was only a degree less than within the Empire. Germans did not displace Celts from central Europe, any more than the Anglo-Saxon invaders were later to drive them from the English plain. They lived on as a substratum until they were assimilated by the dominant peoples.

The Germans. We owe much of our knowledge of the Germans to the writings of Caesar and Tacitus.[34] It was a stratified society that they portrayed, with its chiefs, nobles, and serfs, as well as its mass of free tribesmen. Its organisation was tribal, and Caesar represents the Germans as still semi-nomadic, and eager above all to maintain a frontier of forest around each tribe, separating it from its neighbours. Hostility between tribes was normal, and war frequent. 'They love indolence,' wrote Tacitus, 'but they hate peace.'

Tacitus enumerated the more important tribes, and sketched their movements. Though his knowledge of the physical geography of Germany was less than adequate, the account which he gives of the ethnography of the German tribes is 'one of the most valuable records of [its] kind'.[35] It is evident, however, that tribal groupings were loose, and that tribal unity was manifested only occasionally, and then chiefly for the purpose of attacking or resisting the Romans. Several of the tribes mentioned by Tacitus disappeared from history within a short period of time. Tribes were fluid; one would hive off from another, as the Batavi from the Chatti, and would as suddenly suffer defeat, and be dissipated and lost, like the Bructeri.

Tacitus was familiar with those tribes which bordered the Empire: the Frisii and Batavi, the Usipi and Tencteri; the Chamavi and the Angrivarii, who had defeated and probably absorbed the Bructeri. Several tribes had been taken into the Empire: Vangiones, Triboci, Nemetes, Mattiaci, and Ubii. The Batavi lived amid the marshes of the Low Countries, largely within the boundary of the Empire. The Hermanduri – *civitas fida Romanis* – lived in south-eastern Germany; other tribal areas are defined with considerable precision: the Chauci in the plain of Lower Saxony, the Marcomanni in Bohemia, and the Quadi in the lowlands of Moravia. It is when he turns to the tribes of the interior of Germany that Tacitus becomes a less certain guide. Here, within territories which he is at no pains to define, lived the Cherusci and the war-like Chatti, a federation of the Suebian tribes which together occupied 'more than half Germany', and included the Semnones, who stretched towards the Oder. Other Germanic tribes, vaguely located beyond the Elbe, were no more than names to Tacitus, and their significance to us is scarcely greater.

Tacitus recognised that the Germans were a cultural group. That the Cotini and Osi were not Germans 'is proved by their languages'; whereas the Venethi and their neighbours 'are to be classed as Germans, for they have settled houses . . .'. In this Tacitus was almost certainly incorrect, for these settled peoples of the east may well have been the ancestors of the Slavs.[36] The Germans were certainly not conscious of any ethnic unity. Though tribes occasionally agreed on common action – usually at the expense of the Romans – there is no evidence that the German 'nation' had any concept of itself.

Still less are the ancestral Slavs and the Baltic peoples to be thought of as 'national' groups. They were tribes, whose size and cohesion probably diminished as their territory became less fruitful and more sparsely inhabited. Most of the Celtic peoples had been included within the Empire, and were in very slow process of assimilation within the all-embracing Roman culture. The Cotini of the Carpathian foothills constituted, apart from those of the highland zone of the British Isles, one of the few Celtic tribes to be found beyond the limits of the Empire.

The Slavs. East of the Germans lived the Slavs, though none of the classical writers clearly differentiated between them. They must, however, have established a very open pattern of settlement, through which, as if by some kind of ethnic osmosis, Germanic and other peoples were able to pass without disrupting their way of life significantly. There can be little doubt that the Venethi or Venedi of the classical writers[37] were Slav, though how many of the other peoples loosely associated with them in the classical texts were also Slavs is not known. In all probability these ethnic groups did not each inhabit large and compact areas. The total population of central Europe was small, and the terrain was forested and studded with lake and marsh. Under these conditions peoples of several different ethnic origins could live in small, self-contained communities scattered amongst one another. It is not a question of whether Slavs or Germans lived between the Oder and the Vistula; they both lived there, with the addition perhaps of residual groups of Celts, Scythians, Sarmatians and even Balts.

Jażdżewski shows the Slav peoples extending from a little to the west of the Oder and Neisse eastwards into the Ukraine, and from the Baltic coast southwards to the Sudeten and Carpathian Mountains. From the latter, they are shown, though in smaller numbers, southwards across the Pannonian plain to the Danube. Tacitus represents the 'Pannonian' Osi and the Celtic Cotini as inhabiting the Carpathian Mountains and the middle Danube plain. The Lugii, a name covering 'a multitude of states', were spread across southern Poland. They may have been Slavs, but were more probably the ancestors of the Germanic Vandals.[38] The southern limit of the Slavs during the second century is thus far from clear. Their more advanced elements *may* have penetrated the Carpathians, but it is highly improbable that they had an exclusive control even of southern Poland. Jażdżewski probably exaggerates.

The dominant people within the Pannonian plain were the *Iazyges*, a Sarmatian people who had recently come into the area from the south Russian steppe. These predominantly pastoral and nomadic peoples were Sarmatian, and like the Scythians whom they had followed on the Russian Steppe, they were of Iranian origin and spoke an Indo-European language. Their chief settlement area within the Pannonian plain was along the Tisza, up the valley of its tributary the Körös, and over the Puszta between the Tisza and Danube. A settlement of the imperial period at Ozd, in northern Hungary, was inhabited by a mixed Celto-Illyrian people, and such must have been most of the population of the middle-Danube plain.

East of the plain lived the Dacians, the dominant people of Transylvania and the Bihor and eastern Carpathian mountains. They were related linguistically to the Thracians, and, like the latter, were divided into a number of tribes, which were united only rarely in some common cause. Over a century of intermittent war had ended in A.D. 107 with Trajan's conquest of part of Transylvania, together with its surrounding mountains and the more westerly part of Walachia.

Beyond the curving line of the Carpathian Mountains lived the Getae, a part of the Dacian 'nation', who occupied the plains of eastern Walachia, across the Danube from Lower Moesia. To the north of them were the Roxolani, like the Iazyges, a branch of the Sarmatian people, who had during the first century A.D., driven south-westwards from the steppe towards the Danube. Also like the Iazyges, they were a horse-riding, partially nomadic and mainly pastoral people.

Along the eastern shore of the Baltic Sea Tacitus located the Aestii,[39] and Ptolemy, the Galindai and Soudinoi,[40] ancestors of the Baltic peoples. The Romans had fuller and more accurate information of the Aestii than of the Slavs and east Germans, probably because they were the chief suppliers of amber.[41] The Aestii 'cultivate grain and other crops with a patience quite unusual among lazy Germans'. Tacitus also noted that 'they seldom use weapons of iron'; the Iron Age was in the second century only beginning to dawn in the Baltic lands.

To the north of the Baltic peoples the age of metals had still not begun. Here Tacitus located the Fenni, 'astonishingly wild and horribly poor. They have no arms, no horses, no homes. They eat grass, dress in skins and sleep on the ground. Their only hope is in their arrows, which, for lack of iron, they tip with bone.'[42] These were parts of the Finno-Ugric people who lived scattered thinly through the northern forest from Scandinavia to Siberia, practising at this date and for a long period to come their Stone Age hunting and collecting economy.

Population

No economy can be understood without some appreciation of the size, structure and density of the population. The kinds of evidence upon which estimates may be based has been touched upon in the previous chapter, and their inadequacies discussed. The data for the Roman Empire is no more abundant or reliable than that for the Hellenic period. At best it measures a single element in the population; at worst, it is merely an estimate derived from the supposed extent of cultivable land and the level of urban development. Nevertheless, it is probable that most figures that scholars have derived from the sources are within a factor of two of the actual totals, and this level of accuracy, as Hollingsworth has suggested,[43] is not altogether without value.

Since the great age of Greece, the population of Europe had unquestionably increased, and there is no reason to question Boak's statement[44] that it was greater in the middle years of the second century than it had ever been, or, perhaps, was to be again before at least the later Middle Ages. The urbanisation of Roman society, the rising level of welfare, the extension of agriculture, and the volume of specialised production and trade, are both evidence and explanation of this increase. At the same time there had in recent centuries been a greater mobility of population than previously. Earlier nomadism had largely vanished from the Empire, but the development of cities necessitated mobility of a different kind, a movement from rural areas to cities. The recruitment of the Roman armies and the subsequent foundation of military *coloniae* for the retired legionaries, brought about further mixing of the imperial peoples.

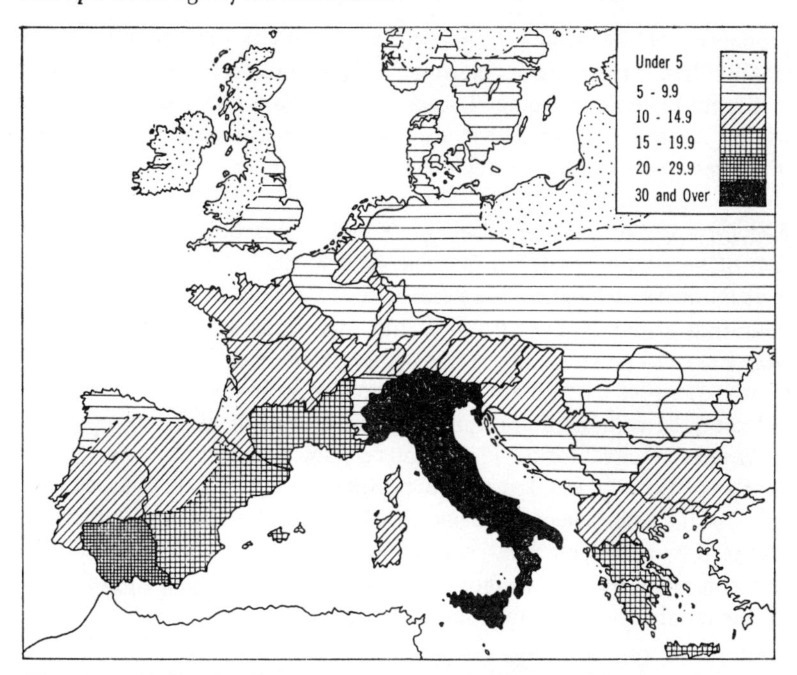

Fig.3.5 Population density per square kilometre in Europe at the time of the Roman Empire (highly conjectural)

Roman population. It had been the practice of the Romans to hold periodic censuses, the results of which were in part recorded by Livy.[45] During the last century B.C. the totals had risen sharply from a little over four million in 28 B.C. to nearly six million in A.D. 48.[46] These figures represented the citizen body, but it was not made clear whether they included women and minors. It is not known to what extent the increase represented, on the one hand, natural growth, and, on the other, the wider extension of citizenship, nor is it clear what proportion of the citizen body actually resided in Italy. Frank argued that in A.D. 48 over 80 per cent were Italian, and, making allowance for families and slaves, claimed that the total population of Italy 'was doubtless over 15,000,000'.[47] On the other hand, a contributor to the *Oxford Classical Dictionary*[48] claims that the census totals included women and children, so that the population in A.D. 48, if the census was even approximately correct, might have been somewhat over six million.

One may look at the problem from the opposite direction: what order of population density could have been supported by an economy such as that of peninsular Italy. The region is mountainous; good cropland is restricted, and pastoralism was growing in importance under the Empire. Except in areas where viticulture had been developed on a large scale, agriculture was probably not intensive. The higher population estimate would give an overall density of almost 50 to the square kilometre; the lower, about 20. If the agricultural land, including pasture, of imperial Italy were approximately the same as that of today, there would have been about 1.3 hectares per person with the higher population estimage; 3.25 with the lower. The area of agricultural land was probably not the same as that of today, and is more likely to have been smaller than greater. An indeterminate amount of food was imported, primarily to supply the city of Rome. The population–land ratios are therefore only the wildest approximations, but if a choice must be made between these two sets of figures, one must accept the lower.[49]

The City of Rome. Rome itself presented at this time the greatest concentration of population that Europe had yet known. Yet it is difficult to say more of it than that it had grown in size with the spread of the Empire and the immigration of non-Roman peoples, and was in the second century a very large city. Less is known of the size of imperial Rome than of Periclean Athens, though attempts to estimate it have been far more numerous. These have been based on the volume of grain imports, the area of the city and reputed numbers of *insulae* or city blocks. These are very uncertain guides. The north-eastern quarter of the city and Trastevere, beyond the Tiber, were characterised by the villas of the rich, where the density of housing was very low. Other areas were occupied by apartment houses. Evidence for their number and size is confusing and contradictory and cannot be made to yield even an approximate figure for the number of their inhabitants.

Estimates of Rome's population, derived from one or another of these sources or from a combination of them, range from a low of a quarter of a million to a high of two million. In reality, the population grew during the early Empire and declined during the later. It *may* have reached its peak in the second century but for this there is no evidence, and those whose estimates lie between 800,000 and 1,200,000 *may* be right, but the figures they suggest are nothing more than intelligent guesses.

For Ostia, the port of Rome, evidence is more complete and more convincing, because a large part of the city has survived. It was a 'miniature Rome' in terms of its building, and its growth and decline must have reflected closely the fortunes of the city which it served. Meiggs, on the basis of its area, estimated its population as at most 50,000 to 60,000.[50] More recently Packer, after a careful estimate of the town's floor-space, has given it as his opinion that the population could never have exceeded 27,000.

Sicily enjoyed a high reputation for its fertility and wealth. According to Pliny there were five *coloniae* together with 63 towns with their regions. Cicero had claimed that the *civitas* of Centuripae, inland from Catania, contained 10,000 citizens. This, together with the evidence for wheat production in Cicero's Verrine Orations, provides the only evidence for the population of the island. It is clearly inadequate. From the number of *civitates* given by Pliny, Scramuzza suggested a population of 800,000;[51] from the wheat figures Beloch derived a total of 600,000.[52] Neither figure is well supported, but it does not follow that the order of magnitude which they suggest is necessarily wrong. Indeed, a density of between 24 and 32 persons per square kilometre, which is what these figures imply, is consistent with what is known of the terrain, the soil and the practice of agriculture. Evidence is almost wholly lacking for Corsica and Sardinia, but their subsequent history suggests that their population was scanty and their economy underdeveloped.

For the Gallic provinces, data consists of little more than Caesar's enumeration of the tribes and his estimates of their numbers. The latter were subjective and may have been exaggerated; they may also have been corrupted in the transmission of the classical text. Both Beloch and Jullian have attempted to derive totals from this grossly inadequate data.[53] Beloch suggested a total of 5,700,000, and an overall density of a little over nine to the square kilometre, while Jullian even suggested that the population could have been little less than that in the time of Louis XIV. If the totals which they suggest have little or no value, it is nevertheless possible that they may give some idea of relative densities of population in Roman Gaul. Narbonensis *may* have had twice the density of northern Gaul, and other areas may have had densities between these extremes. A relatively low density in south-western Gaul is consistent with the low level of fertility and the comparative scarcity of prehistoric sites. It is to be assumed that densities were greater in the second century than when Caesar wrote, but for the extent of the increase and its variation between one province and another there is not a shred of evidence.

For Spain the only figure is Pliny's total of 691,000 for the three north-western divisions of Tarraconensis.[54] From this highly inadequate evidence Beloch generalised for the whole of Spain, and proposed a total of six million, with an average density of about 10 to the square kilometre.[55] Kahrstedt considers this estimate somewhat low, and suggested nine million for the second century.[56] The rate of economic growth may well have been faster in Spain than in Gaul, and classical writers indicate a high level of wealth and prosperity in southern and Mediterranean Spain.[57] Van Nostrand, indeed, suggests that the population may have doubled within this period,[58] an improbably high rate of increase. Much of Spain became highly urbanised, and archaeological remains suggest that even in non-Mediterranean Spain the towns were large and prosperous.

Several estimates have been made of the population of Roman Britain, based almost exclusively on the archaeological evidence. Collingwood considered that 'a round million' would not be too wide of the mark.[59] Later estimates have ranged considerably higher. Wheeler put the total at about 1.5 million,[60] and most recently Frere has argued[61] that the rural population must alone have been at least a million, and the total 'almost two million'. This would mean a density of about 13 to the square kilometre. If this figure can be substantiated, it may lead to a revision upwards of the population estimates for Gaul, which might have been expected to have had a greater overall density than Britain.

Danubian and Balkan provinces. With the exception of Achaea (peninsular Greece), these were less developed and less populous than the western provinces. Except in Pannonia and the plains of Lower Moesia and Thrace, the land was mountainous, and capable of supporting only a sparse population. Urban development was recent; the network of roads was thin, and much of the region was primitive and self-sufficing. Nevertheless, mining was highly important in parts of Noricum and Dalmatia; the needs of frontier defence had led to the growth of towns, and agriculture had expanded to supply them.

Pannonia and doubtless also the lowland regions of Noricum came to be well peopled and relatively prosperous. There is no reason to suppose that the military occupation had not brought about a comparable development in upper and lower Moesia, though the archaeological evidence is slight.

Beloch estimated that the population of the Danubian provinces at the time of Augustus was about two million, and Kahrstedt suggested three million during the second century. These are nothing more than guesses, and are based on neither literary nor archeological evidence. They are probably too low, because these provinces were in fact more developed than Beloch and Kahrstedt appear to have thought.[62]

The Roman occupation of Dacia was short, compared with that of the other Danubian provinces, and the population probably remained sparse. But the most thinly peopled regions of the European provinces lay in the interior of the Balkan peninsula: Dalmatia, Thracia and Macedonia, together with the more southerly parts of Moesia. There was a fringe of Graeco-Roman settlements along the Dalmatian coast, between the sea and the Dinaric mountains. In the interior mining had been developed (see below, p. 156), especially in the area of Bosnia and western Serbia. Roads, with a few small towns along them, followed the major valleys, and towns deriving from the pre-Hellenic period lined the Black Sea littoral. But, by and large, the Balkans were one of the least developed and least urbanised parts of the Empire, and its overall density of population was less, probably a great deal less, than that met with in the Danubian provinces.

The province of Achaia, which comprised much of peninsular Greece, had declined sadly during the previous several centuries. Its cities had decayed, and of many of the smaller 'nothing remained but their names'.[63] Augustus had himself refounded Patras, and Corinth, destroyed in the wars of the second century B.C., had been revived only a few years earlier. Athens was a small, sophisticated, provincial town. The population was almost certainly much less at the time of Augustus than it had been during the Hellenic period. There was some revival under the Principate, but there is no basis for any kind of an estimate of its size in the second century.

The rest of Europe. Beyond the limits of the Empire, archeological evidence for the density of population is scanty, and literary and epigraphic wholly lacking. We are dependent almost entirely on the presumed area of cleared land and the supposed level of the economy and intensity of land use. It may be assumed that urban communities, at least as understood by the Romans, did not exist; that the numbers employed in mining and commerce were negligible, and that the population was engaged almost wholly in agriculture, hunting and fishing.

Tacitus pictured the German tribes as practising a shifting agriculture and as insulated from one another by wide expanses of forest. Proximity to the settled peoples within the Empire, coupled with the market for agricultural produce provided by the frontier towns, influenced the structure of Germanic society and may have led to an increase in private property holding. The area of cleared and cultivated land was relatively small, less in all probability than that shown on the well known map of Otto Schlüter. Agriculture was extensive, and much of the cleared land was probably used for grazing. Under these circumstances, it is unlikely that, even in the most favourable areas, local densities could have exceeded 20 or at most 30 to the square kilometre, while the overall density was probably below ten, perhaps even less than five.

The argument from the estimated number and size of the Germanic tribes is of very little value. The number of tribes is not known; well under half of those named by Ptolemy, for example, are mentioned by Tacitus.[64] Nor is it possible to compute the average size of a Germanic tribe; certainly, there is no good reason for extending to Germany the orders of size attributed by Caesar to the Gallic tribes. Bury estimated[65] that at the time of the *Völkerwanderung*, a small tribe, like the Burgundians, had only about 25,000, and that a large tribe, such as the Ostrogoths, amounted to about 120,000. If one could assume that there were between 30 and 40 tribes – a total which can be obtained only by adding up all the tribal names that are known – there might have been from three to four million Germans. Their density in any case would have been less than 10 to the square kilometre.

East of the Oder, the area of land cleared and settled was a smaller fraction of the whole. The scanty archaeological evidence suggests that crop-husbandry was relatively less important, and few tribes are known to us by name. One may say with some degree of confidence that Poland, western Russia and the Baltic region were appreciably less populous than Germany itself, though the number of known prehistoric village settlements makes it clear that this was very far from an empty wilderness.

Densities in southern Sweden were probably similar to those of north Germany and Denmark. The poverty of archaeological finds suggests that the Småland hills were very thinly peopled, but in central Sweden a great many small hill-top forts were being built, indicating a settled and perhaps growing population. Over 250 of these forts have been mapped within an area of about 62,500 square kilometres. It is not known how numerous or how large were the villages associated with each, nor the extent to which they were occupied at the same time.[66] It is difficult, however, to think of one such fort as the handiwork of a community of much less than 1,000, and many villages, like the Vallhagar site on Gotland, were not associated with a fort. It therefore seems improbable that *central* Sweden could have had a population of less than 500,000, or a density of much under 10 to the square kilometre.

Small farming communities occupied the shelves of cultivable land around the fjords and coasts of Norway. Tacitus portrayed the far north-east of Europe as the home of the savage Fenni, who knew neither agriculture nor metals, and the 'hunt provides food for men and women alike'. Stenberger described the forests of Norrland as a region of cultural isolation, in which 'the old Stone Age culture continued'. The population of sub-arctic Europe can have amounted to only a few thousands, and its density to less than one to the square kilometre.

How populous, then, was Europe in the age of the Antonines? There is no real answer to this question. One can only suggest the outside limits of probability, figures between which the true total almost certainly lay. It is with this reservation that table 3.1 and the population map (fig. 3.5) have been prepared.

Table 3.1 *Population in the age of the Antonines*

Roman Empire in Europe	*Range of population*	*Range of density* per km²
Italy	6–9,000,000	20–30
Sicily	600,000–1,000,000	23–40
Sardinia and Corsica	300,000–600,000	9–18
Gaul	6–10,000,000	10–17
Spain	7–12,000,000	12–20
Britain	1–2,500,000	7–20
Danubian provinces	3–6,000,000	17–34
Balkan provinces	3–6,000,000	6–12
Rest of Europe		
Germany	3–5,000,000	6–10
Eastern Europe	1–3,000,000	1–3
Northern Europe	500,000–1,500,000	less than 1–2

It is generally assumed that population declined during the later centuries of the Empire, but one cannot say when the decline began or from what level. Nor is it clear whether the decline was general, or whether some provinces were spared. There is evidence both from Asia Minor and from the west of *agri deserti* during the later Empire, and one cannot argue, as Boak has tried to do, that changes in population density were a major factor in the 'fall' of the Western Empire.[67] That a changing population was part of a complex of economic factors cannot be doubted, but it is very improbable that it was an independent variable. Whatever changes there may have been in population density, they were dependent on other forces.

SETTLEMENT

Classical civilisation was essentially urban; and that of Rome was no exception. The foundation of cities was part of the process of romanisation. The conquest of the Empire and the defence of its frontiers were accompanied by the creation of towns. Celtic tribal capitals were replaced by Roman cities, and their adornment attracted the munificence of the wealthy and the bounty of the Emperor. The cities became the expression of the wealth of the Empire; it was here that sophisticated people desired, or were at least expected, to live. Yet it is by no means easy to define a town, and the distinction between it and a *vicus* was not in all cases self-evident.

The Roman towns

Towns are usually thought of as larger and more concentrated than other forms of settlement. It is usually assumed that their economic function is different, with a greater stress on manufacturing and on trade and other tertiary occupations than is to be found in rural and predominantly agricultural settlements. Towns, lastly, are conceived of as distinct in legal status and administrative structure. In short, they are able to run their own affairs within certain limits. While many, perhaps most, towns of the Roman Empire would clearly have qualified as such by all three criteria, there were some whose credentials were more than suspect. There were towns whose citizens were probably less numerous than the inhabitants of many a village; there were those where the manufacturing and trading functions had never really developed; there were towns whose chief – almost only – function was to provide services for a nearby military base, and there were settlements with developed crafts and trade and a relatively large body of inhabitants, which possessed no truly urban status during much of the imperial period.

Problems of definition are compounded by the tendency of towns to change through time. In Italy and many areas bordering the Mediterranean, towns developed or were founded under the Republic. Elsewhere they came later, as the conscious creations of the Romans. In Britain Agricola 'gave private encouragement and official assistance to the building of temples, public squares and private mansions . . . And so,' wrote the jaundiced Tacitus, 'the Britons were gradually led on to the amenities that make vice agreeable – arcades, baths and sumptuous banquets. They spoke of such novelties as "civilisation". . .'[68] Few were established after the first century A.D. In the third a number of them became visibly smaller, and many of those not previously defended were fortified with walls. In the later Empire the citizens, overtaxed and afflicted with the burdens of municipal office, merely drifted away and lived on in the countryside. There is mounting evidence that in some parts of the Empire town life wilted.[69] Urban status remained, but urban functions and urban population slowly disappeared.

The norm in most parts of the Empire was for the town to be the focal point of a discrete area – *territorium* – for which it served as market and as administrative centre. In a few instances one can reconstruct the probable limits of the *territoria.* The *territorium* contained its own rural settlements, *vici* and *villae*, but its inhabitants, unlike those of the *polis*, rarely, if ever, had full rights of citizenship in the town of the *territorium*, though conversely the town in many aspects of government administered the whole of its region. The term *civitas* was commonly used for the *territorium* together with its central place, and 'the cities (*civitates*) were the cells of which the empire was composed'.[70]

Outside Italy it was usual for the *civitas* to be a tribal territory and its town was heir in some way to the earlier tribal centre. This was so in much of Gaul and Britain, but there was not everywhere a convenient tribe, settled on its distinctive tribal area. In such cases an imperial estate or an arbitrary area with no evidence of tribal unity could be designated as a *civitas*, and a central-place be made its 'town'.

The foundation of new towns, most of them connected in some way with the military, continued actively until the reign of Hadrian. Thereafter, new foundations were few. The period of the Antonines may thus be regarded as the climax of the urbanisation of the Empire. Towns had ceased to grow in either number or size, but their decline had not yet become significant.

There was a variety of status among the towns of the Empire. The tendency, however, was for their distinctions to become blurred, so that in the second century there was in fact little difference between them. The *coloniae,* which enjoyed the greatest prestige, were originally bodies of citizens sent from Rome to occupy and defend outlying and exposed parts of Roman territory. Later they were used to settle veterans from the army and the proletariat from the city of Rome. The sites of *coloniae* were chosen not only with a view to the availability of cropland, but also to the need to encourage the romanisation of the native communities. The actual foundation of new *coloniae* ceased under Hadrian, and, though the title continued to be conferred, it became wholly honorific. In contrast with the *coloniae* stood the *municipia.* These were in origin the self-governing cities of Italy which had in certain respects placed themselves under the jurisdiction of Rome. The term came later to indicate any self-governing Italian city, other than those established by the Romans themselves, and in the Principate was extended to include native communities which had achieved a certain level of romanisation.

Municipia remained rare in the northern and north-western parts of the Empire, where the degree of romanisation remained somewhat superficial. Here its place was taken by the *civitas*, the least privileged of self-governing local communities. Typically the *civitas* was itself an urban centre with a surrounding *territorium*, but it developed from a system of tribes or of associated groups of villages. In some *civitates*, a distinct urban centre had not appeared, and a village appears to have been chosen as *caput* of the *civitas*, in the hope that it would eventually acquire the appearance as well as the functions of a local capital.[71]

Strabo was aware of the problem of definition. The *civitates* of Baetica were cities functionally and in law: 'many' wrote Strabo, 'are the points . . . from which and to which the people carry on their traffic, not only with one another but also with the outside world.' They clearly combined two, and probably all three of the criteria outlined above. On the other hand, in his treatment of the Meseta (*Celtiberia*), Strabo wrote that 'those who assert that there are more than one thousand cities (πόλεις) in Iberia seem to me to be led to do so by calling the big villages cities; for, in the first place, the country is naturally not capable, on account of the poverty of its soil or else on account of the remoteness of wildness of it, of containing many cities, and, secondly, the modes of life and the activeness of the inhabitants . . . do not suggest anything of the kind'.[72] Here he rejects the urban claims of the so-called towns of central Spain. They were small and agricultural, and comparatively few had acquired an urban status, though many may have served as some kind of administrative centre. The fact is that some parts of the Empire had few places which combined all three of the distinguishing features of towns, and their administrative centres had not risen economically above the status of villages.

One last category of cities owed its existence directly to the needs of the military. Some legionary camps, notably those on the frontier in Britain and along the Rhine and Danube, not only became large and permanent settlements, but also attracted communities of merchants and service retainers. These inhabited the *canabae* or civil settlements, which, with the military, made up a city of considerable size. Its status, however, remained in many instances that of a non-self-governing *vicus*, though in terms of function, it may well have been far more 'urban' than many of the *civitates* listed by the classical writers. On the other hand, the *canabae* sometimes gained municipal privileges, *Corstopitum* (Corbridge), near Hadrian's Wall, is such a case; so too were *Mogontiacum* (Mainz), *Brigetio* (near Komárno) and *Aquincum* (Buda) on the Rhine and Danube frontiers. At *Carnuntum* (Petronell), on the Danube below Vienna, a civil town grew up about two miles to the west of the military station. Not only along the frontier did towns originate in the needs of the military. In the civil provinces military needs dictated in numerous cases the original site of a settlement. 'It can be shown,' wrote Webster,[73] 'that almost all the towns and small settlements [of midland Britain] occupied sites adjacent to the earlier [legionary] forts and must thus have originated as civil settlements outside them, supplying some of the basic needs of troops garrisoned in newly occupied territory.'

In much of Britain and Gaul and also parts of Spain the town was heir to a fortified settlement, commonly called an *oppidum*, of the Celts or Iberians. The latter, usually located on rising ground, was capable, as the Romans knew to their cost, of prolonged resistance. For this reason, and also to emphasise the break with the past and to integrate them with the road net and the commercial pattern of the Empire, they were relocated generally on lower and less defensible ground. The inhabitants of *Bibracte* (Mont Beauvray), metropolis of the Aedui, were in 12 B.C. transferred to the settlement of Augustodunum (Autun), newly established in the nearby plain. Those of Gergovia went to populate *Augustonemetum* (Clermont). *Aquae Sextiae* (Aix-en-Provence) replaced the nearby hilltop fort of Entremont. In Britain, Durnovaria (Dorchester) became the capital of the Durotriges, replacing the impressive Belgic *oppidum* of Maiden Castle, a mile or two to the south-west. Other British examples of a town replacing an *oppidum* are *Verulamium* (St Albans) and *Corinium* (Cirencester).

But a great many tribal towns were not moved. Bourges today still occupies the site, protected by marshes, of *Avaricum* of the Bituriges which Caesar had captured. Langres, high on its plateau, was once the capital of the Lingones, and the Roman Lutetia continued to occupy the island in the Seine, which the Parisii had made their centre. Many, perhaps most, of the cities which the Flavians founded in Thrace were built on the sites of earlier tribal capitals. In Greece also there was a marked continuity of settlement from Hellenistic times. Towns, even those refounded by the Romans, occupied sites hallowed by history and tradition. Time had dealt harshly with some of them, and Roman soldiery with others. Augustus had taken a group of decaying cities of western Greece, and in one grand act of synoecism, had merged them all into the city of Nicopolis, which he founded at the narrow entrance to the Ambracian Gulf. In the same way the small towns of the Roman province of Achaea were subordinated to the refounded and enlarged city of Patras.

The Urban Map. Fig. 3.6 is an attempt to show cartographically the number and distribution of cities within the European provinces of the Empire. Where the extent of the built-up area can be measured, some estimate can be made of its possible population. But these cases are few. In many instances even the

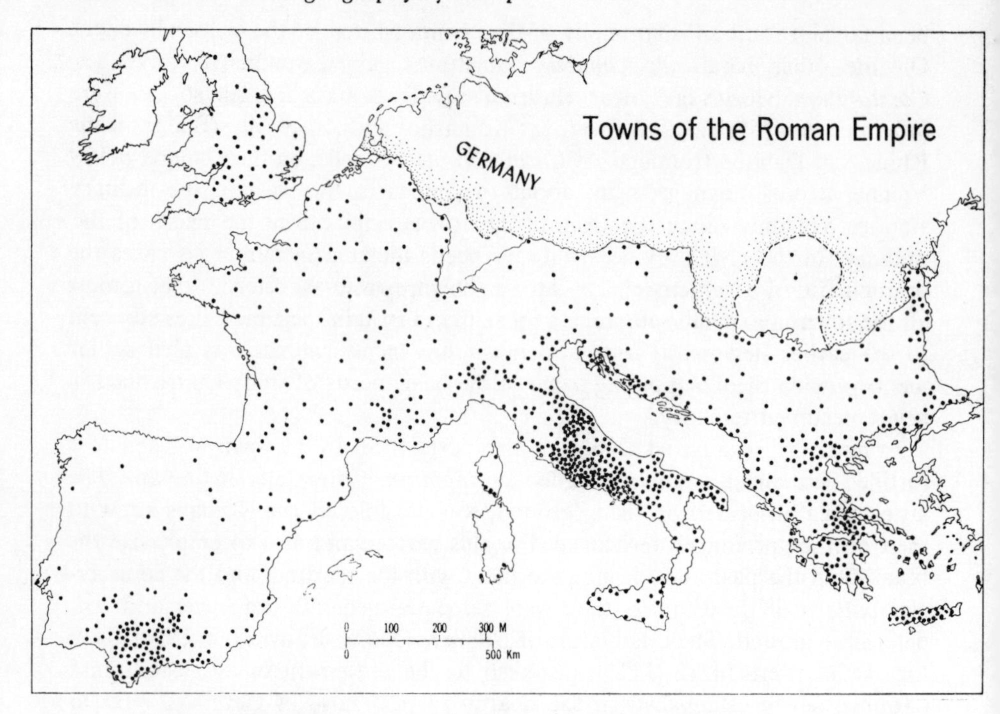

Fig.3.6 Urban development in Europe during the Roman Empire

plan of the Roman town is unknown, and there is no certainty that many places listed by Pliny and Ptolemy or named in the road-books were in form and function anything more than villages.

The cities of northern Italy were in general large. Strabo tells us as much,[74] and his evidence is confirmed by archaeology. Verona covered an area, within the line of its walls, of about 49 hectares; *Augusta Taurinorum* (Turin) had about 53 hectares, and many others were but little smaller. It is difficult, if not impossible, to convert these areas into population,[75] but it would be surprising if this did not represent a total urban population of at least 10,000. Many of the cities of Gaul covered a yet larger area:[76] *Nemausus* (Nîmes), 220 hectares; *Vienna* (Vienne) and *Augustodunum* (Autun), about 200 each, and *Aventicum* (Avenches, in Switzerland), 150 hectares. These were all large cities by any European standard except that of the nineteenth and twentieth centuries.

On the other hand, many – perhaps the majority – of Roman towns were diminutive.[77] *Arelate* (Arles) covered only about 18 hectares; *Caesaromagus* (Beauvais) about 11, and *Condate* (Rennes), *Augustomagus* (Senlis) and *Lapurdun* (Bayonne), less than 10. Some of these areas represent the extent of the walled enclosure under the later Empire when the growing danger of Germanic invasion led many cities to place themselves hastily in a condition of defence. This resulted, in some cases, in a sharp contraction of the walled area: that of Autun was reduced from 494 acres to 25; that of Nîmes, 550 to 20. These however, are extreme cases. It cannot be assumed that the reduction in size was a simple factor of diminishing urban population. The earlier towns were in all probability only partially built-up, like London and *Calleva Atrebatum* (Silchester), and the later towns *may* have been intended to serve as a kind of *Fluchtbürgen,* places of refuge for the surrounding countryside.

It is apparent that many of the towns of the Empire, like the central places of many of the Greek *poleis* six centuries earlier, had populations that could be numbered in the hundreds rather than thousands, and were primarily agricultural in their function. In the spectrum of human settlements there was many a village (*vicus*, Κώμη) that was larger in population and less heavily agricultural in its economic base than many legally constituted towns. The towns tended often to shelter a *rentier* class. Strabo described how the Allobroges 'till the plains and glens that are in the Alps, and all of them live in villages, except that the most notable of them, inhabitants of *Vienna* (Vienne) (formerly a village, but called, nevertheless, the 'metropolis' of the tribe), have built it up into a city'.[78]

In the province of Baetica Pliny enumerated 175 *civitates*,[79] and fig. 3.6 based very largely on the identifications of Thouvenot,[80] shows over a hundred. In the Po valley there were 78 cities in an area, after allowing for the non-urbanised and thinly peopled mountainous areas, of about 100,000 square kilometres. The average *territorium* was about 1,500 square kilometres, larger in the case of the cities bordering the mountains; very much smaller in that of the cities of the plain, where, in Chilver's estimate, it was about 1,000 square kilometres.

The area of Baetica was about 60,000 square kilometres. A similar line of reasoning gives less than 600 square kilometres as the average size of a *territorium*, whereas those of the Guadalquivir valley cannot possibly have had more than about 350 square kilometres each. This is relatively close spacing of urban centres, not very different from that which we find in parts of north-west Europe during the later Middle Ages. In the latter case we know that the towns were very small, having in general not more than about 2,000 inhabitants each. Those of Baetica, with a few significant exceptions, cannot have been larger.

Beloch contended[81] that the number of cities per unit area was a measure of the economic advancement of the region, and on this basis claimed that Cisalpine Gaul was relatively undeveloped. Apart from the fact that this is firmly contradicted by both Pliny[82] and Strabo,[83] it does not accord with what we know of the areal extent and probable population of the north Italian cities. Wealth in any event accrued mainly from the land. The surplus passed into the possession of the landowners who employed much of it in urban building and other forms of conspicuous consumption. A well-built and decorative city clearly implied a developed resource base to support it, even if some of its buildings were erected with loans. But surplus wealth had not necessarily to be employed in town building. Rivet has pointed out that in Britain villas were clustered around secondary towns far more than around the more important tribal capitals.[84] Does this mean that in the one case the surplus was invested more readily in villas, in other in town houses?

The pattern of towns was closer in the plain of the Guadalquivir, in *Baetica,* than in northern Italy, but it is probable that they were on average much smaller in the former. The density of the urban net cannot be taken as a measure of economic growth, though towns could not have arisen without an economic surplus. Towns, it should be noted, were not generally to be found on imperial estates, though it cannot be assumed that these did not yield a surplus that could have been used in this way. The towns may in some instances, as, for example, *Calleva Attrebatum*, have been little more than closely spaced groups of villas, whose agricultural function was dominant. Apart from this, towns were, as often as not, centres more of consumption than of production; they were evidence that there was production elsewhere, indicators rather than pace-makers in economic growth. Of no city was this more true than of Rome itself. In this lies the chief difference between the ancient and the modern town.

Fig. 3.6 shows the distribution of places that meet any of the criteria for towns. The map is necessarily incomplete, because too little is known of many sites. Italy is, not unexpectedly, the most highly urbanised province of the Empire. In Gaul there was a relatively strong urban development in the south and south-east. In northern France, despite its agricultural wealth, cities were thinly scattered and most were small. Only along the Rhine are large cities again encountered, and here their primary function was to house and supply the legions established along the frontier. Certain conclusions may be drawn from the distribution of towns. First, their number and size bear no simple relationship to agricultural productivity and to other forms of natural wealth. Secondly, we may find in comparable regions either a small number of large towns or a dense scatter of small. The distance between towns is so variable, even in regions which one might expect to be fairly densely and evenly settled, that one is obliged to question the significance of the cities themselves as market centres. Thirdly, towns in frontier regions and in areas politically insecure were often far larger and more numerous than could be justified by the local resource base. They must therefore have been supported by a larger area than their immediate *territorium*, perhaps a legionary district, or in the last resort by the taxation of the whole of the Empire which they helped to protect.

The Roman town was, first and foremost, a place for civilised and urbane living, where the wealthier provincials might live in comfort, perhaps taking some part in local administration, and above all, posing no threat to the authority of Rome. The smallest town necessarily had some commercial function, if only because its citizens had to be supplied and fed. Larger towns must have served for the collection and dispatch of the agricultural surplus of their local regions. Baetica made not inconsiderable shipments of foodstuffs and metals (see below, p. 164), and this called for a local apparatus of merchants and markets. The small town must also have had a craftsman or two, perhaps unspecialised, like the one described in the *Cyropaedia*, and satisfying only local urban needs (see page 80).

The medieval town (see below, p. 351), however, small, was the focus of a distinct area for which it provided a market and services. In very few cases did it have jurisdiction or administrative rights over its surrounding area. The classical town was in some respects the reverse. The market and service function was not, as a general rule, strongly developed but the town (*urbs*) had jurisdiction over its whole city-region (*civitas*). It was the vehicle for the collection of taxes (in the Middle Ages this function was more often performed by the parish), for the administration of justice, and for many other governmental functions.

Roman towns and town planning. The cities of the Empire offer a contrast between the extremes of orderly planning and the utmost urban confusion. Both aspects had been present in Italy from its earliest urban development. Whether derived from Etruria or from the hellenised cities of the south, the idea of a planned lay-out was disseminated by the army and regularly employed in its camps. From the fourth century B.C. Greek concepts of planning begun to infiltrate the Italian world. Whether directly or through the medium of soldiers' camps and veterans' *coloniae,* the idea of basing a city's plan on two streets crossing at right angles gained ground. Pompeii was one of the earliest non-Greek cities to be laid out on the basis of straight streets intersecting at right angles. Nevertheless, even if we deny that the old Italic cities – among which Rome must be counted – were planned, we have to admit, in Boëthius' words 'not only a general disposition to discipline but also . . . explicit features in those untidy cities which suggested regular planning'.[85]

But only government could establish and maintain a planned development in a city. The *coloniae*, established in the first instance for veterans, seem always to have had a grid-iron pattern, but planning all too frequently ended when the governmental control was lifted. At *Alba Fucens,* near the Fucine Lake in central Italy, the centre of the city has a regular plan, but the outer area is irregular and conceptually independent of the original *coloniae.* Many similar examples can be found among the north Italian cities, notably Verona, Piacenza and Pavia, of the breakdown of planning beyond the limits of the original *colonia.*

Beyond Italy and the Mediterranean coastlands, the plannned lay-out of cities was sufficiently widespread to be regarded as normal. Here, of course, the cities were founded *ab initio* by the Roman government and usually on virgin sites. The grid-iron plan was general, and is today recognisable in cities as widely separated as Cordoba and Ljubljana; Nîmes and York. In general, the Roman plan survives most completely in southern Europe, where there was a more marked continuity of settlement. In northern Europe, where the cities may have been temporarily abandoned or even destroyed, the former street pattern has been distorted and even obliterated. The street patterns of *Venta Icenorum* (Caistor-by-Norwich) and of Silchester, which were abandoned in the fifth century and never re-occupied, have been almost wholly recovered, whereas that of the Trier, re-occupied after the Germanic invasions, has been very largely lost.

City buildings varied greatly in style and in their density per unit area. The Italic cities were closely built. Along their streets, often narrow and winding, were *tabernae*, rooms with wide openings, combining the functions of door and window, facing on to the street. These were the shops and the workshops of those craftsmen, who were accustomed to work under the eye of the public they served. Only a little altered through eighteen centuries, *tabernae* continue to characterise the older quarters of Naples and other Italian cities. Ruined but ossified, the *tabernae* of the first two centuries A.D. can still be seen at Pompeii, Herculaneum and Ostia. They were built of masonry; sometimes they served as living as well as working quarters; usually they opened into another room in the rear, and wherever the pressure of population on the limited space made it necessary, a second floor was added. It is doubtful whether Italian town buildings often ran to more than two storeys; the monstrous tenement blocks described by the Italian writers are not known outside Rome and Ostia. But it seems likely that, at least in the larger towns, there was an attempt to concentrate the working population near the centre, and the price of this was, of course, apartment blocks.

Local stone was generally used, and Italy fortunately had no lack of travertine, tufa and limestone. It was, however, the introduction of lime-mortar in place of clay that permitted the addition of a second floor to town houses.[86] Even so, urban architecture took grave structural risks and was often none too secure.[87] During the second century the use became widespread of a cement construction, faced with bricks, usually set diagonally in the style known as *opus reticulatum.* This mode of building, which was durable, waterproof and immune to fire, was just beginning at Pompeii on the eve of its destruction; it was at the height of its popularity in the age of the Antonines, and is best seen in the second-century work in Ostia.

Despite the congestion which existed in many towns, especially the Italian, there was space for large houses, built on an atrium plan and inhabited by the rich. They were numerous at Pompeii; they are found at Ostia, and even in Rome itself they formed an 'aristocratic minority' among the huge tenement blocks.[88]

In Cisalpine Gaul, and even more in the provinces, city houses were spaciously laid out along wide streets. Buildings of two or more storeys were increasingly rare, and some variant of the *atrium* plan was common. The larger cities, in which commerce and crafts were relatively important and, doubtless, many of the small had *tabernae* in those parts which constituted the central business district. But around the periphery of the larger and over much of the area of the smaller, various types of courtyard house predominated. Some of the smaller towns, like Silchester, were in effect 'garden cities'.

All provincial cities, whether large or small, were the medium through which Roman civilisation was introduced to the provincials. Their public buildings, forum and basilica, temples and baths, theatre and amphitheatre, were designed to make the town attractive, and no doubt they played some part in inter-city rivalry. They covered a large part of the central urban areas, and in a small city like Caerwent must have dominated the whole settlement.

The amphitheatre was as regular a feature of the Roman city as its lineal descendant, the bull-ring, is of Spanish towns today. Often it was placed at a distance from the city centre and outside the walls, where the riots to which its spectacles sometimes gave rise could be more easily controlled. A theatre, at which one assumes a more sophisticated entertainment was offered to the provincials, was to be found in most cities. Like the amphitheatre, it was often located outside the city itself, and for the same reason.

Even more common than theatre and stadium was the temple. The religious cult of the Empire came to be that of the genius of the emperor and of his deified ancestors of the Julian house. It transcended the local cults of the Empire; it had Hellenistic and Middle Eastern overtones, but spread throughout the western Empire. Every town had its temple to the imperial cult; and in all probability most *vici*, their altars to the emperor. The emperor was a charismatic figure who had brought peace to the Empire; a pledge of allegiance to him – worship in a rather crude sense – was a symbol of loyalty to the unity of the Empire. In the western provinces the cult was 'a convenient adjunct to effective administration ... [rather] than ... an expression of real religious policy'.[89] Other religious cults were tolerated as long as they were consistent with the unifying cult of the emperor.

The temples which began to appear in the provincial towns during the first century A.D. conformed in general to a familiar classical form: rectangular in plan, with colonnaded portico, such as the Maison Carrée at Nîmes or Temple of Augustus and Livia at Vienne,

No less characteristic of Roman cities, except those in the humid north-west of Europe, were the engineering works needed to maintain a water supply. In this respect they differed sharply from the Hellenic, and the epidemics which the Greeks endured were part of the penalty which they paid for their ignorance or their neglect. Whereas the Athenians appear to have derived an inadequate supply from possibly contaminated springs and wells within the city, the citizens of Rome tapped mountain sources far to the east and brought it to the city by an elaborate system of aqueducts.

The Romans, with their numerous fountains and bathing establishments, made heavy demands on any local water supply, and there were few cities that did not need to draw from distant sources. In the dry-summer regions of Italy, Spain and southern Gaul this frequently necessitated very long aqueducts and presented complex engineering problems. Arles was supplied from the Chaîne des Alpilles by aqueducts of which very few fragments survive. Nîmes drew its supply from near Uzès by an aqueduct which dropped only 17 metres in a course of about 50 kilometres and included the Pont du Gard and three tunnels. Fréjus (*Forum Julii*) can also show today impressive remains of its aqueduct. Even more elaborate was the water supply of Lyon. Four separate aqueducts, with a total length of 178 kilometres, brought water from the hills to the north of the city and from the Monts du Lyonnais to the west. Of these, the Gier aqueduct alone extended over 70 kilometres from near St Etienne. Paris had an aqueduct of 24 kilometres; Metz reached out 22 kilometres, and Evreux, 19 kilometres. But the most impressive of all these engineering works today is that whereby Segovia, on the dry Meseta of northern Spain, drew its water from the Sierra de Guadarrama. Scarcely less ambitious, though less well preserved, are the aqueducts at Merida and Tarragona. Even in Britain where water supply might have been expected to present little difficulty, Roman waterworks were more ambitious than any to be found again before the nineteenth century.

Not all cities had defensive walls in the time of the Antonines. Military camps had always been protected by bank, ditch and palisade, and in general the *coloniae*, peopled by veterans, followed the same pattern. The possession of a town wall became a mark of status as well as a means of protection. Many

Italian towns had walls in the first century B.C. and during the following century new towns were usually walled. The period of peace which followed allowed towns to be established without walls, and doubtless many of the older defences were allowed to crumble. Only in areas that were politically insecure, such as Britain, and those which lay near the frontier were town walls maintained or strengthened. During the third century unsettled conditions returned; the danger of Germanic invasion increased, and the cities of the Empire hastened to build walls or to repair those that had fallen into decay. From this period date many of the fortifications that remain in Gaul and Spain.

During the prosperous and peaceful years of the first and second centuries some cities had built walls of more generous proportions than they were now able to maintain and defend. Foremost among these were *Augustodunum* (Autun), whose original walls, almost 6 kilometres in length, enclosed an area of about 200 hectares. How much of this area was built up is uncertain, but the whole appears to have been divided up by a grid-iron pattern of streets. In all probability much of the urban area was dotted with the villas of the romanised aristocracy of the Aedui.

The drastic reduction in the fortified perimeter of several Gallic towns may well have been accompanied by a contraction of the population. More likely the reduced area was what, in the changed conditions of the latter Empire, 'it was economical to defend'.[90]

The walls which the Romans built, even in time of peace, were intended for military use. Towers and gates were designed to give the maximum cover, and the line of the walls made the greatest use of the undulations of the terrain, though there were some city entrances, among them the *Porta Nigra* of Trier, which were more decorative than defensive. To the Greeks, the walls seemed often to have been of secondary importance, mere lines of masonry thrown around an existing community, whereas the Romans appeared to have regarded the walls as dominant, anterior in planning if not also in construction, to the settlement itself, at least under the Republic and early Empire.

City of Rome. To most of these generalisations the city of Rome was an exception. 'Old Athens and Old Rome,' wrote Boëthius, 'were both shapeless and irregular.'[91] Rome resembled the old Italic cities, clustered on their hilltops, but its site was a lowland one, and Rome must, indeed have been one of the very few valley cities that were not regularly planned. In the age of the Antonines Rome was at the height of her prosperity. Its population *may* have approached a million (see above, p. 112), and it was unquestionably the largest city Europe had yet known.

Growth during the Principate had been unusually fast, and the spread of buildings was subject to but little control. The city authorities, it is true, regulated through the censors the location and construction of public buildings, but private buildings were, in the main, erected by speculative builders and were maintained in good condition only as long as it was profitable to do so. It was not until the Principate of Augustus that an attempt was made to restrict the dangerous heights reached by some of the apartment blocks. Further restrictions were placed on builders under Nero and Vespasian, but these related only

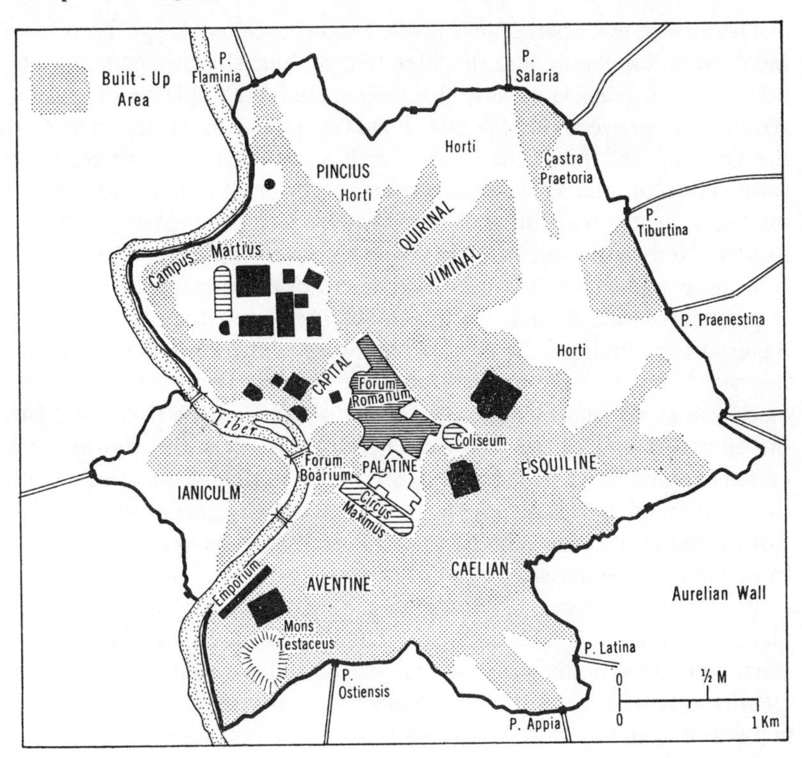

Fig.3.7 Rome under the Empire

to the dimensions of large dwelling houses and the materials of which they were built. No attempt was made to limit their density or to provide wide, straight thoroughfares between them.

By the end of the Republic the Servian wall was ruinous, and at all points the city had spread beyond it. Rome was, apart from the camp of the Praetorian Guard on its north-eastern periphery, an open city, until the Aurelian wall was built in the third century.

Augustus had divided the city, including part of the highly urbanised areas surrounding it, into 14 *regiones*. This administrative division of the city lasted unchanged through the imperial period. By the time of the Antonines very little of the area embraced by the 14 regions was free of buildings, and in the following century the wall of Aurelian was built to enclose more than three-quarters of it. The *regiones* covered an area of 1,783 hectares, or almost 7 square miles. Of this an excessive proportion – at least 40 per cent – was occupied by public buildings. These were scattered over all parts of the city, with the exception of the fourteenth region, which comprised the Ianiculum beyond the Tiber. Public buildings, especially if we include among them the palaces of successive emperors, and the many monuments and temples, occupied much of the area of the central regions, numbers VIII, X, and XI. Even in outlying and mainly residential regions, the baths of Titus and Trajan alone covered almost a square kilometre.

There were few open spaces in the central regions except those provided by the *Forum Romanum*, and the later *fora* of Augustus, Vespansian, Trajan and others, the *Circus Maximus*, the *Stadia*, and other places of entertainment. Around the periphery of the city, however, especially on the higher ground of the Quirinal and Esquiline to the north and east, and also west of the river, were the luxurious villas and gardens of the rich. They had formerly been extensive nearer the centre of the city, but had given place to the spreading *insulae*. Nero had built his *Domus Aurea* on rising ground to the east of the *fora*, on a site which had been laid waste by the fire. He was followed by a large part of the Roman aristocracy, driven from their traditional seats by imperial rebuilding of the city centre. Some built gracious homes out beyond the new house of Nero[92] or even outside the city limits, in the hills by Tivoli or on the coast near Ostia.[93] It is doubtful whether the parks and gardens of the emperors and of the rich gave much pleasure to the crowded proletariat or contributed to the food supply of the city, though there are instances of the more public spirited giving their gardens to the people on their death. But by and large, it was the baths which provided the open spaces and garden-like courts for the masses.

The number of *domi*, of self-contained homes, commonly built on an *atrium* plan, was small, probably less than 2,000.[94] They were, as might be expected, least numerous in the central regions, and those that had survived in the vicinity of the *fora* must have been excessively small. The growth of urban population, without any commensurate development of a system of public transport, led inevitably to an appalling congestion in the central regions, made worse by the short-sighted policy of the earlier emperors in clearing private housing to make room for public buildings. Rows of houses of the *tabernae* type, which under the Republic had risen at most two storeys, were made to support a third and even a fourth. Construction of mudbrick, strengthened with wooden joists, or even of lathe-and-plaster, was unstable and liable to burn. What suburbanite at Praeneste or Tivoli, wrote Juvenal, 'was ever afraid of his house tumbling down . . . here we inhabit a city supported for the most part by slender props: for that is how the bailiff holds up the tottering house, patches up gaping cracks in the old wall, bidding the inmates sleep at ease under a roof ready to tumble about their ears'.[95]

This was the Rome which burned one dry July day in A.D. 64. The fire spread from the southern edge of the Palatine. 'The city's narrow winding streets and irregular blocks encouraged its progress.' Rome was devastated. Of the 14 regions, 'only four remained intact. Three were levelled to the ground. The other seven were reduced to a few scorched and mangled ruins.'

'Nero profited by his country's ruin,' wrote Tacitus, 'to build a new palace.' This was the *domus aurea*; he was also suspected of having set the fire because he was 'ambitious to found a new city to be called after himself'.[96] A new city was indeed built; the rubble was cleared and carried down the Tiber by the barges that had brought up the city's grain supply, and was dumped in the marshes at Ostia. Streets were widened and straightened; masonry construction, either of tufa or travertine, was prescribed. The dangers inherent in the architecture of the old Rome were reduced, but the problems of overcrowding and the intense summer heat were increased.[97] Nor was the height and congestion of the tenement blocks reduced. Perhaps the greater part of a million Romans lived in these *insulae*. Six storeys were not unusual. On the street level were the *tabernae*, often with masonry vaults; above living quarters rose floor on floor. Their external walls were often of concrete, faced with brick; internal walls were generally of masonry, and the use of timber, now becoming increasingly scarce, was kept to a minimum. Sometimes such buildings enclosed small courtyards, and usually balconies projected over the streets. They were plain but functional; the danger of conflagration was kept to a minimum, and, given the number of baths and other places of public entertainment, life in them was probably no more uncomfortable than in the slums of our modern cities. Suetonius spoke of the *immensus numerus insularum*,[98] but how many there were is not known. Many had been built before the fire, and some of these evidently survived the conflagration, and continued to be inhabited.

The ground-plan of the *insulae* is known from the *Forma Urbis*, the city plan of the third century, engraved on marble, which in a very fragmentary form, has survived.[99] The *insulae* were grouped around the Capitoline and Palatine, and spread over the flat land of the *Campus Martius* to the north-west, where they suffered severely from flooding.[100]

The *insulae* were a consequence of Rome's peculiar problems of overcrowding. In a sense, however, they remain with us and have become a feature of the west European city. They contributed to the dispersion of business and crafts through much of the city area. The *tabernae* were the shops and workshops, whereas in Hellenistic cities of the Middle East, the marketing functions were concentrated in a bazaar quarter which was not at the same time residential.

The supply of food and water, of building and industrial materials to the city of Rome posed a problem which the Romans alone would have been able to solve. The surrounding Campagna had long since ceased to be the main source of essential foodstuffs. Wheat was imported, much of it from Sicily, Egypt and North Africa, and handled at the Port of Ostia (see below, p. 132); it was brought up the Tiber by barges, and again unloaded at Rome's emporium, the dock quarter that lined the waterfront below the Aventine. An enormous fleet of barges was employed, and it is estimated that 6,000 barge-trips were made each season. Tacitus recorded in A.D. 62 the destruction by storm and fire of no less than 300 barges.[101] Warehouses, or *horrea*, lined the river downstream from the *Pons Probi*, the lowest Tiber bridge, and 'a vast system of stevedores, grain measurers, and warehouse guards ... received and stored the city's food'.[102] Upstream from the grain warehouses to the *Forum Holitorium*, and accessible only by the smaller barges, were warehouses and open-air markets for all varieties of foodstuffs. Much of the wine and olive oil consumed in the city was sold here; so also were the exotic foods described by Martial; marble and ornamental building stones; metals, wool and cloth. Vegetables and fruits, grown in the Campagna, were brought into the city by cart or pack animal, and animals were driven down from the mountains to the city markets, and lumber for builders and craftsmen was floated down the Tiber from the Apennines or up the river from Ostia.

The maintenance of a steady flow of food to provision a city of a million had

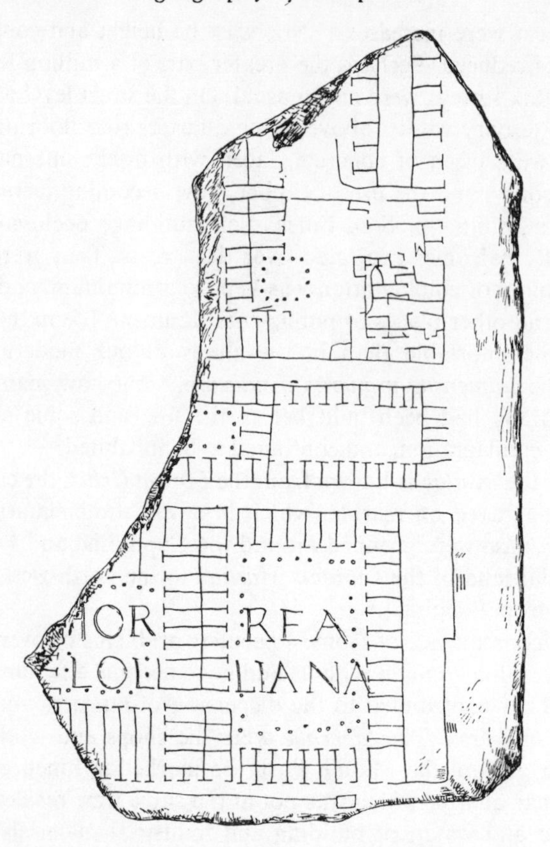

Fig.3.8 Fragment of the *Forma Urbis*, showing the *Horrea Lolliana*

been handled under the Republic on a temporary and *ad hoc* basis. It formed at first part of the duties of the *aediles*. Under Augustus it was entrusted to an office, that of the *praefectus annonae*, created for the purpose.

The supply of water presented problems of even greater difficulty in the Italian climate of long, dry summers. Mention has already been made of the aqueducts which the Romans built to supply provincial cities of even modest size. All these yield before the size and complexity of the works built to bring water to their capital. To Strabo they were among the most valuable of Rome's public works, bringing water 'in such quantities that veritable rivers flow through the city and the sewers; and almost every house has cisterns, and service-pipes, and copious fountains'.[103] The oldest aqueduct, the *Aqua Appia*, had been constructed late in the fourth century B.C., and between that date and the early second century A.D., when Trajan's aqueduct was put into service, no less than nine separate works, with a total length of 264 English miles, were completed. All, except one, drew their water from the Anio and from other left-bank tributaries of the Tiber. The exception, the *Aqua Alsietina*, was built by Augustus to deliver water from the crater lake of *Lacus Alsietinus*, 17 miles north-west of the city. Frontinus questioned the wisdom of its construction since its water is 'positively unwholesome, and for that reason is nowhere

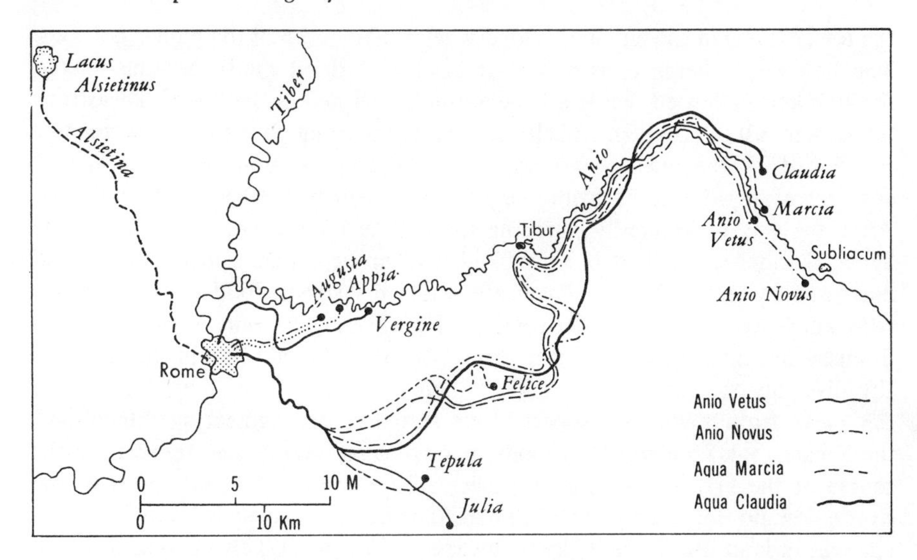

Fig.3.9 The water supply of Rome (The Aqua Felice is a Renaissance work which appears to have followed the Aqua Alexandriana, constructed by the Emperor Alexander Severus.)

delivered for consumption by the people'. It was generally used for watering the gardens and supplying the fountains of the city, and only in time of dire shortage was it used domestically, and then only by settlements on the right bank of the river.

The earlier aqueducts, completed when the city was still relatively small, came in at a low level. As buildings spread up over the higher ground to the east, higher aqueducts were constructed to supply them. The earlier channels had followed the contours of the land from the Apennine foothills, frequently buried a short distance below the surface. The later aqueducts were built on or above the surface, and the last of the major works, the Claudian aqueduct, took its water from the Anio valley far above *Tibur* (Tivoli) and strode across the Campagna on giant columns, the ruins of which still stand. Even in their abandoned and ruinous condition the aqueducts of Rome are an impressive monument to the practical genius of Rome. Frontinus took a justifiable pride in the works which he was called upon to supervise: 'with such an array of indispensable structures,' he wrote, 'compare, if you will, the idle Pyramids or the useless, though famous, works of the Greeks.'[104]

Rome could not have developed without the Tiber. The ancient writers were fully aware that the river had provided protection for the infant city against the Etruscans, that its crossing points had canalised trade through the city, and that it had brought commerce, both upstream and downstream to the markets of Rome. Yet it was a dangerous and difficult river. It did not, fortunately for Rome, dry out in summer, but in winter and spring was liable to particularly severe floods. The *Campus Martius* and all the low-lying land near the Palatine is known at some time or other to have been devastated by flooding. Dio Cassius described a flood[105] which 'rose so high as to inundate all the lower levels in the city and to overwhelm many even of the higher portions. The houses ... being constructed of brick (i.e. dried mud), became soaked through and collapsed, while all the animals perished in the flood'. Disastrous floods were all too frequent. Their reason lay, not so much in the excessive rains mentioned by the ancient writers, as in the deforestation, soil erosion and consequent rapid run-off from the volcanic country to the north of Rome. Augustus was undoubtedly pursuing the correct course when he 'cleared the Tiber channel which had been choked with an accumulation of rubbish and narrowed by jutting houses'.[106] There was a disastrous flood under Tiberius, with much loss of life and destruction of property. The Senate was moved to discuss the matter, and it was proposed that floods should be checked 'by diverting the streams and lakes which nourished them', but, 'because of the pleas from towns, or superstitious scruples, or engineering difficulties', the Senate took no action.[107] Floods, nevertheless, became less frequent in the course of the first and second centuries. The removal of obstructions to the river's flow no doubt helped to improve the situation, but it is likely also that changes in land use in the Tiber valley (see below, p. 147) also played a role in reducing run-off.

The port of Rome. The rise of Rome owed little to its navigable communication with the sea. Though tradition ascribed the foundation of Ostia to the period of the kings, archaeology can find no evidence of a settlement at the river's mouth before the fourth century B.C. and then it existed more to protect Rome from piratical attack than to handle the city's commerce.[108] Puteoli, near Naples, was the first port of Rome, despite its considerable distance from the city. Smaller sea-going vessels sailed up the 24 miles of the lower Tiber, but the shoals of the lower river and the rapid accretion along its banks made navigation hazardous. Julius Caesar even proposed to cut a canal from the city across the Pontine marshes to Terracina to facilitate the transport of grain from the more southerly ports. The volume of food imported by Rome continued to grow; attempts by Augustus and Claudius to make a port at the Tiber mouth for the transshipment of grain met with only qualified success, and it remained for Trajan to create a network of ports and harbours of refuge and to make Ostia 'not merely the harbour of the world's largest consuming centre, but an important line ... in the great trade route from the east to west'.[109]

Ostia was rebuilt and became a populous city (see above, p. 112) of brick-faced *insulae*, rising in many instances to four or five storeys. The transshipment of cargoes took place at first in the river off the city. Caesar had proposed a harbour at a distance from the river, and thus free from silting. Claudius, however, constructed a port opening off the Fiumicino, a small artificial distributary of the Tiber, and Trajan added the hexagonal dock basin whose outline is apparent in the fields today two miles to the north of the ruined town. This became the harbour for the bulk-handling of grain, lumber and building materials. The old town of Ostia seems to have retained the trade in wine and perhaps also in luxury goods, many of which were probably sent on to Rome by road rather than river.

Rural settlement

The study of rural settlement in the age of the Antonines is complicated by the dichotomy which continued to exist between the huts and hut clusters of Iron Age provincials and the pattern of settlement imposed by the Romans themselves. 'Throughout the Roman occupation,' wrote Crawford, 'the natives of southern Britain lived very much the same life as before.'[110] So must have also been the case with the natives of many parts of Spain and the Gauls, and of the Danubian and Balkan provinces. It is, however, impossible to think of these provincials, however remote they may have been from the centres of Roman power, as wholly uninfluenced by the civilisation of Rome. They carried on some form of trade with Rome. Artifacts and tools of the Romans made their way to their villages, and they were subjected to pacification and disarmament, which passed as the *pax romana.* They had, perhaps under pressure, abandoned their hilltop forts, and no longer erected palisades for defence around their hamlets.

The settlement pattern in all parts of the Empire was in process of change. Almost everywhere, except near the Rhine–Danube frontier, peace and security had led not only to the abandonment of entrenched *oppida,* but to the break-up of village settlements and the formation of dispersed farms. This process was as prominent in Etruria as in Cranborne Chase. But throughout the more romanised parts of the Empire two contrasted types of settlement and land tenure were to be found. On the one hand, there were the small holdings, held in some form of freehold, if one may use that term for so early a period, and, on the other, the estates. It is generally assumed that the latter grew in number and size under the Principate. Unquestionably they had become common much earlier, and Toynbee traces the decline of the 'traditional peasant economy' to the Second Punic War and the devastation of much of peninsular Italy.[111] A new plantation agriculture appeared, and for its practitioners Cato, already in the second century B.C., had written a handbook.

The new farming units were relatively large. If the owner lived regularly on the estate, he occupied a villa; if he did not, they would probably have been a *villa rustica,* a sort of grange which constituted the focus of the estate's productive activities. One cannot visualise such estates as vast compact blocks of land; some certainly were, but others were made up of scattered parcels of land. Nor is the antithesis between villa-estate and scattered settlement-small-holding necessarily valid. Many estates were operated wholly or in part with unfree labourers, but there were instances of free peasant farmers becoming tenants on estates. Whether they contributed regularly or occasionally to the labour force of the villa is unclear. It is entirely possible that they anticipated many of the labour relationships met with in the middle ages. The villa-estate is rarely found without evidence of hamlets and small isolated settlements (cf. Picardy, below, p. 137) which *may* have been the farms of a free and independent peasantry.

Most extensive of all estates was that of the emperor himself. It was found in all parts of the Empire. It had been acquired in a variety of ways: confiscation, purchase, or in satisfaction of tax obligations; it was managed on behalf of the emperor by bailiffs (*procuratores*) like any other large estate, and, like other estates, it was made up of lands worked directly with servile labour and of those leased to free tenants.

One looks forward to the time when archaeological exploration will have made more progress than hitherto; when much of the Empire will have been mapped, and it will be known with only a small margin of error where every farm and villa, every military *canabae* and roadside *vicus* were located. We have this data for only a few small areas, and for the rest, it is a race to excavate and record before bulldozer and tractor scatter and destroy the evidence forever.

Southern Etruria. For no part of the Empire is the evidence more satisfactory than for the area lying north and west of the Tiber. This is largely due to the work of the British School at Rome, which has set itself the task of recording every artifact and site that can be ascribed to the ancient civilisations. The area is one of volcanic tufa, horizontally bedded, soft and easily eroded. Older and harder rocks protrude through the tufa, forming, for example, the serrated limestone ridge of Monte Soracte. It was the erosion of this country which contributed to the silting and flooding along the lower Tiber and to the advance of the coastline at Ostia.

Much of this area formerly lay in the Ciminian Forest, and it was not until the third century B.C. that extensive clearing, settlement and cultivation began. By the second century A.D. little primeval forest remained, and the landscape was, by and large, man-made. The construction of a fairly close net of Roman roads, coupled with the peace and security of the Principate, had led to a dispersion of settlement. The Romans had themselves contributed by breaking up and dispersing urban settlements, like *Veii*, *Falerii* and *Lucus Feroniae*, and by the planned settlement of peasant farmers: 'around Veii,' wrote Ward-Perkins, 'small farmhouses . . . are so thick on the ground, and so evenly distributed, that there can be very little doubt that they are the product of some such scheme.'[112]

The area around *Sutrium* (Sutri) and the *Ager Capenas* may be regarded as typical of the region as a whole. Sutrium lay on the southern margin of the Ciminian hills, between Lakes Vico and Bracciano. It was of Etruscan origin, but the surrounding area appears not to have been opened up until after the Roman conquest, with settlements 'penetrating and opening up the woods and forest land, until eventually, at the peak of the Roman period, Sutri . . . had become the centre of a thickly populated countryside'.[113] Most of the buildings attributable to the Roman period were farmhouses of small or moderate size. There was probably never any great number of villas and large estates in this area. The expansion of settlement, in so far as it can be traced by means of the finds of datable pottery, shows the spread of farms outwards from Sutri, following in general the flat interfluves between the deeply incised valleys of this region.

Campania. South of Rome lay the volcanic Alban Hills, and beyond them the Pontine Marshes. This dreary and scarcely habitable tract[114] terminated where the mountains run down to the sea at Terracina. Between the marshes and the sea lay a belt of sandy soil; here, facing the Mediterranean, the villas of the rich lined the shore and, with occasional villages, looked from the sea 'like a number of cities'.[115] Inland, however, malaria depopulated the marshlands, and, except for fisherman's huts and the winter shelters of the herdsmen looking after the transhumant flocks and herds, there were no settlements. South of Terracina, however, lay the most densely peopled countryside in Italy. Its rich volcanic and alluvial soil had long been cultivated when the Romans brought this area under their rule. Although cities had declined in size and importance, the countryside remained rich and populous. But even here change was taking place. It would have been surprising if the heavy imports of grain into Rome had not affected the pattern of agriculture in nearby Latium and Campania. Everything points to the drastic reduction of wheat farming and its replacement by vineyards and olive groves. Domitian had attempted in vain to check the trend,[116] and Campania was increasingly important for its wine, oil and fruit production. This shift in the agricultural base brought with it a change in the settlement pattern, Viticulture and, above all, olive growing were highly capitalised, and on a commercial scale were beyond the capacity of the peasant farmer. The excavation of the site of villas overwhelmed by the eruption of Vesuvius in A.D. 79 throws an important light on this. It is evident that, though small farmers survived, settlement was becoming increasingly concentrated in large villas. These varied somewhat in style and function. Some belonged to well-to-do farmers who lived on their lands; others served primarily as summer residences, and yet others – *villae rusticae* – were the centres of large, slave-run estates. Whatever their organisation, their staple products were wine and oil, and the larger the villa, the more specialised its production. Detailed information relates primarily to villas destroyed and yet preserved by mud and volcanic dust. One assumes that Vesuvius merely interrupted a pattern of economic life, and that the same kind of agriculture and settlement were resumed when the region came to be re-occupied. It was a region of estate-villages, known as villas, relatively closely spaced, because their dominant crops made an intensive use of the land, with the small scattered farms of the surviving peasant proprietors between and around them.

The centuriated landscape. The practice of centuriation provides perhaps the extreme case of the superimposition of a Roman pattern of settlement on the pre-existing landscape. All publicly owned and regularly surveyed land 'was, at least in theory, divided into rectangular plots marked off by roads, paths, or other visible signs'.[117] The establishment of the *centuriatio* was the usual prelude to the establishment of a *colonia.* It characterised thinly settled areas brought under Roman control and settled in the late Republican period and the early years of the Principate.

Centuriation consisted, in short, of dividing the land into squares by a pattern of straight roads or paths intersecting at right angles. The rectangles thus formed were customarily of 2,400 Roman feet (773 yards), and covered about 123 acres. The size of the farm holdings in each *centuria* varied, with the terrain and the fertility of the soil, from 25 to 100 acres.

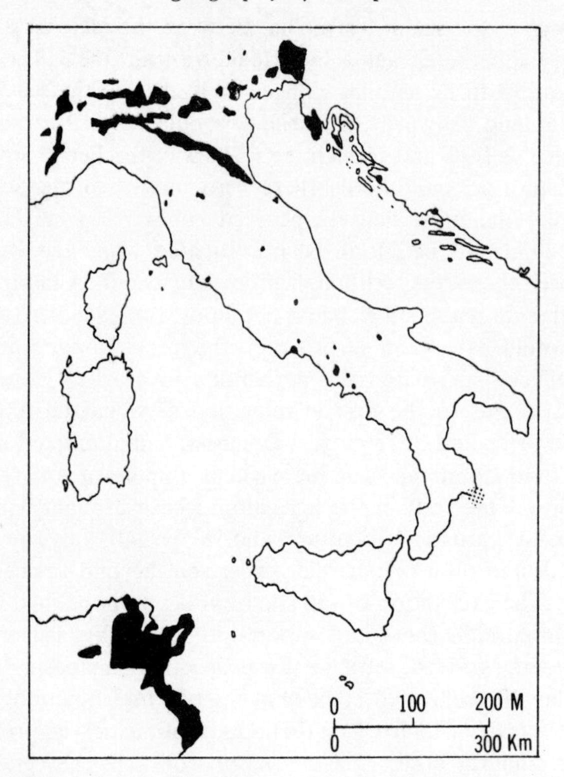

Fig.3.10 The distribution of centuriated field-systems

The evidence of classical centuriation is readily destroyed by later societies, and tends to be preserved best where drainage and irrigation works following the boundaries of the *centuriae* have had the effect of fixing it. There were scattered areas of centuriation in peninsular Italy, but settlement was in general too dense and too well established for there to have been scope for its development. In northern Italy conditions were different. Two broad belts of *centuriae* extended on each side of the marshy central area of the Po plain, from *Aquileia* and *Ariminum* in the east to the neighbourhood of *Augusta Taurinorum* (Turin) in the west. There is evidence of centuriation in Apulia; it is well seen in southern Istria, behind the port of Pola, and along the margin of the Dalmatian Karst, near *Iader* (Zadar) and Salona.

Centuriation is less frequently recognised in the Gauls, and Spain, and scarcely at all in Britain and the Danubian and Balkan provinces. There was, however, a centuriated landscape around *Arausio* (Orange) and to the east of Valence in the Rhône valley. Clear evidence of centuriation has also been found in Greece, especially near Augustus' settlement of Nicopolis (see above, p. 119) and at Dyme in Aetolia. Less certain evidence has been found in Macedonia.

The villa system. The villa system was, in the main, a Roman settlement pattern, superimposed upon an earlier pattern which it neither obliterated nor entirely superseded. The term *villa* is itself by no means easy to define. It

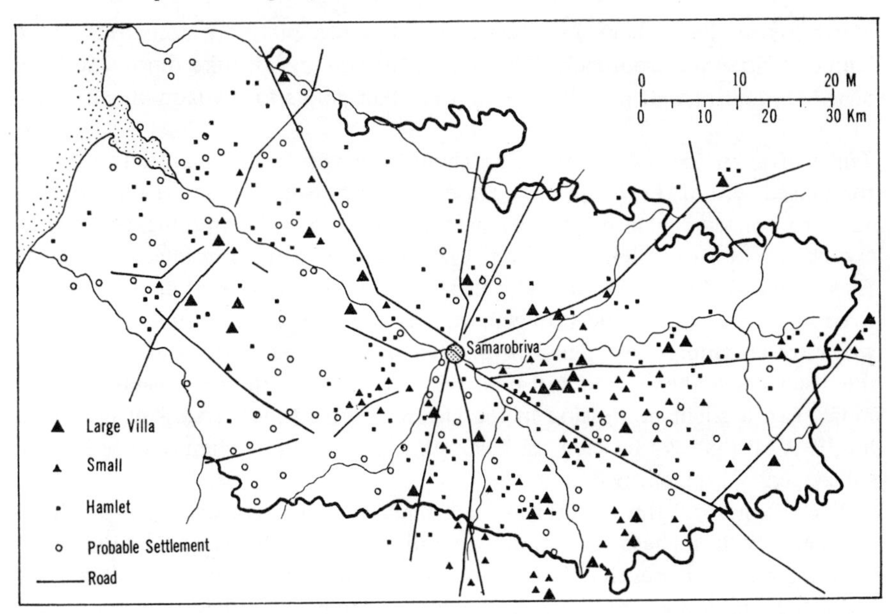

Fig.3.11 Settlement in Roman Picardie (Samarobriva is the modern Amiens.)

implies a farmstead of above average size and sophistication. It was commonly, though not always, built of masonry; it frequently possessed a hypocaust, and may have had mosaic floors and plastered and painted walls. It may have had provision for the weaving, fulling and dyeing of cloth, and for metal-working, but essentially it was a farm. No doubt some *villae* were built by Romans, or at least Italians, who had settled in the provinces, and a few may have served for the holiday amusement of the rich. But most were built by the provincials, whose ancestors may once have helped to defend a nearby *oppidum* against Caesar's army.

On the outskirts of Mayen, in *Germania Superior*, a villa was excavated which had been continuously occupied for three centuries, during which it had grown from a small thatched hut of the pre-Roman Iron Age to an elaborate, if not luxurious mansion.[118] There can be no doubt that its owners and occupiers were increasingly romanised provincials. Grenier claims that most villas in the Gauls succeeded to less sophisticated native settlements.[119] Sometimes the villa stood surrounded by the huts of provincials. Some villas contained the quarters of bailiff, farm labourers and craftsmen; yet others were no more than the family homes of comfortable farmers.

A map of villas is needed, but ambiguities in nomenclature and the uncertainty that attached to the archaeological record until quite recent years makes its compilation a matter of extreme difficulty. Certain areas of western Europe were well colonised with villas. In the Gauls, they were numerous throughout the areas which came later to be known as Burgundy. They spread down the Seine and Yonne towards Paris, and down the Loire towards the sea. They surrounded the more important cities, like Amiens (see fig. 3.11), Autun and Trier. In only five cantons to the north of Autun there were nearly 300 villas.

North-east of Paris there were fewer, but they became more numerous as the Rhine frontier was approached. No doubt the provincials, like those who lived near Mayen, did well by selling food and other goods to the Roman garrisons.

The native settlement pattern. One should not speak of romanised and non-romanised within the Empire, because it is impossible in the second century A.D. to conceive of any peoples or settlements wholly uninfluenced by the civilisation of Rome. The late Iron Age settlement pattern in much of western Europe was a dispersed one. The tribes had hilltop *oppida*, but most people lived in villages and hamlets. In general they were built of local materials; wood was most commonly used, and the absence of stone foundations has resulted in the discovery and excavation of relatively few of them. In many instances, little more than a slight depression in the ground, formed by the sunken floor of a hut,[120] is all that can be seen, and even this remains only when the site has not subsequently been cultivated or used.

The second and third centuries were marked by the construction of a large number of villas. These were built, generally on low-lying ground, by the native peoples, who had presumably lived previously in primitive huts. Evidently, we are witnessing in the age of the Antonines the progressive abandonment of huts and hut-circles in favour of modest *villae* and small hamlets, which were during the following century enlarged and elaborated.

The dwellings of the native peasant farmers were evidently more sophisticated in some than in other areas. Intensive fieldwork in Picardy, supplemented by aerial survey, has revealed a cluster of villas around *Samarobriva* (Amiens), with much of the intervening territory filled out with other settlements.[121] The latter consisted generally of groups of from three to five buildings, mostly rectangular in plan, with two or three compartments and perhaps masonry foundations. The hamlet was evidently becoming the most common form of rural settlement in much of the western Empire.

In the Rhineland and *Agri Decumates* most areas of fertile soil appear to have been occupied by closely spaced *villae rusticae*, which may have been able to contribute to the food supply of the military forces. Small self-sufficing peasant farms were few. A similar pattern of settlement, with a predominance of *villae rusticae*, appears to have spread over the Swiss plateau in *Germania Superior*.

Iron Age settlement in Spain had been dominated by *castros*, hilltop forts, walled with masonry and containing a cluster of small, generally rectangular, huts (see above, p. 118). The Romans, in Strabo's words, 'reduced most of their cities to mere villages'. The Roman policy was evidently to destroy the Celtiberian and Celtic *oppida* and to scatter their inhabitants, and then in time to recombine the villages into cities on the Roman pattern. The Turditani of southern Spain had thus been converted 'to the Roman mode of life, not even remembering their language any more . . . And in the present synoecised cities, *Pax Augusta Julia* [Béja] . . . *Augusta Emerita* [Merida] . . . and some other settlements, manifest the change to these civil modes of life'.[122] This twofold process had advanced least in north-western Spain, where it is probable that the fortified *castros* had not been wholly abandoned.

There has been neither intensive archaeological study nor adequate aerial survey of most of the Danubian and Balkan provinces, and our knowledge of the settlement pattern is slight. Cities were few, and the impacts of Roman and before it Hellenistic civilisation on the native peoples were superficial. Most of the urban settlements of the interior had developed from military camps; those along the Black Sea coast derived mainly from Archaic Greek settlements, while the more important of those on the Dalmatian coast derived from Roman colonies. In the vicinity of most of them villas were in time built by the romanised provincials. In Pannonia, for example, remains have been found of more than 50 villas. They formed a line parallel with the Danube, as if sited to provision the garrisons that kept watch across the river. They surrounded Lake Balaton, then, as now, a resort area, and they were found along the roads leading west to Noricum. Some of the villas achieved a high degree of sophistication. Elsewhere in Pannonia and Noricum lived Illyrian and Celtic peoples, only superficially romanised. Such evidence as there is points to small, self-sufficing villages or hamlets.

In the Balkans villas were fewer than in Pannonia, and the assimilation of the native Thracians and Illyrians to a Roman way of life had made even less progress. The bas-reliefs on Trajan's column show us small villages of closely spaced wooden huts, with high, steep roofs of shingle or thatch. The material and details of the design would have varied, but this must have been the characteristic settlement throughout the Balkan peninsula.

Settlement beyond the Roman frontier

One would expect little significant difference between the pattern of settlement in the remoter areas of imperial territory and that beyond its limits. Both were established by essentially non-romanised and tribally organised peoples. Among their neighbours and enemies, the Romans knew most about the Germans. The latter did not, wrote Tacitus, live in cities (*urbes*). Instead 'they live apart, dotted here and there . . . Their villages are not laid out in Roman style, with buildings adjacent and interlocked. Every man leaves an open space round his house.' Buildings were of wood, plastered with clay. They also, he added, 'have the habit of hollowing out caves underground and heaping masses of refuse on top. In these they can escape the winter's cold and store their produce.' One must think of small, loosely nucleated settlements scattered through the forests of northern Europe, the more permanent encircled by a defensive palisade. In south Germany, Bohemia and Moravia, Celtic *oppida* may have continued to serve as residence and refuge. Some had been abandoned, for even outside the limits of the Empire small, permanent hamlets appear to have been established; others were destroyed by the Romans. In marshy areas of northern Europe, clustered crannogs and pile-dwellings continued to be inhabited, and in the forests of the Baltic region were small hut-clusters, located on defensible sites and sometimes enclosed by a palisade. It seems to have made little difference to the plan and pattern of settlements whether their inhabitants spoke a Germanic, Slavic or Baltic language. Throughout the mountains to the north of the Danube, small, open villages, like those shown on the columns of Trajan and Marcus Aurelius, prevailed, sometimes with a hilltop *oppidum* nearby to serve as a refuge.

Settlement varied in density, but was everywhere thin. Finds in central Germany dating from the first three centuries show a number of clusters, separated by areas in which finds have been few. The largest of these clusters (fig. 3.19) lie in Saxony, along the lower Elbe and in Brandenburg, in eastern Jutland, and in Silesia, northern Bohemia and Moravia. Clusters, less clearly differentiated, are found along the upper Vistula and the middle Elbe, and in the plain of Bohemia. The areas where archaeological finds have been most numerous show a rough correspondence both with areas of good – in part loess – soil and with areas shown by Schlüter as cleared in early historic times.

In Scandinavia careful archaeological work has made possible the reconstruction of some villages and the location of the sites of many more. The Danish peninsula and islands, central Sweden, Gotland, and Oland appear to have been quite populous. In Jutland, not only the infertile and sandy west, but also the areas of ground moraine further to the east, were settled and cultivated. At Vesterup, in northern Jutland, a village site of the Roman Iron Age has been excavated.[123] Two rows of huts, not unlike, in all probability, the 'black-houses' of the Hebrides, were exposed, forming together an early example of a street-village. Here and elsewhere settlements appear to have been relatively permanent. At Borremose, also in north Jutland, a village of about twenty elongated huts has been found.[124] This was much larger than most, and it is possible that not all the huts were inhabited at one time. On the islands off the east coast of Sweden were similar settlements. Farms lay in groups of two or three; less frequently of four or more. The single, isolated house was rare. Sometimes, especially in Oland, a number of such clusters formed 'a kind of scattered village', with the houses surrounded by small, compact and somewhat rounded fields.[125]

It is, however, Vallhagar, on the island of Gotland, that provides the clearest picture of a Scandinavian settlement of the Roman Iron Age.[126] The village lay near the south-west coast of the island, and was located on ground moraine in an area of mixed woodland. It consisted of five or six 'long' houses with their ancillary buildings, and was surrounded by a complex of small, subrectangular fields. The houses were walled with stone taken from the boulder clay, and a roof of turf or thatch was supported by wooden posts. This was a permanent settlement. It was well established in the second century, and lasted until the period of the migrations. Similar settlements have been noted in southern and western Norway. In Denmark, the plan of the settlement is usually broadly similar, but walls were more often of turf, and the roof was supported by posts. In the north European plain, masonry construction was rare.

At this time central Sweden was probably the most developed and populous region of Scandinavia. A belt of small hilltop forts stretched from the Kattegat to the Baltic, most of them probably occupied in the Roman Iron Age. A few such settlements lay along the Swedish coast of the Gulf of Bothnia, but they ended abruptly where the hills of Norrland rose from the lowlands of central Sweden. The map of these Iron Age strongholds in Sweden conforms with the distribution of articles of Roman origin.

North of the hill forts of central Sweden and the villages of southern Nor-

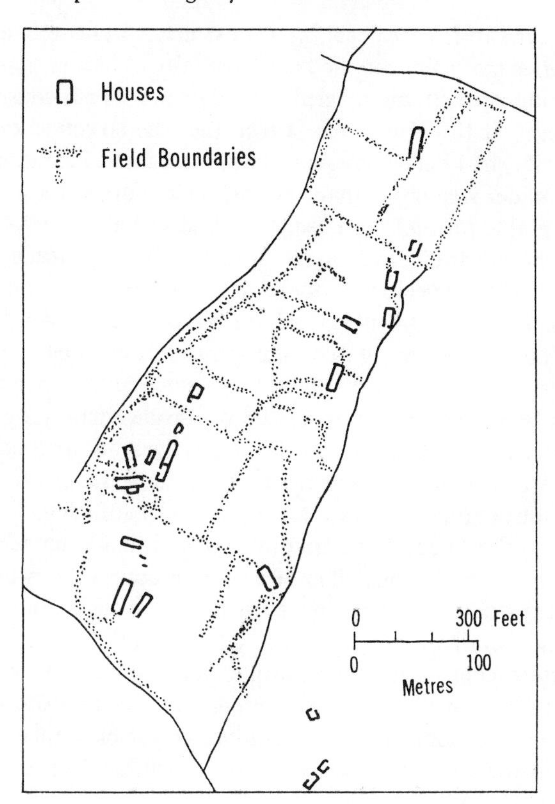

Fig.3.12 Vallhagar settlement

way, there were fewer people in all probability than there had been six centuries earlier. The deterioration of the climate had to some degree emptied this wilderness. The existence of a hunting, nomadic people can be demonstrated for only part of this area, and there is a complete lack of archaeological finds over most of it. One must accept the general accuracy of Tacitus' remarks about the Fenni (see above, p. 110): 'astonishingly wild and horribly poor'.[127]

AGRICULTURE

It is impossible to exaggerate the importance of agriculture in the economy of the Roman Empire. Despite the local prominence of crafts and the pervasiveness of commerce, 'by far the greatest part of the national income of the Roman empire was, so far as we can estimate, derived from agriculture'.[128] the number of large and quasi-feudal estates was undoubtedly increasing at this time, but the foundation of agriculture was still the small, free peasant, who cultivated his own farm with the aid of his family. Vast quantities of grain were handled commercially in order to supply towns, above all Rome, and the garrison towns along the frontier. Some parts of the Empire had adopted a specialised agriculture, such as viticulture and olive growing in Campania, and thus came to depend upon distant markets. Nevertheless, over most of the Empire agriculture was at a near-subsistence level, and most areas were self-sufficing except for the few luxuries that were supplied commercially.

Much has been written about an alleged soil depletion and consequent decline of arable farming at this time.[129] It is true that the structure of agriculture, especially in Italy, had been changing; that grazing had increased, and that larger areas were under tree crops than formerly. Cato urged the planting of olives as a very profitable form of land use,[130] and he and also Varro[131] and Columella[132] stressed the advantages of planting vineyards and garden crops. But these observations related specifically to Italy, where the development of a large urban market combined with subsidised grain imports to make the cultivation of specialised crops more profitable than grain. But even here specialised agriculture was restricted to a few favoured areas, such as Campania.

Wheat continued to be grown over most of the peninsula, mainly by small peasant farmers, to satisfy the needs not only of themselves but also of those who lived in the towns or served in the army. Outside Italy the *civitates* and other comparable territorial areas were in most respects self-sufficing.

During the second and third centuries agriculture was probably more widely practised in the territory of the Empire than at any time before the twelfth or thirteenth century. Local studies indicate the recession of the forest and the reclamation of marsh and heath on a scale unprecedented hitherto. Iron tools were now in general use except in northern Europe, and encouraged the clearing of new land. Evidence from Etruria, for example, suggests that this period marked the culmination of a long period of gradual forest clearance. There was thus some truth in the rhetorical flourishes of Tertullian (see above, p. 95).

The crops

Cereals continued to provide the bulk of the diet of most people. Wheat was the most important food-grain in Rome, and probably also in Sicily and much of Italy. In Greece, however, and probably also in the Balkans, barley remained the dominant cereal. Barley was important also in western Europe, but in northern Europe yielded place to rye and oats. Barley cropped more heavily than wheat, and for this reason continued to be preferred in countries, like Greece, where cultivable land was relatively scarce. In Italy, however, the natural advantages of barley were less marked, and the more palatable wheat had in recent centuries greatly increased at the expense of barley.

Several varieties of wheat were known and cultivated. Foremost among these were the wheats of the emmer group, particularly the hulled variety (*Triticum dicoccum*). This was not a good bread-wheat owing to its low gluten content, and the glumes, which were attached to the grains, made milling difficult. It was being replaced at this time by a 'naked' wheat which was easier both to thresh and to mill. There were several varieties, but that which appears to have gained most ground was durum (*Tr. durum*), a naked variant of emmer, followed in importance by poulard (*Tr. turgidum*).

Other wheats cultivated at this time were einkorn (*Tr. monococcum*), a poor quality cereal of small and diminishing importance, and the spelt group. Apart from certain generic differences, the spelts differed from the other wheats in their superior baking qualities, and this explains why, under the Principate, spelt went far towards replacing einkorn and competed with emmer. Like emmer the spelts were known in both naked and hulled varieties and the naked spelt (*Tr. vulgare* or *compactum*) was very highly favoured for the light bread which it made. The hulled spelt (*Tr. spelta*) was a crop of the regions lying north of the Alps (see p. 205). Its cultivation may have extended as far south as the Po valley, but certainly no farther; it was important in western Germany, southern Scandinavia and Britain, and locally also in northern Spain and the Danube valley.

The choice of a bread grain was not determined only by its suitability for bread-making. It had also to be adapted to the climatic peculiarities of the region. The duration and intensity of the summer drought over much of the Mediterranean basin made it desirable to sow cereal crops in the autumn. On the other hand, few varieties were at this time hardy enough to withstand the winter. It is probable that naked spelt yielded the best results in Italy, but southern Italy and Sicily were probably too dry for it, and here it was replaced by the quicker growing emmer.

Outside Italy barley was more frequently grown. It was suited to calcareous soils, which were widespread in western and southern Europe; it grew and matured quickly, and was thus better adapted to spring sowing than many of the wheats. Lastly, it was also tolerant of the cooler and damper conditions of north-western Europe. On the other hand, it contained little gluten and could not be made to 'rise' like the bread-wheats. It was important in south-eastern Europe and was grown in Spain and France. In Britain, where the evidence of crop remains has been carefully studied, barley was cultivated only to a small extent.

Other cereal crops were less important within the Empire. Rye was a crop of the Germans, but was also grown in Britain; oats were known but were not widely cultivated, except in northern Europe. Pliny regarded them as the foremost 'disease in wheat',[133] in other words as a pernicious and ineradicable weed of the wheatfield. They were, however, a staple of the Germans, and were gaining favour in Britain. Applebaum, in his fanciful account of villa farming, represented oats as a crop with which the more progressive farmers were experimenting[134] because it provided a good fodder and stood up well in poor soil and wet climate.

Maize was unknown. Millet was known, but was chiefly grown in areas where summers were both hot and wet, such as the plains of the Po. Rice was also known, but its heavy demand for water restricted its cultivation to a very few areas.

The lack of statistical data of grain impressions and carbonised plant remains makes it difficult to form any quantitative comparison between the cultivated grains. It appears, however, that in Denmark at this time barley was about four times as important as all other cereal crops, and that oats were next in importance, followed by rye and wheat. This is supported by the evidence from Gotland where barley was also the dominant crop.

Rye came into these northern regions, along with oats, as a weed of the wheatfield, and established itself as an independent crop by its power to thrive in soils and under climatic conditions where wheat did increasingly poorly. It was first cultivated for its own sake in central Europe. From there it was taken to Sweden, and during the Roman period was establishing itself as the chief bread-crop of the Germans and Slavs.

Varieties of pea and bean were known to the Romans, and, if the space given to them by the classical writers is any guide, were extensively cultivated. In addition to providing food for human beings, they were used as fodder and it was also held that they contributed to the fertility of the soil. Lucerne and vetch were grown for the stock. Turnips, cultivated chiefly as human food, were also fed to the cattle in Gaul.[135] Sesame and several varieties of vetch and clover were grown for the stock, and flax and hemp were widely grown for making rope and coarse fabrics.

Such was the range of field crops commonly grown on a large farm by a well-to-do farmer. The peasant farmer, even in Italy, probably grew little more than essential bread-crops, and even so limited himself to those that cropped most heavily in his own district.

The vine and the olive. The Roman writers continually stressed the importance of these twin pillars of classical husbandry. They were writing for a Roman, or at most an Italian, public, and were read in regions where the vine and olive were both climatically suited and generally profitable. Wine and olive oil were necessities of civilised living. The Roman public was a discriminating one, with distinct preferences in both. The superior varieties thus entered into long-distance commerce, and were in fact second in importance only to grain. We have for instance already noted the wine markets and cellars at Ostia and along the riverside in Rome, where the Monte Testaccio, beside the Tiber, was built of broken amphorae that had been used to carry the wine and oil.

Classical allusions to the vine and viticulture are so numerous and relate to so extensive an area that it is possible to compile a map of the crop during our period. Not only do we have the reasoned accounts of the geographical writers, notably Pliny, Strabo, Diodorus Siculus and Columella, but also the allusions by poets like Virgil, Martial and Ausonius, the letter-writers such as the younger Pliny, Seneca and Julian, and novella-writers and symposiasts like Apuleius and Athenaeus.

In some respects the distribution of viticulture was strikingly modern, with its concentration in Apulia, Campania and Latium, southern Spain and the south of France. Strabo reports, however, that it had not reached the more westerly parts of the Spanish peninsula,[136] and there is no evidence that it was practised at this time in the valleys of the Loire, Moselle and Rhine. There can be no doubt, however, that the Rhône valley was important for its vineyards, and that wine was transported northwards into regions still without vineyards of their own. Dion has persuasively argued that the famous vineyards of Côte-Rotie and l'Hermitage owed their start to the Romans, who may have produced here the wine needed for their garrisons in Britain and along the Rhine. In the third and fourth centuries the cultivation of the vine was extended yet farther, to the Moselle valley, to Bordeaux and to the Paris region. At the same time, vineyards were planted along parts of the Dalmatian coast, and attempts were made, with but slight success, to introduce the grape-vine into Pannonia, Noricum and Rhaetia.

The olive had a more restricted distribution than the grape-vine. It was cultivated extensively in peninsular Italy, as well as in Greece, and some places, such as Venafrum, had achieved a high reputation for their oil. Outside Italy the most productive regions of the Empire were southern Spain, which seems regularly to have supplied the Roman market, and Greece, which also had an exportable surplus. It is uncertain whether the olive had yet been introduced into Provence, which became in the later years of the Empire, to judge from the fragments of presses that have survived, an important source of oil. The olive could spread only slowly. The establishment of a grove of olive trees required a large capital outlay on which there could be no return for many years. It was effectively beyond the capacity of the peasant farmer and its cultivation was a feature of large *villae rusticae* such as that at Francolise. The olive, furthermore, was less tolerant of climate than the grape-vine. Its inability to resist frost restricted it to the coastal regions of the Mediterranean Sea, but within this area it often exceeded the grape-vine in importance. The classical agronomists gave a great deal of attention to the varieties of olive and their physical requirements and to methods of crushing and pressing them in order to secure the best oil and of preserving them in brine. The olive was the only source of vegetable fat, and with the relative scarcity of animal fats, oil played an important role in cooking and diet.

Animal husbandry

The rearing of farm animals appears to have been almost as insignificant in early Rome as it had been in classical Greece. Large-scale animal husbandry in a region of Mediterranean climate required some degree, however limited, of transhumance, and this was difficult if not impossible, as long as the land was politically fragmented between a large number of small tribes. With the extension of Roman authority over the Italian peninsula, animal husbandry became of greater importance. Routes (*calles*) were established from the *saltus hiberni,* or winter pasture, in the plains of Campania, Latium or Apulia up to the *saltus aestivi* in the central and southern Apennines. The *Lex Agraria* of 111 B.C. had even provided for broad *tratturi*, with free roadside pasture for the flocks and herds,[137] as they migrated between their summer and their winter homes. Those who tended the animals – *pastores* – lived in temporary huts (*casae repentinae*) during the summer, and their permanent homes, in so far as they had any, were in the lowlands. In many instances the winter grazing was on the burned stubble of land which had been cropped the previous year. Silius Italicus mentioned 'the multitude of fires that the shepherd sees from his seat on Monte Gargano [Apulia] when the grazing lands of Calabona are burned and blackened to improve the pasture'.[138]

The transhumant animals were primarily sheep and cattle, though doubtless goats and mules took part in the semi-annual migrations.[139] They moved in flocks and herds, which numbered up to 1,000 head, and their passage through cultivated valleys must have occasioned no little disturbance. It is known that in one second century instance tolls were levied on the migrating sheep.[140] We know also that the animals did not always keep to the *tratturi*, and an inscription in the southern Apennines complained that the flocks sometimes crossed cultivated land to the damage of standing crops.[141]

In central and northern Italy the summer drought is less extreme, and transhumance therefore less necessary. On the other hand, mountain grazing was restricted by winter snow which forced the animals to move to the lowlands. Pastoralism, whether transhumant or not, was important in Etruria and the northern Apennines. The younger Pliny praised the meadows of his Tuscan villa, where water was never lacking.[142] In the mountains much of the land provided only rough grazing, which seems mostly to have been given over to sheep. The formation of such 'ranches' had been begun under the Republic, and probably developed after the Second Punic War.

Little is known of animal husbandry in other parts of the Empire. Unquestionably it was relatively more important in the more northerly regions of year-round rainfall, and it may locally have been more significant than crop-farming. It is evident, too, that meat formed a conspicuous part of the diet. A Pannonian villa revealed a great quantity of the bones of animals which had been consumed, with cattle predominating. Similar though more fragmentary evidence exists from numerous sites in Gaul. One might perhaps generalise by saying that in the non-Mediterranean regions of Europe animal farming was important, and that in mountainous regions and in such damp lowlands as the Belgian plain and parts of Britain, the agricultural emphasis was upon stock almost to the exclusion of crop husbandry. Cattle, with their more diversified uses, proved to be the most common. Swine, which appear to have been the most numerous animal in the Neolithic and Bronze Ages, were now, as a general rule, greatly exceeded in numbers by sheep. The most probable explanation of the shift in emphasis lies in the reduction in the area of forest and its replacement by open country more suitable for the latter.

Villas, even the most specialised of them, carried on some kind of mixed farming. The olive groves of an Italian *villa rustica* were probably grazed and manured by flocks of sheep, and there were in all probability few villas, apart from those which served as luxury homes of the rich, which did not produce some crops. Applebaum's typical villa carried on a kind of mixed farming, in which manure from the animals was fed to certain of the crops. Crop-farming must also have been associated with animals close to the frontier if only to satisfy the needs of the army.

Land use

Most rural communities were self-sufficing in the basic necessities of life, and the volume of agricultural produce which entered into long-distance trade must, relatively at least, have been very small indeed. Bread-grains provided the staple diet in all parts of the Empire except those where a pastoral economy prevailed. Yet agriculture was far from static. Fluctuations in the total population and in popular taste led to a changing pattern of demand. The availability of farm labour itself influenced the practice of agriculture, and the shortage which characterised the later Empire may have led to an increase of pastoral activity at the expense of crop-farming.

The area under bread-crops is likely in the aggregate to have increased during the first two centuries A.D., and it is probable that this expansion took place in most parts of the Empire. Grain cultivation appears to have been extended in southern Gaul, while the area under tree and field-crops increased greatly in the Guadalquivir valley of southern Spain.

An extension of the area under crops is mostly clearly demonstrable in the frontier regions of the Empire. It was clearly necessitated by the need to provision the garrison towns with the bulky foodstuffs that could not easily be transported great distances. It is no less evident in the increasing number and elaboration of the villa farms built within a day or two's journey from the imperial boundary. It is in Britain, however, that the clearest picture emerges of the changing pattern of agricultural production under the impact of military occupation. Here there was a considerable increase in grain production to supply the legions which were, in the main, based on areas of low agricultural fertility. It seems likely that many frontier regions, notably Pannonia, became virtually self-sufficing in spite of the large body of troops stationed there.

In Italy, however, crop-husbandry declined before pastoral, and there was a tendency for small holdings to be concentrated into larger estates, owned in many instances by absentee landlords. Pliny's judgment[143] that great estates had been the ruin of Italy undoubtedly exaggerated the situation. Italy had not been, at least for several centuries, predominantly a land of peasant proprietors, and large estates certainly antedated the Principate. Columella's account of farming presupposed large farms, and the villas of the younger Pliny and of his neighbours were also, in effect, estates.[144]

It is not easy to form an estimate of the size of such estates. Döhr considered[145] that peasant holdings ranged up to 80 *iugera* (50 acres); that medium-sized estates – a category which would have embraced most of the *villae rusticae* of Campania – had from 80 to 500 *iugera* (50 to 310 acres), and that large estates had over 500 *iugera* or upwards of half a square mile. The estates of the Younger Pliny greatly exceeded this limit, and probably amounted to at least 8,000 *iugera*.

A distinction must be drawn between an estate such as that of the Younger Pliny and the 'ranches' that had developed especially in southern Italy. It is probable that the Romans understood only the latter by the term *latifundia*. The estates of the wealthy Roman landowners did not in general constitute continuous blocks of land. On the contrary their owners most likely acquired them piecemeal, like, for example, the monastic estates of the middle ages, and they each comprised a number of relatively small, scattered holdings. Nor were they necessarily run by gangs of servile labourers, though slaves may well have been important in their labour force. It is likely that many of the individual holdings were leased or rented to tenant farmers, who would in all probability have practised some form of mixed husbandry, and whose self-sufficiency would have been tempered only by the need to obtain a cash income to pay their rent.

On the other hand, the vast grazing estates of southern Italy – the *latifundia* in the narrower sense – are more likely to have formed compact blocks of territory, whose origin probably lay in the disposal of public land after the Second

Punic War. Their owners clearly did not live on them, and they were most likely to have been worked by slaves. The Veleia inscription,[146] which relates to an extensive area in the northern Apennines, suggests that large holdings tended to predominate in the areas of forest and rough grazing, whereas small holdings survived in the agriculturally more productive valley lands. On the rich lands of Campania, at least, there was no strong movement towards the formation of really large estates, and it seems probable that the latter were established mainly on lands that were marginal or sub-marginal for crop-farming, particularly in the Apennines and in the south of Italy.

If there was no strong tendency to consolidate peasant holdings, there was, nevertheless, a change in the character of land use, accompanied perhaps by a decline in the rural population. This was what shocked the moralists like Seneca and the elder Pliny. It is doubtful whether central Italy could have supplied the food needed by a city the size of Rome, and during the later Empire other large towns had to rely on imported grain. Much was imported by sea, and some was given free to the populace. But it was only a minority that received this grain issue; the rest had to buy in the open market at prices which fluctuated with the season. There was always a market in the city for grain produced in central Italy, and despite the increase both of pastoralism and of intensive wine and oil production, mixed farming continued to be practised by both villa-owners and peasant farmers.

Crop-farming remained important on the estates of Veleia and over the plain of the Po, but in parts of central Italy grain was replaced either by vineyards and olive-groves or by grazing land. Wine and oil could still command a good price in the Roman market, and demand had probably been rising up to the mid-second century. The Campanian villas destroyed by Vesuvius in A.D. 79 had been engaged mainly in viticulture. Animal husbandry was also relatively profitable. Demand for wool and hides was probably increasing, and the large, transhumant flocks and herds called for extensive grazing lands but little labour.

The formation of estates appears to have gone on in most of the European provinces. Cicero's friend, Atticus, owned an extensive estate in Epirus,[147] and many other Romans had broad acres outside Italy. In Greece in the second century 'the wealthy must have held a larger share of Athenian lands than ever before',[148] though it is entirely possible, even probable, that they leased or rented it to small farmers. In Gaul, the Rhineland, Britain, and even Pannonia, the large and affluent villas which have been revealed by excavation presuppose an income from very extensive estates. It cannot, however, be assumed that there was outside Italy a similar movement toward replacing cropland with vineyard, orchard and grazing. The simple fact that most of the food consumed in the European provinces was produced within Europe would have prevented too great a change.

Despite the overall economic growth of the Empire there must have remained areas where conditions deteriorated and population became impoverished. Such an area was the Euboea of Dio Chrysostom's seventh discourse. The author pictures himself as shipwrecked upon the rugged east coast of the island and seeking refuge among the pastoral people of the interior. The latter described their plight in terms which are realistic enough, but probably not wholly applicable to Euboea. There is no evidence that Dio ever visited Euboea, though he spent a considerable time in the Balkans while banished from Rome. There were enough such depressed areas, even in southern Italy, for his picture to carry conviction. 'Almost two-thirds of our land', he represented the Euboean herdsmen as saying, 'is a wilderness because of neglect and lack of population',[149] and he described – very improbably – only pastoral pursuits right up to the walls of the towns.

Farming techniques

Despite the writings of the Roman agronomists, surprisingly little is known of the farming technique over much of the Empire. Columella's treatise, which was in all probability written for the guidance of the owners of estates in Italy, is at great pains to describe the accepted methods of planting, pruning and grafting, of vines and olive, fig, nut and fruit trees, but nowhere does he mention the method of ploughing or the type of plough used. His readers would not have been interested in such peasant matters. Knowledge of the field systems in use is fragmentary. Vergil has left two short, poetic descriptions of ploughing,[150] which are often taken to refer respectively to the light and heavy ploughs. Pliny's description of the heavy, wheeled plough[151] which, he claims, had originated in Raetia, is more to the point. It seems possible that the heavy plough, fitted with a coulter and capable of turning over the sod by means of a curved mould-board, originated in south Germany. It may have spread southwards into the Po valley, where Vergil encountered it. It was probably taken to Britain, perhaps by the Belgae, and there it was probably used during the period of the Roman occupation. A number of iron plough coulters has been found, most of them on villa sites, suggestive of the use of a heavy plough. Varro also decribed the coulter, when enumerating the parts of a plough, but made no mention of a mould-board.[152]

The introduction of the heavy plough was of great significance because it influenced the pattern of fields and systems of land tenure, and also enabled the heavier soils to be cleared and cultivated. Such land could, of course, be tilled without a heavy plough, but the large amount of labour involved makes it unlikely that this was done on more than a small and local scale. Scholars have generally assumed that the light plough (*aratrum*), drawn usually by a single draught animal and capable even of being carried by the ploughman, gave rise to small and roughly square fields. The heavy plough, with its large team of oxen, was a great deal less manoeuvrable. It was turned as infrequently as possible, and thus tended to give rise to long, narrow parcels of land (see below, p. 283). It should be noted, however, that there is no conclusive evidence in Roman Europe for the existence of strips such as characterised the medieval open fields. Indeed, all the fields that can be ascribed with any degree of assurance to the Roman period are small and compact. It must be concluded that use of the heavy plough was by no means widespread, and that in most parts of the Empire fields continued to be tilled with the light plough.

Some light is thrown on the possible use of the heavy plough by that of the *vallus*. This was a cart-like implement, which was pushed by a draught animal through the grain field. Teeth along its forward edge tore off the ears and allowed them to fall back into the cart. This was the earliest harvester. It was described by Pliny[153] and Palladius,[154] who both regarded it as peculiar to the Gauls, and it is illustrated on bas reliefs in France. It does not appear to have been widely adopted, and was probably suited only for use on large estates with extensive level land.

It is very difficult indeed to say to what extent cropland was being reclaimed from forest, heath and marsh. The picture given by Tertullian: 'cultivated fields have overcome the forests; the sands are being planted . . . the swamps drained',[155] may have been overdrawn, but in all probability was not fundamentally wrong for some areas. A major reason for the extension of agriculture was the need to provision the armies, and this, as has already been noted, led to the intensification of crop-farming in areas that might otherwise have been considered marginal. In eastern Gaul large areas were brought under cultivation and numerous estates created. Pressure on the land appears to have justified a number of attempts to drain marshland, and Cassiodorus addressed the inhabitants of the Po marshes: 'Here after the manner of waterfowl have you fixed your home . . . here are seen . . . your scattered homes, not the product of Nature, but cemented by the care of man into a firm foundation. For by a twisted and knotted osier-work the earth there collected is turned into a solid mass'.[156] There were also more ambitious and less practicable projects. Julius Caesar planned to drain the 'marshes by Pomentium and Setia (south-east of Rome) and to create a plain which would be cultivated by many thousands';[157] Claudius drained the Pontine marshes and also the Fucine lake in the Abruzzi, and Nero and Trajan undertook reclamation projects.

The Romans not only made use of manure, but were even familiar with the practice of marling. Nor were these practices limited to the well-to-do Italian farmers for whom Columella wrote; Varro described their use in northern Gaul.[158] In many parts of the Empire one gains the impression from texts, inscriptions and bas reliefs of a class of wealthy men who regarded the acquisition of land as the most respectable as well as the safest form of investment, who read the farmers' manuals and experimented with crops and farming methods. They probably made no great advances in farming technique, though some have credited them with the three-field system and even something resembling the bipartite manor of the Middle Ages. The use of slaves on the land was not calculated to stimulate innovation or the introduction of labour-saving techniques. There was nevertheless a wide gulf between the farming practice of those who owned the *fundi* and read Columella and of the peasant proprietors with their small and self-sufficing holdings. Of the latter almost nothing is known.

Agriculture beyond the imperial frontier

Progress was far from uniform within the Empire, and there were places where the standards of the pre-Roman Iron Age had scarcely been exceeded. The practice of agriculture beyond the frontier was on average lower than within its limits. The Germanic peoples of central Europe and Scandinavia, as well as the proto-Slavs, were primarily farmers, but their flocks and herds may have been even more important than their cropland. Tacitus' description of German agriculture is far from unambiguous. He admitted that Germany was 'fertile in grain crops', but 'unkind to fruit trees'. Cultivable land was evidently abundant; 'they change their plough-lands yearly, and still there is ground to spare'.[159] This enigmatic phrase has been taken by some to mean that the Germans practised a system of fallowing; it is more likely that there existed some form of shifting cultivation – *Feldgraswirtschaft*, as the Germans called it. On the other hand Germanic agriculture had almost certainly been improved by the example of the Romans since the time when it was described by Caesar

The Germanic peoples of Scandinavia seem to have cultivated small and approximately rectangular fields. There is evidence from scratch marks that they cross-ploughed, thus suggesting that they used a light plough. One can no more generalise regarding the field-systems and agricultural tools in use among the Germans than one can regarding their settlements. This was a time of change in all these respects, and the picture is as confusing to us as it seems to have been to Tacitus.

The Germans are represented as growing a number of grain-crops and a considerable range of vegetables. In southern Scandinavia barley appears as the leading grain; in Germany barley, rye and several varieties of wheat were known, and farther east, in northern Poland, the bread-wheats (*Tr. dicoccum*, *compactum* and *vulgare*) as well as barley have been found at settlement sites of this period.

Though the Germans and their eastern neighbours were primarily tillers of the soil, animal husbandry played an important role in their economy, and their animals were a source of pride. Cattle were used to pull the plough and also for meat and milk. Germany was 'rich in flocks',[160] though sheep were apparently numerous only in lightly wooded areas, and pigs were more common in the woodland. At Vallhager, remains of sheep and pigs were roughly equal, and this is in line with what one knows of the extent of clearings on Gotland. Natural meadow was scarce, and fodder crops were unknown. How to keep farm animals alive through the winter posed a major problem. Many were probably slaughtered in autumn; it is possible that others were turned loose to fend for themselves during winter and their survivors were rounded up again in spring. In hilly areas of central Europe mixed farming gave place to pastoralism. Except for a few fertile basins within their limits, the Taunus, Westerwald and Vogelsberg were inhabited mainly by small groups of pastoral peoples. The proto-Slav and Baltic peoples practised a similar agriculture to that of the Germans and, according to Tacitus, with a greater assiduity. Cattle were numerous, and they kept sheep, pigs and horses.

A pattern of small villages, each surrounded by its square fields, tilled without doubt with a light plough, stretched eastwards into Byelorussia and north-eastwards into Latvia. Beyond this frontier of settlement, where the climate became increasingly harsh and the soil unrewarding, lived only the Finno-Ugric hunting peoples. The limit of permanent, agricultural settlement in the east Baltic lands, as well as in Scandinavia itself, may have retreated with the deterioration of climate.

South-eastwards from the north European plain the forests merged gradually into the Steppe, and with the change in the vegetation went a transition in the economy, from mixed farming of central Europe to the nomadic pastoralism of the steppe itself. The Sarmatian Alans and Roxolani, kinsmen of the Iazyges of the Pannonian plain and of the Getae of Walachia, were horse-riding, pastoral nomads. Their wealth was in their cattle, but they probably obtained plant foods from settled cultivators, analogous to the 'Forest Scyths' of earlier centuries (see above, p. 78).

MANUFACTURING AND MINING

Though the economic basis of the Empire was agriculture, manufacturing industries were nevertheless necessary and important. Their scale, however, was small; they catered for local needs, and regional specialisation and distant trade in the products of industry were at a minimum. Indeed, it could be said that agriculture showed a higher degree of specialisation and contributed far more to the total volume of trade than manufacturing.

Very few of the cities of the Empire contained manufacturing industries of more than local significance. The processing of raw materials was much more a rural than an urban function, and in this the villas were probably more important than the towns. No less than fifteen of the relatively small number of villas in Belgium contain evidence of iron-working, and seven had pottery kilns. Rome, in particular, lacked craft industries of more than local importance, and produced little for export.

The milling of grain, the pressing of grapes, the crushing of olives and the tanning of leather were carried on wherever these crops were grown and animals were reared. Cloth manufacture was also widespread, and most communities were able to satisfy their own needs. The western provinces even had a small surplus for export, but the high costs of transport limited trade and restricted it to fabrics of luxury quality. Only the Middle Eastern provinces can be said to have had a cloth industry.

Other manufacturers, however, were more localised by the nature of their raw materials. These were the manufacture of pottery and building materials, the smelting of metals, and the extractive industries. It was these which, after agriculture, supplied most of the goods entering into long-distance commerce.

Ceramic and glass industries

These are, from the nature of their products, better documented than most other manufactures. Potsherds are found on almost all Roman sites. Most are of a coarse local ware, but some, of more refined craftsmanship, came from centres of more specialised manufacture. Indeed, *terra sigillata*, a smooth, red-brown pottery often with a raised pattern, was used by all in the west who could afford something better than a coarse ware. In Italy its most noted centre of production was Arezzo, but there were others making a similar type of ware in northern and southern Italy. The manufacture of arretine pottery spread to Gaul where it tended to displace the native types. In the first century A.D. it was established successfully at several places in southern Gaul, of which Lezoux and La Graufesenque (Aveyron) were most well known and productive.

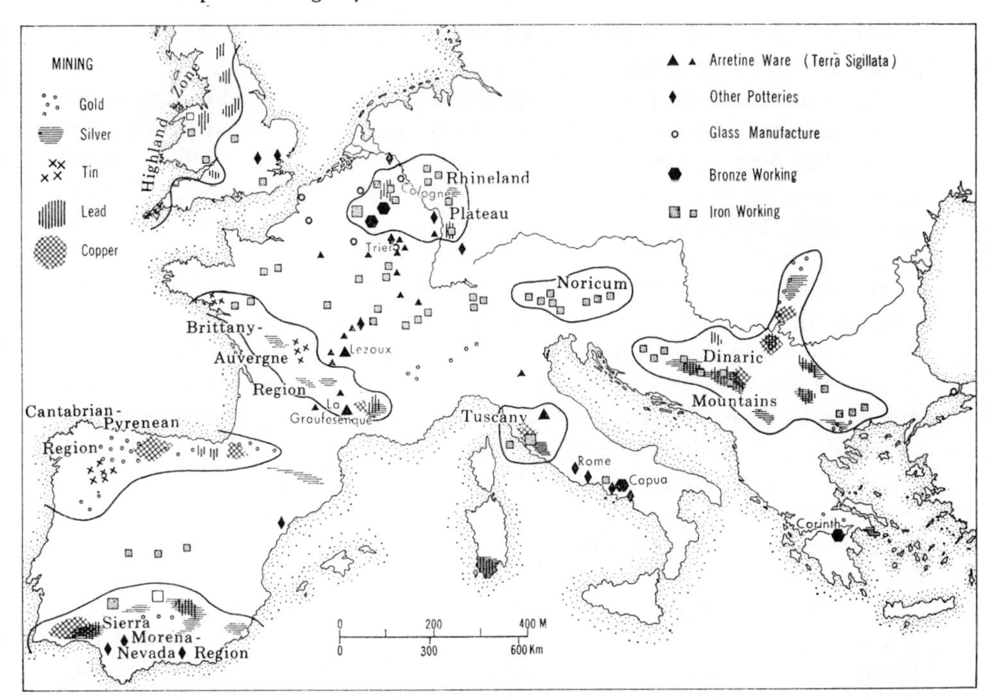

Fig.3.13 Manufacturing and mining centres in the Roman Empire

In time, rivals to the manufacturing centres in southern Gaul arose farther to the north, where demand was growing amongst the romanised provincials. In the second century the technique made its appearance in the Rhineland. Inevitably this wide diffusion of the manufacture of arretine pottery led to some decline in quality. It nevertheless remained in great demand throughout the western Empire, and was marketed from Scotland to southern Italy. The chief centres for the manufacture of *terra sigillata* are represented in fig. 3.13. Of these the most important were Lezoux, near Lyon; in Lorraine, and along the Moselle and Rhine. Clearly the availability of transport by water was an important factor in localising the pottery manufacture. Potteries were also to be found in Spain, though they never achieved either the distinction or the wide markets of those of Gaul.

The kilns at Rheinzabern and Lezoux must have approximated closely to conditions of mass production, but their wares never entirely displaced the native Celtic crafts. The physical difficulties and high cost of transport were in part responsible. Many were the areas where the pots in use represented a Celtic rather than a Latin tradition. In the third century – perhaps earlier in some areas – the Roman pottery began to lose its appeal, and demand increased for the products of traditional and mainly Celtic craftsmen. There was 'a re-awakening of the Celtic genius'.[161] This was associated with the decline in the importance of towns under the later Empire and the diversion of both demand and production to the rural areas, where Celtic crafts had survived.

Related to the manufacture of *terra sigillata* was that of oil-burning lamps, a relatively specialised product which was distributed widely, and also the making of terra cotta figurines and decoration.

The manufacture of glass had been developed in the Middle East, and was introduced into Italy during the first century A.D. From here its manufacture spread, like that of arretine pottery, to Gaul and the Germanies. *Colonia* (Cologne) came in the second century to be particularly noted for its glass manufacture, but there were many other centres in the western provinces of the Empire, particularly in *Belgica.*

The manufacture of bricks and tiles was related to the ceramic industry, but was less discriminating in its requirements. Bricks were too bulky and of too low a value to be transported far, and there must have been brick kilns in the vicinity of almost every town of the Empire, at the headquarters of every legion, and on the estates of many landowners. Many bricks were stamped with the name or mark of their maker, perhaps for some fiscal object. They were produced in a variety of shapes and sizes and for many purposes. They were made for flues and hypocausts, drains and patterned walling. Tiles came flat, flanged and curved like pantiles.

Mining and metal-working

The Romans inherited the scattered workings of the Iron Age Celts, and subjected them to a degree of government control. Most ores, except probably those which could be dug from surface workings, were the property of the state, and were often worked by gangs of slaves or condemned criminals, under the control of an imperial procurator. In general, however, the right to work a mine was leased to private operators, or *conductores.* The Aljustrel tables, found at a mining site in Lusitania, preserve the regulations which governed such a working.[162] The government received a fixed proportion of the ore or metal, and probably had the right to pre-empt the remainder if it wished. The mining communities were on this occasion extra-territorial, that is, outside the jurisdiction of their *civitas*, and were placed directly under the imperial procurator. In general the mining lessees had to exploit their mines with proper care and attention to the continued profitability of the undertaking.

A bas relief, found at Linares in the Sierra Morena, shows a gang of miners, carrying picks as they proceed to their work (fig. 3.14). Conditions of work were hard and skeletal finds in mines of the Roman period suggest that mortality was high. Mining techniques remained rudimentary, but it is doubtful whether much advance was made on Roman mining techniques until the end of the Middle Ages. Although the Romans acquired a useful knowledge of the ores, their understanding of the geological structures was slight, and they showed little skill in finding a lode displaced by faulting. Much of their mining was in well-jointed rock in a dry climate, and they thus experienced less difficulty with water in the mines than many medieval miners were to do (see below, p. 395). Outside the Mediterranean region, however, they had to resort to drainage adits, and to water wheels and the archimedean screw to lift water from the workings.

Apart from iron-ore, the most important metals obtained were gold, silver, lead, copper and tin. Italy itself was not an important source of these metals.

Fig.3.14 Relief of Roman miners, from Luiares, Spain

Alluvial gold was known in the Alpine valleys as well as in the northern Apennines. Sardinia was a source of lead, from which silver was obtained as a by-product, but southern Italy and Sicily had lost what little significance they once had possessed, and the metal industries of Campania used imported materials.

Spain, by contrast, contained the most developed mining regions of the Empire at this time. The whole peninsula, it seemed to Pliny, abounded in metals.[163] Most important were the mines of the Sierra Morena and of the northern Meseta and Cantabrian Mountains. Gold was obtained both from placer deposits and also from lodes. Silver was in the main a by-product of lead, and the chief commercial ores were probably lead and copper, though tin was important in Galicia.

The resources of Gaul were less concentrated than those of Spain. Iron-ore was probably the most important, but gold was obtained from a number of sites in the Massif Central and the Pyrenees, and silver from Melle, in Poitou. On the other hand, the sources of tin, opened up in Brittany during the Bronze Age, appear to have been abandoned.

The Rhineland provinces of the Empire were richer than Gaul. Iron-ore was widespread, and the smelting industry, doubtless encouraged by the needs of the army, was well developed. The lead ores of the Eifel were worked, and copper was probably mined.

The Danubian provinces were notable chiefly for their iron-ores. Those of Noricum, obtained mainly from the Huttenberg district of Carinthia and the Erzberg of Styria, were amongst the best in the Empire, and provided bar-iron for much of Italy. Mining appears to have been more sporadic in the eastern provinces. Iron-ore was worked and smelted in Pannonia, where silver was also mined on a small scale. The Balkan peninsula itself is highly mineralised. Iron-ores are widespread, and were worked in a broad belt extending from north-west to south-east within the Dinaric mountain system. The scale of mining was probably small, but Cassiodorus nonetheless praised the iron mines of Dalmatia, and implies that they provided tools and weapons for Italy.[164] In Bosnia silver-lead ores were mined, and gold was panned in the valleys of Illyria, the Bihor Mountains and the Transylvanian Alps.

It is more than probable that gold, silver, lead and iron and possibly also copper were mined in the Stara Planina of Moesia and the Rhodope of Macedonia and Thrace. But the classical writers were extremely vague; the physical evidence is difficult to date, and archaeological investigations have been inadequate. The gold and silver mines of Thrace, which had been important in the early classical period, had largely been abandoned, though the panning of alluvial gold continued in the hills of the interior. In Greece itself there were few ores worth mining, and even the Laureion deposits of silver-lead had been abandoned or were only feebly worked.

Lead was in great demand as a roofing material and for making water-pipes, essential in the large bathing establishments which the Romans were building. Copper was generally alloyed with tin to make bronze and sometimes with calamine to produce brass. Though the sources of copper are known, it is far from clear where the Romans obtained their tin. The Breton mines appear to have been little used, and the stream-workings of Cornwall acquired some small importance only during the later Empire.

The metal industries were far more widespread than the mining of the ores upon which they depended. Unfabricated metal was relatively valuable, and well able to bear the cost of transport to distant markets. The metals in greatest demand were bronze and iron, the former for decorative articles of all kind as well as for pails and jugs; the latter for tools, weapons and armour But lead was also much used in building construction and there was a perpetual demand for silver and gold both for currency and also for ornament and decoration.

Smelting and metal-working were normally practised in larger workshops than those used for manufacture of pots, and required a bigger capital investment. Their heavy demand for fuel ensured that they were carried on close to forested areas. Italy had formerly been important for its metal industries. The iron-ore of Elba continued to be smelted in the forests of Tuscany, and Capua was still noted for its bronzes, but the Italian metal industry in general had declined in the face of competition from the provinces.

The focus of iron-working had moved to Noricum, where the ores were rich and abundant, and the forests extensive. Iron-ores were widespread in Gaul and Spain, where they had given rise to a great many manufacturing centres. The Pyrenees were noted for their ironworks, and slag heaps of the Roman period are met with in many parts of central and eastern France. The ironworks were in general small and ephemeral, and were abandoned as soon as the ore or fuel in their immediate vicinity had been exhausted. There is no evidence that the numerous centres were ever at work at the same time. In some instances, it is claimed, iron-working even formed the economic foundation of the villa.

In Spain, also, iron-working was widespread, though it may here have yielded in importance to the non-ferrous metals. The iron of Bilbilis, in the north-eastern Meseta, was well-known, and the vast reserves of Vizcaya were also being exploited. Less is known of iron-working in the eastern provinces, though it was carried on in Illyricum as well as in the Carpathian Mountains of Dacia, in the Stara Planina and in Macedonia and Thrace. Every legionary fort must have had the means of fabricating iron goods, if not also of smelting the ore, as weapons and armour needed continual repair and replacement.

Bronze-working was less widespread and produced goods of much higher value. Capua was the most noted centre in Italy, but was probably eclipsed by the scale of manufacture in southern Spain and Gaul. The Meuse valley in particular was already noted for its metal work, a reputation which it was to retain – or revive – during the Middle Ages.

The making of jewellery, enamelled ware and artistic metal goods in gold, silver and bronze was important, and every wealthy citizen probably possessed many pieces. It is far easier, however, to show where such goods were used than the workshops in which they originated. It is clear, however, that much of the art-work of this type originated among the craftsmen of Alexandria, Antioch (Syria) and Rome, where there are, amongst the inscriptions, numerous memorials to gold- and silversmiths.

Quarrying was of great importance in all parts of the Empire. The Romans used stone in their public buildings, and this practice was imitated in many of their private homes. No town, and few villas, were without a quarry. Difficulties of transport necessitated heavy reliance on local stone, unsuitable as it may sometimes have been. Great use must have been made of the navigable waterways for rafting stone to the building sites. The use of marble, especially coloured, in the more ostentatious works of the emperors called for stone from Greece, North Africa, and elsewhere, and ships wrecked on the Sicilian coast have been found to contain large quantities of marble, destined almost certainly for the buildings at Rome. The Romans developed great skill in quarrying, cutting and dressing the hardest igneous rocks, using a combination of saws and wedges. Softer stones were chosen with great care and allowed to weather before being used.

Little is known of classical salt manufacture. It was obtained by evaporation in salt-pans along the Mediterranean coast, notably on the shores of Latium and Etruria, of Apulia and Sicily, and of Spain and Greece, and was also obtained from mines. Doubtless also the brine from inland salt-springs was evaporated over fires and used to supply places remote from the sea. There is no evidence that the salt deposits, worked from the early Iron Age in Upper Austria, continued in use during the Roman period.

Peoples living outside the boundaries of the Empire were greatly influenced by the example of Rome. Roman artifacts were sold in Germany, but at greater distances from the frontier the tendency was to manufacture rather than to buy. In southern Poland, near Kraków, there appears to have been an 'industrial combine', with 'dozens of two-chamber pottery kilns, dating from the Roman period'.[165] Iron was widely smelted in the northern plain, with a few concentrations of great size. Near Kielce a group 155 iron furnaces was discovered, and throughout the Holy Cross Mountains of southern Poland there were ironworks at this time, some of them serving it would appear, quite distant markets.

The weaving of wool and flax was widely practised, but has left little trace. Finds from the Rhineland, Gaul and Britain show that complex weaves were used and the fabrics were dyed with bright colours. Similarly crafts were practised in Scandinavia and the Baltic region, becoming simpler in their techniques and cruder in their products with increasing distance from the Empire.

TRADE AND COMMERCE

In the mid-second century A.D. Aristeides represented the city of Rome as the commercial hub of the known world: there are always ships putting into or sailing out of harbour, and the whole world's products could be seen in Rome. He was correct, at least for the summer months, but he omitted to add that most of those leaving carried no cargo, for Rome had nothing to export. The grain, wine and oil which the ships brought to the Tiber mouth were paid for by the taxes levied on the provinces. The Roman populace which consumed them contributed very little to the economy, and if it fabricated goods for its own use, it made none for export. The trade described by Aristeides was illusory: it supplied the vast, parasitical city of Rome, and disguised the fact that in most of the Empire there was little long-distance commerce.

The Empire was made up of self-sufficing communities. Every provincial town was in large measure rural in function; it was the focus of a region from which were derived all the foodstuffs which it was physically possible to grow there. In the town locally produced articles of metal, clay and wood were sold. There was little inter-regional trade, and what there was consisted largely of luxury goods, wine, oil and other locally specialised foodstuffs.

A reason for this was the great difficulty and high cost of transport by land. It has been estimated, on the basis of Diocletian's price edict, that a land journey of 300 miles would approximately double the price of grain. Most bulk transport was by water, and Rome would have been inconceivable without its seaborne trade. It appears that the road-net which the Romans had spread across their Empire was of no great commercial value, whatever may have been its military significance; indeed, it is by no means clear that merchants had the *right* to use the road system, at least without authority, though it is clear that the routes across the Alpine passes were regularly used by them.

The navigable rivers played perhaps the largest role in internal trade. The most important of the commercial towns were accessible by boat, and its wealth of navigable rivers was not least among the factors which made Gaul the most flourishing province of the Empire. Bas reliefs depict the traffic on the Rhône and Moselle, and inscriptions attest the companies of boatmen on these and several other Gallic rivers. At least one Roman emperor contemplated a canal that would link the Rhône system with the Moselle,[166] and a waterway was actually cut in the Fenland of Britain, providing perhaps a route all the way from Cambridge to York.

The Po appears to have been much used, and Cassiodorus represented ships

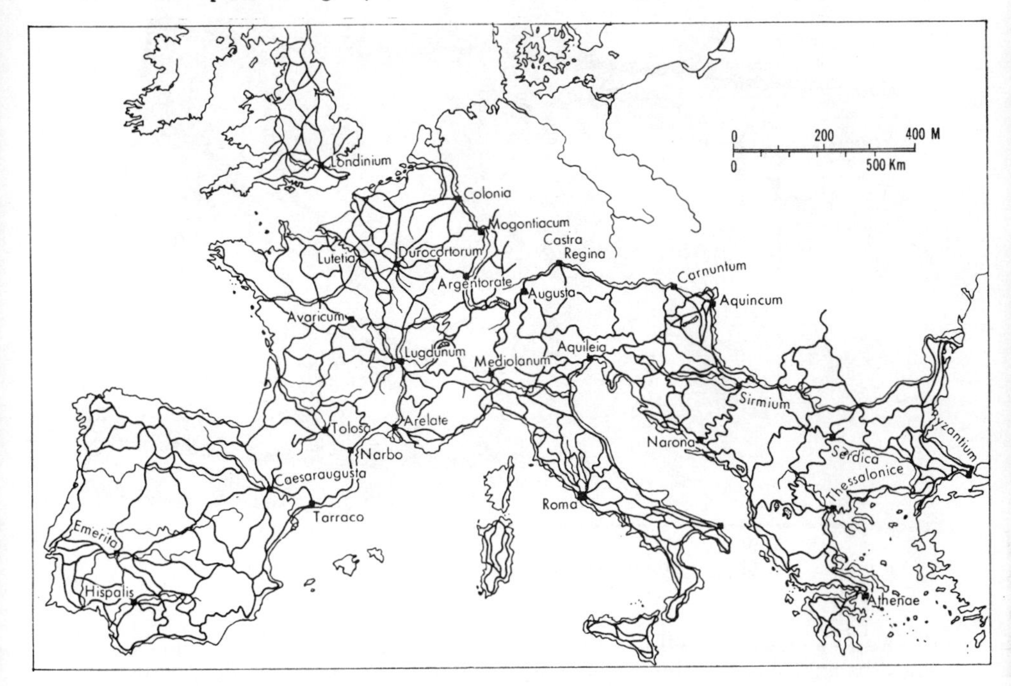

Fig.3.15 The Roman road system

as being towed through the channels of its delta rather than face the storms of the Adriatic.[167] In Spain the *Betis* (Gualaquivir) was the commercial artery of Baetica, and the Marica was navigable throughout Thrace. The Rhine was probably the most used river of Roman antiquity. On its banks lay some of the largest provincial cities, and supplies for the Roman garrisons must in part have been distributed by river.

Yet waterborne transport was not without its difficulties. The Romans never readily sailed the Mediterranean in winter, and the annual grain supply for Rome had to be brought to Ostia during the summer months. Few Mediterranean rivers were easily navigable, least of all the Tiber, and on many the upstream movement of boats was difficult or impossible. Only the smallest sea-going vessels were able to make the journey up to Rome, and grain and other imports had to be transshipped at Ostia to small barges, of which many hundreds were in regular use.

Much has been written about the unity of the Mediterranean basin and the supposed ease with which commerce and communications could be maintained between its opposite shores. There was, indeed, a greater volume of trade within the Mediterranean basin during the early Empire than at any previous time. It declined during the later Empire. The tendency was for seaborne trade to concentrate on a relatively small number of important ports, while much more numerous small ports were used by coasting vessels and fed goods to the larger entrepots. Most of the seaborne commerce was carried in ships with a capacity of between 150 and 500 tons. There were very much larger ships, but the majority were probably smaller. They ranged downwards to boats capable

Fig.3.16 Shipping on the Danube (relief from Trajan's column)

of carrying no more than five or six tons. The larger ports were equipped with dock basins, masonry-built quays and warehouses; many of the smaller remained open roadsteads, with little more equipment than those of Homeric Greece.

No quantitive study of the ports of the Roman Empire is possible, but it is evident that the largest in the volume of goods handled were Ostia, Alexandria and Carthage. Ostia served mainly for the import of food for Rome, and Alexandria and Carthage were primarily export ports for grain and oil for Italy. Both, however, imported some foodstuffs, particularly wine, as well as fabricated goods.

There was also a vigorous trade between the Italian ports – principally Ostia, but also *Tarracina* (Terracina) and *Puteoli* (Pozzuoli), and Gaul and Spain. In Gaul *Arelate* (Arles) and *Narbo* (Narbonne) were the most active ports. In Mediterranean Spain, commerce had come to be concentrated in *Tarraco* (Tarragona) and *Nova Carthago* (Carthagena), while that of the province of Baetica passed through the ports of *Gades* (Cadiz) and *Hispalis* (Seville).

The legacy of the Greek cities was important in the commerce of southern Italy. The chief ports on the mainland were *Brundisium* (Brindisi), the most important link with Greece, and *Tarentum* (Taranto). In Sicily, the ports which handled most of the grain export were *Catana* (Catania) and *Syracusae* (Syracuse) on the east coast, and *Panormus* (Palermo) on the north.

The Adriatic sea had gained in commercial importance because its ports gave access to the Balkan and Danubian provinces. Most important of these was *Aquileia*, at the head of the Adriatic, from which it was but a short journey to Noricum and Pannonia. *Salona* (near Split) was probably the most important port of the Dalmatian coast, but *Narona* (near Metković) on the navigable *Naro* (Naretva) served as an outlet for the mining region of Bosnia and Serbia.

Small ports were numerous along this mountainous coast, but in most instances their hinterlands were mountainous and difficult to penetrate and their commerce small.

Dyrrhachium, on the coast of Epirus, lay at one of the shortest crossings of the Adriatic. From it the *Via Egnatia* ran eastwards through Macedonia and Thrace. It thus controlled the shortest route between the western and eastern parts of the Empire. *Patrae* (Patras), refounded by Augustus, was the chief port of western Greece, through which the Romans maintained their communications with the province of Achaia. The coastline and islands of the Aegean were dotted with small ports, between which, as in Hellenic times, there was a busy commerce, carried on in very small ships. In the second century Byzantium was still a small town, and its trade had not yet begun to dominate the Aegean and Black Sea. After its refounding by Constantine in A.D. 324 it grew to be the equal of Ostia and Alexandria.

Outside the Mediterranean basin and southern Spain, seaborne trade was of little importance. Ships penetrated the Black Sea and carried on a trade with the coastal towns of Thrace and Moesia Inferior and with Bithynia and Pontus. In the extreme north-west of the Empire *Bononia* (Boulogne) was the most important port for Britain, and a few trading ships threaded their way through the waterways of the Low Countries and reached Denmark and the Baltic, but their importance was small.

Several of the more important ports lay at the mouths of the navigable rivers, which provided their chief link with their hinterland. Thus Puteoli was replaced by Ostia as the port of Rome. Massilia, the port of the Greek settlers, declined before *Arelate* (Arles), situated on the Rhône and accessible to seagoing ships. But others, including Aquileia, the ports of North Africa, south-eastern Spain, Asia Minor and the Levant, had no such natural link with their hinterlands, and goods must have been moved over the road network.

Road system

The magnificent network of well engineered and surfaced roads was designed to serve military and administrative purposes rather than commercial, but it would be a mistake, however, to assume that they were not used by merchants. In any case, they were supplemented by a complex of secondary roads, less substantially constructed, if indeed they had been 'constructed' at all. They linked the *vici* with their nearest town and served as 'feeders' to the rivers and ports. The nexus of local roads has been established for only restricted areas. It is fairly well known for Etruria and for parts of Gaul and Britain and for the provinces of Noricum and Pannonia. Elsewhere little work has been done, and the data for the reconstruction of the road system may, in fact, no longer exist.

Apart from the physical remains of these roads, the most important evidence for their extent is to be found in the late Roman roadbooks, the Antonine and Burdigalia *itineraria* and the Peutinger Table.[168] The map, fig. 3.15, is derived from these sources. It shows a pattern of roads radiating from Rome to all parts of the Italian peninsula. The net is less dense in Sicily and in the Lombardy plain, where it consists essentially of roads along the northern and southern margins of the lowland.

The most frequently used Alpine crossings were the coast road to Arelate and the Monte Génèvre, the Little and Great St Bernard Passes; the Maloja, Septimer and Julier; the Reschen-Scheideck and Brenner, and the group of low and easy routes between the head of the Adriatic and the Danube basin. The last would appear to have been of the greatest commercial and probably also military importance, and they focussed upon the busy port of Aquileia.

Lugdunum (Lyon), approached from Italy by way of the Rhône valley as well as by three mountain routes, was the focus of the roads of Gaul. There was, over most of the provinces, an open but fairly even road net. In Spain, the coastal cities of the south and east were connected; roads ran into the interior, and were linked at a number of route centres, of which *Emerita* (Merida) and *Caesaraugusta* (Saragosa) were the most important. Apart from Léon, in the northern Meseta, the north-western part of the peninsula was not really opened up.

The road system of the Balkan peninsula was even less well developed. It consisted essentially of four elements. A route with several branches followed the Dalmatian coast from Aquileia to Epirus; a much used road, the *Via Egnatia*, crossed the region from the coast of Epirus, through Macedonia and Thrace to Byzantium. From the latter route a road ran up the Vardar and down the Morava valleys to the Danube below Sirmium, and the fourth important route linked this road at *Naissus* (Niš) with the shores of the Sea of Marmara. This road pattern was to a large degree dictated by relief, and, with only minor variations, has continued to provide the primary transport net until today. It was, however, supplemented by other roads, which penetrated the Dinaric Mountains, followed the Black Sea coast, and linked it with the Danube frontier in Moesia and Pannonia. The road system of Greece in general followed the coastline of the peninsula, and few routes penetrated its mountainous interior.

A highway ran parallel to the northern frontier of the Empire all the way from the delta of the Danube to that of the Rhine. It linked the garrison towns and provided a means of rapid movement between them. Between the latter lay smaller forts and the marching camps of the legions, as they patrolled the frontier. Stretching back into the imperial hinterland were the roads by which the garrisons were maintained and supplied.

The *itineraria* do not represent all the roads that were in regular use, though they may well show most of the surfaced, all-weather highways. In the more densely settled areas, such as Etruria, there was evidently a very much more dense net of secondary roads and tracks, deriving in many cases from the pre-Roman pattern of movement.

Travel was assisted by the provision, at least on the more important roads, of *mutationes* where the horses could be changed, and of *mansiones* for overnight lodging. Very rapid speeds could be made, and instances are quoted of up to 200 miles in a day. About five Roman miles an hour was more usual for most couriers. Merchandise was carried on carts, each fitted with four large wooden wheels, and drawn by horses, or on the backs of pack animals (fig. 3.17).

The Roman Empire was no free-trade area, and the movement of merchan-

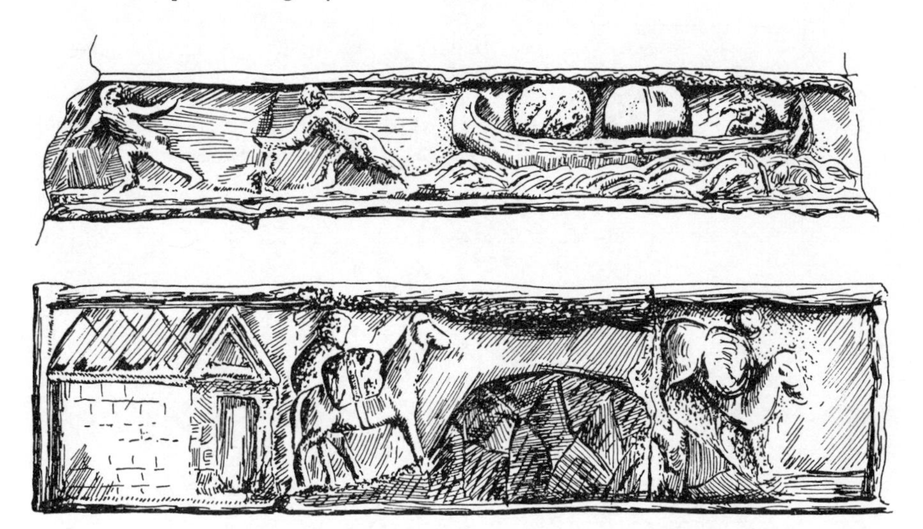

Fig.3.17 Transport by boat and pack animal (frieze from the Igel column)

dise by both sea and land was burdened with tolls. These were generally small – a 2½ per cent levy was normal – but on a long journey merchandise might be called upon to pay several times. Toll stations, or *portoria*, were set up near the boundaries separating provinces; they were numerous along the Rhine and Danube, where they served not only to levy tolls on incoming goods but also to prevent the export of 'strategic' goods – wheat, salt, iron – to the Germanic peoples, and they occurred in at least the more important ports. They were found at the approaches to the more important Alpine passes and at a few places in the interior of provinces.

The tolls levied were a significant source of income to the imperial treasury, or *fiscus*. It had been re-organised under Tiberius and appears to have undergone little subsequent change, though very little is known of its operation under the later Empire. The emperor could grant exemptions from it, and Domitian relieved veterans of the legions from the obligation to pay, but, apart from such instances, the obligation was universal under the early Empire. The effect of the *portoria* must have been in some degree to make goods more expensive and to reduce their circulation.

The articles of trade

The commerce of the Roman Empire was largely made up of two categories of goods: foodstuffs, and manufactured goods of light weight and relatively high value. The former were dependent overwhelmingly on waterborne transport, and consisted largely of the movement of bread-crops to Rome, to the military centres and to the larger towns. Cargoes of grain were supplemented by wine, oil and a few luxury foodstuffs. Climatic conditions restricted the olive to the Mediterranean coastlands, and it is doubtful whether the grape-vine had spread deeply into Spain, Gaul and the Balkan peninsula by the second century. Oil and wine were not only shipped to supply Rome but were also distributed over

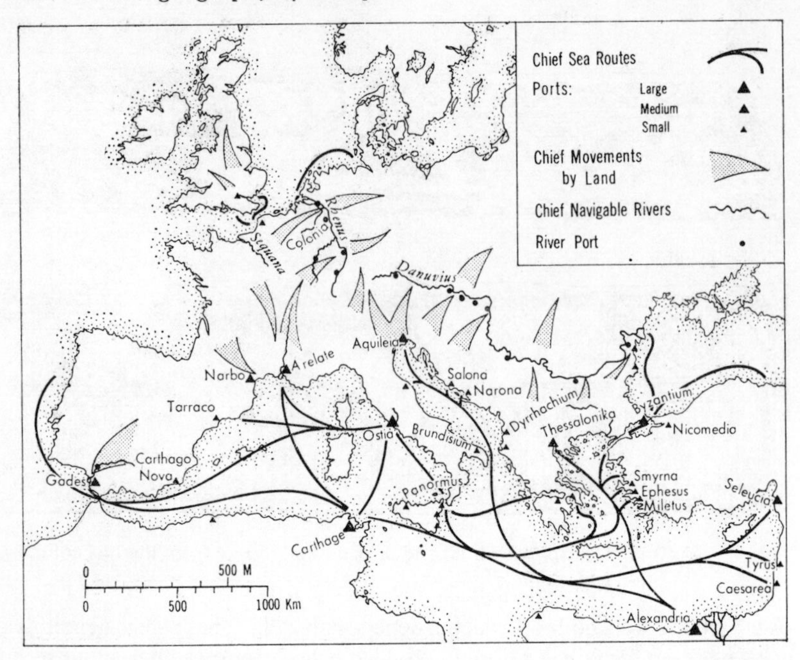

Fig.3.18 The commerce of the Roman Empire

areas which did not produce them. The bas relief showing the boat-load of wine barrels being rowed against the current of the Rhône is an eloquent tribute to this trade. At first the wine was Italian, carried in amphorae from central Italy; later it was Gallic, shipped northwards in wooden casks.

The volume of the trade in essential foodstuffs cannot easily be measured. Egypt is said to have provided twenty million bushels each year for the city of Rome, and van Berchem estimated that the total needs of the city amounted to sixty million – a figure which one cannot but regard as excessive.[169] Central Italy had long since ceased to produce more than a fraction of the wine and oil which it consumed. Most of the imports, as is demonstrated by the shapes and stamps of the amphorae in which they were transported, came from southern and south-eastern Spain. Most of the fragments of amphorae in 'the gigantic mass of the Monte Testaccio', have been shown to be of Spanish origin. Gaul was probably of only slight importance in provisioning Rome, perhaps because of the restrictions placed on Gallic viticulture in the interests of the Italian.

Specialised foodstuffs – the cheese of Trabula, honey of Attica, smoked and salted fish of southern Spain, lobsters of the Hellespont – were probably neither large nor significant items of trade, however important they might have seemed to the wealthy Roman who consumed them.

Amongst manufactured goods pottery appears to have been amongst the most important in long-distance trade. By the second century Gaul had become the chief source, and immense quantities of Samian ware were shipped out from the factories of Lezoux and La Graufesenque in southern France and later from Rheinzabern in the Rhineland, bringing great wealth to the owners of the potteries and employing large numbers of workmen and distributors.

Samian ware from Gallic factories made its way to Britain and Spain. It is found in Italy, where a crate of such goods was discovered in excavating Pompeii, overwhelmed by the volcano before it could be opened. It was also shipped eastwards to the Danubian provinces. Italy itself continued to export some of the more specialised forms of pottery and earthenware, including lamps and terracotta tiles, and the Campanian bronzes still commanded a market in the western Empire.

The legionary forts along the northern frontier constituted a market for foodstuffs and consumers' goods second only to the large cities of the Empire. One of the advantages of a river boundary was that water transport was available in time of peace. Nevertheless the roads that ran northwards to the rivers must have been well used by those whose task it was to provision the armies.

It is very difficult to form a clear picture of the trade in cloth. Most communities were self-sufficing, but there was a market at least for fabrics of the better qualities. Juvenal[170] and Martial[171] mentioned the coarseness of Gallic cloth, but at least they knew of it. But the quality of Gallic workmanship improved, and there is good reason to suppose that in the second century cloth from Gaul was welcomed in Rome. It must not be forgotten that the impressive Igel monument, near Trier, was erected by a cloth merchant and portrayed the manufacture of and trade in cloth.[172] It is even possible that the silk industry had been established in the Rhône valley.

Metals must have been prominent in long-distance trade, especially lead, copper and tin. They moved from Spain to satisfy the plumbing needs of Rome and to supply the bronze-founding industries of Campania. Large quantities of marble were shipped from Greece to Italy, but were now beginning to meet with competition from the white Carrara marble from Tuscany. At the same time copies of Greek statuary were made to adorn the villas of the rich in many parts of the Empire.

After surveying the evidence – literary, epigraphic and archaeological – for the trade of the Roman Empire, one cannot avoid the conclusion that its volume was very small. For most communities trade with the outside was of marginal importance. Most of the goods in long-distance trade seem, apart from the food supply of Rome and perhaps military supplies, to have satisfied the whim and the appetite of a very small and very rich minority. The excavation of the British city of *Calleva Atrebatum* (Silchester), which was abandoned at the end of the Empire and not again occupied, revealed a very short list of articles from beyond the local area.

Lack of sources precludes a quantitative study of the volume and direction of Roman trade. It would appear that the largest flow of goods, apart from the grain trade, was between Spain and Italy. 'With no country in the world,' wrote West, 'did Rome have so extensive and constant a trade as with Spain.'[173] It is likely, however, that by the second century Gaul ran a close second, but Greece, Britain and the Danubian provinces probably contributed little to the flow of goods within the Empire.

It is by no means certain that a cash economy had spread into all parts of the Empire. Coin was, in some of the less romanised parts of the Empire, used primarily as a means of storing or hoarding wealth and of paying taxes, and during the later Empire, even taxes tended to be discharged in kind, and coin served little purpose.

Trade beyond the frontier

If much of the commerce carried on within the Empire consisted in luxury goods, its exports to the Germanic and other tribes were almost exclusively so. There is little evidence, other than archaeological, for the nature and extent of this trade. There was an export from the Empire not only of coin, but also of bronze, silver and glassware, and of large quantities of arretine pottery. The chief points of contact between those who peddled wares far into northern Europe and the Empire were *Trajectum*, the port of the Rhine mouth, *Vetera* (Xanten), *Moguntiacum* (Mainz), *Carnuntum* (Petronell), and *Olbia* and other ports on the Black Sea coast. From these points the German imports followed water routes – the North Sea coast, the Lippe, Main, Morava and Dniestr, towards the north and east. Pottery commonly made the shortest journey; it has been found most abundantly near the imperial frontier, and was probably much used by the more romanised of the tribes. It is, however, also found in Poland. Bronze statuary and vessels were of very much greater value, and were spread more deeply into northern and eastern Europe. They were the prized possessions of the tribal chiefs, in whose graves they were sometimes buried.

It is likely that many of these more distant finds had been distributed by the northern sea-route. The Danish islands appear, from their great number of finds, to have been an emporium for this trade, from which goods were distributed to Norway and the Baltic region. Not all such materials, however, came from workshops in the Rhineland. The greater number probably originated in Italy, and were transported by way of Aquileia to Carnuntum and other Danubian market centres, and thence distributed to northern Europe. The evidence of datable finds suggests that this trade was at its most prosperous in the second century, before it was interrupted by the frontier wars.

Trade between the Roman and the non-Roman world was a balanced trade. Imports from the Empire were requited by the export of cattle, forest products and, above all, slaves. Slaves were obtained in the course of tribal wars; perhaps the wars were themselves undertaken in order to obtain prisoners for the slave trade.

Cattle were driven from the Danubian frontier into northern Italy. It is possible, however, that Germans in Roman service may have remitted or taken home coin as well as goods. The sphere of Roman trade embraced at least the southern half of Scandinavia, and the Baltic littoral (fig. 3.19). Trade between the Empire and areas lying beyond its borders was probably carried on most vigorously from the Middle Eastern provinces, but that, important though it was in the overall economy of the Empire, lies beyond the scope of this book.

CONCLUSION

The Empire of the Antonines underwent little outward change for a century, until, in fact, the province of Dacia was abandoned under the Emperor Aurelian.

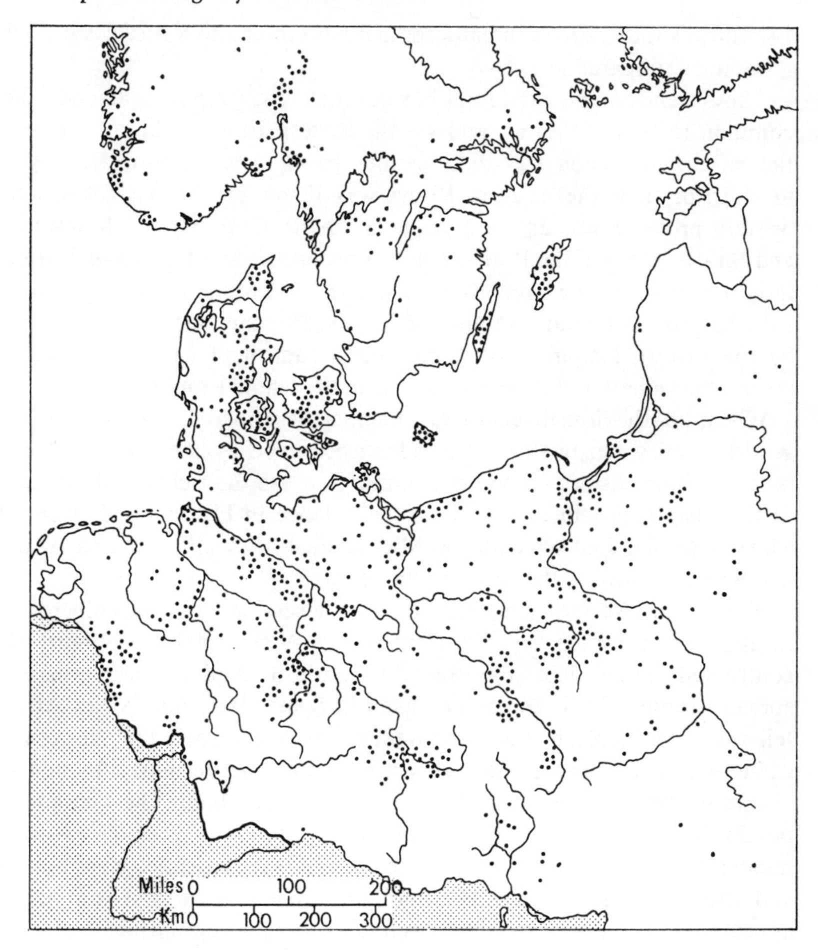

Fig.3.19 Roman finds among the Germans

In 301 Diocletian divided the Empire into an eastern and western part by a line which ran from the Danube, near Sirmium, southwards through the mountains of Montenegro to the Adriatic coast, and at the same time introduced far-reaching administrative reforms. His actions, however, provided no remedy for the deep-seated ills which plagued the Empire. The pressure of Germanic tribes intensified along the frontier of the western Empire; the total population probably began to decline, and, in order to maintain the production of essential goods, the cultivator was bound to the soil and the craftsman to his trade. The movement of commodities declined, and the self-sufficiency of local communities was gradually intensified. At the same time the level of taxation increased with the mounting burdens of the Empire.

Late in the third century, the Germans broke across the Rhine and raided deep into Gaul. Walls were thrown around exposed settlements, and the speed with which this was done is apparent today in their poor construction and the use in some of them of materials from nearby buildings. The 'Saxon Shore' of

Britain was increasingly threatened, and here too human energies were diverted to military preparedness.

These trends alone would not have brought the Empire to an end. They were common to both divisions, and yet the eastern Empire did not fall. No doubt the military situation was more serious in the western Empire; it was easier to overrun than the eastern. Rome was taken by the Visigoths, and every western province was exposed to the Germans. Gothic tribes, followed by Slav and Bulgar, ravaged the Balkans, but Constantinople remained inviolate, owing, in part at least, to the strength of its site and its coastal location, which allowed it to be provisioned and reinforced by sea. But the chief advantage possessed by the eastern Empire was its continued control of its Middle Eastern provinces, the richest and most prosperous in the whole Empire.

When, in the fourth century, Libanius described[174] Antioch, it was as a wealthy and sophisticated city, undisturbed even by the rumour of barbarian invasion. The eastern emperors continued to use the riches and the manpower of the eastern provinces to redress the balance in Europe, and, even after the Moslem armies had overrun the Middle East, they still retained Asia Minor, and with it control of the surrounding seas.

Not so the Western Empire. There a succession of weak emperors proved incapable of pursuing any consistent policy. A more difficult military situation confronted leaders less competent to handle it. And so the Germanic tribes spread through Gaul, and settled even in Spain, Italy and North Africa; rule fell from the feeble hands of Romulus Augustulus, and the Western Empire came to an end in name – in reality it had long since terminated – in 476.

Life continued in western Europe on a lower plane. The Germans settled beside the romanised provincial, and in time the families of the two were assimilated to one another. Villas, burned during the invasions, were not restored, and their place was taken by more primitive structures. Exotic goods disappeared from the homes of the rich, and literacy diminished until it became the monopoly of the church. Here and there a provincial, like Symmachus or Ausonius, tried to maintain the standards of the vanishing Roman culture, but in time their heirs and successors gave up the uneven struggle.

A partible tribal kingship developed among the Franks of Gaul, with all the conflict and confusion that this entailed. The political history of France became one of sanguinary conflict between the descendants of the Frankish leader, Clovis, until at last the Merovingian line became extinct, and the crown passed to the more vigorous house of Pippin.

Germanic tribes had settled and German kings were enthroned in Spain and Italy, the Visigoths in the one; the Ostrogoths and then the Lombards in the other. If their history was less turbulent than that of the Frankish Merovingians, it was no more successful. The Visigoths were overthrown by the Moslem Berbers, under Arab leadership, who invaded the Spanish peninsula at the beginning of the eighth century. In Italy the Byzantines had succeeded by virtue of their fleet and their command of the sea in retaining a few footholds around the coast, but in the interior of the peninsula successive German invaders had so weakened the society and the economy of the Empire that conditions were little better than those in contemporary France.

Through these centuries the successors of Diocletian continued to rule the Eastern Empire from their capital beside the Bosporus. They had natural advantages denied to the western emperors, and to these they added an appreciably higher level of competence. Their Balkan hinterland was ravaged by the Goths, overrun and settled by the Slavs, and invaded by the Bulgars. The effective frontier of the Empire in Europe retreated from the Danube and then fluctuated in the Balkans as in turn it yielded to barbarian pressures and then reasserted itself and regained much of what it had lost.

Such, in political terms, were the fortunes of Europe during the declining years of the Roman Empire and the following centuries. The record was not, however, entirely one of political disintegration and economic decline. Two external forces, both of them of Middle Eastern origin, were exerting a profound influence on Europe. The less important was the rise of Islam. Armies of mixed origin, but led by Arabs and imbued with the new-found faith of Islam, spread around the Mediterranean, from Syria to Spain. They never completely controlled the sea itself, nor did they interfere except occasionally with the feeble seaborne commerce that had survived. Their danger was more imagined than real, but in the minds of contemporaries the Mediterranean became a dangerous, if not a hostile sea.

The second influence was Christianity itself. In the age of the Antonines it had been just another oriental cult, slowly disseminated through the territory of the Empire. But it proved to be more enduring than its rivals. Under Constantine, early in the fourth century, it became the official religion of the Empire, and few succeeding emperors did anything to check its progress. It was at this time an urban cult. Its early ministers, or bishops, became important figures in the cities of the Empire, and their churches were established beside the temples of the deified emperors and Roman gods. Roman gods and procurators disappeared in the fifth century, leaving the Christian bishop as the chief local dignitary and his religion as the only local cult. After the cultured Roman aristocrat had disappeared from its streets and *forum*, and the flow of traders had diminished to minute proportions, the church, with its bishop, nevertheless provided some continuity in town life, bridging the gap in urban history between the declining Empire and the Middle Ages.

4

Europe in the age of Charlemagne

The early years of the ninth century were, compared with those which preceded and followed, a period of relative peace. In the east, the Arab attack on Constantinople had failed, and, though the Middle East was finally lost, Asia Minor had been retained for the Eastern Empire. So also had the imperial footholds in southern Italy and around the shores of the Adriatic Sea. The First Bulgarian Empire straddled the Balkan peninsula. Its rulers at intervals threatened the Byzantine Empire, but they were imitators of the emperors, rather than barbarian invaders intent only on destruction.

The Western Empire had disintegrated more completely than the Eastern. Its line of emperors had ended in 476, and the splendour of the imperial court and the unity of its administrative system were but memories. But on Christmas Day, in the year 800, Charles, son of Pippin the Short and King of the Franks, was crowned emperor by Pope Leo III. The meaning of this act continues to be disputed. But to most people who knew of it, it symbolised the power and continuity of imperial rule; it gave hopes of a new period of peace, prosperity and internal harmony, and it represented Charles as picking up the reins of power that had fallen from the ineffective hands of Romulus Augustulus, and establishing himself as the new Constantine.

Charlemagne, as he has become known to posterity, was indeed fortunate. The Moslem threat to western Europe had receded two generations earlier. Charlemagne himself reduced the warlike Saxons to submission and completely defeated the Avars of the Danube valley so that 'to perish like the Avars' passed into the common speech of the Slavs.[1] The period of Charlemagne was one of relative peace and stability between the turmoil and ineptitude of the Merovingians and the divisions and quarrels of his feckless descendants. It was marked by a short-lived and abortive renaissance of art, literature and learning. The scale and significance of this revival may be questioned, but there can be no doubt that its level of creativity was above that of the centuries which preceded and even of that which was to follow.

Charlemagne was undisputed leader of the Franks. Germany eastwards to the borders of the Slav lands acknowledged his authority, and his rule extended into Spain, southward through Italy to the March of Benevento, and down the Danube to the Hungarian Plain. The reality of his power was recognised when he was crowned emperor, successor to the Caesars.

POLITICAL GEOGRAPHY

The pretensions of pope and emperor could well be disputed. Over eight hundred miles to the east, in Constantinople, there ruled another emperor whose claim to be the heir to Diocletian and Constantine could bear more careful scrutiny than that of any upstart ruler in the west. The political geography of Europe at this time was dominated by the two empires: in the west the Frankish Empire upon which posterity bestowed the epithets of 'Holy' and 'Roman', and in the east the waning splendour of the Byzantine, decadent but remarkably tenacious of life, its principal organs almost invulnerable behind the fortifications of Constantinople. The two empires had little contact. They conducted only a frigid diplomacy, and for the rest were separated by the lands of Slavs and Avars, into which their merchants, missionaries and soldiers made occasional forays.

In one important respect the empires both of west and east differed from that of Rome to which each claimed to be heir. The latter had embraced the whole Mediterranean littoral. Pirenne has argued, persuasively if not altogether convincingly, that all parts of this empire were linked with one another by seaborne commerce. This system of trade, he argued, was only damaged by the Germanic invasions; it was the expansion of Islam which destroyed it. It had, in fact, been declining during the later years of the Roman Empire, but, in diminished volume, it actually survived the Moslem conquests. European seaborne commerce in the Mediterranean never entirely ceased, however restricted it may at times have become. Pilgrims continued to journey to the Levant, and small quantities of merchandise continued to be imported into western Europe from the eastern and southern shores of the Mediterranean Sea. But travel by sea was fraught with danger; piracy was rife, and the Moors even had the temerity to establish a base on the Provençal coast, at Garde Freinet, from which their war-bands raided far into France.

The trade which was thus jeopardised had been neither vigorous nor important in the western Mediterranean. In fact only the Byzantine Empire continued to practise any significant seaborne commerce. This focussed as certainly on Constantinople as that of the Roman Empire had on Rome. Indeed, some have seen in the restrictive commercial practices of the Byzantines rather than in the Moslem conquests the chief reason for the general decline of trade in the early Middle Ages.

The Carolingian Empire. The lands which Charlemagne inherited were substantially those which had been conquered and settled, however sparsely, by the Franks. They extended roughly from the Pyrenees to the lower Rhine, and beyond the upper Rhine they reached to Bavaria and the present Thuringia. This vast domain had no precise boundaries. Royal power had begun to fade out long before the boundaries, as shown in historical atlases, were reached. Even within the limits of the empire there were areas where the authority of the Frankish king was scarcely known or recognised. Brittany lived to itself and remained for practical purposes beyond the limits of the Carolingian Empire.

The nucleus of this empire was the Rhineland and the area stretching westward to the Meuse. There was no capital in the formal sense. The seat of the government was wherever the emperor happened to be; the officers of his government were his domestic officials, and governmental records accompanied the king in wagons as he travelled. Nevertheless, the emperor passed much of his time at Ingelheim, near Mainz; at Nimeguen, where 'he began two palaces

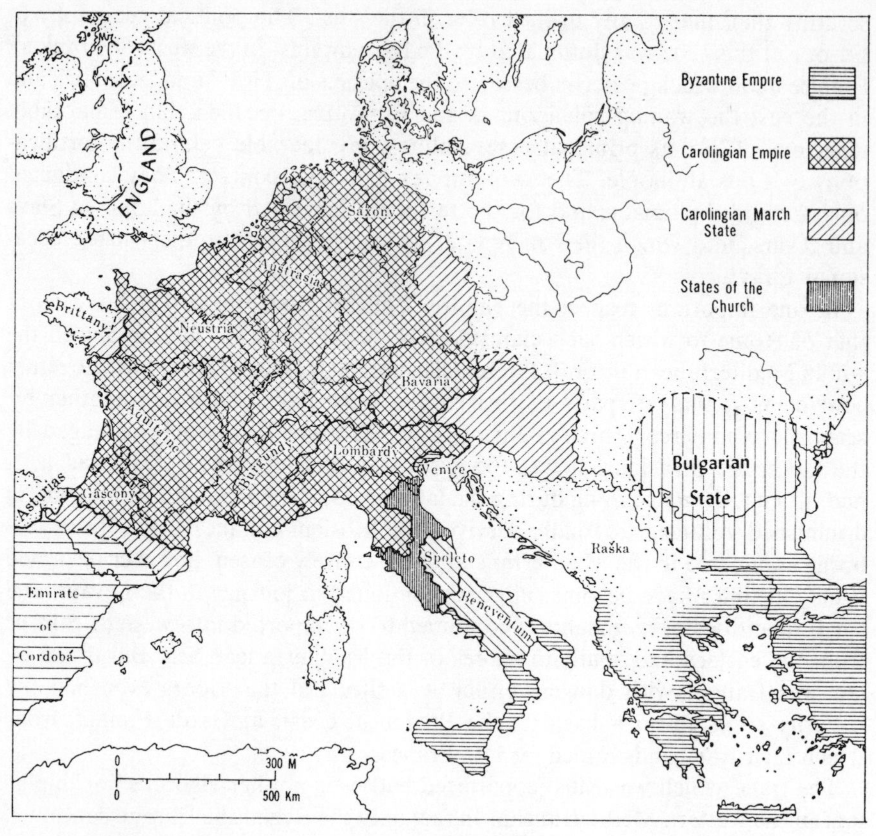

Fig.4.1 Political map of the Carolingian Empire

of beautiful workmanship', and at Aachen where his palace was adorned with marbles brought 'from Rome and Ravenna'.[2] It was from these palaces and from nearby Heristal, Thionville and Frankfurt that most of the *capitularia* were promulgated. An exception is the 31st *capitularium*, which was recorded as having been published when the emperor happened to be in Maine: *quando in Caenomanico pago fuimus*.[3] It is evident that, except when campaigning, Charlemagne did not stray far from Austrasia. 'The Rhineland was the geographical and educational centre of Charlemagne's Empire; and Aix (Aachen) was, in a greater degree than any other place, the administrative capital.'[4]

Much of his empire Charlemagne never visited, and the ties which bound the remoter provinces, like Aquitaine and Bavaria, to himself were slender in the extreme. The memory of recent independence remained strong, and it took the most careful and tactful administration not merely to retain their loyalty, but even to prevent open revolt. Even close to the centre of Frankish power the cohesiveness of the empire was little more apparent than at its periphery.

Yet public administration was for its period not only enlightened but also efficient. The corpus of *capitularia*, the ordinances and instructions promulgated by the emperor and his servants, constitutes a monument both to the solicitude and the administrative skill of Charlemagne. As if to weld this diffuse and polyglot empire into a cohesive whole, Charlemagne divided it into some 300 counties, *pagi* or *Gaue*. These were in large measure traditional territorial units. In each was a count, *comes* or *Graf*, who represented locally the power of the king. The *comites* appear generally not to have been drawn from the local aristocracy, who might have tended to reinforce the particularist tendencies of the county. Furthermore, they held office only during the pleasure of the emperor, and were visited at intervals, generally irregular and sometimes frequent, by *missi dominici*.

The latter were charged to inquire into abuses, as well as to exercise a general oversight over local administration and to provide a channel of communication between the emperor and his local officials. By such means Charlemagne created a semblance of unity, and his efforts were successful at least to this extent that the territory which he ruled was never again fragmented to quite the same degree that it had been under the Merovingians.

If imperial rule bore lightly on some of the lands of the Carolingian Empire, it was even less effective in the areas which his armies had recently conquered. The Spanish March, which embraced the Pyrenees and extended an indeterminate distance towards the Ebro, was established in 778. The mountain valleys remained immune to the raids of the Moslems and also probably to the injunctions of the emperor. South of the mountains lay a contested zone where Frankish authority was occasionally asserted and, lastly, the Ebro valley where at this time Moslem control remained secure.

The Frankish experience in the Breton March was similar. Desultory war between Franks and Bretons appears to have been terminated only by the creation of an ill-defined March, within which the Franks established a defensive line of forts, or *guerches*.

The greatest danger to the Frankish state came from the east. 'Except in a few places,' wrote Einhard, 'where large forests or mountain ridges intervened and made the bounds certain, the line between ourselves and the Saxons passed almost in its whole extent through an open country so that there was no end to the murders, thefts, and arsons on both sides'.[5] Reprisals had no effect and, their patience exhausted, the Franks undertook the conquest of the Saxons. The war began in 772 and was effectively concluded in 785. It resulted in the annexation of the plain of north-western Germany from the Rhine eastward to the Elbe. The base of the Danish peninsula was also brought within the empire, and southward from the middle Elbe the boundary appears to have followed the Saale river into the forested uplands of Thuringia. From here a frontier zone ran the length of the Šumava, or Bohemian forest, to the hills of southern Bohemia, and thence to the Danube.

For over a century the Pannonian basin had been occupied by the nomadic Avars, whose raids up the Danube valley had presented a grave threat to Bavaria. In a series of campaigns in 791–6 their armies were destroyed and their power broken. Charlemagne at the same time brought Noricum and Carinthia within his empire, and created a march, or border zone, designed militarily and administratively to resist invasion from the Danubian plain. The limit of the empire in this direction is obscure. It is likely to have been co-extensive with that of the former Roman Empire, but it is doubtful whether

Charlemagne's authority in fact extended into either the Pannonian plain or the Dalmatian Karst.

The frontier thus established appears to have included most of the Germanic peoples and to have excluded the Slavs. In terms of the security which it gave, it was scarcely more successful than that which it replaced. Charlemagne built forts along the Elbe and Saale, especially at Magdeburg and Halle, to restrain Slavic attacks, but was nevertheless obliged to lead campaigns into Slav territory to demand nominal submission and to exact tribute.

The Slavs consisted at this time of a large number of tribes, closely related in language and culture. These groups varied greatly in size and in their readiness to resist the Franks. According to Einhard, Charlemagne encountered resistance from the Welatabians, the Sorabians, the Abotrites and the Bohemians, or Czechs; 'he had to make war upon these; but the rest, by far the larger number, submitted to him of their own accord'.[6]

Einhard probably exaggerated the extent of Charlemagne's campaigns. It is highly improbable that his military journeys took him as far as the Oder, and beyond the line of the Elbe and Saale rivers and of the Šumava the empire of Charlemagne was only nominal. Acknowlegment of his suzerainty lasted only as long as his armies were present, and the Slav tribes appear never to have paid any form of regular tribute.

The Franks reacted more vigorously against the Avars than against the Slavs, because the former constituted the more aggressive foe. They were a Ural-Altaic people who had entered the Pannonian plain from south Russia and had there formed an aristocracy among the Slav and Celtic peasants who occupied – though very sparsely – the whole middle Danubian basin. Armies under Frankish leadership penetrated to the plain, and there had captured the Ring, or fortified enclosure of the Avar khan. The destruction was complete, and the site of the khan's palace became 'a desert, where not a trace of human habitation is visible'.[7] To the west and south of the plain, Slavs had moved up the Alpine and Dinaric valleys. Here they mingled with the Germanic Bavarians, and with the latter were brought within the Carolingian Empire.

Italy. The Italian peninsula had long been a political vacuum, under the nominal authority of the eastern emperors. The city of Rome, abandoned by its rulers, had fallen under the sway of its bishops. The latter had inherited not only the obligations of imperial rule, such as that of feeding the Roman proletariat and maintaining the water-supply, but also its perquisites. With its acquisition of secular control of much of central Italy, the Papacy became indeed the 'ghost of the deceased Roman empire', as Hobbs described it, 'sitting crowned upon the grave thereof'. The pretensions of the Papacy were resisted by the Lombards, a Germanic people who had in the mid-sixth century come into northern Italy from central Germany and had constituted an *élite* among the Latin-speaking provincials. Their conversion and assimilation had not been accompanied by any high regard for papal pretensions, which in general they resisted. It was ostensibly in reply to the Pope's request for aid against them that the Franks had first invaded Italy.

The documents which were submitted in order to enlist Charlemagne's sympathy and help had been forged. So also was the Donation of Constantine, to which he was induced to give his approval. This was to be the legal basis of the pope's claim to territorial sovereignty over Italy and the foundation on which the States of the Church were established. But Charlemagne, having in 773 defeated the Lombard king, assumed the title of King of the Lombards. In this new role he proved less ready to concede the utmost claims of the Roman bishop. Thus the Patrimony of St Peter was restricted to a strangely shaped territory which embraced in the north the Ravenna Exarchate, and, in the south, Etruria and the Campagna. These territories, both inherited in fact from the possessions of the Byzantine emperor in Italy, were joined by a narrow corridor along the upper Tiber valley. They excluded the Lombard duchies of Spoleto and Benevento, the former of which came to be loosely attached to the Carolingian Empire.

Such then was the empire over which Charlemagne claimed to rule. If one excludes the States of the Church, within which Charles made no claims, the Slav and Avar lands in the east, Brittany, and the Duchy of Spoleto, within which his claims were rarely respected, we have an area of about 460,000 square miles. Within this area Charles sought to create a sense of unity and cohesion. But its sectionalism – better described as tribalism over much of it – was never overcome within the Carolingian period. The administrative machinery proved inadequate; the *comites*, communicating only intermittently with the seat of government, came to identify themselves more and more with regions which they had been sent to administer, and within a relatively short period were to become hereditary local rulers.

Scandinavia. The British Isles, Scandinavia and the whole of eastern Europe, amounting together to more than half the area under review, lay outside the limits of both the Carolingian and the Byzantine empires. Neither empire had much contact with these areas, but there were stirrings in the north which threatened ill for the Western Empire.

By 800, Vikings had already visited the northern shores of the British Isles, but it was almost a generation before their raids on western Europe were to begin in earnest. A century or two earlier, however, Scandinavian peoples had begun to cross the Baltic Sea, some to settle among the Slavs of central Europe; others to occupy the east Baltic shores and thence to press into the interior of European Russia. The well-documented site of Vallhagar was abandoned in the sixth century, and evidence from other sites points to a migration from Scandinavia at this time. This movement was accompanied and perhaps in part caused by the creation of larger political units within Scandinavia itself.

The nucleus of the Swedish state lay in central Sweden, around Lake Mälaren. It passed under the rule of the *Svea* kings as early as the sixth century. Swedish rule had then been extended southward into Vestergötland, so that the southern boundary of the Swedish state about 800 ran in somewhat indeterminate fashion through the Småland highlands. Towards the north, Swedish authority faded out in the almost uninhabited hills of Norrland.

The Danish state emerged later than the Swedish. In 811, a treaty with the Franks established its southern boundary on the Eider river, where it was to remain until the nineteenth century. The Danish peninsula and islands were inhabited by tribes loosely knit together under a common king. The focus of political power lay at this time in southern Jutland, and the Danish islands and southern extremity of the Scandinavian peninsula were probably no more than nominal members of the Danish state.

Norway was politically more backward than other parts of Scandinavia. At the beginning of the ninth century the Danish king exercised a vague suzerainty over the lowlands around the Oslo Fjord, but in west and north Norway, the communities grouped around the fjords were in fact separate and autonomous units, each under its king or chieftain, which formed a changing pattern of alliances and federations.

Eastern Europe. The chronicler, Adam of Bremen, wrote in the eleventh century that 'Slavia is a very large province of Germany . . . In breadth it extends from . . . the Elbe River to the Scythian Sea. And in length it appears to stretch from our diocese of Hamburg . . . towards the east . . . in boundless expanses'.[8] The Slav lands of Europe impressed the Germans as an area of immense extent, and yet they underestimated their size. The Slavs were at this time the most widespread if not also the most numerous people of Europe. They were also among the least developed politically. Except on the borders of the Byzantine Empire, their organization was still tribal. The Poles, Czechs and Croats were, nevertheless, on the eve of those momentous developments which were to weld together related tribes and create states with some form of centralised government.

Many of the tribal names of the Slavs are known from writers of the following century. The Russian chronicle of Nestor, written some 300 years later, echoed the conditions of this period. Wrongly ascribing the origin of the Slavs to the Danube Valley, the chronicle tells how 'from among these Slavs parties scattered throughout the country and were known by appropriate names, according to the places where they settled'.[9] The chronicler mentions by name some fifteen Slav tribes, and adds 'thus the Slavic race was divided, and its language was known as Slavic'. The emphasis throughout was on tribal organisation. It is doubtful whether there was any high degree of unity and cohesion even within the tribes. The basic social unit, was as it had once been in the Celtic west, the village community, centring usually in its fortress, *gród* or *hrad*.

The expansion of the Slavs had by about 800 almost enveloped the Baltic peoples (see below, p. 185). King Alfred's informant, Wulfstan, had visited this region, which he called 'Estonia', and found there 'many towns (*manig burh*), and in every town there is a king'.[10] The 'towns' were the forts which crowned the hillocks in the moraine or occupied islands in the lakes. The basic political unit was the village community, though a passage in the Life of St Ansgar, suggests that these were sometimes associated into larger states.[11]

By the beginning of the ninth century the Slavs had spread eastwards to the headwaters of the Volga, and perhaps also to the lower Don and Sea of Azof. Though early writers gave lists of the Slav tribes of Poland and western Russia, the highest level of political organisation was in fact the village community or group of villages. This basic political unit was common to most of Europe.

'Whatever the subsequent divergences of Russian and Western history,' wrote Clarkson, 'there can be no question but that, at the starting point of Russian history, its institutional equipment was a piece with that of the West.'[12]

At the beginning of the ninth century bands of Swedes had already reached the Slav settlements to the east of the Baltic. Their role in the political evolution of these communities continues to be hotly debated. The extreme 'Normanist' view that the Russian state was the exclusive achievement of the Swedes, like the opposite opinion that it was the unaided creation of the Slavs. has to be rejected. Undoubtedly the latter had developed some rudimentary form of city-state in the pre-Viking era, and had laid the foundations of the later commerce between the Baltic and Black Seas. It is, however, no less clear that it was the Scandinavians who finally 'united the scattered tribes of the Eastern Slavs into a single state based on the Baltic-Black Sea waterway'.[13] If the political developments lay in the future, it is none the less apparent that Scandinavian influences were already at work in the east Baltic lands.

In northern Europe, beyond the indefinite limits of the Swedes, Balts and Slavs, lived the thin, scattered population of Proto-Finns. Their level of political organisation was primitive, being based essentially on the matrilinear family. To Jordanes they were 'the most gentle Finns, milder than all the inhabitants of Scandza',[14] perhaps because they posed no threat to their neighbours. They were in fact pastoral nomads, hunters and fishermen, whose technological level was little, if at all, above that of the Neolithic.

By and large, the Slavs had settled the zone of deciduous and mixed forest which tapers eastwards towards the upper Volga. To the south lived another and less peaceful complex of nomadic peoples – the Ural-Altaic tribes of the steppe. The Indo-European Sarmatians had disappeared either by migration or absorption into other peoples. The Slavs from their forest cover to the north-west intruded at times into the steppe, but generally were driven back by the military power of the steppe-dwellers.

The political geography of the steppe is difficult – even impossible – to reconstruct in detail, because its peoples were in almost continuous movement. They belonged to one or other of the two divisions of the Ural-Altaic peoples: the Finno-Ugric and the Turkic. These two groups had lived in close association for so long and had borrowed so extensively from one another that it is sometimes difficult to draw any distinction between them. The most powerful people of the steppe at this time were the Khazars, whose empire stretched to the north-east of the Black Sea. Constantine Porphyrogenitus referred to the 'nine regions of Chazaria',[15] which may indicate a ninefold tribal organisation. Nomadic though they were, the Khazars nevertheless possessed several towns, of which Itil, in the Volga delta, served as a kind of capital.

North-east of the Khazars, between the Volga and Ural rivers, lived the Pechenegs or Patzinaks, and to the east of them the closely related Guzes and Kumans, or Polovtzi. These were restless peoples. Their way of life was nomadic, and at times under pressure from their neighbours, they made long-distance migrations. The Guzes, for example, joined forces with the Khazars to attack the Pechenegs 'and prevailed over them and expelled them from their country, which the ... Uzes have occupied till this day. The Pechenegs fled and wandered round, casting about for a place for their settlement; and when they reached the land which they now possess and found the Turks living in it, they defeated them in battle and expelled and cast them out, and settled in it, and have been masters of this country, as has been said, for fifty-five years to this day'.[16]

The 'Turks' who were thus expelled by the Pechenegs were in fact the Magyars, a Finno-Ugric people who had acquired some of the linguistic traits of the Turkic peoples. In the early ninth century the Magyars inhabited the western steppe from the Dniepr to the Pruth, whither they had moved from the middle Volga region. Constantine Porphyrogenitus referred to their organisation into a number of tribes, of which one *του Μεγερη*, may have given the whole people their Magyar name. These peoples appear in general to have been forest-dwellers, and their traditional home was in the forests of the Kama and middle Volga. They became, perhaps under Turkic influence, pastoral nomads, without at the same time abandoning entirely their rudimentary agriculture. The Arab traveller, Ibn Rusta, described them as living 'during the summer on the steppes, moving with their tents wherever they found a better pasture for their horses and cattle. They even tilled some land. But with the coming of winter they went to the river to live by fishing.'[17]

The Balkans. The true heir to the Roman Empire was the Eastern Emperor, who, behind the formidable defences of Byzantium, had been able to withstand the attacks of Slavs, Bulgars and Arabs and to maintain a direct continuity from the empire of Diocletian and Constantine. The empire had, however, lost much of its territory in both Europe and Asia. Slavs, Bulgars and Avars in the one; mixed peoples of Middle Eastern origin, under Arab leadership, in the other, had stripped away its outlying provinces and reduced its European extent to the hinterland of Byzantium, together with the coastlands of Thrace and Macedonia, the Greek peninsula and a few scattered possessions which the emperors had been able to retain only by virtue of their command of the sea (fig. 4.2).

By the year 800, Slav tribes had overrun much of its area. By their occupation of Illyria they had thrust a wedge between the western and the eastern divisions of the empire, with profound consequences for both. Henceforward communication between them had to be largely by sea, and even this link had become tenuous under Arab maritime attack.

The basic social organisation of the early Slavs was the village community, which was usually closely knit. On the other hand, social and political association among these communities was loose. There are instances of Slav communities uniting before 800 to form a rudimentary state, such as that of the Antes, the White Croats, and Samo's Moravian state. But these were short-lived. Indeed, the only Slav state which at this time possessed any element of permanence owed its origin and direction to Bulgar, that is, non-Slav leadership.

The Bulgars were a Turkic people, who in the seventh century had moved into the Balkans from the Russian steppe. Some of their number even made

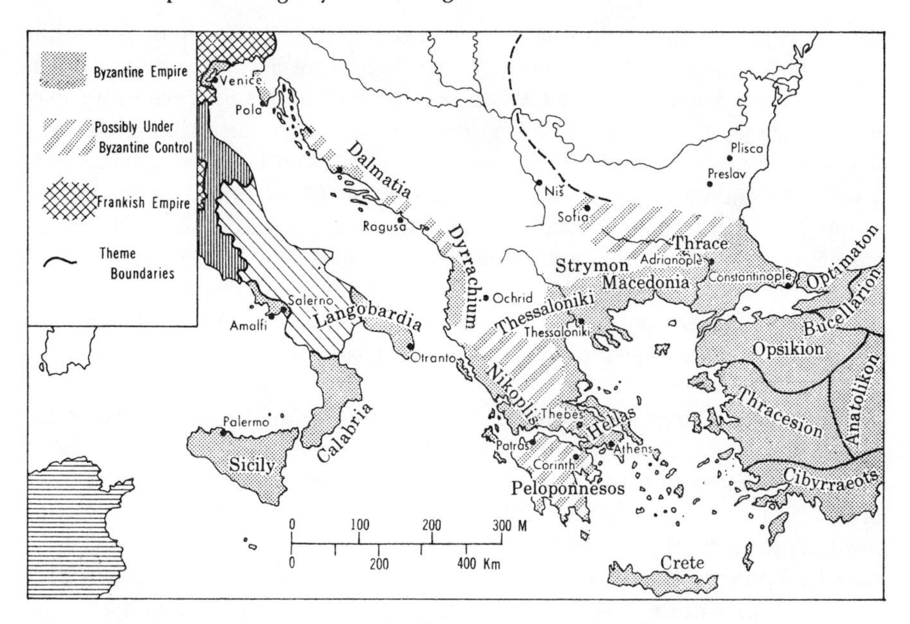

Fig.4.2 The Byzantine Empire and the Balkans in the ninth century

their way as far as Greece and southern Italy, but the main body crossed the Danube near its delta and settled in Dobrudja and north-eastern Bulgaria. They do not appear to have been particularly numerous, and constituted little more than a Turkic aristocracy which lorded it over the larger Slavic and Thracian masses. The nucleus of the Bulgar state was the platform which lies between the Danube and the Black Sea, and here they founded their first capital, Pliska. By about 800 the social and linguistic distinction between Slav and Bulgar had begun to break down, and to this the Bulgar khans had themselves contributed by elevating well-to-do Slavs to an equality with the Bulgar nobles. The Bulgars failed to maintain their indentity. Though the two languages continued to be spoken in the eighth century, during the ninth Bulgar yielded to Slavic, to which it contributed a substantial number of loan words.

The Bulgarian state was not established without strong opposition from the Byzantine emperors. Intermittent wars had failed to repel or even to halt, except temporarily, the Bulgar advance. The early years of the ninth century witnessed the consolidation of the Bulgar state under Khan Krum. The destruction of the Avars removed his chief rival, and Bulgarian armies even defeated the Byzantine forces and killed the emperor in battle. A treaty of 815–16 settled for a time the boundary between the Bulgars and the Byzantines. It ran from the Black Sea at Develtus inland to Microlivada on the Maritza river, and from here north to the Balkan mountains, which it followed north-westward to the Danube. Though the strength of the Bulgar state lay south of the Danube, it exercised – at least after the destruction of the Avars and until the advent of the Magyars – a kind of overlordship over the plains north of the river.

The Bulgars made extensive use of forts and linear earthworks to protect their frontiers. A line of fortifications had previously been built along the line of the Balkan mountains, and a bank and ditch were drawn from here northward to the Danube to protect Plisca on the west. Subsequent defensive lines marked stages in the expansion of the Bulgar state. The delimitation of a boundary from Develtus to Macrolivada was followed by the construction of a yet more ambitious earthwork, the Erkesiya, or 'Great Fence', between these points. This line, which can still be traced at some points, consisted of a bank, surmounted by a palisade, with a ditch on its outer side. So elaborate a work must have been constructed with the consent of the Byzantine authorities, and probably it served a similar function to that of Offa's Dyke between Mercia and Wales; not so much a defensive line as a visible separation of two jurisdictions.

The Bulgar Khan was apprehensive, after the defeat of the Avars, of the eastward advance of the Franks. Twice he made overtures, suggesting that a mutual boundary might be agreed upon, but the Franks were unwilling to be drawn into such negotiations. It was probably already too late. The Slavs who lived between the Bulgar state and the Friulian March of the Frankish Empire were beginning to form political organisations, which would present a barrier to the further expansion of both Bulgar and Frank. These independent Slavs were organised into territorial and political units known as *zhupe,* each under its chief or *zhupan.* At the beginning of the ninth century the *zhupan* Vishelav was beginning to unite the *zhupe* of the Raška region and to create the nucleus of the Serbian state. To the west the Croats had a similar social and political organisation to that of the Serbs, and despite Frankish overlordship, a similar coalescence of *zhupe* was taking place here. The Dalmatian Croats combined to defeat Charlemagne's Friulian margrave, and a Slavic duchy of Dalmatia appeared with its centre at Nin. A similar development was taking place beyond the Dinaric mountains, where a rudimentary Croat state was taking shape under the *zhupan* Ludevit, with its capital at Sisak – the Roman *Siscia* – on the Sava river. On the other hand, the Slavs of the eastern Alps were falling increasingly under the control of the Franks.

The Byzantine Empire. The emperors had been able to retain control in Europe over only Thrace, Macedonia and Greece, and even here were threatened by Bulgars from without and dissident Slavs within. Slav tribes in the Peloponnesos 'neither obeyed the military governor nor regarded the imperial mandate, but were practically independent and self-governing'.[18] The prevailing insecurity was reflected in the administrative divisions of the empire. The civil divisions, or Eparchies, of the Romans, had been replaced early in the seventh century by *Themes* (Θέματα). These were military, and consisted of unit-areas, each occupied by an army corps and controlled by a *strategos* or general. Inevitably provincial government became military and autocratic, and civil government disappeared.

The number of *Themes* changed frequently and tended generally to increase. Their boundaries are not clearly known, but in the European parts of the empire there seem at this time to have been about a dozen *Themes,* most of which bore regional names, such as Thrace, Macedonia, Strymon, Thessalonika,

Hellas, Peloponnesos, and the Aegean. The *Themes* of Dalmatia, Dyrrhachium, Nicopolis and Kephalonia lay along the Adriatic coast of the empire. Cherson in the Crimea formed a *Theme* as long as it was in Byzantine hands, and imperial possessions in Italy made up the *Themes* of Langobardia and Calabria (fig. 4.2).

The Byzantine Empire seemed to have few elements of cohesion or unity; it was never 'a true national state with an ethnically homogeneous population',[19] and it never shrank from admitting barbarian peoples of every race and language. Greek, nevertheless, remained the official language of the empire, and the only tongue common to all its provinces. It was the language of the well-trained, conservative and in general efficient civil service as well as of the Church. Language and religion were the chosen instruments of hellenisation, by which the imperial administration tried, though with little success, to achieve some unity among the discordant elements. It was through the Orthodox Church that the Eastern Empire achieved such unity as it was ever to know, and religious orthodoxy became the empire's peculiar kind of nationalism.

It must not be forgotten that the Byzantine was a sea-based empire, and that, threatened as it might be by the Saracen fleet, it was able until well into the ninth century to command much of the eastern Mediterranean. In addition to the Aegean islands, the Eastern Empire controlled Crete, and the Ionian islands, which made up the *Theme* of Kephalonia. The agreement of 812 with Charlemagne confirmed that the Dalmatian coast, with its virtually independent Slav *zhupe* and self-governing Roman cities, should continue to be at least nominal parts of the empire. Venice was in a somewhat different position. The small communities, gathered around its lagoons, served as commercial intermediaries between east and west, and were torn between their allegiance to the Eastern Emperor and their commercial need for outlets westward to Lombardy. The ultimate failure of the land-based Frankish armies to annex a state which could be continuously supported and supplied by the Byzantine fleet, not only explains the survival of Venice, but forms an interesting commentary on contemporary military logistics.

Byzantine control over Sicily and southern Italy seemed secure. Charlemagne's authority was limited to the area lying north of the Papal States. South of Rome, the Lombard Duchy of Benevento, though admitting allegiance to the Frankish Empire, was in fact independent of both the Western and the Eastern. The Campanian cities and Apulia, Calabria and Sicily were under the administration of the Byzantine Empire, but they were so penetrated and interrupted by enclaves of Lombard territory which the Franks were too remote to control and the Byzantines too feeble to absorb, that communication between them had to be maintained by sea.

Naples, together with the Campanian cities which she controlled, played an ambiguous role, somewhat like that of Venice, between west and east. Here the hellenising influences of the Eastern Empire struggled for mastery with the Latin civilisation of the west, and lost in the final round. The position of Sicily made it an outpost of the Byzantine Empire against the African world of Islam. The first Saracen attacks had begun in the seventh century, but the island was held without difficulty by its Byzantine *strategos* until external attack coincided with internal revolt.

The Byzantine emperors had been able to retain one other overseas possession, Cherson in the Crimea. It was, like Venice, essentially a commercial city, and, also like the latter, essentially self-governing, but dependent upon the empire for the security of its sea lanes and the defence of its territory against Khazars and Magyars.

The Iberian peninsula. Within Europe the forces of Islam had established themselves only in Spain. They had failed to hold their conquests north of the Pyrenees, and within the belt of mountains there was little to attract them. During the early years of the ninth century the frontier lay to the south of the Pyrenean and Cantabrian Mountains and was in 812 established along the Ebro. The Pyrenean region comprised the Spanish March. To the west lay the mountainous Kingdom of Asturias.

A frontier zone of variable width and sparse population separated the Christian states from the Emirate of Córdoba, which occupied the rest of the peninsula. There a small Arab *élite*, backed by a large body of immigrant Berbers, ruled the fundamentally Celto-Iberian masses (see below, p. 187).

POPULATION

For no period is it more difficult to form an estimate of population than for the early Middle Ages. Literary sources are fragmentary and the evidence drawn from place-names and settlement patterns highly unreliable. It cannot be doubted that the last two centuries at least of the Roman Empire formed a period of diminishing population, nor that population probably continued to decline during the period of the barbarian invasions and of the tribal kingdoms which followed. 'The period from A.D. 543 to 950,' wrote Russell, 'probably marks the lowest ebb of population in Europe since the early Roman Empire.'[20] It is sometimes held that there was a momentary recovery during the relatively peaceful years of Charlemagne, followed by a further recession, but for this, as Russell observed, 'it seems singularly difficult to find evidence'.

The most important source for the study of population at this time are the polyptyques, of which that of Abbot Irminon yields the most valuable data. It consists in its present form of surveys of twenty-five separate domains belonging to the Parisian abbey of Saint-Germain. Most lay within a few miles of Paris. All were large, well managed and relatively densely populated. Furthermore, their location in the Paris basin, in general a region of good soil and early settlement, must inevitably have given them a density of population well above the average for France. Any estimate of population density derived from conditions on the estates of Saint-Germain is likely to be too high. Ferdinand Lot's estimate of 14 to 15.5 million for France as a whole must, therefore, be rejected. Levasseur, after allowing for the greater extent of woodland than today, suggested that the whole of France had then from 8 to 10 million.[21] He added furthermore that those areas which had been cleared and were under cultivation supported more people than in the late nineteenth century when he wrote. This, however, has not been demonstrated. The food production was a great deal less in the ninth century from the same area of land, and only a handful of people can have inhabited the towns.

It is possible that even the more modest estimate of Levasseur is too high. A simple projection backwards of fourteenth-century estimates (see below, p. 328) suggests a much smaller total, perhaps no more than 6 million.

The amount of tribute paid by the later Carolingian emperors to the Norsemen throws some small light on the question. The sum was based on the number of *mansi*, and varied according to whether these were seigneurial, free or servile. On this basis, the 150,000 *mansi* of Francia and Burgundy should have had a population of about 900,000. This area amounted to about 18 per cent of the area of France, so that the total might be put at about 5,000,000.

The evidence for the population of the Iberian peninsula is even less adequate, and consists of a single fragmentary tax roll. An estimate can be made of the population of those areas for which the amount of tax collected is known. The district of Córdoba is said to have had about 113,000 houses, representing a population of perhaps 500,000, and the whole of the former province of Baetica may have had 1,200,000. This would appear to suggest that the whole area lying south of the Pyrenees may have had about 4,000,000, of which, in all probability, no more than half a million would have been in the non-Moslem regions of northern Spain.

For Italy there is no statistical evidence. The sources speak only of depopulated cities and abandoned fields, offset in some small measure by villages newly founded in the security of hilltops. Russell estimated[22] that the population of Italy was less than 4,000,000, but there is no formal evidence either to support or refute this figure.

Any estimate of the population of Germany and Scandinavia can be little more than intelligent guesswork. We know from the evidence for settlements (see below) that the cultivated area was relatively small, and the insecurity of the times probably kept the population even below that which the cleared land was capable of supporting. Russell has suggested 4 million for the area as a whole. This seems reasonable, and is consistent with the totals suggested for western Europe.

For the Slav land of eastern Europe even informed guesswork is impossible. Everything points to a small population, thinly spread across the country. 'Living apart one man from another,' wrote Procopius, 'they inhabit their country in a sporadic fashion. And in consequence . . . they hold a great amount of land.'[23] Both Polish and Hungarian historians have adopted a density of 5 per square kilometre or less as consistent with this semi-sedentary mode of life, and a total population for the three historic Polish provinces of 670,000 has been suggested. For Poland within its present boundaries the total might have been about 1.25 million, and densities in other regions of eastern Europe would have been comparable.

The Byzantine Empire is somewhat better documented than most other parts of Europe, but the evidence is, nevertheless, conflicting and inconclusive even for the city of Constantinople itself. For the rest of the empire one can only conjecture. Russell has suggested a total of 3 million for the Balkans, a figure with which, for lack of evidence, one cannot quarrel.[24] The population of the Eastern Empire had unquestionably suffered severely during the invasions of the fifth and sixth centuries. Tax receipts suggest that it may have reached its lowest point in the sixth century, and that it was slowly rising in the seventh and eighth.

The data presented in the previous pages may be tabulated thus:

Greece and the southern Balkans	2,000,000
Iberian peninsula	4,000,000
France	5,000,000
Italy	4,000,000
Germany and Scandinavia	4,000,000
Slav lands	6,000,000
	25,000,000

To this perhaps 1.5 million could be added for the British Isles.

Ethnic composition

It is fortunately easier to construct an ethnic map of Europe for the early years of the ninth century than a map showing population density. There had, of course, been large-scale and momentous migrations during the preceding five hundred years, and it is sometimes assumed that the ancient population had been driven out from many areas of Europe. This is far from the truth. Destructive though the invasions were, there is no evidence at any time for the wholesale obliteration of tribes or peoples. Nor do the Germanic invaders of western and southern Europe appear to have caused much movement in the peoples among whom they intruded. The invading tribes do not, in general, appear to have been large. Even one of the largest of the invading peoples, such as the Visigoths or Ostrogoths, is unlikely to have exceeded 120,000, and the smaller were probably no more than 25,000.

Population was spread so thinly throughout much of Europe that tribes such as these could have passed through many regions and not even have been seen by many of the local inhabitants.

The Avars, who, according to the Russian Primary Chronicle, disappeared completely after their destruction at the hands of the Franks, were probably only a ruling aristocracy whose defeated remnants merged with the Slavs of the Danube basin. The Bulgars, also an aristocracy of invaders from the steppe, after ruling the Slav masses for a period were eventually merged with them.

There is abundant evidence that although the *Völkerwanderungen* brought about far-reaching changes in language and culture, there was little modification of the racial structure of the population. D. T. Rice wrote of the population of Byzantine Asia Minor that the Slav and other invaders 'can hardly have affected the population as a whole, and just as the inhabitants of the uplands were in the main Anatolians of Armenoid extraction, so those on the coastal fringes remained Greeks, stemming from the Mediterranean racial group.[25] Very little anthropometric work has been done that might throw light on strictly racial changes in Europe during the historical period, but what there is suggests that the complexities of racial composition derive from the pre-

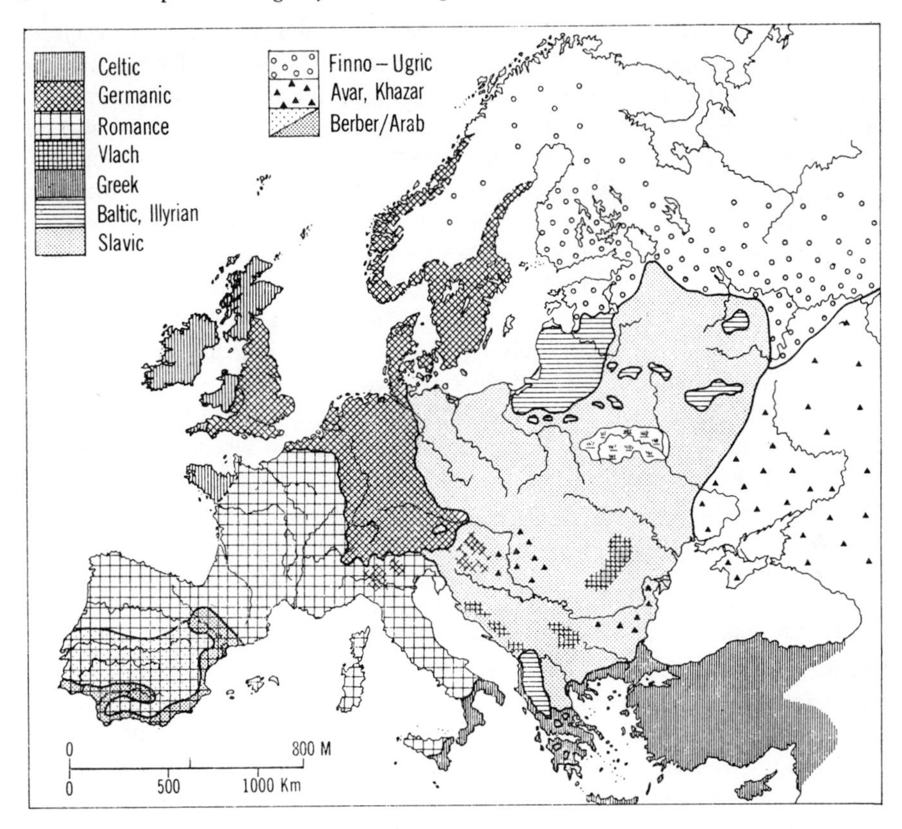

Fig.4.3 Ethnic map of Europe in the ninth century

Roman rather than the post-Roman period. Morant, for example, concluded that the Anglo-Saxon conquest of England had no appreciable effect on the physical characteristics of the population. The same author's studies of the racial composition of central and east European peoples have shown that the measurable physical characteristics, such as cephalic index, stature, and pigmentation, are of very little value in distinguishing between cultural groups. 'If no other evidence had been available, we might have concluded that Germans, Magyars, and Rumanians in central Europe are practically undifferentiated by physical characters.'[26] In other words the establishment during the early Middle Ages of the present linguistic pattern was not accompanied by the creation of any parallel racial boundaries. The bearers of new languages seem in most cases to have been few in number and were racially absorbed, leaving little measurable inheritance in the physique of the population.

Except in a few areas, such as the Pannonian basin, Transylvania and the eastern borders of the German realm, the linguistic pattern of Europe had been in its broad lines established by the early ninth century. The languages were very far from assuming their modern forms, but the basic division between Romance, Germanic and Slavic languages had appeared.

Within the western provinces of the Roman Empire the Latin tongue had, in some form, become the normal vehicle of communication. Not everyone spoke Latin; Celtic languages continued to be used in parts of Roman Britain and Gaul, and the Basque and probably also the Iberian languages remained of some importance. During the period of Germanic invasions the frontier of Latin speech had retreated somewhat in the east before the advance of Germanic. At the same time, vulgar Latin was slowly developing into the later dialects of France, Italy and the Iberian peninsula. In 813 the Council of Tours permitted the use of the 'rustic' or 'vulgar' tongue for preaching, instead of Latin, which had ceased to be intelligible. The first appearance of French as a distinct language, however, is in the Oath of Strasbourg of 842, sworn by two of Charlemagne's grandsons, Charles and Lewis, in French and German, so that it could be understood by their respective followers.

There must have been at the beginning of the ninth century some sort of boundary stretching from the North Sea to the Alps and separating an area of rudimentary French speech from one of Germanic. It may not at this time have been as sharp a divide as it subsequently became, but there is good reason to suppose that its general course has changed little in the course of the last thousand or more years. It was at one time supposed that the language boundary marked the extent of the Germanic – that is, Frankish – invasion of Gaul, and physical obstacles were sought which might have set a limit to the German advance. Kurth, Pirenne and others saw this barrier in the *Silva Carbonnaria*, the great forest which was assumed to have extended across Flanders from the Boulonnais to the Meuse. Des Marez emphasised the significance of Roman fortifications in restraining the invaders along approximately the present language boundary,[27] and Dubois argued that the Roman roads were used to demarcate the boundaries of Germanic settlement.[28] It should be said, however, that neither the *Silva Carbonnaria* nor any other physical obstacle served to restrain earlier Germanic invasions and that there is no evidence whatever that in the dying years of the Roman Empire any attempt was made to hold militarily any particular line of defence.

In contrast with the essentially French view that the Germanic invasions were, except for small bands, limited by a line, corresponding roughly with the present language boundary, is the standpoint of Franz Petri and the German school.[29] Petri claimed that neither forests nor roads set a limit to the invading tribes, but that they fanned out across France, settling the land in diminishing numbers with increasing distance from the former boundary of the empire. He cited the Germanic place-name elements in France and the archaeological finds of Germanic origin. The former he regarded as the more significant, indicating as it did that the invaders had settled and established deep roots in the land. However exaggerated Petri's interpretation might be, there is no doubt that the Franks and other Germanic tribes penetrated far beyond their present language boundary. The Germanic Burgundians are known to have been settled in Savoy, where they were completely assimilated – perhaps before the ninth century – by the local Romano-Celts. Only Germanic place-names survive as evidence.

There is, lastly, a suggestion that the language boundary in Belgium and the Ardennes was established before the end of the Roman Empire: that it represented a pre-Roman division between Celtic and Germanic tribes, and was not significantly changed by the invasions. St Jerome had listened to the Celtic-speaking Treveri;[30] if these people were so conspicuous in the fifth century, it is at least possible that their language continued to be spoken in the Moselle region in the ninth. Indeed, it could be argued that the language boundary in such regions as the Ardennes and the Vosges was formed by the encroachment of both French and German on an area of Celtic speech.

It is not clearly known what languages, Celtic or Ligurian, had survived in the recesses of the Alps. If a Celtic language had lasted through the Roman imperial period in the Lombardy plain, one cannot doubt that it lasted much longer in the mountains.

French dialects were slowly taking shape within Francia, but there is no reason to suppose that the Pyrenees constituted any kind of a divide; Basque was spoken on both sides of the range, and it is improbable that Iberian and Celtic languages had disappeared, especially from the north and west of the Spanish peninsula, nor Gothic from north-eastern Spain.

In the Spanish peninsula at this time Arabs had settled along the Mediterranean coast and in the Guadalquivir valley, and Berbers in the mountains of the interior. Their languages were spoken throughout these areas, though not to the exclusion of dialects derived from Latin and earlier languages. North African Berbers, it has been suggested, moved into the more mountainous regions in which they felt most at home.

In Italy the Latin of the later empire was slowly developing into Italian. The change was primarily in the spoken tongue which diverged more and more from the Latin which remained for several centuries the vehicle for written communication. Italy had been invaded by successive Germanic peoples: Visigoths, Ostrogoths and Lombards. Their languages failed to survive, though it is unlikely that the Lombard dialect was wholly extinct by the early ninth century.

The triumph of Italian was less easy in southern Italy. Since early classical times this region had maintained close cultural ties with Greece, and it is even possible that the Greek tongue had never disappeared completely from Magna Graecia. In the sixth and seventh centuries it was reinforced by the immigration of Greeks from the Peloponnese before the attacks of Slavs and Avars. They are described as settling in Calabria and Sicily, where some of them were able to 'preserve their own Laconian dialect'.[31] One assumes that the Byzantine administration also served to strengthen the Greek language, but it seems to have run into opposition from Latin-speakers, and this may have been a factor in the Neapolitan revolt against the Eastern Empire.

Scattered through western Europe, generally in less accessible areas, were peoples who belonged culturally to the pre-Roman period. Similarly there were other minorities who owed their presence to Roman imperialism. Gregory of Tours[32] mentioned the Syrian colonists in France, and there were similar groups in Rome, Verona, Ravenna and Naples.

The eastern boundary of Germanic speech was probably even less precise than the western. Though the period of most active German advance into Slav lands had not yet begun, there was already a forward movement, especially along the lower Elbe and in the Danube valley. This progress was, however, slow until after 1100 (see below, p. 248). The language boundary in Bohemia is obscure. Slav sources generally locate it within the encircling ring of mountains, but these probably formed at this date an uninhabited frontier between the two peoples.

The former Noricum had become largely Germanic with the eastward movement of the Bavarians. Population in the eastern Alps was probably mixed, while the western margin of the Pannonian plain was mainly Slavic, with only a few scattered Germanic settlements. Slavs had also penetrated the Dinaric Mountains, had reached the shores of the Adriatic, and had destroyed some at least of the classical sites of the Dalmatian littoral. To the south, the migration had taken them to the southernmost peninsulas of Greece. They followed the valleys of the Morava and Vardar. Thessaloníki, which received the brunt of their attack, was besieged several times. The Slav bands may have infiltrated the mountain valleys, but few penetrated the mountains of Albania or reached the coast of Epirus. The Slavs in the Peloponnese, however, were numerous enough to establish their independence of the Byzantine Empire for a time, and two of their tribes in the Taygetus Mountains were subdued only by a military campaign. An eighth-century life described Monemvasia, on the coast of Laconia, as 'in slavonica terra',[33] and Slav settlements were to be found at the very 'gates of Athens'.[34]

The original Greek population remained dominant in most coastal locations, and in the Chalcidice peninsula and Thrace. The commercial cities of the Black Sea coast remained largely Greek, while Adrianople became partly Slav, and Macedonia predominantly so.

The eastern and north-eastern limits of Slav settlement are less certain. For a long period – perhaps considerably more than a thousand years – Slavic-speaking peoples had been advancing eastward across the territory of European Russia. Their unchronicled movement had taken them by two routes, separated by the Polesian marshes which lay around the upper reaches of the Pripet river. The more northerly had extended to the region of Novgorod, and had thus enclosed the main settlement area of the Baltic peoples. The Slavs had thus become the dominant people within the belt of broad-leaved and mixed forest extending as far east as upper Volga valley. Their advance south and south-east into the wooded steppe and the steppe proper was slow and was, temporarily at least, halted by the steppe peoples. The Antes, for example, had in the fifth century formed some kind of empire which had reached into the grasslands. It was destroyed by the Tatar peoples. The early ninth century, however, saw a weakening of the Tatar empire of the Khazars; part of their territory was occupied by the ancestors of the Magyars, while the Scandinavians and Slavs were pushing down the Russian rivers and laying the foundations of the first Russian state.

Almost a third of the land area of Europe was at this time settled principally or in part by Slavic peoples. Their spectacular spread during earlier centuries had not altogether been achieved at the expense of the pre-Slav inhabitants. Many of these lived on, like the Baltic communities, surrounded and isolated, but none the less resistant to assimilation by the Slavs. This was pre-eminently the case with the Greeks, who sought refuge in mountain villages, but were able within a few generations to absorb and hellenise the invaders. In the Albanian mountains, also, the Illyrian peoples, hellenised or romanised in some small degree, were able in the main to resist the encroachments of the Slavs as later they were to do those of the Turks.

The Dalmatian coast had been romanised. There remain considerable traces of centuriation in its fields; cities were numerous, and at Split the Emperor Diocletian had built his palace. According to Constantine Porphyrogenitus, the Ρωμανοι (Romani) 'used to extend as far as the river Danube', but in the face of the Slav advance, 'the remnant of the Romani escaped to the cities of the coast and possess them still . . . the inhabitants of which are called Romani to this day'.[35] It is doubtful whether all the Romani found refuge in the cities. Constantine speaks of them also as being dispersed through the hinterland, where they lived by agriculture and were constantly harassed by the Slavs.[36] In these peoples we can see the forbears of the pastoral Vlachs of the Balkans (see below, p. 250).

Among the pre-Slav inhabitants of eastern Europe were groups of Germanic peoples, Goths and Gepids in particular, who had passed through this region and probably contributed to its population. Nor is there reason to suppose that the Sarmatian Jazyges had disappeared completely from the Danubian plain. The numbers of these pre-Slavic peoples were probably small, and diminished gradually as they were assimilated to the dominant Slav society.

The post-Slavic invaders are better documented. There were three such groups: in order of their appearance, the Avars, the Bulgars and the Magyars. The Avars, an eastern people of Turkic or Finno-Ugric linguistic affinities, had entered the Pannonian basin from the south Russian steppe in the sixth century. Their primary area of settlement lay between the Danube and the Tisza, though they also spread westward into the Dunántul (Transdanubia), and raided far into central Europe. Such skeletal evidence as is available suggests that their numbers were relatively small, and that those who did not perish at the hands of the Franks were absorbed by the Slav majority. Their language, whether Finno-Ugric or Turkic, was in all probability still spoken in the early ninth century, though it may have disappeared before the Magyar invaders reached the Pannonian plain.

The Bulgars, who had crossed the Danube into the Balkans late in the seventh century, established themselves on the low Dobrudja plateau. Their densest settlement was around their capital, Plisca, to the north-east of the Balkan Mountains. They were a relatively small, conquering people who settled as an *élite* among the local Slavs. The Bulgar leaders made no attempt to maintain the racial separateness of the two peoples. There was intermarriage between them almost from the first, and though the Turkic language of the latter continued to be spoken in the ninth century, it disappeared soon afterwards, leaving slight traces in the Slavic language of the Bulgars.

Skeletal remains are said to show almost no change from the pre-classical to the medieval period in Bulgaria, and it is assumed from this that the Bulgar invaders were too few to have brought about any great change in the racial composition of the people and, furthermore, that the Slavic invaders cannot have differed greatly in race from the autochthonous Balkan peoples.

The third group of post-Slavic invaders of eastern Europe, the Magyars, had at this date appeared only recently in the south Russian steppe. Here, until their ultimate expulsion by the Pechenegs, they lived on the western margin of the empire of the Khazars. It was not until the end of the ninth century that they crossed the Carpathian Mountains and spread over the Hungarian plain.

SETTLEMENT

The pattern of human settlement within the former boundaries of the Roman Empire had undergone a radical change. For this the barbarian invasions were largely though not wholly responsible. The fundamental reason was the insecurity of the times and the changing pattern of social relations. The breakdown of central authority on the one hand, the invasions and the depredations of bands of marauders on the other, made small settlements unsafe and cut off the commerce on which the fortunes of the larger were founded. This in turn tended to make the local communities as self-sufficing as possible.

Cities

It is assumed – in general correctly – that urban life decayed during the later centuries of the Roman Empire, and that over large areas of imperial territory it disappeared completely during the barbarian invasions. Many provincial cities were at best little more than villages; others drastically reduced their built-up and enclosed areas in the third and fourth centuries. During the succeeding centuries many were abandoned. The Anglo-Saxon poem, *The Ruin*,[37] presents a picture of a ruined and abandoned Roman city:

> the work of giants moldereth away.
> Its roofs are breaking and falling; its towers crumble
> in ruin. Plundered those walls with grated doors –
> their mortar white with frost. Its battered ramparts
> are shorn away and ruined, all undermined
> by eating age.

Some cities of the empire were abandoned and never again occupied, like *Uriconium* and *Calleva* in Britain, or reduced to the status of a village, like *Civitas Diablintum* and *Civitas Albensium* in Gaul. But as a general rule their walls, however ruinous, conferred some advantage in a period of insecurity, and probably continued through the Dark Ages to shelter a handful of people. Indeed, Isidore of Seville regarded the walled city as essentially a place 'quo esset vita tutior'.[38] London, for example, probably never ceased to be inhabited after the withdrawal of the legions, and in southern Gaul and Italy the continuity of settlement in some form can be demonstrated in most cases.

Many Roman cities had been laid out on a regular plan, and in some of them the rectilinear street pattern has largely or wholly disappeared. Does this indicate that the city was abandoned for some period of its history,[39] or can the distortion of the regular pattern be attributed to 'the more positive disrupting influence of a continuous and comparatively dense occupation through a period of indifferent civic discipline?'[40] In general, the Roman plan seems to have survived best in those towns where evidence of continuity of settlement is most marked.

Commerce for which the Roman cities had served as centres unquestionably declined in volume and importance, but whether it disappeared from the markets and roads has long been questioned. Evidence is accumulating for the continuance through the Carolingian period of both maritime and overland commerce. The evidence is fragmentary but none the less adequate to demonstrate that, despite their depleted population, reduced area and no doubt half-ruinous condition, many cities nevertheless retained that essentially urban function, the conduct of long-distance and local trade.

To this vestigial commercial role the cities of the Carolingian era had added another which in numerous instances was destined to contribute more than trade to their future importance. The Roman Empire had been a congeries of self-governing urban communities, each exercising jurisdiction over its surrounding *territorium*, the whole being known as a *civitas*. Each of these civil jurisdictions tended to become, well before the end of the empire, the sphere of a Christian bishop, and the city itself became a cathedral town. There must have been many instances of the institutions of the Church spanning the gulf between the Empire of Rome and the commercial and urban revival of the Middle Ages. A similar though generally less consistent role was played by monasteries. It became almost normal during the Merovingian period to establish monasteries within the limits of cities, and we may assume that Roman materials were sometimes used in the new constructions.[41] If a decadent Roman city failed either to attract a colony of monks or to become the seat of a bishop, its chances of survival and revival were slender in the extreme. This transfer of function was most strongly marked in Italy and France. In the Spanish peninsula, the Moorish invasion of the eighth century created a hiatus between Roman and later medieval times, and, in Britain, Christianity was re-introduced into the urbanised parts of the province almost two centuries after the withdrawal of the legions.

In all cities, with the exception perhaps of Rome and Constantinople, the citizens were engaged to some degree in cultivating the surrounding fields, and in most of them this agricultural function was dominant. Most were in the ninth century self-sufficing in the basic requirements of human life.

Only in their urbanised landscape did they differ from rural settlements. The city in the ninth century was walled, and without defences it was no city. The monks were attracted to it perhaps primarily because it gave security, and the merchant used it because there he could store his wares more safely than in the open country. The country magnate might have a town house, and rural communities could here accumulate supplies and take refuge in an emergency. In some cities there was a Jewish quarter, and in others, a Syrian.

The cathedral in episcopal cities was heir both to the location and to the architectural style of the Roman basilica. It was centrally placed and, at least in its later developments, dominated the skyline of the city. Monastic settlements were generally more remote and secluded, forming generally a walled enclosure within the city or a defensible annex beyond its walls. St Denis was built on the outskirts of Paris; St Riquier close to Abbeville, and St Remigius

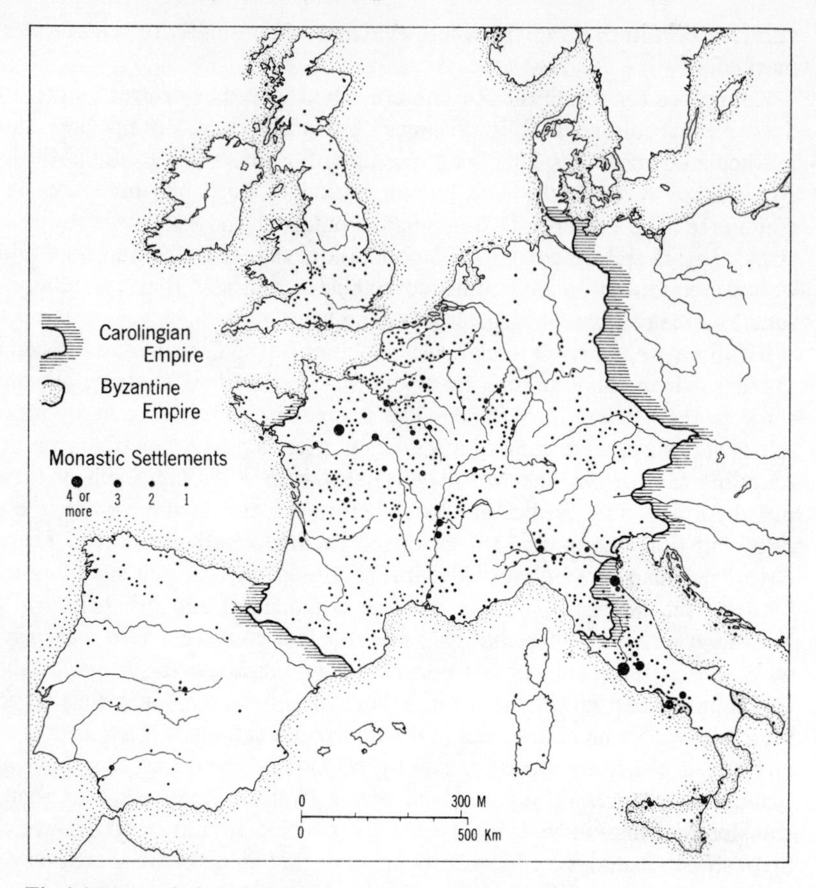

Fig.4.4 Monastic foundations

just outside the walls of Reims. Little survives of the cathedral and monastic buildings of this period, but the fragments suggest large and impressive structures that foreshadowed the romanesque of the tenth and eleventh centuries.

Most Roman towns of Spain had fallen into the hands of the Moors. Islam has always shown a propensity for urban life, and under Moslem rule the Roman cities of southern Spain became populous and prosperous, and were probably among the largest in Europe. Those of Italy, on the other hand, had suffered from the Gothic, Lombard and Frankish invasions. Although some trade continued, most were primarily places of refuge. Urban life continued only where there were walls to protect it. Many hill settlements, abandoned since the time of the Republic, were reoccupied. Rome itself was no exception. Its fortifications, built by Aurelian, were continuously repaired and restored, though their perimeter was far too great for the diminished population that found refuge within. The Capitoline and Palatine hills, and the Roman *fora* had been abandoned. Agriculture was practised within the walls. The diminished population lived mainly on the low ground of the Campus Martius, to the north of the old city, where most of the buildings and the charitable institutions of the Christian Church had been established. It is curious that the monks, in their search for solitude, had withdrawn to the low hills which had constituted the nucleus of the first Rome.

Small as it was, Rome of the Dark Ages was still too large to subsist on the food that could be produced from the surrounding fields. Grain was imported from Sicily during the Gothic Wars, and the popes continued to organise the grain supply until the Moors in the eighth century put an end to the commerce of the ports of Latium. One presumes that the population further declined until, like that of the primitive city, it could be fed from the fields of the Campagna.

The cities of Italy were reaching the nadir of their fortunes while at the same time a new vigour was becoming apparent in those of north-west Europe.

The larger cities of the Eastern Empire never underwent the eclipse that overcame most of those of the Western. Beset though they were by barbarian invaders, the eastern towns nevertheless retained their urban institutions far into the Middle Ages and continued to perform in some degree the urban functions of industry and commerce. An important factor was the centralised imperial authority, which through its *themes* and its *strategoi* retained some control over them. Another was the continuance of trade and the peculiar geography of the Byzantine Empire which gave great importance to whatever power continued to control the sea. Above all, Constantinople itself had advantages denied to Rome. Its situation was naturally strong, and its walls almost impregnable. It could import food and maintain contact with many parts of its empire, even when under attack. And when the Moslems made the Aegean unsafe, it could always turn to the not inconsiderable resources of the Black Sea coast.

Constantinople was at this time the largest and almost certainly the most prosperous city in Europe. It covered the peninsula, roughly triangular in shape, which extended eastward to the Bosporus between the Golden Horn and the Sea of Marmara. Its site was a low plateau, which rose steeply from the water. On the west the city was protected from the Golden Horn to the Sea of Marmara by a wall of a size and complexity that aroused the admiration or the envy of all who saw it. The area enclosed by the Theodosian walls, the last of the series to be built across the neck of the peninsula, was about 2,880 acres. Little is known of the interior arrangements of the city at this time. Churches were larger and probably more numerous than in Rome, and the visitor who came by sea was greeted by the majestic pile of Hagia Sophia which towered above the Hippodrome and the congestion and squalor which must have marked the older quarter of the city near the tip of the peninsula. Two wide thoroughfares, interrupted by squares or *fora*, ran respectively north-west and south-west from the old quarter to gates in the Theodosian wall. The homes of the rich were commonly built on an atrium-plan, as in Italy, but lay scattered amid working-class and industrial quarters. Water was supplied by a masonry aqueduct from hills west of the city and stored in elaborate underground cisterns, some of which have survived till the present.

The main harbour of the city was the Golden Horn, where most of the imports of the city were handled; the poorer quarters extended over the north-facing slopes above the docks. Several small harbours lay along the southern

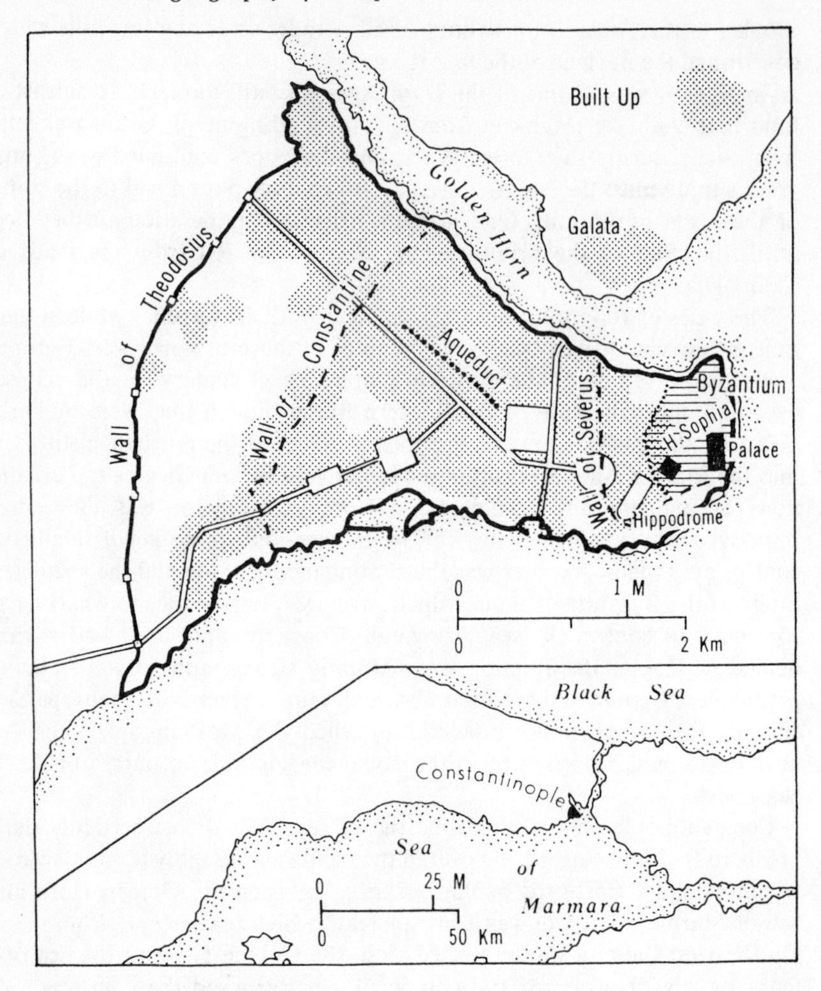

Fig.4.5 Plan of Constantinople

or Marmara shore, too small for regular use, but valuable when adverse winds made it difficult to reach the Golden Horn.

The large area enclosed by the Theodosian walls has led to excessive estimates of the population of the city. Under Justinian Constantinople has been said to have had a million people and it is claimed that the population never fell much below this figure until the thirteenth century. On the other hand, the city was far from built up. The area between the now destroyed wall of Constantine and that of Theodosius had never been fully occupied, and the urban area as a whole was, in Runciman's words, 'a conglomeration of townships, each complete in itself and joined to its neighbours by orchards and gardens'.[42] This open character of all except the oldest quarters must greatly reduce any estimate of its population. In any case, the accepted total is too large to have been regularly provisioned, even when Byzantium had command of the sea. It also implies that far too high a proportion of the population of the empire lived within its capital city. Russell has suggested that the population under Justinian might have been 160,000.[43] We can be reasonably sure that it did not exceed a quarter million.

Constantinople was a more significant focus of routes than any other European city at this time. Sea routes radiated over the Black and Aegean Seas, to which the arc of islands from Rhodes, through Crete, to the Peloponnese gave some protection from Arab raids. The *Via Egnatia* linked the city westward to Macedonia, and other roads ran north-westward to the Maritza valley and the Danube. From the opposite shore of the Bosporus roads ran to the provinces of Asia Minor, which the Byzantine emperors prized above all other parts of their empire.

It is no accident that the next largest cities in the European provinces, Thessaloníki, Corinth and Athens, lay close to the coast and were dependent on the trade of the Aegean. With its Macedonian hinterland and its trade with the Slavs, Thessaloníki was probably the largest and most important. Corinth, though smaller than it had been in classical times, still had some commercial and industrial importance, but Athens was but a ghost of the Hellenic city. The Valerian wall enclosed a mere 22 acres to the north of the Acropolis, but the repair under Justinian I of the more extensive Themistoclean wall suggests that there may have been a change in the city's fortunes. Excavation has demonstrated extensive building from the sixth to the eleventh century, and it is probable that Athens had regained some of its lost prosperity. Patras, on the coast of the Peloponnese, was able to beat off the attacks of the Slavs, and urban life continued in Spalato (Split), Ragusa and in a few other cities of the Dalmatian coast, 'because they obtained their livelihood from the sea'.[44]

Cities had never been numerous in the interior of the Balkan peninsula. Those sufficiently strongly protected survived the successive waves of barbarian invaders. Some, like Stobi in the Vardar valley to the north of Thessaloníki, had even been strengthened to face the invasions and thus had a new lease on life. Stobi, which may still have been inhabited in the ninth century, was a walled enclosure of only about 20 hectares, but contained paved streets, courtyard houses for the richer townspeople, small homes for artisans and tradesmen, a small Jewish colony and perhaps a synagogue. Justinian similarly developed the Macedonian village in the Kosmet, where he had been born, surrounding it 'with a wall of small compass ... in the form of a square, placing a tower at each corner', and building an aqueduct, as well as stoas, market-places, fountains and baths.[45] Other towns are less known, because the continuity of urban life has either hidden or obliterated the evidence. The cities of *Scupi* (Skopje), *Naissus* (Niš), *Serdica* (Sofia) and *Hadrianopolis* (Edirne) probably retained some elements of urban life, and the small Roman town of Singidunum, overlooking the junction of the Sava with the Danube, was already inhabited by Slavs, who had perhaps already begun to call their new home Beograd, the 'White Fortress'.

Justinian is credited with founding or at least fortifying an immense number of towns in the Balkans. In Macedonia there were no less than 46, and in Epirus, both old and new, he restored 49 and founded 44.[46] When all allowance is made for exaggeration in Procopius' list, it nevertheless represents an ambitious programme of building, far more so than that undertaken by the children of King Alfred in order to restrain the Danes. In the end, however, it was unsuccessful. Many of the sites listed by Procopius are unidentifiable, and it is doubtful whether many of these fortified towns continued to be occupied three centuries later.

Rural settlements

The collapse of central government during the later years of the Western Empire led to a profound change in the pattern of rural settlement. Most cities contracted, but remained within the framework of their fortified walls. Rural settlements had no such protection, and all that rural population could do was to gather wherever possible into larger nucleated settlements for mutual help and defence. The extent of the change in the settlement pattern in Italy and Sicily has already been noted. The chronicle of Monemvasia described the similar changes wrought in the Peloponnese by the Slav invasions: some fled to southern Italy; 'others found an inaccessible place by the sea-shore, built there a strong city which they called Monemvasia . . . Those who belonged to the tenders of herds and to the rustics of the country settled in the rugged places located along there'.[47] Nowhere else can the change be documented as it is in these areas, but the simple fact that settlements of the Roman period, ranging from villas to groups of huts, were abandoned, implies that their inhabitants found refuge elsewhere. There is evidence for the emergence of fortified villages and of castle-like structures in which the previously dispersed population could gather.

The change did not happen at the same time throughout the European provinces of the empire. It may have begun with the Germanic invasions of Gaul in the third century, continued with the Gothic incursions into Italy in the fifth, the Slavic into Greece in the sixth and seventh, and culminated with the raids of Moors and Vikings from the eighth to the eleventh centuries. It happened not infrequently that the most strongly built and most easily defended building within a rural settlement was the church. In many parts of Europe and over a period of many centuries, the fabric of the church became a village castle, *ecclesia incastellata.* Such familiar examples as the church of Les Saintes Maries in the Rhône delta, the many fortified churches of the Auvergne, and the defensible churchyards of Transylvania belong to a somewhat later age but similar structures were probably built in the ninth century.

Throughout western Europe there was some degree of change in the pattern of rural settlement. Koebner has distinguished three zones of barbarian settlement.[48] The first included Italy and the highly romanised provinces which bordered the western Mediterranean. Here the weight of Roman civilisation was overwhelming; the invaders could not set it aside, even if they had desired. Their numbers probably were not great, and they settled 'in place of or beside the Roman landlords'.[49] Place-names indicate that continuity was normal. Only on the borders of this region, as in Visigothic Aquitaine, do we find that the place-name evidence for Germanic settlement begins to be abundant, thus suggesting that there was here some kind of hiatus between the late Roman and the early medieval patterns of settlement.

Beyond this region, in lowland Britain, the rest of Gaul and the provinces bordering the Rhine and Danube, lay an area where German settlers were more numerous and Roman civilisation weaker. Place-names and the archaeological evidence suggest that the newcomers not only took over existing settlements, but also established their own. The third zone lay beyond the boundaries of the Roman Empire, where the influence of Roman civilisation had been slight, and the movement and settlement of one Germanic tribe could only be at the expense of another. The settlement pattern of this zone is discussed later.

It is tempting to try to assess the roles of the Roman inheritance and of the Germanic invaders in shaping the later pattern of settlement in western Europe. The debate on this issue has been long and inconclusive. The extreme view on the one hand regards the villa as the normal form of settlement and land exploitation during the later years of the Roman Empire, and assumes that it survived the invasions to become the villa estate of the Carolingian period.[50] Villa life was described by a number of late Latin writers, and the literary evidence tempts one to assume that the villa system was more widespread than was probably the case. For the Carolingian period itself, the evidence consists primarily of the polyptyques of monastic estates and a very few documents relating to the personal lands of the emperor. The latter are either inventories of holdings, or precepts regarding their administration. Their evidence, it has been argued, showed that the large estate, analogous to the late classical villa, was at this time the normal unit of landholding. This argument is not supported. The documents sometimes mention specifically the existence of small free peasant holdings.[51] Furthermore, only the rich and powerful had the ability, the need, and even the desire, to prepare such inventories. The holdings of independent peasants went unrecorded. The *Capitulare de Villis* and the *Brevium exempla*, both of them capitularies originating in the Chancery of Charlemagne, present a somewhat idealised situation.[52]

The opposite viewpoint postulates little or no continuity from the structure of settlement and landholding of the later empire. It is no more supportable than the views of the extreme Romanists. One must conceive rather of both villas and small peasant communities existing side by side at the time of the invasions, when groups of invaders displaced the romanised provincials, or, more probably settled among them. It cannot be said that villas did not survive as discrete economic units, but it seems highly improbable that they underwent no change or fragmentation.[53] The invaders are said to have avoided the towns of the Romans, and even the villa and village sites from which the Gallo-Romans had fled. Their own villages were often established nearby, as if they were unwilling to disturb the spirits of the Romans by settling on precisely the same sites, but this can hardly be regarded as interrupting the continuity of occupation of the site.

Two lines of argument have been used to demonstrate either the continuity of settlement or the lack of it from Roman times through the Dark Ages: place-names and the typology of the settlements themselves. The evidence of place-names must be used with great care. The mutations which they undergo, together with the human tendency to imitate the typonomy of earlier generations, introduces a degree of uncertainty into their value. Names, for instance, incorporating 'ville' or 'court', usually as a suffix, are relatively common in eastern France. The rest of the word is made up of a personal name, always modified but usually of recognisably Germanic origin. These names, Longnon claimed,[54] indicate settlements made by the Germanic invaders, and the suffixes themselves are no more than translations of 'dorf' and 'hof'. To this Bloch[55] and Lot[56] replied that these were romance names, and must have been given by speakers of a dialect or derivative of Latin. It is possible that such place-names came relatively late, many generations after the initial Germanic settlement, when Germanic personal names were borne by people who knew not a word of German,[57] More nearly contemporary with the Frankish and Alemannic invasions are the names made up of a Germanic personal name together with the adjectival suffix *-ius* or *-iacus,* which in time became *ingue*, *-ange*, *ens*, or *-ans*, implying the people or clan of the person named.[58] These names are distributed deep into France, suggesting that the Germanic bands penetrated to the centre and even the south, settling very irregularly, wherever they were tempted by the richness or the emptiness of the land. The relative isolation of the Germanic communities from one another would explain their early cultural divergence and their rapid assimilation by the Gallo-Roman majority.

It was a thinly peopled land into which the Germans had come. Population during the later years of the Roman Empire had been declining; farms were abandoned and fields allowed to revert to waste. It is doubtful whether the late Roman density of population and settlement was regained before the ninth century, so that it is possible that in many instances the Germanic settlers merely re-occupied and renamed earlier settlement sites, or reinforced the feeble complement of peasants who had survived. In any event the Germans probably only took over the fields of the Gallo-Romans. There is no evidence whatever that they came to Gaul to undertake the arduous work of forest clearance and land reclamation; they came for better, not poorer conditions of living. It is therefore unlikely that their occupation of the land brought about any fundamental change in the settlement pattern of Gaul; in Marc Bloch's words, 'place-names alone cannot help us to resolve the problems of settlement'.[59]

The typology of settlements, especially if studied in conjunction with place-names, throws some light on the problem. It was formerly the practice to relate the geographical pattern of the settlement itself to the ethnic origins of the people who founded it. We no longer equate nucleated settlements with Germanic settlers, but it seems probable that the morphology of each settlement bore some relationship to the social and economic conditions of its founders. The laws of the Salian Franks, codified by Clovis at the end of the fifth century, depict a society not unlike that portrayed by Tacitus. It consisted essentially of small communities of free men, whose settlements must have consisted in the main of loose clusters of huts, or *Weiler.*[60] Sometimes a larger group, presided over by the head of a clan, might give rise to a village made up of houses loosely 'thrown together', or *Haufendorf*.

One may presume that the settlements of the Germans conformed to these two types, but one cannot look in any village today for a perpetuation of its medieval plan. Village houses, built of wood or of clay and wattle, and roofed with thatch, were combustible and short-lived. The positions and interrelations of the buildings were continuously changing, and the site of the village itself might even shift slowly through the centuries. Furthermore, with growing population, the settlements became larger and their component buildings denser. The *Weiler* would be transformed into a nucleated village, while new, *Weiler*-type settlements might be founded. While it is possible, during the later Middle Ages, to relate certain settlement forms to particular waves of settlers, one cannot in general project backwards the later village plans and arrive in this way at their original forms.

One can, lastly, approach the question by examining the kinds of sites occupied by the new settlements of the Dark Ages and comparing them with those of the Roman period. The dating of settlements is very difficult and for this reason attempts to formulate and present cartographically a sequence of settlement patterns are open to criticism.[61] Nevertheless, those that have been made suggest that the Germanic invasions brought about no immediate and fundamental change in the geographical distribution of settlements, at least in continental Europe. A study of settlement in a region of south-western France[62] shows the Dark Age sites, here defined as *-ingas* place-names, interspersed on the river terraces among the Gallo-Roman; clearly no great expansion of settlement accompanied the invasions. A similar study of Touraine,[63] a region in which Gallo-Roman settlement had been concentrated in a few areas of light or alluvial soil, shows Dark Age settlement filling out the approximate area of Roman cultivation and advancing a little beyond it.

Gradmann has studied[64] the settlement pattern of Württemberg, east of the Rhine, but within the Roman *Limes*. Both the typology of settlements and the presumed date of their foundation are plotted on maps. The pattern of Alemannic (Germanic) settlement conforms very closely with the Roman and pre-Roman and also with the distribution of nucleated and *Weiler* settlements. The areas of post-Dark-Age settlement, on the other hand, are marked almost exclusively by forest villages (see p. 362) and isolated settlements. In Burgundy, Déléage found little change between the distribution of settlements in the third century and that of the early Carolingian period.[65]

The Byzantine Empire. Little is known in detail of the pattern of settlement in the Byzantine Empire. Rural settlement seems largely to have consisted of the nucleated villages of free peasants, surrounded by their gardens and vineyards, their cultivated fields and their grazing lands. Population was increasing after the losses of the invasion period, and villages, becoming overpopulated, were sending out groups of settlers to found dependent hamlets. Town life remained more significant than in most parts of the west, and the relationship of village to the nearby town, which served as market centre, was closer. Villages were, in consequence, less self-contained than in the west, and social mobility and communication were probably greater.

Central, eastern and northern Europe. Beyond the former confines of the Empire of Rome there had been little change or development in the pattern of settlement. Tacitus presented a picture of Germans living in small agricultural villages, which they were ready to abandon at a whim. This had not greatly changed. They still lived in clusters of small, rectangular huts, grouped irregularly in forest clearings.

The density of settlement can have increased little if at all in the previous eight centuries and locally Germanic settlement may have retreated before the Slavs. On the other hand, some progress had been made in clearing at least the margins of the forested Eifel, Taunus and Westerwald. Settlements were probably more numerous and extensive in Hesse than they had been a century or two earlier. But over much of Germany there had been little change. East of the Rhineland population was small and the economy primitive and self-sufficing. Thietmar, writing in the following century, described the Thuringians as 'a nation of swine-herders, feeding their hogs upon the mast in the forests'.[66]

It is unlikely that the Germanic settlements at this time had permanent or elaborate defences. This was not the case, however, among the Slavs, who appear to have had communal defences which were elaborate for the age. The Moslem traveller, Ibrāhim Ibn Ja'kub, noted the rounded 'castles' which the Slavs built in the damp valley bottoms by throwing up earthen ramparts.[67] Over 250 such *grody* are known in Poland alone; they are numerous in eastern Germany, and an indeterminate number have been found eastward into Russia and the Baltic provinces. Most were built on the low and often sandy elevations which bordered the flood-plains of the rivers of the northern plain; some on the alluvium of the valleys themselves. All were intended primarily for defence, and a few even had crescent-shaped courts, also enclosed by earthen banks and palisades, below the *gród* itself, possibly for corralling cattle.

These *grody* or *Herrenburgen* are most numerous in areas of good soil and presumably denser settlements. They probably served as refuges for the inhabitants of the surrounding villages. Of the latter very little is known. They consisted mainly of crude pit dwellings – *ziemianki* – roughly roofed with wood, but simple rectangular huts were beginning to appear. Procopius wrote of the Slavic Antae, that they 'live in pitiful hovels which they set up far apart from one another . . . every man is constantly changing his place of abode',[68] but in the more westerly Slav lands rudimentary oval and street villages (see below, p. 255) were probably beginning to appear.

A similar pattern of *grody* and villages existed among the Baltic peoples, but to the north, roughly where the mixed forest yields to coniferous, and forest soils to podsol, there was a change in culture. Finno-Ugric hunters took over from the Slavs and Balts, and the temporary shelters of nomadic peoples replaced the villages and forts of their southern neighbours. One cannot for this period speak of a geography of settlement in northern Europe.

An exception must, however, be made for southern Scandinavia, where there was a large, settled population. There had been widespread clearing in the plains of central Sweden, and forts were clustered closely around Lake Mälaren, in Uppland, and near the west coast. These refuges each served as a focal point for a group of hamlets similar to that excavated at Vallhagar, on the island of Gotland (see p. 140). The huts were here elongated and roughly rectangular. Their low walls were commonly built of large stones taken from the boulder clay. In the hills which bordered the lowlands were the isolated huts (*fäbodar*) used in summer by transhumant pastoralists. Swedish settlement had been extended thinly along the Bothnian coast, but had probably made little or no impression on the forested waste of Norrland.

Along the Norwegian coast, over the narrow coastal plain of south-east Norway, and on the strandflats of the fjords were isolated farms and small hamlets, where a scattered population won a living from its fields, its herds and from the richer and more reliable harvest of the sea: 'Their cattle and their crops yielded . . . food, and . . . they caught fish and shot venison and game in the forests . . . but the corn-fields were small, and the food generally ran short at midsummer.'[69]

The farmsteads consisted each of a dwelling-house, usually elongated like those of Vallhagar, together with barn, cowhouse, sheepshed and perhaps other outbuildings. It was inhabited by a patriarchal family of perhaps 20 to 30 persons. Most were overcrowded. Though additions were made to the farm-house and its subsidiary buildings, expansion was restricted by the shortage of cultivable land. Sooner or later a part of the *Grossfamilie* hived off to found another settlement elsewhere.

Such was the settlement pattern described in the Sagas and illustrated by Norwegian place-names. The period from A.D. 400 to 800 was one of expanding population and of broadening settlement. The place-name element *-land* is considered by Olsen to indicate a new settlement of this period; *-land* settlements 'belong to the outskirts . . . They developed out of pieces of *land* in the more or less distant surroundings of older farms, and their names have advanced in rank from what students of place-names usually call "field-names".'[70] Almost a quarter of the place-names in Viet-Agder are said to incorporate this element. The Norwegians, like the Swedes, must also have had summer dwellings, or *saetars*, on the high fjeld, where they occasionally made contact with the nomadic Finno-Ugric peoples.

AGRICULTURE

Any study of agriculture in the Carolingian period must rely heavily on the evidence of the polyptyques. This is unfortunate, because these estate records, emanating mainly from the larger monasteries, are apt to give a somewhat one-sided picture. The polyptyques were registers of the lands, tenants and dependants of the monasteries, and the most detailed of them provide a remarkably full record of land use, cropping and population. But the estates which needed such records and were capable of producing them were far from typical.

The lands to which the polyptyques relate lay almost wholly between the Loire and the Rhine, in an area which had been both occupied by the Romans and settled by the Franks. It was one in which the villa had been a normal unit of settlement, and it would be reasonable to suppose that many of the settlements recorded in the ninth century already existed in the second.

Only seven of the surviving polyptyques relate clearly to the ninth century and can be said to reflect conditions existing under Charlemagne and his immediate successors. All were monastic, and all described estates that were large and

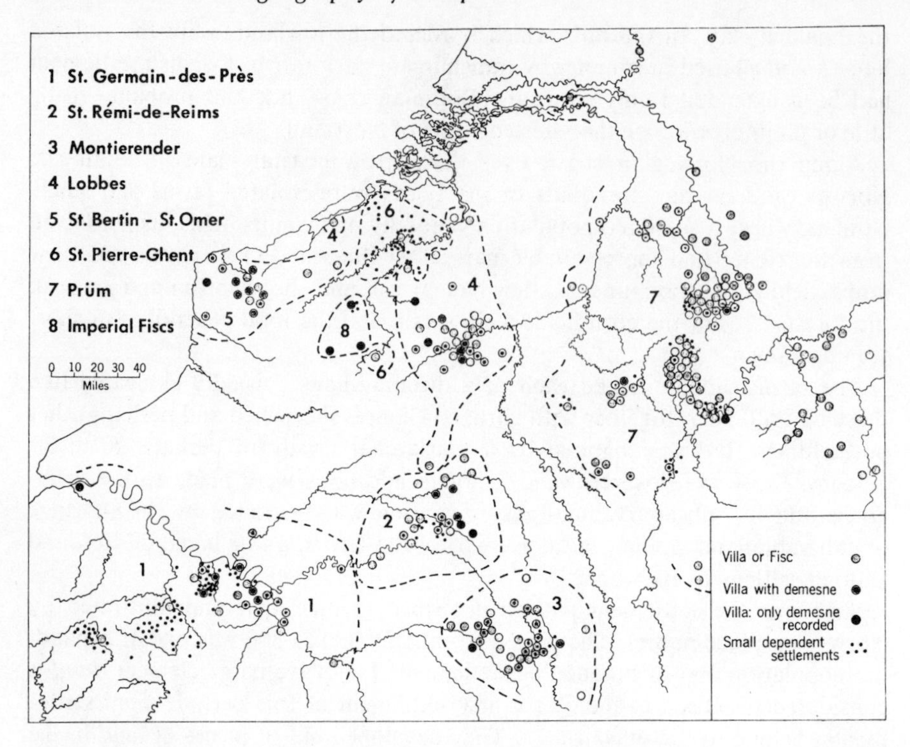

Fig.4.6 Bipartite villas on monastic estates in the ninth century

widely scattered. Nevertheless, they are not strictly comparable. That of the Parisian abbey of Saint-Germain gives not only the area but also the names of the tenants under the monastery's control. Others are less detailed and consistent. Saint-Germain possessed perhaps 75,000 acres of land. Of Prüm, probably the next largest, it can only be said that the monastery possessed wholly or in part no less than 119 named settlements. Saint-Rémi of Reims, Montierender, Saint-Bertin of Saint-Omer and Lobbes possessed lands smaller in extent than those of Prüm and Saint-Germain, but nonetheless extensive.

There is no doubt that these were very large estates, and it has sometimes been suggested that this was the normal method of land tenure in the Carolingian period. This view has been vigorously contested by Dopsch and more recently by Verhulst,[71] who claimed that small-holdings must have made up a large part of the cropland. Some of the calculations by which Dopsch demonstrated the small average size of holdings are more than suspect, but small-holdings, owned and cultivated by free peasants, must clearly have been numerous, and may have comprised the greater part of the agriculturally productive land. Indeed, it is from such small peasant holdings that part at least of the large monastic estates had been built during earlier centuries. The polyptyque of Saint-Pierre of Ghent was, in effect, a catalogue of donations of land, most of them quite small, to the monastery. Estates were growing in size at this time; the number of small, independent peasant holdings was probably diminishing with their absorption into large estates, but they had not yet disappeared from north-western Europe.

The picture of settlement and land use which emerges is very far from simple. In each of the surveys, the land possessions were grouped into *villae* or fiscs. A majority of these were bipartite manors of the kind long familiar to economic historians. The land was divided into (a) the demesne, owned by and operated on behalf of the monastery, and (b) dependent tenures held by the peasants of the monastery in return both for labour service on the demesne and also for payments in kind. In other cases there was no demesne, and the peasants, while owing labour services, were not called upon for ploughing and harvesting obligations. Instead they performed a variety of carting duties, cut and prepared timber, wove flax and wool, and paid rents in kind.

The bipartite manor is relatively common (fig. 4.6) from the Seine valley to north-eastern France. It is absent from Flanders, and infrequent in the Prüm lands which were spread over the Ardennes, Eifel and Rhineland. The map suggests that the bipartite manor was relatively common only in areas where a *villa* organisation had existed under the Roman Empire.

The map also shows that hamlets and isolated settlements were numerous, especially to the west of Paris, in Flanders and in the Ardennes and Eifel. These were areas where it seems unlikely that the Roman villa-system had ever been important or widespread. It would, however, be premature to relate the bipartite *mansio* exclusively to the areas occupied by late Roman *villae*.

Agricultural land seems usually to have been divided into strips which were relatively long and narrow, and were in turn grouped into *culturae*, or open fields. The shape and size of the strips is commonly and probably correctly regarded as a concession to the demands of the heavy plough. It does not follow that each *cultura* bore a single crop in any one year. The volume of the different grains harvested, where they are given in the records, is good evidence that so simple a practice was not generally followed. It seems more likely that the cultivator was still left with a certain freedom of choice, and that the full rigour of the *Feldzwang* had not yet been imposed.

The relationship of the demesne, when it was present, to the tenants' lands is far from clear. Occasionally the demesne embraced a whole *cultura*, but one is obliged to assume that in general it was scattered among the dependent tenures. Nowhere does the untidy evidence of the polyptyques point to the existence at this date of a *system* of landholding; instead, lands acquired piecemeal continued to be cultivated in their individual and peculiar ways. It was from this complexity that later field-systems emerged.

A three-field system seems to have been widely used in north-west Europe. In several polyptyques autumn and spring ploughing duties were exacted from the tenants, but only on the lands of Saint-Amand do we find the classical three-field system in its textbook simplicity.[72] The areas sown on each of four manors were listed, as shown in table 4.1.

Nowhere else can a system of three almost equal fields be derived from the figures given for ploughing, sowing and cropping. Very often one finds a disharmony between the seasonal labour obligations of the peasants and the areas actually ploughed and cropped. At many places in the Ardennes and

Table 4.1 *Areas sown on the manors of Saint-Amand* (*in bonniers*)

Manor	Autumn sown	Spring sown	Fallow	Total
A	5	6	(5)	16
B	10	10	(10)	30
C	16	16	(16)	48
D	5	5	(5)	15

Eifel the only field-crop was oats, which presumably alternated with fallow; at Aldenselen in the Netherlands was a tract of land which took oats 'whenever it was sown',[73] and near Ghent there was land which yielded a crop of oats every third year.[74]

The polyptyques contain a great deal of rather unsystematic data on the crops sown. On the lands of Saint-Rémi of Reims,[75] most of which lay in Champagne, the amounts of grain either sown on the demesne or received as rent from tenants were as shown in table 4.2.

Table 4.2. *Amounts of grain sown on the demesne or received as rent on the lands of Saint-Rémi of Reims* (*in modii*)

	Sown on the demesne	Percentage	Received from tenants as rent	Percentage
Wheat (*frumentum*)	150	1.7	55.5	2.3
Rye (*sigillum*)	586.5	6.6	53.5	2.2
Barley (*ordeum*)	6	0.1	271.5	10.9
Spelt (spelta)	7,256	81.0	2,024.0	81.6
Oats (avena)	..	..	15.5	0.6
Grain (annona)	988	11.0	60.0	2.4
Totals	8,986.5	100.0	2,480.0	100.0

In this area spelt was the most important bread-crop, as indeed, it probably was throughout north-eastern France. On the Lobbes estates,[76] most of them in Hainaut, spelt shared the soil with oats and barley. On five Lobbes manors, for which figures are given, the total grain sown was:

Spelt	2172 *modii*
Oats	1028 *modii*
Barley	630 *modii*

These totals are consistent with a three-field system, with winter-sown spelt alternating with spring-sown oats and barley. This, however, is not so with the Rhineland manor of Duisburg, where dues payable in kind by the tenants were as shown in table 4.3.[77]

Table 4.3 *Dues payable in kind by the tenants of the manor of Duisburg (in modii)*

Mansio	Rye	Barley	Oats
A	16	15	14
B	12	31	..
C	16	12	10
D	16	12	10
Total	60	70	34

The amounts paid as dues do not necessarily reflect the cropping pattern. In this instance the volume of the superior winter-sown grain (rye) is so much less than that of the less desirable spring-sown crops (oats and barley) that it is very difficult to assume that a system of three approximately equal fields prevailed.

Spelt, dominant grain-crop in the Paris Basin, was replaced in the Rhineland and Germany by an association of rye, barley and oats. The grain-crops comprised the greater part of man's diet, but they were supplemented by vegetables and herbs. A capitulary of Charlemagne required that these should be grown on all royal manors, and added a list of those thought desirable.[78] Mustard and hops (*umblones*) were widely cultivated, and flax and hemp were, locally at least, of great importance and contributed significantly to the revenue of some monasteries.

The chief drink, other than water, was beer, which was brewed even in important wine-growing areas. The polyptyques mention very large numbers of breweries and rents were sometimes paid in part in malt or beer. Almost any grain could be used for malting, but oats, in general the least valuable, were employed most frequently.

Wine by contrast was the specialised product of a few regions especially favoured by soil and climate. Foremost amongst these, at least according to the evidence of the polyptyques, was the Seine Valley near Paris. This had been a wine-growing region under the Romans, and its importance continued until the later Middle Ages. Its chief advantage appears to have been its navigable rivers which assisted the transport and distribution of the wine itself. Next, at least in north-west Europe, came the area around Reims and Laon. This was the most northerly of the important wine-growing areas, and for this reason several northern monasteries had acquired possessions here in order to ensure their own supply of the wine. In the ninth century Lobbes had a small but exclusively wine-producing estate near Laon, and the imperial estates in Artois controlled the wine-growing *villa* of Thiel (*Treola*), in the Seine Valley near Paris.

A third wine-growing region lay along the banks of the Moselle and Rhine, as far north as Bonn, where the abbey of Prüm had extensive vineyards. Viticulture had probably not spread far to the east of the Rhine, and the Verdun agreement of 843 gave Lewis the German the Worms region *propter vini copiam*.[79] The carting of wine was an important autumn task of the peasants.

Some of those on the estates of Lobbes, which had few vineyards of its own, 'faciunt ad vineas carra II in festivitate S. Remigii',[80] presumably the long journey into Champagne, where the wine harvest was associated with the feast of Saint-Remigius (Saint-Rémi) on 1 October.

Each autumn, after the grapes had been pressed, there must have been a busy traffic in wine by cart and boat. But transport of so bulky a commodity was not easy, and many monasteries tried to produce their own, however unfavourable their local conditions. The most northerly vineyards recorded were near Lille[81] and at Ghent.[82]

Very few Carolingian manors in north-west Europe were without meadow, which, in the absence of fodder crops, provided hay, the only winter feed for the animals. It was usually small in area, and lay along the valley bottom. Sometimes it was measured in acres, but usually in terms of the amount of hay which could be harvested from it. Only in Flanders, where it was more abundant than elsewhere, was it regularly used for grazing.

Rough pasture was rarely mentioned, perhaps because it was so abundant that its value was negligible. The abbey of Saint-Bertin had extensive grazing rights, probably over the downs of Artois, and the Flemish monastery of Saint-Pierre (Ghent) grazed its sheep in the meadows and marshes of the lower Scheldt. One area of meadow was here described as capable of supporting sheep in winter as well as in summer.[83] Most, it is presumed, would have been too damp for winter grazing.

Pigs, however, were the most numerous and probably the most important farm animals recorded. They were reared everywhere where there was woodland, and the most frequent measure of woodland was the number of swine it could support. *Silva grossa* was hardwood forest, containing oak and beech, which provided the chief food for the swine. *Silva minuta,* recorded not infrequently, but never associated with swine, was made up of varieties which yielded neither acorns nor beech-mast. Through the winter days the forests could provide little subsistence for the herds. Large numbers of swine were slaughtered in the late autumn, and their carcases salted. Swine bred fast, and their expectation of life was short. On the imperial *villae* near Lille there were, at the time when the *Brevium exempla* was compiled, 1,025 swine grazing in the woods and 645 carcases (*baccones*) freshly salted down.[84]

Acorns and beech-mast were not the only products of the forests. Wood formed the only fuel, and among the obligations placed on the tenants were those of trimming and cutting trees and carting the wood to the monasteries and granges. The greater part of the timber was consumed as fuel, but it was also used for building. Tenants were assigned the task of cutting and carrying large beams, of preparing wooden shingles (*scindulae*) and rafters (*axiles*) and, in areas where viticulture was important, the cutting of barrel-staves and vine-poles. Resinous woods were sought for the manufacture of torches (*faculae*), and oak bark (*durastauuae*) was gathered for tanning.

The polyptyques are strangely silent on most farm stock. Cattle are rarely mentioned though they were the chief draught animal. Dairying was at a discount and in none of the polyptyques is there a clear reference to dairy cattle.[85] Sheep may have been numerous, but were mentioned only in northern France and Flanders. It was the enormous herds of swine which provided most of the meat. On the other hand, poultry were very numerous, and poultry and eggs were among the commonest of the dues in kind paid by the peasantry. The writer estimates that the monastery of Saint-Rémi received yearly from its tenants almost 2,000 chickens and 8,000 eggs, in addition to whatever it may have produced on its own demesne.

The Carolingian capitularies suggest that broadly similar conditions must have existed on the imperial and, one assumes, other lay estates. But what of the farms, owned and worked by free peasants? One can only assume that they resembled the small *villae* of Prüm or the more westerly lands of Saint-Germain (fig. 4.6), where there was no demesne and a handful of peasant families operated some rudimentary field system and shared rights in the meadow and forest.

Except for this restricted area of north-west Europe information is scanty. Prüm possessed several manors beyond the Rhine. None had a demesne; all were small, and their obligations to the monastery were minimal and discharged in kind. East of the Rhine shifting cultivation, or *Feldgraswirtschaft*, was practised, and this method of clearing a tract of land, cultivating it for a year or two, and then allowing it to revert to the waste, seems to have been even more common in Germany itself. When permanent cultivation and the organisation of the land into a two- or three-field system came to Germany is far from clear. Lamprecht and others appear to place its origin well before the Carolingian period,[86] but there is little compelling evidence for this. The change is likely to have been necessitated by the rising population and made possible by the introduction of the heavy plough. One may, in any case, question whether it had made much progress by Carolingian times.

In the Drenthe region of the Low Countries open-fields are said to have appeared in the sixth and seventh centuries, with the introduction of the heavy plough. At the same time a 'pre-manorial' system, with small fields and scattered settlements, still prevailed over much of the lower Rhineland. In south Germany a developed bipartite manor is recorded on the Bishop of Augsburg's lands at Staffelsee,[87] and a similar manorial organisation may have been found also in Bavaria and Swabia.

There is no reason to suppose that the crops cultivated deep within Germany differed greatly from those in the Rhineland. Rye, oats and barley were the chief cereals; spelt and wheat were probably of minor importance. Millet, peas, beans, lentils and flax were grown as field crops. Pastoral activities, on the other hand, were unquestionably more important. Pigs were, in general, the most numerous animals, but the ratio of cattle to pigs was higher than in north-west Europe. Skeletal remains from early medieval settlements suggest that sheep and goats predominated in areas of heath and mountain, and that in the north German plain cattle outnumbered all other domestic animals, including pigs. The dependence of the latter on forests of beech and oak may explain their relatively small numbers in this region.

In Scandinavia population had been increasing; settlements had grown larger, and subordinate settlements had been established wherever the land permitted, but the general pattern of agriculture had changed but little. Large

Grossfamilie farms were each made up of a cluster of buildings, surrounded by its small enclosed fields. There was no field system in the west European sense. Hardy cereals, especially oats and rye, were grown, but the greatest reliance was on the farm stock, and as one progressed northwards so pastoralism increased in importance and crop-farming gradually disappeared. The Norse sagas, which relate principally to Norway and Iceland, present a society which relied almost wholly on pastoralism and fishing. Cutting, drying and bringing in the hay were the most critical activities of the farmers' year and it was for very good reason that the sagas described with such care the weather at the time of the hay-harvest.[88] Pastoralism was transhumant, as, in fact it has since remained. Cattle were summered on the hills and high fjeld, and brought down to the lowland farms only after the hay-harvest was in. Occasionally the herds were trapped on their high grazing lands by an early snowfall, unable to make the descent.[89]

Fishing was important in the coastal regions of Scandinavia, and towards the north tended to replace crop-farming as an adjunct to pastoralism. Several centuries later even the cattle were fed in winter on seaweed and fish-waste, and such may well have been their diet in the ninth century. It has been generally assumed, following the arguments of Sernander, that the climate deteriorated during the Scandinavian Iron Age, and that summer temperatures, reduced by as much as 3°C, made the cultivation of grain-crops difficult in central Norway and impossible in northern. In the opinion of Magnus Olsen the local population adapted itself to the change by an increased dependence on pastoralism and fishing. This climatic deterioration must have reduced the food-producing capacity of the land and contributed to the movements of the Viking age.

Most of the fields in western Europe, cleared and cultivated in the age of Charlemagne, had first been brought under cultivation in the period of the Roman Empire or even earlier. The process of bringing fresh land under the plough was making very little progress. Abbot Irminon reclaimed a few tracts of waste belonging to the abbey of Saint-Germain, but on the lands of Prüm were several *villae* described as *absi* – vacant.[90] The great period of medieval land reclamation still lay in the future, but it is probable that in north-western Europe the downward trend of population had been checked, if it had not everywhere been reversed.

It is doubtful whether this was also the case in southern Europe. Lands, ravaged by Germanic, Slavic and Arab invaders, had been abandoned; the population was decimated and driven to find refuge in fortified villages amid the hills, while the cultivated plains gradually reverted to grazing land and malarial swamp. The Campagna became 'a wide tawny plain . . . the pastures of which gave nourishment to thousands of cattle and other domestic animals, for the soil was too water-logged to be tilled to any large degree'.[91] In the south of the Campagna the Pontine Marshes gradually took shape, and in southern Italy and Sicily similar changes were hastened by Arab attacks which even made the grazing lands insecure.

Yet there were exceptions to this gloomy picture. In southern France rural life must have continued very much in the way described by the late classical workers. The Roman *villae* and *vici* continued, as a general rule, to be inhabited and their fields to be tilled with the light *aratrum*, sown with wheat and barley, and fallowed every other year. That the grape vine was widely cultivated is apparent, not only from the late classical writers, but also from the survival of classical vineyards.

Yet in the ninth century southern France must still have borne the scars of recent invasions. Southern Aquitaine was 'une immense solitude'.[92] Life was precarious in the coastal regions, and parts of Provence were intermittently in the possession of the Arabs. South of the Pyrenees conditions were worse. The wars had depopulated large areas, and in the ninth century settlers from France were being encouraged to migrate to the valleys of the Spanish March.

On the Meseta the new settlers were in part Berber, accustomed to the semi-arid conditions of North Africa. Their agriculture was based mainly on wheat and transhumant sheep. Conditions were different in the coastal regions, where the Romans had developed irrigation agriculture. The Arabs took over the land, and repaired and re-activated the irrigation works so effectively that they earned the reputation of having created them in the first instance.

The Arabs were probably the foremost agriculturalists in Europe at this time. Their treatises on agriculture were the first significant works in this field since Columella, and nothing comparable was again to appear until the Renaissance. They improved on the Roman irrigation works, and greatly extended the irrigated area in southern Spain. The cultivation of fruit and vegetables reached a level not previously known. The cultivation of the grape-vine declined in this Moslem land, but the olive remained important. Temperate fruits, known to the Romans, continued to be grown, and the Arabs themselves introduced many sub-tropical plants, including the banana, orange and peach, as well as numerous vegetables and herbs, which they had brought from the Middle East. Industrial crops, such as flax, hemp and dyestuffs, were grown, and perhaps at a later date, cotton appeared. At some indeterminate period also the sugar-cane was introduced, the mulberry tree planted, and the silk-worm brought in to feed on its leaves.

The post-classical economic decline was most marked in Italy. The balance of agriculture had been tipped even more towards pastoralism. Nevertheless, in the eighth century the Popes attempted to reclaim some of this land. They established in and near the Campagna a number of model manors, *domus cultae*, to provide food for both the papal court and the city of Rome. These were essentially *latifundia*. Doubtless crops were grown, but even these papal estates seem to have been primarily pastoral, and by the ninth century the enterprise was faltering. The Campagna, and with it much of coastal Italy, was reverting to 'a sodden wilderness into which shepherds with their flocks already were descending . . . from the Abruzzi . . . agriculture had already all but disappeared from the ancient *ager romanus*'.[93] To this the plain of Campania – the hinterland of Naples – was an exception. Its rolling terrain prevented it from degenerating into marsh, and its volcanic soils continued to attract the farmer.

While southern Italy was thus slipping into that condition of backwardness and depression which was to characterise it for centuries, in northern Italy there was economic expansion and growth. The polyptyque of the monastery of Bobbio,[94] one of the very few to come from Italy, describes a large and evidently well managed monastic estate of the northern Apennines. There was a large demesne, and no less than 650 dependent tenancies; it received annually 14,000 *modii* of grain, 1,600 loads of hay and 2,000 *amphorae* of wine, not to mention chicken, eggs and sheep. Over 5,000 swine fed in its forests in the Apennines, and, since Bobbio lay high up in the mountains, it had lands beside Lake Garda, which produced olive oil. It was a monastic estate comparable in every way with those of north-western Europe.

The monastery of St Julia at Brescia also possessed very extensive and varied lands in the north Italian plain and the bordering region of the Alps.[95] They were made up of at least fifty-five *curtes*, most of them small but some bipartite, and all extremely varied in their resources and revenues. Wheat and rye were the principal grain-crops, but oats, barley and millet were grown. Vineyards were widespread and olive groves numerous; there was extensive meadow, and even the production of edible chestnuts (*castanea silva*) was recorded. Farm animals were varied. Swine were reared in the woodlands, measured, as in north-west Europe, by their capacity to provide pannage. There were goats, sheep, dairy cattle, oxen and immense numbers of poultry. The practice of transhumance is to be inferred from the possession of alpine pastures (*alpa*), which yielded a rent in cheese. The monastery possessed boats which brought salt to Brescia, from, presumably, the lagoons at Commachio, and the use of iron tools was specifically noted.[96]

It cannot be said that such estates were typical in Italy, but large landowners seem, nevertheless, to have been numerous, and to have leased their lands, broken up into small holdings, to free peasants on a share-cropping basis. Much of the land of Bobbio was tilled in this way. The northern plain of Italy bore a very different aspect from that of today. The course of the Po and the lower courses of its tributaries were bordered by marshes, and it was several centuries before a vigorous effort was to be made to control the rivers and drain the marshes. But on the low terraces which bordered the flood-plain, the fields, more or less as the Romans had left them, continued to be cultivated with the same crops and by the same methods. The farmers may have spoken with Gothic or Lombard accents but in all other respects they had been assimilated into the Mediterranean way of life.

For the territory of the Byzantine Empire sources are even more scanty than for the Italian and Spanish peninsulas. Only the so-called Farmer's Law,[97] belonging to a somewhat earlier period, throws light on rural conditions. It presents a picture of free peasants, living generally in village communities, and cultivating their land sometimes with the help of either slaves or dependent free men. The law regulated their disputes about field-boundaries and the damage done by animals, and controlled the clearing of forest. It presupposed enclosed gardens and vineyards, but cultivated fields were intermixed and separated only by balks or markers that could easily be destroyed by ploughing. It regarded the rearing of cattle and other farm animals as normal, but also safeguarded water rights for irrigation.

This society of free peasants had come into existence only recently, with the break-up of the *latifundia* which had characterised the later Empire.[98] The Slav invaders had probably played an important role in this process. But this regime was to last only a century or two before succumbing to a creeping feudalisation.

How widespread, one may ask, were the conditions portrayed in the Farmer's Law. Like the *Capitulare de villis*, it probably related to a particular region or province. According to Setton, it illustrates the way of life of the Athenian peasant,[99] but it contains no mention of the olive, though there are numerous references to viticulture. The allusion to irrigation (cap. 83) is somewhat non-Attic, and the prescriptions regarding forest clearance can have hardly applied to the dry and almost treeless conditions of eastern Greece. The circumstances portrayed in the Law fit more closely the physical conditions in Macedonia, Thrace and the Maritza valley, and it is to this region that it can most easily be related.

INDUSTRY

Long-distance trade declined to minute proportions during the centuries which followed the collapse of the Roman Empire in the west. Each locality became self-sufficing, not only in foodstuffs – which, by and large, it had always been – but also in manufactured goods. What could not be produced locally most men had to do without. Only luxury goods, together perhaps with salt and some metals, entered into long-distance trade. Village crafts were pursued as never before. The carpenter, mason, iron-worker and potter provided for local needs. Their activities, for the most part unchronicled, perpetuated and even developed local styles; but only rarely did their wares travel far beyond the limits of the local community.

On this, as on agriculture, the polyptyques throw an uncertain light. Monastic settlements had of necessity to include craftsmen, who not only built and maintained the fabric of the church and conventual buildings, but also supplied the monks with their daily needs. The tenants of Prüm and of other monastic houses paid their dues to the monastery in part in constructional materials. Some were required to build so many perches of walling or to work at the granges and stables, while their wives wove woollen and linen cloth.

Iron was needed in windows and in locks and hinges, as well as in domestic equipment. Servile tenants of one of the manors of Saint-Germain owed 100 pounds of iron, presumably a small bloom.[100] The Prüm polyptyque contains no mention of iron-working, but the lands surveyed stretched across an area which is known to have produced iron at this time. Of the contemporary organisation of iron-working almost nothing is known, though iron was evidently a material of some value and not always easy to come by. Even greater quantities of iron were needed by the laity for arms and armour; axes, knives, and the tools of agriculture. To what extent Roman metallurgy survived into the Carolingian period to supply these needs is uncertain. It is not even known whether Noricum remained active as a producer. It might, however, be noted that not inconsiderable amounts of iron were listed among the possessions of St Julia of Brescia.[101] Iron was smelted and fabricated at this time in many parts of central and western Europe, and, if the finds of artistic ironwork and the needs of building and agriculture are any measure, it was produced in quantity. The Moselle region was unquestionably active in this field, and, though finds have not been dated earlier than the tenth century, the Siegen area may already have become important.

Iron was probably worked in the mountains of northern Spain, as well as in the British Isles. The laws of King Ine[102] of Wessex spoke of the nobleman, accompanied by his smith, as if ironworking was something that could be carried on almost anywhere by the hired retainers of the rich. Many of the known ironworks were associated with the monasteries and many a prelate, including St Dunstan of Glastonbury, was regarded in popular folklore as a worker in iron. Yet iron was not everywhere available. Lupus, abbot of Ferrières, seems to have had difficulty in obtaining a supply for his monastery.[103] One must assume, however, that unchronicled and unrecognised bloomeries were at work in many areas of the continent.

Even less is known of the mining and smelting of the non-ferrous metals. Lead, it is claimed, was at this time increasingly used for roofing large ecclesiastical buildings, and Lupus of Ferrières appealed to King Ethelwulf of Wessex for the material with which to cover his own monastery.[104] Since a subsequent letter from the abbot is concerned with difficulties of importing it through Etaples, one assumes that his appeal was successful. There is evidence also that lead was mined at Melle (Deux-Sèvres), but the mines in the Harz and Erzgebirge had not yet been opened.

There is, on the other hand, abundant evidence that salt was worked at brine springs and salt-pans, and that it entered into long-distance trade on perhaps a larger scale than any other mineral substance. There were salt-pans on the Mediterranean coast in Italy and near Narbonne, where the brine was evaporated by the heat of the sun. The salt-pans of the Atlantic coast were however more extensive and better known, perhaps because more accessible to consumers in northern Europe. They stretched along the coast from the Gironde to Morbihan, and salt was distributed by Irish traders to the ports of north-western Europe. Some of the salt-pans were owned by inland monasteries. Coastal salt-pans were also to be found on the coast of the Adriatic Sea, near Venice, and salt was not only an important cargo on the river Po but also the basis of the early prosperity of Venice itself. The abbey of Bobbio probably obtained its salt from this source. Other salt-pans lay on the coast of the Maremma and Laguna marshes, and of Macedonia. Much of Europe, however, was supplied with salt from inland salt-springs, where, except in the heat of summer, fires had to be set to evaporate the brine.

In Lorraine there were salt-springs at Vic and Marsal. The Archbishop of Sens is recorded to have purchased a load of salt from Lorraine, and a share in the salt-pans was acquired by the abbey of Prüm and described its polyptyque.[105]

Pottery of good design and competent craftsmanship had been produced in many parts of the western Empire. This manufacture ceased with the Germanic invasions. Locally at least the art of turning pottery on a wheel had disappeared, and a coarse ware, made for local needs and from local clays, took its place. Early in the ninth century, however, a stoneware of higher quality began to be made in the Rhineland, between Bonn and Cologne, where good pottery clay was to be found. It is not without significance that this revival occurred in an area previously noted for its Roman pottery manufacture.

The manufacture of glass was technically more difficult and more demanding in its materials than that of pottery, and was in consequence more localised during the Roman period. The industry survived the invasions, however, and in the Carolingian period was carried on in northern France and the Rhineland. Monasteries in Britain obtained much of their glass from France, and glass from this source even made its way to Poland.

Building was mainly in wood, though stone was used for the larger ecclesiastical buildings as well as for imperial and probably other residences. Roman materials were re-used where they were available, but major constructions were in general of rubble masonry with facings and coins of freshly quarried stone. Newly made brick cannot be recognised in buildings which survive from this period.

Most widespread of all crafts at this time was the manufacture of textiles. There were weaving sheds at many of the monasteries, where cloth was woven and vestments made, but usually it was a domestic occupation. There can have been very few cottages without the apparatus for spinning and weaving. Silk was woven in the Byzantine Empire, and cotton may have been spun and woven at this date in Spain, but in north-west Europe cloth was made from wool and flax. Bundles of prepared flax and even pieces of finished linen were contributed by the tenants, especially those of the monasteries of Lobbes and Prüm, as part of their obligations. Wool and woollen cloth were also produced, but were less conspicuous amongst the peasants' obligations.

The monastery of Prüm alone must have received at least 200 pieces of cloth each year, in addition to the fabric woven in its own *gynaecia* from flax provided by its tenants. On the most generous computation, this was greatly in excess of the needs of the monastery, and one can only assume that the surplus was sold into the market.

Pirenne has argued that even at this early date cloth was made in Flanders and sold by Frisian merchants.[106] The polyptyques of Saint-Bertin (Saint-Omer) and Saint-Pierre (Ghent) show that large numbers of sheep were kept in Flanders and Artois, and many monasteries besides Prüm and Lobbes produced cloth. The monastery of Saint-Gallen was receiving *pallia fresonica* brought up the Rhine by Frisian merchants, and, as if to show that this was no isolated event, Ermoldus Niger, in his fanciful dialogue between the Rhine and the Vosges Mountains, makes the river boast[107] that by its help merchants transported garments (*toga*) upstream.

The reality of this trade is not in doubt. The question is rather the source of the cloth. It is highly improbable that cloth of a sufficient high quality was produced by the Frisians themselves. It has been suggested that it came from England,[108] especially as Charlemagne had himself requested a continuation of this trade.[109] Pirenne, however, saw this as the beginning of the specialised cloth industry of Flanders and northern France. The only cloth manufacture certainly large enough to provide a significant surplus was that organised by the monasteries, and these are likely to have been the source of the *pallia frisonica.*

The existence of a cloth manufacture and even export in southern England cannot be questioned, but its scale was probably small. In Bavaria and perhaps also in other parts of Germany cloth was more than adequate for local needs.

When all is said, however, the volume of specialised manufacture in north-western Europe remained very small, and in northern and eastern it was unquestionably smaller still. Only in the Byzantine Empire did the crafts of the later Roman Empire survive. Constantinople was at this time by far the largest centre of manufacturing in Europe, and urban crafts were more widely distributed and more vigorously pursued in the territory of the Eastern Empire than anywhere in the west.

Towards the end of the ninth century the police regulations of the city of Constantinople were codified. The so-called *Book of the Prefect* was the result;[110] it prescribed in minute detail how the craftsmen might organise themselves, purchase their materials and sell their products. The greatest prestige attached to the silk-weavers and dyers. The sale of their best quality manufactures was an imperial monopoly, and only goods of lower quality were sold in the market. The linen-weavers, leather-workers and saddlers, the makers of soap, candles, and perfumes, the butchers and bakers – all were numerous enough to have their own corporate organisations. The industrial structure of Constantinople was highly regimented; it was protected from competition, and the export of everything that had strategic value was forbidden. Under so strict a supervision and control, it is surprising that manufacturing industries were as successful as in fact they were.

Most industries were pursued by craftsmen on a small scale, and only those which were an imperial monopoly, such as the manufacture of armaments and the weaving of the finest silks and brocades, were conducted in large workshops. The latter were clustered near the imperial palace, but other crafts were grouped in their own quarters of the city.

The larger cities of the Byzantine Empire were not, like most of those in the west, primarily agricultural in function. They had craft industries, some of which were conducted on a scale far larger than was required by local demand. Even the little town of Stobi in Macedonia seems to have had a small weaving and dying industry. Corinth was reviving as an industrial and commercial town, and there is evidence that the volume of coin in circulation was considerable. The larger European cities of the Empire: Thessaloníki, Athens, Thebes, Patras, had a larger industrial development than was to be found anywhere in western Europe. Luxury goods, it is true, dominated production, especially silk, brocade, jewellery, enamels and carpets. One manufacturer of cloth and carpets in the Peloponnese at about this time is said to have employed no less than 3,000 slaves. Crafts were, however, practised in general by free men; only the large workshops, whether imperial or private, relied heavily on slave labour.

TRADE

The vigour of manufacturing activity in the Byzantine Empire, in contrast with its absence except on a local and domestic scale in much of the west, points to a major difference in economic development between the two parts of Europe. Despite barbarian attack and invasion the Eastern Empire had been able to preserve much of the economic life which had existed during the Roman Empire. Seapower undoubtedly played an important role in this. Constantinople retained control of the sea routes which converged on it from the Black and Aegean Seas, and, since these had always been its most important avenues of supply, the city was able to preserve in somewhat attenuated form its earlier pattern of industry and commerce. Not so the Western Empire. The character of the land mass itself made it inevitable that land-borne commerce would be the most important. A century or two earlier Rutilius Namatianus, returning from Rome to his estates in Gaul, was obligated by the insecurity of the roads to travel by sea,[111] but in Carolingian times the sea routes were probably as insecure as the roads. It is impossible to overestimate the importance of the breakdown of the central authority and the resulting insecurity in reducing the volume of trade.

Long-distance trade

Before the Arab conquest of Spain and domination of the Mediterranean, vessels had sailed from Mediterranean and even Middle Eastern ports to those of the Atlantic coast and north-western Europe. The life of John the Almsgiver records a voyage from Egypt to Britain.[112] Wine and oil, in vessels of Mediterranean manufacture, found their way to Cornwall. The Celtic 'saints' had moved to and fro between the Celtic promontories of north-west Europe and had without hesitation set out on the long road to Rome. There was a decline in the amount of trade and movement, but they never ceased. The Mediterranean ports of western Europe were exposed to Moslem raids, and their commerce became intermittent. Nevertheless, Byzantine silks and brocades; leather goods from Moslem Spain, and the spices of the Middle East continued to reach western ports by sea as well as to arrive by land routes which extended westward from the Black Sea. Arles and Narbonne were trade centres in the western Mediterranean, though the volume of merchandise that passed through their harbours was certainly very small and probably irregular. Moslem control of the sea was by no means complete; nor was it to the advantage of the Moslems to destroy a commerce from which they derived a not inconsiderable profit.

Long-distance trade was primarily in luxury goods, but within north-western Europe there was a considerable volume of trade in daily necessities. Each of the richer monasteries possessed widely scattered estates, from which they derived income and services. The produce of the demesne on the most distant villa and the rents paid by its tenants had to be transported to the monastery itself. The distances involved and the volume of goods to be transported were considerable. This is reflected in the labour dues required of the tenants. Carting duties were general, and the obligation to make longer journeys,

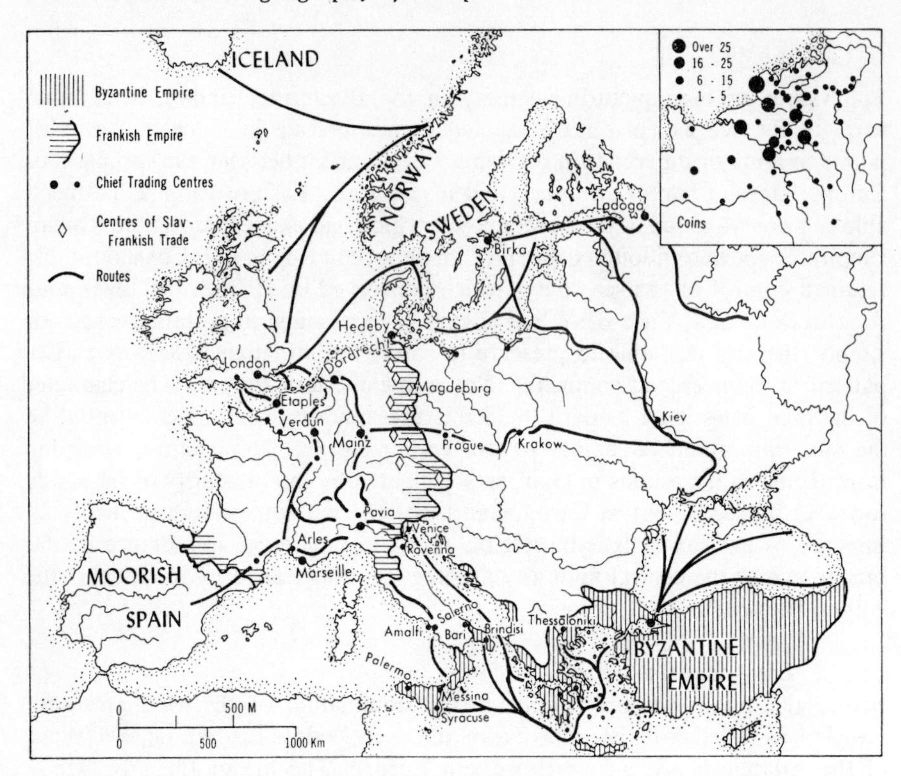

Fig.4.7 Trade in Europe, with inset of coin finds

necessitating absence for three or more days was widespread. Some carting duties were defined more specifically: carrying the hay or grain to the monastic grange and fetching wine from the vineyards which sometimes lay at great distances.[113] Sometimes these journeys were made by boat on the Seine, Moselle or Rhine. Occasional visits to distant cities were provided for; some of the tenants of Saint-Rémi (Reims) had the obligation to make the journey to Aachen, and Saint-Germain (Paris) similarly allocated to specific tenants the duty of travelling to the port of Quentovic when necessary.

Although most is known about those monasteries whose polyptyques have survived, many others must have been faced with the same transport problems. The abbey of Stavelot-Malmédy was in 814 granted exemption from tolls for all its ships on the Rhine and Meuse.[114] Not only was this monastery conveying its produce by river boat, but the total river-borne commerce was sufficient to make the establishment of toll stations worth while.

Many monasteries possessed very distant estates for the purpose of providing a highly localised commodity, such as wine or salt. Lobbes had such an eccentric possession near Laon (see p. 205); Bobbio and a church in Reggio nel Emilia each owned olive groves beside Lake Garda; Prüm had a share in the salt works of Lorraine, and the salt-pans of the Atlantic coast of France and at Comacchio on the Adriatic were owned and operated by a number of distant monasteries. All these distant possessions necessitated the transport of goods.

The roads may not have been crowded, but there can have been no lack of traffic, especially at harvest time and when the vintage was ready.

Not all such movement was within the limits of the monastic estates. The monasteries must have produced a surplus of cloth, wine and perhaps also of artistic and luxury goods. It has been suggested that much of these goods passed into the hands of Frisian and other merchants, who are known to have shipped cloth from the lower Rhine to markets in Switzerland and to have taken Alsatian wine in return.[115] There is no reason to suppose that these were all the goods they handled. The Frisians had colonies at Mainz, London and York; they were to be found at Saint-Denis, and in the small river ports of the Rhine, at the starting points of the alpine routes, and even in distant Narbonne. They were a well organised community of middlemen. It seems unlikely that the products of their own Frisian marshlands contributed significantly to their trade, and they must have visited the monastic houses and fairs, of which that at Saint-Denis was the most important, purchased wine, cloth and wool, and transported them, perhaps by road, but more likely by flimsy river craft.

The area from the Loire to the Rhine was clearly one of not inconsiderable commercial activity. A century ago a coin hoard was found near Amiens. The origins of no less than 493 of the coins were identified. They belonged to the middle years of the ninth century and had been minted at commercial centres within this area.[116] The distribution of their points of origin gives a rough measure of an economic region within which goods and currency circulated.[117] But this region of relatively intense activity nevertheless had external commercial relations.

Commercial links towards the south were probably the least important. Mediterranean trade had declined; the ports of Languedoc had decayed, but trade had not altogether ceased. The merchants of north-west Europe were in touch with those from the shores of the Mediterranean. From Paris and Orléans a routeway extended through Poitiers to the Spanish border; another ran through Toulouse to Narbonne. One followed the Meuse Valley through Verdun, whose merchants – *Virdunenses negociatores* – were well known, to Langres, and thence to Dijon, Lyon and the Rhône valley. The chief emporium in southern France was Arles, from which routes branched towards both Spain and Italy.

Much of the trade between Italy and north-western Europe, however, crossed the Alps to reach Pavia, the chief emporium of the plain of Lombardy. The Mont Cenis, the Great St Bernard and perhaps the Septimer were the most used passes, though there is some evidence for the use of all the better known passes with the exception of the St Gotthard. The passes themselves presented no difficulties other than the arduous ascent and descent. What travellers appear to have feared most were avalanches and the danger of losing the snow-covered track. Not so, however, the approaches to the higher mountains, where the roads ran through the defiles of the foothills. These were the *clusae*, or 'cluses', which were sometimes fortified and often used for the collection of tolls. This could be the most dangerous part of the journey, and the Carolingians generally took special care to see that the *clusae* were under their firm control.

Towards the east routes ran from the Rhineland through Germany and the lands of the western Slavs into Russia. The most important of these appear to have followed either the open country of lower Saxony or the Main valley. The latter linked with both Regensburg and Prague, and continued by way of Kraków and Przemysł to Kiev. Mainz was the chief focus of these routes, and received grain, brought down the river Main from Hesse and Bavaria. Doubtless many other goods came this way to the Rhineland, including spices from the east and skins and furs from the northern forests. This eastern trade was of sufficient importance under Charlemagne to justify an attempt to regulate it.[118] Trading with the Saxons, Slavs and Avars was tantamount to trading with the enemy. Weapons might not be sold to them, and trade was permitted at only a small number of points near the Carolingian frontier: *Bardaenowic* (Bardenwic), *Schezla* (Schesel, near Celle), Magdeburg, *Erpesfurt* (Erfurt), *Halazstat* (Halberstadt), *Foracheim* (Forchheim), *Breemberga* (Bamberg), Regensburg, and *Lauriacum* (Lorch) (fig. 4.7). Slav traders nevertheless penetrated to Fulda and Mainz, and Jewish and Arab merchants travelled over these routes from the Black Sea. Yet the volume of trade which crossed the frontier between the Empire and the Slav lands must have been very small. Few finds can be related to it; very little coin circulated, and the monastery of Corvey was accorded the right to mint its own because so little was available.[119] Nevertheless some goods of Byzantine and Middle Eastern origin must have reached the west by these routes.

It was towards the north and north-east that the external commerce of the Carolingian Empire was most active. The rivers which discharged to the North Sea were regularly used by merchants; here coin circulated, markets were active, and merchants frequented the *porti* (suburbs) that had grown up either at the gates of the monasteries or beside the cities which had survived from the Roman period.

Merchants did not need to operate from established ports: they could load and unload along the river banks and on open beaches. But a port gave protection to them and their merchandise, and furthermore it usually fell under the immediate jurisdiction of the emperor. Much of the trade, then, was canalised through a small number of ports. Of these Quentovic, or *Wicus*, was the most important for the Paris basin. It lay near the mouth of the Canche. The monastery of Saint-Germain did a regular business there (see p. 216),[120] and the volume of coin minted is evidence of the importance of its trade. Rouen, Etaples (*Stapulae*) and Boulogne were also used. The chief port of the Rhine delta and the focus of Frisian commercial activity was Duurstede, which lay on the lower Rhine and commanded routes westward to the Channel crossings and northward to the Zuider Zee and Frisian Islands.

Trade was carried on with England, where the reform of the Mercian currency under Offa is evidence of increased commercial activity. The export of cloaks and lead from England to the continent was balanced by the import of wine and pottery vessels, cloth, glassware, millstones and luxury goods for the nobles and churchmen. The chief English ports were London and *Hamwih* (Southampton). Trade also flowed from the ports of northern France towards the north-east. Frisian merchants had opened up trade with the Danes. They followed a route which kept close to the coast of Lower Saxony and crossed

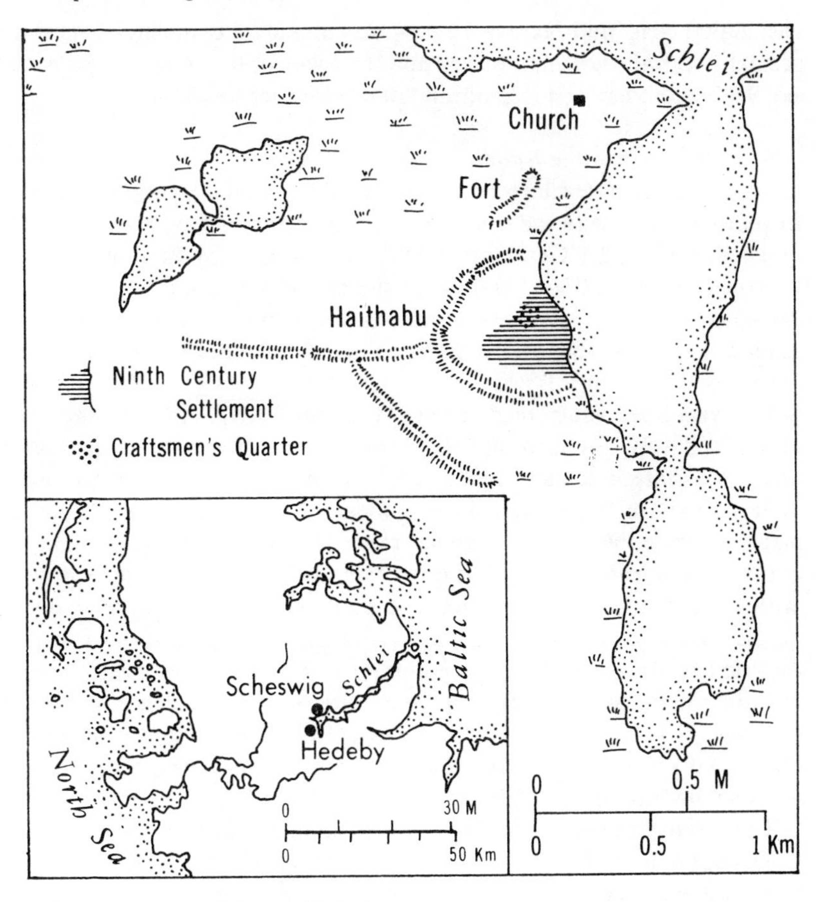

Fig.4.8 Plan of Hedeby (Haithabu)

the base of the Danish peninsula to the Schlei Fjord on the east coast. There, at Hedeby, they made contact with Scandinavian and oriental merchants.

This was a different commercial world from that of north-west Europe. Its ties were eastward, by way of the Black Sea rivers, with the Byzantine Empire and Middle East. Its emporia were Birka, on Lake Mälaren in eastern Sweden, Wolin on the north German coast, and Hedeby itself. But there must have been many small trading ports like the 'Truso', described in Othere's narration.[121] Trade was growing at this time, with the formation of centralised monarchies in Denmark and Sweden. The commercial towns of Hedeby and Birka were relatively large and well equipped with warehouses and shops, and there were numerous smaller trading places around the Baltic coast.

Coins of Byzantine or Middle Eastern origin have been found along the valleys of the Oder, Vistula, Dnepr, Daugava and Narva, thus suggesting the routes by which trade had reached the Baltic. The early traders by these routes were Syrians and Jews, but already the Scandinavian peoples had begun that outward movement which was to take them by way of the Russian rivers to the Black Sea and Mediterranean. Settlements had been founded on the east Baltic coast and Staraya Ladoga was already a commercial station within reach of both the northern forests and the rivers of the steppe, though trade had not yet begun to move in significant quantities over these routes.

The Byzantine Empire

In its commercial development, as in its industrial, the Byzantine Empire stood in sharp contrast with the rest of Europe. Its economy had successfully resisted attack from Arabs, Germans and Slavs. For this Byzantium was primarily indebted to its continued control of the sea and its ability in consequence to provision its capital and its coastal cities. For this reason also the Byzantine Empire had been able to retain the south of Italy and Sicily and exercised at least a nominal control over the rising commercial cities of Italy. On the Black Sea, Byzantine commercial power was unrivalled, though the emperor's authority rarely extended far inland from its coastline. The Byzantine Empire was a sea-state in a more real sense than the Roman Empire had ever been. A large part of the movement of goods within it was by ship, and without maritime trade there would have been very little transport and movement.

Contemporary records speak of the vigour with which trade was conducted within the limits of the Empire. The ports were full of shipping, and it was as easy to get a passage from one Byzantine port to another as it had been in the Mediterranean of the first century. The Rhodian Law, which codified the maritime practice of the Byzantine Empire, is itself evidence of the intensity of trade. Road travel by contrast was of minor importance; bands of Slavs threatened the security of merchants, and the road-books, formerly compiled to guide the traveller, had almost disappeared.

The foremost obligation of Byzantine commerce was to provision the city of Constantinople itself. The immediate hinterland yielded little, and most of its food and raw materials were brought in by sea. Numerous small ports were active not only along the shore of the Bosporus, but also around the Propontis. Bread grains had come from Egypt and North Africa until the Moslem conquest cut off this source of supply. In the early ninth century grain was obtained from Anatolia, Thrace and Macedonia. It was also obtained from the lower Danube valley, but probably not from the south Russian steppe. Wheat and barley were most in demand. The grain trade was a government monopoly, and the government itself maintained warehouses and, by regulating supply, controlled the price.

Other foodstuffs, like meat and fish, and industrial goods, such as skins and hides, timber, metals and coloured and ornamental stones were also brought to Constantinople. Silk production had been introduced into Greece, and raw silk, as well as flax supplied the clothing industries of the capital. The leading cities of the Empire – Thessaloníki, Athens, Corinth – were very much smaller than Constantinople, but their demands were broadly similar, and were satisfied by imports by sea whenever the resources of the local area proved insufficient.

Trade beyond the limits of the Empire was restricted in both volume and variety. Such trade was not encouraged by the government, and was in fact restricted to luxury goods. From Persia and other countries lying to the east came silks, carpets, gemstones, ivory and bullion; cotton and cotton cloth was received in small quantities from Egypt, and some raw silk seems still to have made the long journey overland from the Far East. In the Black Sea ports Byzantine merchants obtained slaves, skins, furs and amber from the northern forests and even dried fish from the Baltic. The *Book of the Prefect*[122] makes it clear that a great number of spices and herbs were available in the shops of Constantinople, and many of these must have been imported from lands outside the Empire.

The Byzantine Empire requited this trade by the export of silk fabric – especially pieces of inferior quality, since the best was reserved for the use of the emperor – of linen and cotton cloth, and of ivory, mosaics and other works of art. The volume of this trade was certainly small; how small there is no means of knowing. It was closely controlled and heavily taxed, while the Byzantine government pursued its ideal of a stable and self-contained economic unity.

The Merchant

Who, lastly, were the merchants who conducted this dangerous and difficult business? In previous centuries Syrians and Jews had dominated long-distance trade, and continued to play a prominent role. Jewish merchants, however, remained important in western Europe though obliged to share the field with increasing numbers of local traders. Foremost amongst the latter were the Frisians and Italians. The latter came to the fore when the Italian ports became the places of exchange for western and Byzantine goods. Greeks operated within the Byzantine seas; Italians lived on the borderland of two empires and belonged clearly to neither. They were 'not irretrievably discriminated against as the Jews, nor irretrievably distant from the Mediterranean centre of Oriental trade as the Scandinavians, nor irretrievably tied to the economic systems of antiquity as the Byzantine'.[123] The future was to be theirs. It was probably Italian merchants who carried the products of the Byzantine Empire and even of the Middle East across the alpine passes, and who met somewhere with the Frisians and other merchants of north-west Europe.

Local trade was generally in the hands of local people. Though some appear to have been well-to-do, many operated on the smallest scale, and were little more than pedlars. Yet others had as their chief task the procurement of foodstuffs for the imperial and royal courts.

A picture emerges of two areas, north-west Europe and the Byzantine Empire, within each of which there was a considerable volume of movement and trade. Contact between them at this time was far from close, and was maintained primarily along a commercial axis which ran from eastern France and the Rhineland across the Alps to Pavia, Venice and other trading cities which lay on the fringe of the Byzantine Empire. A third area of more intensive activity can be recognised in the Baltic region. Its links with north-western Europe were at this time maintained by the Frisians, though the Scandinavians were in the near future to play a more active role. The third side of this triangle lay between the Baltic and Black seas, but the full development of commerce along his route belongs to the tenth century.

Through these three broad routeways was canalised much of the interregional trade in the early ninth century. Its volume is incapable of measurement, but the emphasis placed on luxury goods in an age of poverty suggests that it must have been very small. Commerce was most active within each of the three heartlands of economic development. Of these the Byzantine was most active and developed, but in commercial policy and economic organisation it belonged to the past. It was ossifying in the face of changing conditions. The Baltic development was to be meteoric and short-lived. Only the north-west European region possessed the physical resources, the means of social communication and the human flexibility and adaptability that were necessary for continued economic growth. To it should be added, as a kind of southern appendage, the north Italian cities which, from being intermediaries in a small-scale commerce between north-west Europe and the Byzantine Empire, were beginning to take on the aspect of commercial foci in their own right.

CONCLUSION

Europe in the age of Charlemagne was characterised by two nodes: Constantinople and its neighbouring provinces, and north-eastern France and the Lower Rhineland, the former the core of the Byzantine Empire; the latter, the Frankish state of Austrasia. They had, as has been seen, little contact with one another, and each was a world to itself. Here their similarities end. The Byzantine world was defensive and static. The Frankish world of north-west Europe was vigorous and expansive. Trade was stable, if not declining in the one; expanding in the other. Art-forms had become fixed in the eastern node, and very little was being added to the total either of art or of architecture. A wave of building, on the other hand, was sweeping north-western Europe and spreading, in one direction to southern France and Spain, in the other, across Germany. The most vigorous activity was between the Seine and the Rhine, where the evidence for trade was also the greatest (fig. 4.7).

Charlemagne spent most of his life in this region, and probably left it only when on military campaigns in Italy, Spain or Germany.[124] Almost all his capitularies originated here, and here were to be found most of the seats which he frequented in the course of his travels (fig. 4.9). Already, however, certain nuclei were beginning to crystallise within this broad Austrasian realm, indicative of the future power centres in western and central Europe: the Paris region; the eastern Low Countries, between Liège and Cologne, and the middle Rhineland.

The Carolingian Empire was too grandiose, built on too large a plan. The imperial chancery and servants, efficient as they were, could not effectively control an area almost as large as the present European Economic Community. Even during his own lifetime Charlemagne had delegated power over some of the provinces to his sons. After his death, the Empire fragmented, not merely into three major kingdoms but also into a mosaic of quasi-independent local jurisdictions.

The period of almost three centuries which followed the death of Charlemagne, was characterised by certain movements and trends, all of which had been evident at the beginning of the ninth century. The first of these was the

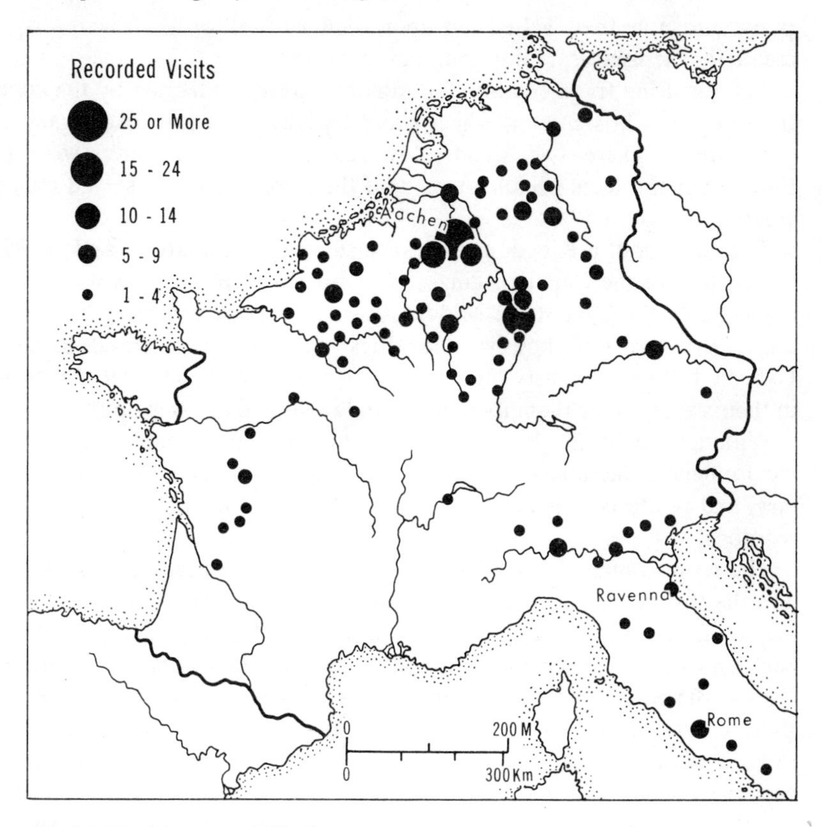

Fig.4.9 The itinerary of Charlemagne

continued pressure on Europe of external forces. The danger from the Moslem peoples of the Mediterranean intensified during the ninth century, with the Moorish occupation of Sicily and other islands which had hitherto been in the hands of the Christian powers. If, in the west, the Christian states of Spain made some headway against the Moors, the Moslem armies of the Turks in the eleventh century actually overran most of Asia Minor, and posed a direct threat to the security of Constantinople. The Avar danger had been eliminated, but another people from the steppe, the Magyars, about 900 invaded the Danube basin and again placed central Europe in jeopardy. The Vikings, whose raids had barely begun when Charlemagne died, intensified their efforts in the ninth and tenth centuries, dominating the Baltic Sea, settling in England and northern France, and, directly and indirectly, bringing about significant changes in the political map of Europe.

A second feature of these years was the political fragmentation of the Carolingian Empire. For a time the descendants of Charlemagne maintained some control over it, but the last members of the dynasty, for whom posterity has been able to devise no more appropriate epithets than the 'Child', the 'Bald' and the 'Simple', died in the tenth century. In the words of Reginon of Prüm, the Carolingian lands 'deprived of a legitimate heir, broke up into their component parts: and now they waited not upon their natural lord, but were disposed to create for themselves a king from their own stock'.

The resulting fragmentation and anarchy were moderated by the concept of the unity of Christendom, which was Charlemagne's principal legacy to later generations. These two trends, towards unity and disunity, were present throughout the later Middle Ages, and their conflict was resolved only by the creation of national states.

The concept of a Kingdom of France survived the anarchy, and in 987 Hugh Capet, first of the Capetian kings, ascended the throne, which was held by his descendants throughout the Middle Ages. The monarchy provided a focus for unity. In Germany, however, loyalties were never polarised in this way. Dynasty followed dynasty on the German throne, and squandered its resources in their struggles for the imperial title and their Italian campaigns.

Within each of the larger political units, effective control rested with the local officials, the dukes, margraves and counts, who made their offices hereditary, and reinforced them with extensive land-holding and a nexus of personal relations.

The more fragmented western and central Europe became, the more refined was the theory of European unity. The concept of empire, revived for Charlemagne in 800, was never allowed to lapse entirely, though some of the titular emperors went unrecognised throughout most of their nominal empire. In 962, Otto I, the Saxon king of Germany, was crowned emperor, and the German Empire was born. At the same time the theory of papal authority was elaborated. In general, the Gelasian thesis was accepted, that 'there are two powers by which this world is governed . . . the consecrated authority of priests and the royal power', and that the former authority should prevail since it has 'to render account at the judgment-seat of God for all men, even Kings'. On the other hand the emperors might themselves claim to share in some measure in this divine ordination, since they had been anointed and crowned by the pope. Thus was the scene set for a conflict of empire and papacy. That it arose in the last quarter of the eleventh century was due to the reform movement within the church and to the character and problems of the emperor himself.

While the states of western and central Europe were breaking apart, new states were coming into existence in eastern. The Slav, Magyar and Vlach peoples were developing from tribes, loosely linked with a particular territory, with chiefs whose role was merely that of wartime leaders, into states, with specific areas of land and established organs of centralised government. Poles, Magyars, Bohemians, Croats, Serbs suddenly appeared in the pages of history. They had strong kings who, in most cases, were the founders of dynasties. Some such states – Great Moravia, and the Vislanian state of southern Poland, for example, failed to survive, and their territories were divided between their more successful rivals. By 1100 a tier of 'proto-feudal' states had emerged along the eastern border of the empire and extending into the Balkans. They were at a stage of political evolution which much of western Europe had reached five centuries earlier and which even Germany had achieved two hundred years before.

These centuries were also characterised in western, central and southern

Europe by the emergence of the politico-socio-economic system known as feudalism. This term is used to denote a complex of human inter-relationships which evolved during the early Middle Ages and some of which lasted into modern times. It is, however, an error to speak of it as a system. It was not a coherent and distinct body of relationships. It arose and decayed piecemeal; its political, social and economic aspects did not necessarily co-exist one with the other. Nor did they emerge everywhere at the same time. They were in decay in some parts of western Europe while they were emerging in eastern. The years between 800 and 1100 were nevertheless the period of the growth and ascendancy of feudalism in the west and of its gradual extension into east-central Europe.

In a political sense feudalism consisted in the delegation of the functions of government to, or their assumption by, the local aristocracy. The fief was the reward for the performance of these duties, but alongside the emergence of fiefs there developed the twin concepts of benefice and vassalage. Both in fact ante-dated the Carolingians. The former consisted in entrusting a piece of land, a tenement, to an individual in return for a service, which may have been onerous but was often merely symbolic. Vassalage, on the other hand, was a personal relationship, by which one person pledged loyalty or commended him-self to another, usually in return for some intangible benefit such as protection or support. A complex hierarchy of such relations could be established.

Benefice and vassalage, in origin distinct, came in time to be merged. The benefice was conferred upon a vassal who in return owed both loyalty and service to his lord. This act was at first *ad hominem*, and the relationship lapsed with the death of either party. In fact, however, it quickly came to be heredi-tary. The counts and dukes were in time assimilated to the ranks of vassals of the king, and their fiefs were conceived of as benefices for which they did homage and owed service. Their own barons stood in the same relationship to them as they did to the king, and the barons and knights had themselves established similar contractual relations with their own dependants. Thus a chain of responsibility, loyalty and obligation extended from the humblest peasant to the king himself.

The system of feudal relationships which had emerged by 1100 was far more complex than any verbal description can suggest. It produced strange contra-dictions, as when a noble found himself the vassal, in respect of one piece of territory of a person who was his own vassal in respect of another. A baron could be a vassal of two or more lords, and was obliged to default in some degree in his obligations to one of them if they should find themselves at war. Feudal society was aware of its own contradictions, and devised what rules it could to reconcile them. Through its emphasis on status, loyalty and obligation, feudalism infused some measure of order into a society that was basically tur-bulent and unstable. It aimed to preserve a balance between king and vassal. When too much power passed into the possession of the latter, the result was likely to be anarchy; if the king became all-powerful, the result was despotism. Feudalism tried to avoid both extremes.

Feudalism came also to be both a legal and an economic system. The legal principles which it evolved were principally those governing personal status and obligation. It also acquired its own mechanism for the administration of justice. Among the obligations of vassalage was that of attending the lord's court. The king judged the cases of those who owed him homage, and the petty lord of the manor administered a rough justice to his own serfs and tenants.

The cement which bound the system together was however economic, the obligation to contribute service or rent in either cash or kind, both in return for the benefice and as symbol of vassalage. The personal relationships of feudalism came ultimately to be attached to the land. The vassal owed military service to his lord roughly in proportion to the amount of land he held, and the dues which he demanded of his own tenants were in rough proportion to the lands entrusted to them.

This system, so simple in its broad lines, so complex in detail, developed slowly in western Europe during these centuries. Yet the feudal structure was never complete. Before it had succeeded in embracing all men it had begun to disintegrate. There were always some people who stood in uneasy and uncertain relationship to it, and with the rise of the towns the number of those who were excluded from its personal and tenurial relationships increased.

These centuries, lastly, were a period of economic growth. About 800 the volume of trade, both local and long-distance had been minute. It was not large in 1100, but it was no longer restricted to a few luxuries. Long-distance trade in bulky goods was growing; a few specialised centres of manufacturing had emerged, and one can for the first time recognise a pattern of trade, a system of routes in regular use, with a flow of goods between established centres.

Underlying this growth in economic activity was an increase in total population. Some time in the ninth or tenth century it took an upward turn, and the evidence points to continued growth through the eleventh. Urban life, which had disappeared from some parts of Europe and had been moribund in most others, began to revive. The towns of Roman origin, most of which had become the seats of bishops, grew in size and others were established at the gates of monasteries and in the shadow and under the protection of castles.

Such indicators as there are suggest that growth intensified during the eleventh and twelfth centuries, and culminated near the end of the thirteenth. The year 1100 has no special significance in this expansion of the European economy, but it comes about half-way between the depression and gloom of the tenth and the splendour of the thirteenth century. A picture of the Europe of about 1100 reveals a still primitive but nevertheless expanding economy.

5

Europe about the year 1100

At the turn of the twelfth century Europe was in a condition of rapid change and growth. The incursions of the Northmen had at last been checked. The last wave of Ural-Altaic settlers from the Russian steppe was well on the way towards assimilation, and in the south the drawn-out struggle with the forces of Islam had turned decisively in favour of Europe. Advance was still slow in Spain, but in the eastern Mediterranean European peoples had gone over to an offensive, both commercial and military.

Within Europe itself, the division between the west, the heir of Rome, and the Byzantine south-east, implicit since the third century, had now come into the open, and the formal schism, which developed in the eleventh century, was to continue through the rest of the Middle Ages. On the other hand, the sphere of western civilisation, as if to redress the balance, was at this time being extended into Scandinavia, the middle Danube basin, and eastward across the plains of Poland.

Population within Europe was increasing more rapidly than at any time in the previous thousand years. The forest was yielding to the settler's axe, and the area of land under cultivation was keeping pace with the rising population. Long-distance trade, a product both of mounting population and of a rising level of welfare, was being developed anew, and from Italy to Flanders and from Spain to Poland, small urban communities were coming into existence under the protection of monastery or castle or within the crumbling walls of former Roman towns. Everywhere there was building activity on a scale not seen since the early centuries of the Roman Empire. Europe was, in the words of Raoul Glaber, 'putting on its white-robe of churches'.[1] It was also assuming its more sombre garment of castles and city walls. Merchants and pilgrims were stirring in growing numbers along its roads and in its ports, but so also were the robbers and pirates who lived at their expense. It was a period of great insecurity, but one of high promise which was to be fulfilled some two centuries hence.

POLITICAL GEOGRAPHY

By 1100 the political map of Europe had assumed in its broad outlines the shape which it was to retain until recent years. Dominated by the German Empire and the Kingdom of France, it was ringed by smaller states from León and Castile to Hungary and Poland, each of which reflected some ethnic individuality. One would be very rash indeed to assume that any form of national feeling had received political expression at this date. The age of the nation-state lay many centuries away, but the political frameworks were already established within which national consciousness and the nation-states were to grow.

Most of Europe was at this time divided into states, whose rulers – kings and princes – claimed, if they could not always exercise, political authority in all parts of their respective territories. Only the extreme north of Europe, together with an ill-defined region which extended eastwards from the borders of Poland and Hungary to the effective limits of Muscovy, did not at this time have an established and relatively stable government. The political map, however, is deceptive in many ways. Boundaries were not as finite and political control was not as uniform as they imply. In each state the king or prince was merely the apex of a feudal hierarchy. His prestige he derived from his title, but his power was based upon those lands and resources which he could personally control. Few rulers had the power to control or coerce the whole of the realm of which they were nominal rulers, and it is doubtful whether they knew, except in restricted areas, what were the limits of this legal authority. The king of England was an exception, but his contemporary on the throne of France was master in less than a twentieth of his kingdom, and was virtually ignored in the remainder.

Not only was the ruler weak within the borders of his own state, but those borders were themselves fluid and uncertain. In but few instances had they been reduced to a finite line. The Crusaders, led by Godfrey of Bouillon, in 1096 crossed the river Leithe and entered Hungary;[2] such precision was rare. More often states were separated by zones of varying width and indefinite political allegiance, in which few people lived and resources were undeveloped. In time boundaries would gain precision, but it was a long, slow and pragmatic process. It was not that the boundaries were in dispute; rather, they ran through areas which at this date were unsettled and not worth disputing. On the rare occasions when it became necessary to define a boundary with any degree of precision, it was customary to ask the local people to which side they belonged or to make the boundary conform with some conspicuous natural feature.

The indefiniteness of boundaries was intensified by the fluidity of feudal allegiance. It was possible, under certain circumstances, for a vassal to transfer his allegiance from one lord to another, from one prince or king to another, and thus – if one may impart a modern concept – from one state to another. In many parts of western Europe boundaries which at a later date can be termed 'national' owned their origin at this time to the rivalries and idiosyncrasies of one petty lord or another.[3]

The boundaries of feudal loyalties and obligation could change because there was no sense of nationalism to weigh them down. People – at least those who travelled – were aware of varieties of language and custom. The chroniclers of the First Crusade made frequent references to peoples of unintelligible tongues and strange and barbarous habits. Godfrey of Bouillon was chosen to meditate the disputes that continually arose amongst the Crusaders, 'because he had been reared on the borders of both races and so was acquainted with the language of each'.[4] Even within any particular language or culture group there was no sense of unity or kinship. Peter the Venerable, Abbot of Cluny, for example, could not visualise himself as a member of a French nation; his monastery lay in a region 'sans roi, sans duc et sans prince'.[5] The former ties of tribal kinship had been broken, and had been replaced by the mutuality of feudal duties and

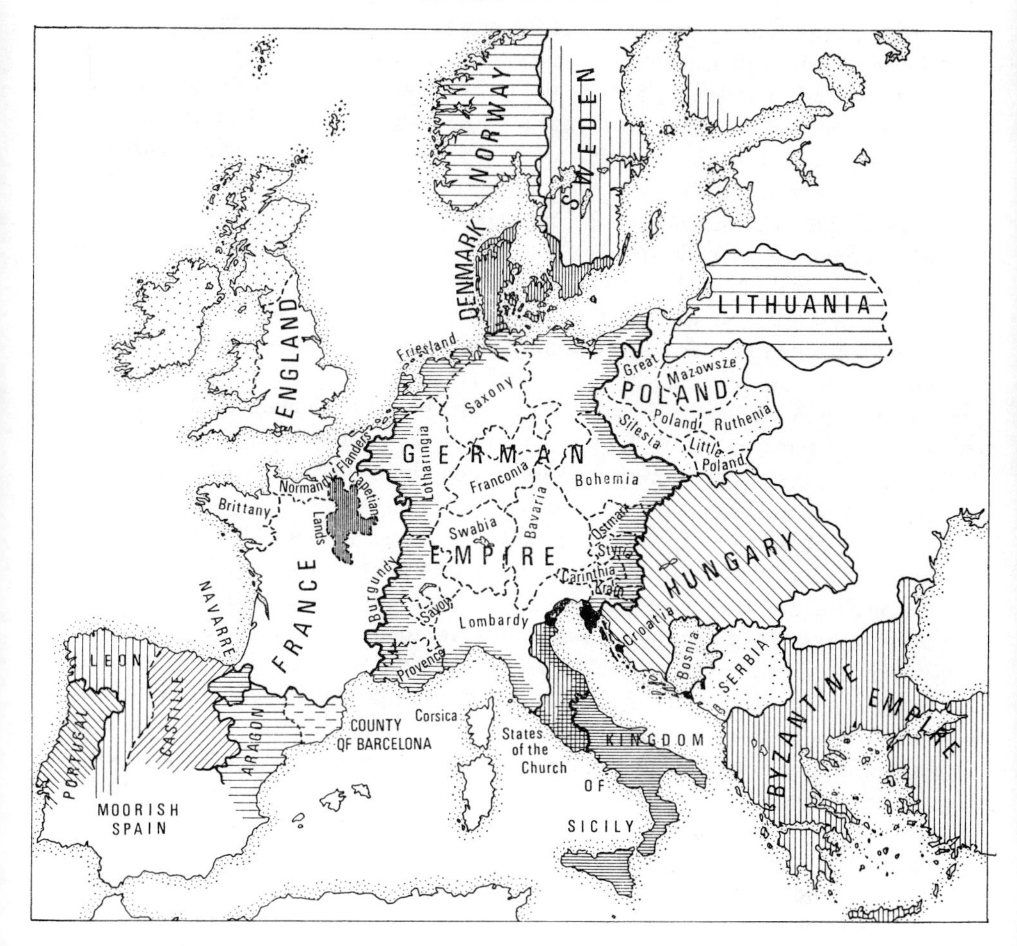

Fig.5.1 Political map of Europe, about 1100

obligations. The compass of people's lives and thoughts was narrow, almost local in its extent. King and prince were to most people remote and ineffectual; what mattered was the occupant of the nearby castle to whom they were bound, according to their status, by obligations to provide labour, to pay rents, or to protect and defend.

The political pattern

However fragmented the Europe of the twelfth century may have been, the fragments were nevertheless grouped into kingdoms and principalities. Their central authority in general was weak; they lacked a machinery that would permit their rulers to supervise all parts of their territory. But these rulers nevertheless had a moral authority, which the Church had confirmed by its act of coronation. A state might dissolve in anarchy, as England did under Stephen, or, on the other hand, its ruler might consolidate his authority and become both lawgiver and guardian of peace and justice.

France. France and Germany were the most extensive and, potentially at least, the most powerful European states at this time. Territorially France was the most westerly of the divisions of the Carolingian Empire which had emerged from the conference held near Verdun in 843, and its boundaries had changed only in detail from those traced by the grandsons of Charlemagne. Within its borders, however, the change was more profound. Authority had passed from the enfeebled Carolingians to the house of Capet, and its focus from the plains of Austrasia to the valley of the Seine. It was an accident of history that Hugh Capet, the first of his dynasty to assume the royal title, was in his own person Count of Paris and master of lands around Paris, with further possessions which extended north-east to Compiègne and south to beyond Orléans. These lands were not particularly rich or valuable, but they were more centrally situated than the personal possessions of the later Carolingians had been. The early Capetians set out to acquire control of the intervening lands, so that by the twelfth century they were masters of a continuous belt of territory from the Aisne valley to Berry.

The royal domains spanned the valleys of the Seine and Loire, and the pattern of rivers, which appeared to converge on the Capetian lands has appeared to some historians to have endowed the expansion of the Capetians with a kind of geographical imperative. 'The rivers in their courses', wrote Haskins, 'fought against the Plantagenets', who at the time were trying to defend their possessions in Normandy. The rivers were much used commercially, but military campaigns were conducted over the intervening plateaus. The wealth of the region lay along the valleys, but these were too damp and too tortuous ever to have guided movement by land, and the roads in fact avoided them. Attractive as possession of these valleys might be, there was no 'movement' along them from the Paris region into the provinces. The expansion of the royal demense until the beginning of the twelfth century had been by the slow erosion of surrounding territories by whatever process – conquest, marriage, inheritance – seemed appropriate.

The demesne was not only a continuous area of land which the king owned and from which he received rents. It embraced also a body of rights and privileges which were at once more immediate and a source of greater profit and power than those which he exercised over the rest of France. Carefully and firmly managed they were a source of power sufficient to permit and encourage their own expansion, but in weak hands they could easily have been dissipated, until power over the royal demesne became as insubstantial as that over the rest of France. The Capets, however, were in the main a strong-minded and vigorous family.

The County of Paris was ringed with similar complexes of lands and privileges which constituted the duchies and counties of France. Down the Seine lay one of the strongest, the Anglo-Norman Duchy of Normandy. To the west lay the counties of Blois, Maine and Anjou; to the east, Champagne, Nevers and Burgundy; to the south, Poitou, Marche and the counties which made up the Duchy of Guyenne. In the extreme south of France, the County of Toulouse extended from the Garonne to the Rhône. Most of these, and of the many other feudal units of France, had begun as administrative areas under the Carolin-

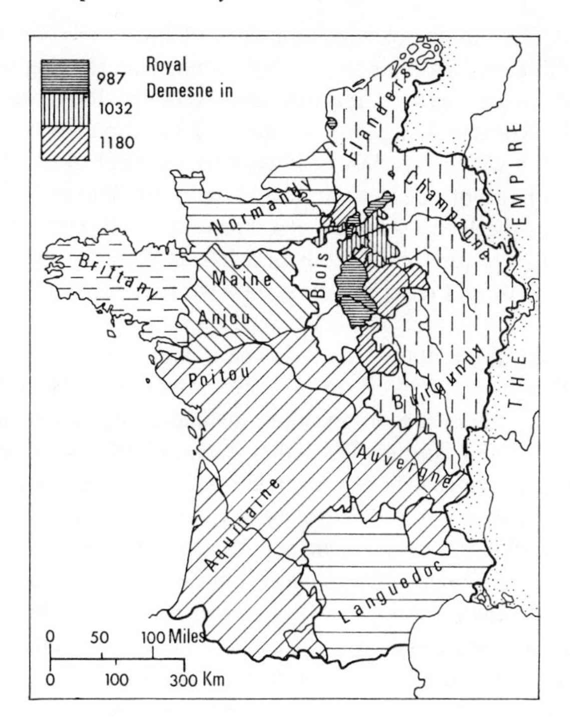

Fig.5.2 Map of France, about 1100

gians and had become hereditary in the families of their appointed *comites.* In each a count or duke was pursuing as diligently as he was able the same policy as the Capets in the County of Paris – strengthening their authority, adding to their rights and privileges and increasing the extent of their own physical demesne. Their success or failure in this is reflected in the changing geography of fiefs and seigneuries.

Within these larger units was an immense number of small, discrete territorial units, with the functions of government, such as they were at this date, discharged mainly at this territorial level. The smaller units nested – though not smoothly and without overlaps – within the territorial boundaries of the larger, and these similarly within the limits of France itself. There was nowhere any sense of French unity. The primacy of the Paris region among them was not inevitable, and at one time would have seemed hardly probable. It was still, about 1100, the petty capital of just one – and that by no means the largest and richest – of the *comtés* of France. But capital cities tend to expand with the expansion of the lands which are effectively controlled from them, and so Paris grew with the power and prestige of the Capets.

Germany. If France lacked both unity and sense of cohesion, the same was even more marked in Germany. The German Empire of the twelfth century derived from the remaining divisions of the Carolingian Empire, the middle realm, or Lotharingia, and the eastern, or Germany in the restricted sense. On the west its boundary was as definite as the eastern boundary of France. To the east it was bordered by an ill-defined frontier in which German settlers mingled with the native Slavs, and in the south it reached into southern Italy, where its marches bordered the Norman Kingdom of Sicily. The empire was a vast, ungovernable area, within which a variety of languages was spoken and a sense of German nationhood had not begun to appear. East of the Rhine and north of the Danube, it lacked the Roman legacy of roads and cities. The Germanic *Stamme* had ceased to have any reality, but had left as their legacy the historic regions – Swabia, Saxony, Franconia, Bavaria – which continued to attract a kind of tribal loyalty.

Political fragmentation in Germany had followed a similar course to that in France. The Carolingian *Gaue*, created for administrative purposes, had been turned into hereditary seigneuries, with, in many instances, the descendants of the original governors as their counts. The tribal duchies had now little more than a nominal existence; the reality of power lay with the feudal states that were developing within them, based on territory rather than kinship.

Amid the growing confusion of the German political map, it is nevertheless possible to discern nuclear areas. These were endowed with more – and better – agricultural land than other parts of Germany; their population was probably more dense, and their towns were larger and more numerous. Wealth was being accumulated more rapidly in these areas than elsewhere, and, since wealth meant power, they were the seats of the most powerful of the German leaders. Foremost among them were the plain of the middle Rhine, home of the Franconian dynasty, and the plains bordering the northern Harz Mountains, from which the Saxon emperors drew their strength. To these should be added the plain of the upper Danube, from Regensburg to Augsburg, and of the Lower Rhine, from Aachen to Cologne.

It is difficult to delimit these regions in the absence of any statistical basis. Perhaps the itineraries of the emperors demonstrate most readily what were the bases of imperial power. Fig. 5.3 shows where the Emperor Henry IV is recorded to have stayed. He was most often in the middle Rhineland – he was, it must be remembered, a Franconian – and, after this, in Saxony and Bavaria.

The German kings were elective and weak. A movement under the Salian dynasty had tried to anchor their authority firmly to a precise tract of land, as that of the French kings was based upon the Paris region. 'What a revolution this would have brought about had it been successful,' wrote Fisher, 'had this homeless German monarchy been firmly brought to anchor in the Harz Mountains, to spread the circle of its power steadily outwards through the German lands.' The attempt failed, perhaps because it came too late, after the territorial seigneuries had become well established; perhaps because it was not pursued sufficiently consistently; perhaps because the emperors themselves were distracted by their contest with the papacy and lured by the mirage of Italian conquest.

The imperial title (*imperium Romanum*) was not altogether an empty one, though it did not, at this time or at any other, designate any kind of power or authority over western Christendom. It was used to indicate the 'German' Empire; there was nothing 'divine and necessary' about it, and in terms of

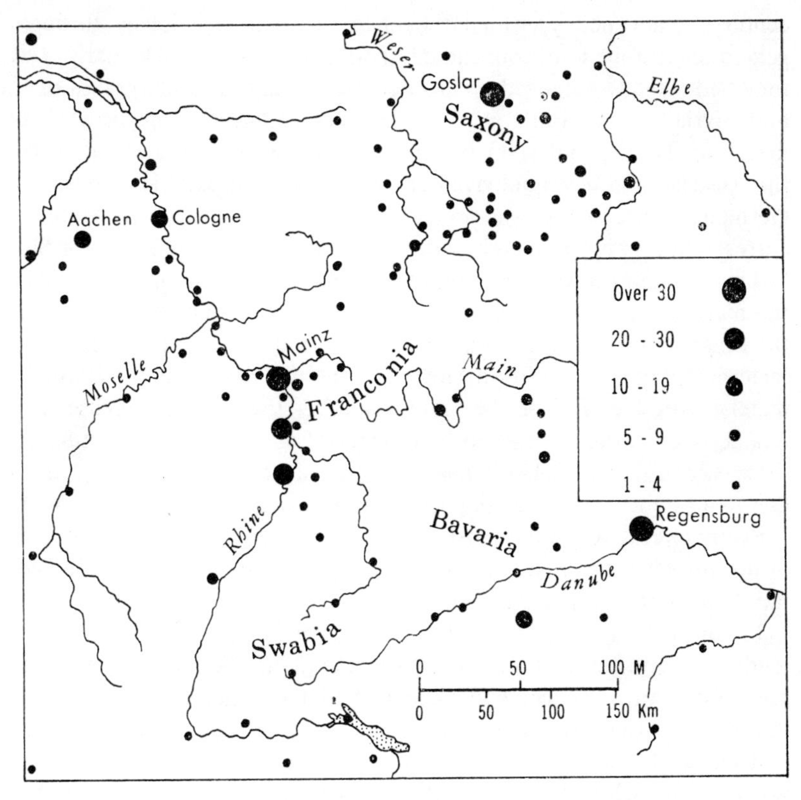

Fig.5.3 The itinerary of the Emperor Henry IV, showing the number of recorded visits to each place

dominion and power it denoted nothing more than an expression like the 'Kingdom of France', and usually a good deal less.

While most of Germany was dissolving into a maze of small states, both lay and ecclesiastical, along the eastern frontier larger and more compact units had been able to survive. These derived from the 'march states of the tenth century. Their nuclei lay along the Elbe valley, but eastward expansion had taken their limits to the Oder and beyond. If the Nordmark, which occupied the base of the Danish peninsula, was by 1100 beginning to break up into small territorial units, the Marches of Brandenburg, Lausitz and Meissen remained large and powerful.

The Kingdom of Bohemia was a Slav state within the German Empire. Its boundaries, drawn through forested mountains, had already begun to assume the shape which they were to retain into modern times. Bohemia remained – what the German duchies had already ceased to be – a state held together by an ethnic bond, and embracing all the Slav peoples living in the plains of Bohemia and Moravia, but is however too early to speak of it as a nation-state.

Bohemia was bordered by the Ostmark, the later Duchy of Austria, still in the possession of the Babenberg family which had been granted it in the tenth century. Their policy, which they had pursued tenaciously through several generations, aimed to consolidate their control of the Ostmark, and in this they had already achieved a considerable success. The margraves of Carinthia and Styria to the south were no less acquisitive and purposeful, but their marches, also dependent on Bavaria at first, were less strategically placed than the Austrian and never achieved an importance comparable with the latter. The German Empire, lastly, was protected on the south-east by the March of Krain, corresponding approximately with Istria and the later province of Slovenia.

The boundary of Germany on the east had no precision; it extended as far as the marcher lords were able or willing to extend it. The western boundary, on the other hand, ran in general through more densely settled territory than the eastern, and greater precision came to be thought desirable. It accorded approximately with the line of the 'four rivers' – the Scheldt, Meuse, Saône and Rhône – and came in time to be thought of as following the actual courses of these streams. The Scheldt basin, at the northern extremity of the Franco-German boundary, was at this time an area of rapid economic growth, and, if it had not yet become the most densely populated and highly urbanised region of north-west Europe, one might nevertheless discern in its existing conditions the shape of its future development.

In contrast with the rest of France's eastern border, the boundary was here tending to diverge *from* its theoretical line along the river Scheldt, and in two areas Germany in effect had expanded at the expense of France. The more important of these was the Ostrevant, a small territory, between Valenciennes and Douai, bounded by the Scheldt and its left bank tributary, the Scarpe. It was a well-developed and valuable region on which the counts of both Flanders and Hainaut cast eager eyes. In the later years of the eleventh century it had become a fief of Hainaut, which was in turn a fief of the empire.

Imperial pretensions to the marshy regions lying roughly between the lower Scheldt and the city of Ghent went back a great deal farther. The land of Waes, which lay across the Scheldt from the future site of Antwerp, together with the Quatre Metiers, which lay to the west, was established in the tenth century as an imperial march, and remained indisputably a part of the empire. On the other hand the County of Alost, which lay to the east of the Scheldt and unquestionably within the German Empire, passed into the possession of the Counts of Flanders, who were vassals of the king of France.

The County of Flanders thus came to span the French–German boundary, and the count himself to be vassal of both the king of France and the German emperor (see fig. 5.4). Flanders was bounded to the south by Artois, beyond which lay the growing power of the king of France. If the count wished to increase the extent of his country and to add to its wealth, this was possible only by expanding eastwards into Brabant and Zeeland. This is, in effect, what the counts were attempting to do and their policy contributed to the prevalent unrest in the Low Countries.

Instead of breaking up into a number of lesser seigneuries, most of Flanders was divided into *Châtellenies,* small districts, each centring in a castle and ruled by a castellan appointed by the count. The castles originally were built, like the 'boroughs' of the English Midlands, to deter the incursions of the

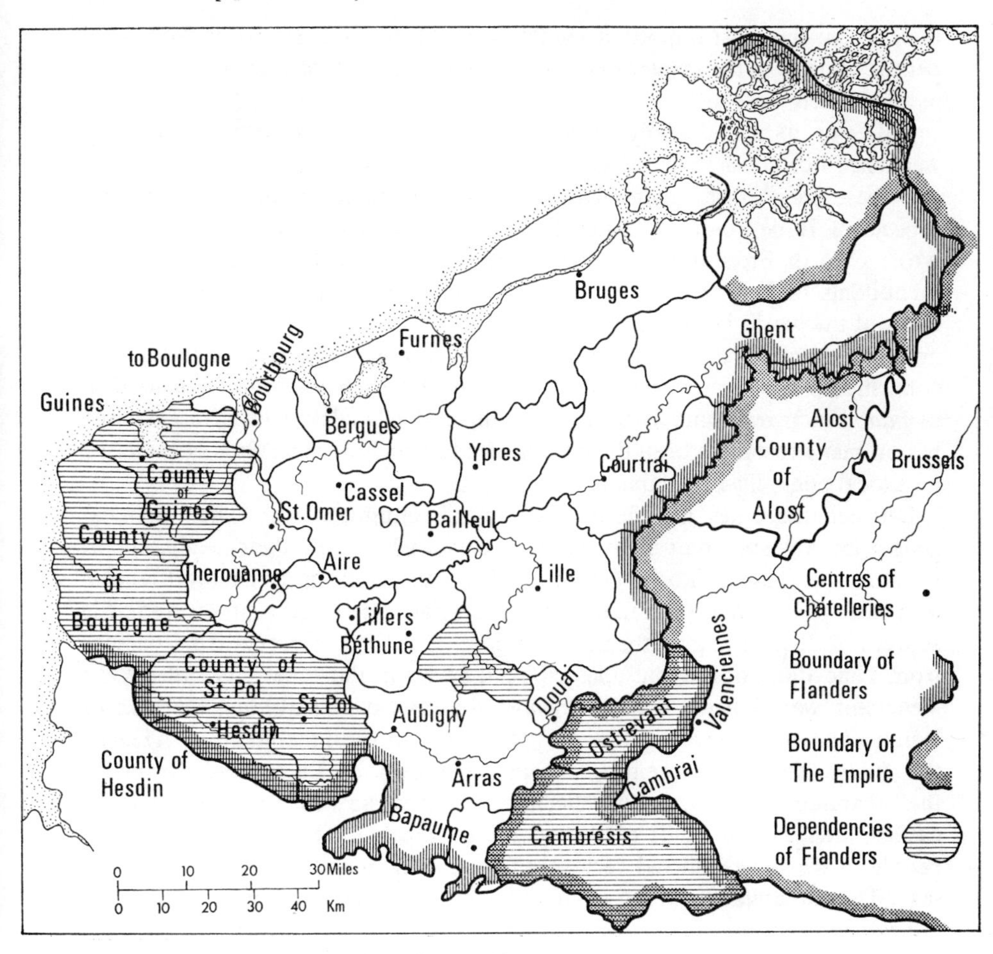

Fig.5.4 Map of Flanders, about 1100

Norsemen; they served also to reinforce the authority of the count and eventually to form the nuclei around which most of the Flemish cities were to develop.

Italy and Burgundy. The German emperors claimed authority over the kingdoms of Italy and Burgundy, but their power was feeble in the former and non-existent in the latter. The emperors were as a rule crowned as kings of Italy with the iron crown of the Lombards, increasing their prestige but adding little to their real power. In Burgundy, however, the royal power had disintegrated with the rise of powerful seigneuries within the kingdom, and the death of the last nominal king in the mid-eleventh century and the inheritance of his pretensions by the German emperor made no difference to the actual situation. Within Burgundy power rested with the Count of Burgundy, whose authority coincided approximately with the later Franche Comté. The Count of Arles was master of much of Provence, and the Count of Savoy in the western Alps. The last was guardian of those alpine passes which were most used and, in virtue of this, one of the most important feudal rulers of the time. It was he who could permit or deny to the emperor admission to his kingdom of Italy, at least by the more westerly passes.

Italy was, as it had always been, a mosaic of city-states. City life had been weakened but not destroyed by the invasions of the Ostrogoths and Lombards, and these peoples had themselves been assimilated to urban life as to other aspects of Italian culture. 'Almost the whole country pertains to the cities,' wrote Otto of Freising late in the twelfth century, 'each of which forces the inhabitants of her territory to submit to her sway... They surpass all other cities of the world in riches and power.'[6] As king of Italy the emperor was the nominal master of the north Italian cities; occasionally he granted to them or confirmed a charter of privileges, but in general each was master in its own *contado*. Such restraints as bishops or imperial officials might have exercised had in many instances been thrown off by 1100, and in each rule was exercised by a local oligarchy or *commune*.

Between the cities was fierce and almost continuous war. Their political geography was such that many could, if they wished, hold their neighbours to ransom, and their territorial growth, their competition for agricultural land to feed their people, and their need for commercial outlets and routes were fertile sources of dispute among them. The record of their wars reads like a passage from Thucydides or Polybius, describing the wars of the Greek city-states. 'The Venetians waged fierce warfare against the people of Ravenna, the men of Verona and of Vicenza against the Paduans, and the inhabitants of Treviso, the Pisans and the Florentines against the men of Lucca and Sienna... the inhabitants of Verona and Vicenza ... laid waste with fire and sword the castles, villages and fields of the people of Treviso. The Venetians and the people of Ravenna inflicted very great damage upon one another by land and sea. The inhabitants of Pisa and Lucca ... waged war with unwearied frenzy ...'[7]

The number of such city-states, even at the beginning of the twelfth century, was considerable. The average distance between them was only from 20 to 25 miles, 'so that even after the *contado* had been wholly subjugated, the conquered territory never reached much further than ten to twelve miles from the city walls. To go beyond was to enter foreign and unfriendly territory.'[8]

Nevertheless, a feudal society of more orthodox type existed in some parts of Italy. Territorial seigneuries, both lay and ecclesiastical, were to be found in the more westerly parts of the Lombardy plain and in its enclosing foothills, as well as in many parts of the Apennines. But the *contado* tended always to absorb both seigneurie and seigneur, and the latter merged with urban aristocracy, exchanging his rural castle for a tall, fortified tower within the city walls and conducting his feuds in the city's streets. Among the areas where the anarchy was intra- rather than inter-urban was the Patrimony of St Peter, within which the feudal overlord was the Pope himself. An elected ruler, chosen from among the turbulent nobility of Rome and holding office usually for only a few years, he was in no position to create a strong government, and barely prevented the city of Rome itself from going the way of the communes of northern Italy.

Southern Italy, beyond the Marches of Ancona and Spoleto, was still, at least in name, a part of the Byzantine Empire, but Apulia and Calabria, as well as Sicily, had been occupied by the Normans, and in Benevento, Salerno, Capua and other towns there were Lombard counts who carried on a desultory war with the Normans. Only around the coast did a few cities continue to recognise the pretensions of the Byzantines for the sake of commercial privileges within the Eastern Empire.

The marginal states

The Kingdom of France and the German Empire, together with its Burgundian and Italian appendages, had very roughly comprised the Carolingian Empire. Bordering them to the south and east was a belt of states which formed the frontier of Europe in this age. European commerce and culture; European methods of civil and military organisation; of land tenure and urban development, were spreading slowly into them as Islam retreated in the south, and Slavs and Magyars in the east were won over to western ways.

Spanish peninsula. In 1099 the Cid, the legendary hero of the wars between the Spaniards and the Moors, died, and an era of Christian expansion was brought to a close. A few years earlier both Toledo, the ancient Visigothic capital, and the port-city of Valencia had been captured by the Spaniards. But their advance was now halted, and Valencia was lost early in the twelfth century. A strategic desert separated the two sides and extended from Portugal to the Mediterranean. Here, in the words of the medieval Spanish epic:

> The land is poor, gaunt and barren.
> Every day Moors from the frontier
> And some from beyond kept watch on my Cid.

Warfare and raiding were continuous:

> As far as Alcala went the banners of Minaya,
> and from there with the spoils they return again,
> up along the Henares and along the Guadalajara.
> Such great spoils they bring back with them,
> many flocks of sheep and cattle.[9]

North of this no-man's land within which settlers from the north were beginning to appear, lay the Christian kingdoms of Spain, alternately fighting against the Moslems and feuding among themselves. In 1072 Leon and Castile, not for the first time, were joined, creating a state of considerable power which extended from the western Pyrenees to the Atlantic coast and from the Bay of Biscay to the Tagus.

There was, it has been argued, a Spanish national sentiment at this time, more developed and articulate than in France.[10] Leon had inherited from the Visigoths and Romans the ideal of an undivided Spain, and claimed primacy among its Christian states. This claim was not undisputed, but tradition and the ever-present forces of Islam nevertheless imposed on Christian Spain a higher degree of unity than was present in France. The King (*imperator*) of Leon and Castile enjoyed in fact a kind of hegemony over the little Pyrenean states. He had for many years been encroaching on the Basque kingdom of Navarra, and at the end of the eleventh century had made a successful attack on Zaragoza.

East of Navarre the Carolingian Spanish March had given rise to a number of small Pyrenean counties. A group of these lying in the central Pyrenees had come to be grouped under the King of Aragon, while to the east the Count of Barcelona had extended his authority over the mountain counties of Cerdaña, Conflent and Besalu, and had expanded south-westward to Tarragona. Beyond the Ebro, the Cid had himself captured Valencia, and at the time of his death is said to have been struggling to hold the city and its surrounding region against renewed attacks by the Moors.

Al-Andalus. Moslem Spain, united and militant in the Carolingian period, had begun to disintegrate even before the Caliphate of Cordoba came to an end in 1031. Thereafter a mosaic of petty states, or *kuwar*, emerged to the number of perhaps as many as twenty-one. They had been units of local administration under the caliphs, and consisted each of a central city (*hadra*) and its surrounding region. Some became cultural centres, and those which lay close to the Mediterranean engaged in seaborne commerce. They were pacific and tolerant, content to co-exist with the Christian states provided the latter would tolerate them. This they would not always do. The governors of the *kuwar* had recourse at the end of the eleventh century to Berber tribes from North Africa, more fanatical and ruthless than themselves and incomparably less civilised. These were the Almoravides, who in the last decade of the eleventh century and the early years of the twelfth, overran, pillaged and united Moorish Spain; regained part of what had been lost in the campaigns of the Cid, and for a generation stemmed the advance of the Christian states of the peninsula. Thus at the beginning of the twelfth century, though only for a short period, a united and fanatical Al-Andalus looked northwards across the desert frontier to a group of weak and divided Christian states.

Southern Italy. In southern Italy and Sicily the situation was more complex, because the forces playing upon the region were more varied. Italy, south of the Papal States, had been, until early in the eleventh century, a group of petty duchies, counties and independent cities, under the control of Lombard rulers. The Byzantine Empire still claimed sovereignty over them, but had little authority beyond what sprang from the goodwill of the coastal cities, whose commerce it encouraged. Most of Sicily was in the hands of Moslem emirs, who quarrelled amongst themselves as violently as the Lombard princes of southern Italy. Into this situation had come bands of Norman invaders, early in the century. They had taken sides in the local disputes, profited from the political divisions, and made themselves masters of Apulia and Calabria. They then nibbled away at the Lombard principalities to the north until they controlled Benevento, Capua and Salerno, and had wrested from Byzantine control the ports of Bari and Amalfi. Late in the century they turned to Sicily, employing the same tactic of intervening in the disputes of local rulers until they could effectively control the whole.

At the beginning of the twelfth century southern Italy and Sicily still formed two separate states under Norman lordship, the former weak and divided; the latter, recently conquered from the Saracens and about to pass into the strong hands of Roger II. Both were divided into lordships, each controlled by vassals of the Norman dukes. Conquests made by the Normans on the opposite coast of Epirus had mostly been lost, though they continued to hold some of the Ionian islands, and their bridgehead of Durazzo was used by one of the crusading bands on its eastward march to Constantinople.

The Byzantine Empire and the Balkans. At the beginning of the twelfth century the Byzantine Empire was ruled by Alexius Comnenus, the first of his distinguished family to occupy the imperial throne. Though threatened in Anatolia by the advance of the Seljuk Turks and occasionally raided by the Pechenegs from the Russian steppe, the empire is generally thought to have had a greater unity and, perhaps, prosperity at this time than it had had for several generations. It covered the more westerly as well as the northern coastal regions of Anatolia, whose interior formed at this time the Moslem Sultanate of Iconium. It included Cyprus and the Greek islands, and in Europe extended northwards to the Danube and westwards to the Ionian Sea and the borders of Zeta (Montenegro) and Raška (Serbia). Here direct rule by Byzantium ended and in the highlands of modern Serbia, Montenegro and Dalmatia the loosely knit state of Zeta was in practice independent. On the other hand, the Bulgarian state, which had formerly extended eastward from the borders of Zeta to the shore of the Black Sea, had collapsed almost a century earlier. The Bulgarian provinces were now under Byzantine occupation, but the memory of the First Bulgarian Empire had not faded, and within the century was to inspire a renewed and successful Bulgarian revolt and the creation of the Second Bulgarian Empire.

The Danube basin. The northern boundary of the Byzantine Empire at this time was formed by the river Danube from its mouth upstream approximately to the junction of the Sava. Beyond the lower Danube, in the plains which today comprise the provinces of Walachia, Moldavia and Bessarabia, lived the Pechenegs and other Tatar tribes, while in the mountains were the remains of the ancient Thracian and Illyrian inhabitants of south-eastern Europe. These were unruly neighbours of the Byzantines, who had frequently to protect the Danube crossings against them. Not infrequently the Pechenegs acted in concert with the Hungarians, whose state at this date filled out much of the plain which now bears their name. For a century they had been a turbulent people, but their raids into both the German and the Byzantine Empires had now become infrequent. On the west, their boundary was stabilised along the Leithe river, where it was to remain until 1918. To the east they had expanded into Transylvania. Within the plain the Hungarians had by and large abandoned their nomadic way of life, and had made the Dunántúl, the region lying to the west of the Danube, the centre of their state. Here, in Székesferhérvár, they established the seat of their government, while their area of settlement and rule extended up the Nitra and Váh valleys to an indefinite frontier with Poland in the Carpathian mountains.

Late in the eleventh century Hungarian power began to extend south-westwards into the hilly country beyond the Drava. Here, with its focus in the upper Sava valley, lay the Kingdom of Croatia, a weak and divided power, with a social structure that was still basically tribal. In 1097 Kálmán of Hungary overran Croatia, and was crowned king. Then, urged on by a desire to reach the sea and to control the ports near the head of the Adriatic, he occupied Dalmatia and most of the offshore islands. To the commercial cities of Zadar, Trogir and Split he granted privileges so extensive that their previous independence was but little curtailed. In the hills of the interior the clan structure of society continued, and effective power over most of Croatia resided, not in the king, but in the *Župani* of the many clan-districts. Hungarian armies also invaded Bulgaria, but in this direction they failed to hold their conquests. It is probable, however, that groups of Magyars settled in the Balkans (see below, p. 249) and were assimilated in time by the Slavs.

Poland. About 1100 Hungary was united under the leadership of the house of Arpád. The same cannot be said of Poland. The Polish state had come into being almost a century and a half earlier. The core from which it developed lay in the region of boulder clay and glacial lakes, almost enclosed by the branches of the Warta, which had since been known as *Wielko Polska*, 'Great Poland'. Its limits were clearly extended to the line of the Oder and Krkonoše hory on the west, and eastward to a broad and indeterminate frontier zone which lay roughly along the valley of the Bug and extended southwards across the headwaters of the Dnestr to the Carpathians. On the south the boundary, also indefinite, followed the course of the Carpathian mountains.[11]

The state thus lay almost wholly in the northern plain of Europe, and it has become a commonplace of historians that the lack of mountain barriers and the conquest ease of movement within the limits of the plain exposed Poland to invasion and rendered her existence precarious. Such a view, however, is not supported by contemporary sources, which continually represent the network of river valleys as marshy and very nearly impassable.[12] It would probably be nearer the truth to regard the repeated partition and division which medieval Poland underwent as, in part at least, a consequence of these physical barriers to movement and communication within its boundaries. It was, in the words of the anonymous Gallic chronicler, so studded with lakes and marshes that none could lead an army through them.[13]

The Germans regarded Poland as a fief of the empire, and if the Polish king paid no homage to the emperor, he at least refrained from calling himself 'king'. Partitioned early in the eleventh century; re-united by Kazimierz Odnowiciel, 'the Restorer', it was again divided between Władysław Herman and his sons. The basis of division was not feudal, as in Germany; it was rather based upon the strength of local feeling within well marked physical regions. Five such regions were distinguishable in 1100, each with its own ruler who rendered little more than lip-service to the Polish prince: Great and

Little Poland, with their capitals respectively in Poznań and Kraków; Mazowsze, with its centre at Płock, and the provinces of Pomorze (Pomerania) which had never been more than nominally a part of the Polish state, and Silesia, to which the Bohemian king had laid claim. Such divisions could however be repaired almost as quickly as they were made; by 1107 most of Poland was again united under Bolesław Krzywousti, or 'Wrymouth', who soon afterwards set out to bring Pomorze more effectively into the Polish state.

Pomerania at this time was a thinly peopled and primitive region where, behind the protection afforded by the marshes of the Oder, Warta and Noteć, the primitive paganism of the Slavs had been little touched by Christianity. By 1121 Polish armies had penetrated its forests, captured its chief city of Szczecin and brought the local princes to submission. Polish command of Pomorze proved, however, to be short-lived, and it was German influences from the west rather than Polish from the south that ultimately overwhelmed it. East of the Pomeranians lived the Prussians, Lithuanians and Estonians, still organised into tribes and almost untouched by western culture.

Scandinavia. The heroic age of northern Europe had ended, and the three Scandinavian kingdoms had already assumed the approximate shape which they were to retain until the seventeenth century. Denmark, economically and politically the most advanced of them, extended northwards from the Eider river to the extremity of Jutland, and embraced the Danish islands and, to the east of the Danish Sound, the provinces of Scania, Halland and Blekinge.

Sweden was more backward. Though Uppland continued to be the most densely peopled and politically the most important part of Sweden, the country was far from united around it. Sweden at this time was 'a collection of semi-independent provinces administering their own laws and recognising little more than the nominal overlordship of the king'.[14] To the south of this core area of the Swedish state – Svealand, as it was called – lay the forested and thinly populated uplands of Götaland and Småland. These regions were at this time being opened up, and a strong rivalry developed between them and the longer-settled lowlands to the north. Around the year 1100 it was Gotland which was dominant, providing the Swedish king and extirpating the last remains of paganism in its rival provinces. To the north of the central Swedish lowlands stretched the forests of Dalarne and Värmland, thinly peopled, remote from the centre of Swedish political power, and inhabited by Lapps, against whom the Christian Swedes were occasionally moved to wage a campaign for the salvation of the former and the profit of the latter.

Norway was at this time more peaceful but hardly less divided. Magnus III ruled from Trondheim a kingdom which in name at least extended from the Gota river to Finmark, reached inland to 'the ridge of the country',[15] and embraced also the Orkneys, Shetlands, the western isles of Scotland, and Man. In fact, however, development had been local or regional, and the units that mattered were the clan-like groups which lived along the shores of the fiords. A larger political grouping first appeared among the Trönders, who lived around Trondheim fjord, where cropland was more extensive and communication easier. This region of Tröndelag was the nucleus of medieval

Norway, but secondary concentrations of population and of power, each with its Christian bishop, were to be found at Bergen, Stavanger and Oslo. Between the small nuclei of human settlement which lay along the fjords the land was, in the works of the twelfth-century writer, Saxo Grammaticus, 'craggy and barren ... beset all around by cliffs, and the huge desolate boulders give it the aspect of a rugged and gloomy land'.[16]

The authority of the Norwegian and Swedish kings faded out in the forests and fjelds of northern Scandinavia, where the scanty Lappish population, still practising a late Stone Age culture, was organised in small clan groups, and lived mainly by the chase.[17] In Finland and the Baltic provinces the institution of the state had not yet appeared; this was politically a no-man's land between the Swedish and Polish states to the west and the remote principality of Muscovy to the east. Society was still in the main tribal, though modified in some degree by the impact of the Vikings, and its economy was pastoral and self-sufficing.

POPULATION

Most of Europe's population at this date lived and worked on the land which provided its sustenance. Crafts were practised and long-distance commerce was carried on, but these employed only a minute fraction of the total population. Yet this was a period of great creativity. Cathedrals, monasteries and humble parish churches were rising from one end of Europe to the other. Kings and princes were perfecting the mechanism of government, while their feudal aristocracy was building castles and their merchants and burgesses were raising walls to protect their developing urban settlements and institutions. This activity was paid for by the labour of serfs and bondmen, by the rents and dues of free tenants, by the exactions for the privilege of using mill and brewhouse; in short, by the surplus that accrued from the successful cultivation and use of the land. The more fertile the soil and the more profitable its use, the greater was the surplus and the more ambitious the building could be.

The natural endowment of Europe was very uneven. Areas of great natural wealth alternated with areas of poor soil and unfavourable physical conditions, and by and large, areas of superior agricultural wealth were regions of denser population and more conspicuous cultural and political growth. Wealth, in general, was consumed where it was generated, because the mechanism for its transfer over more than short distances was undeveloped. A map of building activity is, by and large, a map of the capacity of Europe to generate surplus wealth, and it accords more or less with a map of general fertility.

We thus have a number of hearths or core areas, in each of which the agricultural resources permitted the formation of an agricultural surplus and the creation of more sophisticated institutions of government and the erection of larger and more elaborate buildings. These activities were, with the addition of land reclamation, almost the only forms of investment open to most people. Between these hearths were regions less populous and less developed both culturally and politically, for the evident reason that their agricultural potential was smaller. Not until later in the Middle Ages do we find wealth generated in one area used on more than a very small scale for investment and development in another.

Several such cultural and political hearths are self-evident on the basis of their early urban growth, their building activity, and other symbols of greater productivity and wealth. A list would include parts of the Paris basin, of Burgundy and of the Lombardy plain; Provence, Flanders and the middle Rhineland; the loess region of Saxony; Great Poland, and the plain of northern Bohemia. The political capitals of the time lay with scarcely an exception in areas of developed agriculture, around which and on the basis of which their respective states were built.

Agricultural wealth was of little value without the population to develop it, and there are no sources which provide any kind of a basis for estimating the population of this 'préhistoire démographique'.[18] Everything suggests, however, that population was increasing – perhaps rapidly – during these years. The expansion of urban settlements, the spread of cultivated land at the expense of waste, the movement of settlers from France into Spain and from Germany into Slav lands, and of the Crusaders themselves to the Middle East, all suggest a steady population growth. The size of Europe's population probably reached its lowest point in post-Roman time during the period of Norse and Saracen inroads, and by the late tenth century had begun again to increase. The increase, however, was uneven. Moorish raids probably continued to restrict the population of the coastal regions of Provence, as the Norman conquest must have reduced that of southern Italy. On the other hand, the evidence – wholly circumstantial – suggests a rapid growth in the Lombardy plain.[19] In parts of southern France and in northern Italy the fragmentation of farm holdings had become excessive and rural over-population was considerable before migration, urbanisation and a reform of the tenurial system again brought prosperity to the countryside.

In Burgundy growing prosperity and wealth are demonstrated not only by the extension of cropland but also by the increasing size of cities and the building of churches and monasteries. In northern and western France, where the density had never been high and the population had been reduced by the raids of the Norsemen, new settlements were being created on a lavish scale. Population growth was such that even unpromising heathland, like the Gâtine of Poitou, was being colonised. Here the eleventh century was 'a time of colonization and expansion which dwarfed in importance all other contemporary events. Beginning slowly and sporadically, it gained its greatest momentum at the end of the eleventh century, and then gradually subsided at the end of the twelfth'.[20] There were many areas in Brittany and Normandy, in the Auvergne and in the wooded ridges of Lorraine with a not dissimilar history of settlement and population growth. But there were also regions which continued to resist settlement. Among these was the Ardennes. Here, in the words of the eleventh century Egbert of Liège:

> Ardennam valles et cingunt ardua saxa
> In pluviis heret, nimio sub sole fatiscit;
> Iudicio nostro semper cultore careret,
> Nullus carorum cupiat fieri incola terrae![21]

The valleys of the Vosges and Black Forest were only beginning at this time to be penetrated by human settlement. Further to the east, areas of untouched forest were more extensive, and a belt of unsettled country extended from the Bohemian hills through the Thuringian forest and the Harz to the northern plain. Beyond it the Slav lands of the Elbe and Oder valleys were even less populous than those of western Germany. German settlers were beginning to appear around Leipzig and Meissen, and groups from the Low Countries were beginning the work of dyking and reclaiming the marshy lands along the estuaries of Lower Saxony and Mecklenburg,[22] but the chronicler was undoubtedly exaggerating when he wrote of the 'innumerable multitude of different peoples' who came to the land of Wagria (Holstein).

In the Slav lands hitherto untouched by German settlement population was even more sparse. Otto of Bamberg described the 'nemus horrendum et vastum'[23] through which he travelled for the space of seven days on his journey to the pagan Pomeranians, and doubtless, travellers in eastern Poland and the Baltic region would, if they had left a record of their journeys, have described this wilderness in even stronger terms. Settlements were but islands in a sea of forest, numerous and occasionally large in Great Poland, Silesia, Little Poland and close to the Baltic coast, where Rügen was at this time 'insulam parvam sed populosam',[24] but elsewhere few and small.

Any estimate of the density of population of Europe at this time is at best subjective. That it was low compared with the density in the fourteenth century is obvious; that it was considerably higher, at least over much of the area, than during the Carolingian period is probable, but any quantitative measure is lacking except for England. Lamprecht attempted to derive a figure for the total population of the Moselle region at this time from the increase in the number of settlements recorded in contemporary sources between 800 and 1237.[25] He assumed that the average village size increased from about 20 in the Carolingian period to about 250 in the thirteenth century. At the same time their number rose from 340 to 1380. Lamprecht and those who have manipulated and used his figures[26] have inferred a steeper rise of population than is likely to have occurred both by underestimating the number of settlements at the earlier date and overestimating village size at the later. Lamprecht's data are of such uncertain value for the computation of population densities that it is better not to use it at all for this period.

At this same time the Domesday Book suggests a density of about 10 per square kilometre for England, and Russell, arguing that the French population trend was analogous to that in England, suggests a total of 6.2 million – a density of a little over 11 per square kilometre – in France within its 1794 boundaries.

The population of central Europe was unquestionably very much lower. If Lamprecht's figures are generalised for medieval Germany one has a total of about 2.5 million, a density of 5 per square kilometre. This is consistent with Beloch's suggestion of 3 million for the territory of pre-1914 Germany.[27] For the Slav lands and Scandinavia one can only formulate an estimate on the basis of non-quantitative data; the vast extent of virgin forest, the low level of the economy, and the primitive techniques of agriculture. The population is unlikely to have exceeded 1.5 million in the former and 1 million in the latter.

Polish scholars, working backwards from late medieval data, have suggested that the population of Poland was about half to two-thirds of a million about the year 1000, and somewhat larger in 1100. In any event its density was considerably less than 5 to the square kilometre.[28] The slight evidence for Hungary suggests overall densities that were but little higher.[29]

The most populous areas of Europe at this time were Spain and Italy. The Arab geographers[30] provide evidence for the size and prosperity of Spanish cities, the importance of their commerce, and the intensive agriculture carried on in their surrounding *huertas*. Though this was most marked in the Moslem regions of Spain, there is evidence of expanding settlement in the northern areas recently reconquered by the Christians. Russell has suggested that the population of the whole peninsula was about seven million in 1000; a century later it may have been about 7.5 million.

In Italy the development of population was probably more rapid than in Spain, as demonstrated by the growth of cities and the intensification of trade. Growth was most rapid in the north, where cities were larger and commerce more developed than elsewhere in Europe at this time. A population of about 6 million is consistent with the estimates furnished by Russell. If one adds at most 3 million for the population of the Balkan peninsula, for which, in fact, there are no acceptable data, one has, for the continent of Europe as a whole, a total of about 30 million, an average of about 6.5 to the square kilometre.

The ethnic map

In its broad outline the ethnic map had already taken shape in the ninth century, and there had occurred since then no migrations capable of greatly modifying it, except the invasion of the Magyars. In detail, however, ethnic boundaries had changed, and everywhere the process of assimilating minority groups had progressed. Writers appear to be increasingly conscious of linguistic diversities, and appear to recognise that there were in some areas well-defined linguistic boundaries.[31] Nothing, however, brought home to west Europeans the ethnic variety of the continent more forcefully than the First Crusade. The crusaders met with a number of unfamiliar language groups and the several chronicles of the First Crusade show at least a realisation of the ethnic diversity of the continent.

The Romance language area. The principal language boundaries in western Europe had shown little change. Indeed, it is likely that the division between Romance and Germanic speech had in fact become more definite in France, while the islands of Germanic language which had existed in northern Italy three centuries earlier had by now probably disappeared. The west European region of Romance languages was not expanding, except in Spain, and internal colonisation was able to absorb its increase in population. Colonists from southern France were, however, moving into northern Spain, where the wars with the Moors had devastated and depopulated large areas of land. French-speaking colonies were established in the Ebro valley,

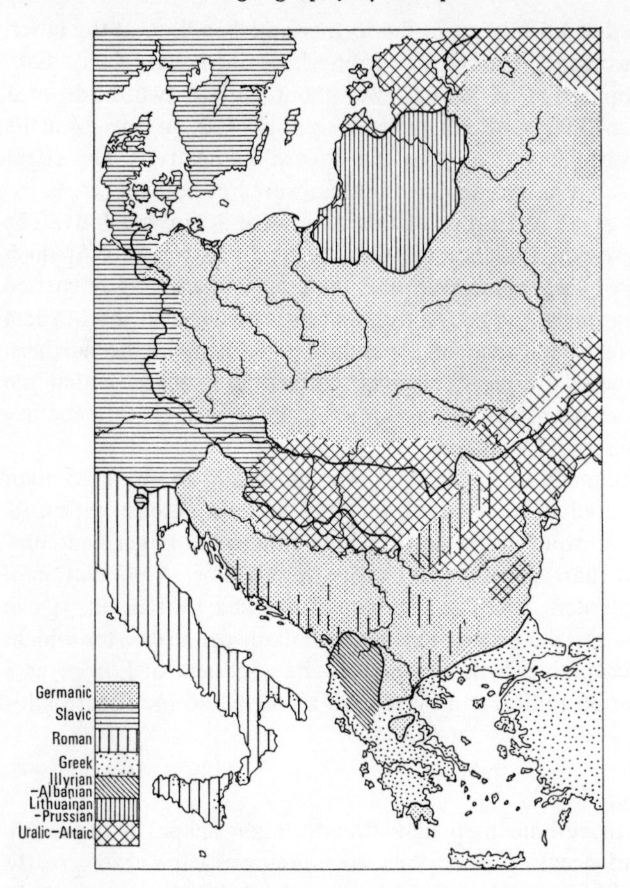

Fig.5.5 Ethnic map of central and eastern Europe, about 1100

especially in Huesca, Zaragoza and Tudela, where they were quickly assimilated by the local population. Frenchmen also settled along the pilgrim route which led Campostella in Galicia, but only in Navarre, where the French settlers had been given the right to use their own language, was their speech able to survive. These were joined after the ninth century by a large number of Mozarabs, Christian and probably romance-speaking refugees from the Moslem region to the south. They were, like most political and religious refugees, *revanchistes* and intolerant, and they contributed greatly to the bitterness and cruelty with which the subsequent struggle was waged.

In Christian Spain the Castilian dialect was developing into a dominant language, though Basque remained the established language in Navarre, and Catalan, with its close affinities with Provençal, was generally spoken in the eastern Pyrenees and Barcelona. South of the *tierra despoblada*, the battleground of Christian and Moor, the situation was more confused. The bulk of the population remained, as it had long been, an amalgam of Iberian, Celt and Goth, of Phoenician, Vandal and Latin, with the language of the last providing the basis of the local speech. The Arabs who had led the Moorish invasion of Spain in the eighth century formed only a small minority, and were greatly outnumbered by the Berbers who had joined their forces during the advance through North Africa. By 1100 the Arabs formed a land-owning but urban aristocracy. They were largely responsible for the cultural and architectural developments of the time, but in any ethnic sense they had little influence on their Berber co-religionists and none on the Mozarabs who retained their romance dialects and their Christian practices. The Berbers – the Maghribins, as they were called – on the other hand came in greater numbers and tended to form a rural proletariat in southern Spain. Their Berber language, spoken by neither the Arabic ruling classes nor by the local population, tended to disappear, and the Berbers were in time assimilated to the Romance-speaking Spaniards. The communal complexity was further intensified by the Muwallads, romance-speaking Spaniards who had been converted to Islam; by the not inconsiderable number of slaves, many of them acquired in the course of warfare with the states of northern Spain, but some deriving from all parts of the Mediterranean world and even sub-Saharan Africa, and by the Jewish communities which were both numerous and large (see below, p. 252).

The Arab–Berber incursions into Sicily and southern Italy were on a smaller scale than into Spain, and their consequences were less far-reaching. Nevertheless, Sicily had fallen, on the eve of its conquest by Roger, under the control of Arab and Berber emirs, who had arrived in numbers sufficient to impose some aspect of their culture if not also their religion. The Moslem impact was less marked in southern Italy. It is possible that Greeks, who formed large colonies in Calabria and eastern Sicily, were in fact of greater ethnic significance than Arabs and Berbers (see below, p. 251).

The Norse incursions into Romance-speaking Europe were on a more limited scale even than those of the Arabs and Berbers. They had, however, been sufficient to create the Norse state of Normandy in the lower Seine valley and to endow the region with a dense pattern of place-names with Nordic affinities. But the Norse settlers, like so many Germanic invaders of the romance lands, remained politically dominant, but became culturally assimilated. It is very doubtful whether the Norse language was represented about 1100 by more than a few loan-words in the prevailing Norman-French speech of the descendants of the Norman settlers.

It was from this Franco-Norse state that settlers moved in the eleventh century to both England and southern Italy, and from the latter to Sicily, Epirus, Macedonia and the Middle East. The band of *condottieri* who followed the sons of Tancred of Hauteville to Apulia early in the eleventh century, and, in the manner of *condottieri*, ended by conquering what they had come to protect, had already become in most respects northern French. But they wore their culture lightly and assumed naturally and easily the brilliant and brittle civilisation – Greek, Latin, Byzantine and Moslem – which they found in southern Italy and Sicily.

The Germanic sphere. The German language area, smaller, less populous and less developed than the Romance, had for many years been expanding towards the east. About 1100 the indistinct boundary of German and Slav ran from the Baltic coast near Schleswig southwards to the Elbe, where Hamburg was a frontier city of the Germans, and thence to the Harz, Thuringian Forest and Šumava. In general, it ran through sparsely populated country, but German settlers had already advanced their settlements along the Börde, founded Magdeburg, and reached the river Saale. They had also moved down the Danube valley, where, in lower Austria, they made contact with the Hungarians. Numerous Slavic enclaves, however, had been enclosed by the German advance, especially in the valleys of Carinthia and Styria, but it is doubtful whether many of them still retained their identity as late as 1100.

The eastward advance of the Germans was accompanied in part by the extermination of the Slavs or by their withdrawal or assimilation. All can be illustrated from the Slavic Chronicle of Helmold. The local population, sparse in any event, was greatly reduced by warfare between the Saxon and the Slav. Helmold, whose narrative is concerned almost exclusively with the territory lying to the north-east of the lower Elbe, described how Count Adolf of Schauenburg, 'as the land was without inhabitants, . . . sent messengers into all parts, namely to Flanders and Holland, to Utrecht, Westphalia and Frisia, proclaiming that whosoever were in straits for lack of fields should come with their families and receive a very good land – a spacious land, rich in crops, abounding in fish and flesh and exceeding good pasturage . . . An innumerable multitude of different peoples rose up at this call and they came with their families and their goods into the land of Wagria.[32] But the Slavs had not wholly disappeared. They were made to pay tithe to the church; they attended the Sunday markets along with the Saxons and gradually became assimilated to a German society. The German–Slav frontier must, throughout its course, have been a zone of mixed population, with the German language and culture gradually triumphing over the Slavic.

But small groups of Germans had already begun to move far ahead of the main body of German settlers. Though these also were in the main peasant farmers, they also included miners and metal workers. Germans had already begun to settle in the metalliferous Spiš region of Slovakia, where tradition puts their appearance a century earlier; and by the early twelfth century they formed a well-established colony.[33]

The Slavs. About 1100 Slav peoples were spread from the German frontier eastwards to the upper Volga and southwards into Greece. Except in a few favoured areas such as Great Poland and the plains of Bohemia, their numbers can only have been few, and it is highly probable that groups of pre-Slavic peoples – Celts, Germans, Illyro-Thracians and Latin- and Greek-speaking peoples – were far more numerous and widespread than the number of their survivors into modern times would suggest. Fig. 5.5, based on Jaździewski, shows the approximate distribution of the Slavic peoples among whom the distinct language groups – Polish, Czech, Serbo-Croat, Russian – were beginning to crystallise.

The Slavs were divided into a northern group, extending from Germany into Russia, and a southern, living mainly south of the Danube and its tributary the Sava. Between were Germans, newly arrived from the west,

Magyars and other invaders from the Russian steppe, and possibly some romance-speaking peoples, the ancestors of the Romanians. While the Slavs in the west were yielding to the Germans, towards the east they were pushing into the Baltic regions inhabited by Prussians, Lithuanians and Finno-Ugrians, and were slowly assimilating pre-Slav minorities in the Balkans. In Romania they had settled among the local Dacians, they had spread through the region of hardwood and mixed forest, had established settlements between the upper Volga and the Oka, and had penetrated the wooded steppe of the Ukraine, and perhaps reached the Black Sea. Balts and Finno-Ugrians and groups of steppe peoples – Pechenegs and Kumans – continued to live in the areas overrun and loosely held by the Slavs. Throughout this vast area the culture of the settled Slavic peoples had much in common, but the southern Slavs, beyond the barrier of German and Magyar, Vlach and Kuman were beginning to go their separate ways.

Between the northern and the southern Slavs, the Germans, moving down the Danube valley, had met the Magyar immigrants from the east and had already established a mutual boundary which followed in part the Leitha river. Both peoples must have imposed their authority on a population which was certainly Slavic, and may in part have been pre-Slav. Both incorporated loan-words from the languages which they found in the Danube basin, and both peoples interbred with the local peoples. In the case of the Germans little genetic change resulted, but in the case of the Magyars the racial characteristics of the resulting population tended to approximate those of the native Slav and pre-Slav peoples. Culturally dominant, the Magyar stock gradually became racially submerged. The Szekely, very closely related to the Magyars, were already established in Transylvania,[34] and the *Anonymous Chronicle* also mentions groups of Magyars who had penetrated the Balkans south of the Danube.

The Magyars were not the only Ural-Altaic people from the steppe to invade and settle in eastern Europe. Other steppe peoples had penetrated the territory of the later Roumania[35] and some had crossed the Danube and settled within the land of Bulgaria. The account of the Third Crusade by Geoffrey de Vinsauf describes how Richard I was in 'the furthest passes of Bulgaria', attacked by 'Huns, Alans, Bulgarians and Pincenates'.[36] This is probably no more than a list of names familiar to the chronicler, but it does suggest that there was some considerable ethnic variety in the mountainous region of what is today north-western Bulgaria. The Pechenegs and Kumans, nomadic, mongoloid and Ural-Altaic peoples of the steppe, exerted a continuous pressure at this time against the Danube frontier. Anna Comnena tells us that it was the policy of the Byzantine emperors to hold the passes of the Balkan range against these invaders, who nevertheless on occasion plundered the Maritza plain and doubtless even settled there. We can probably assume that the plains of Walachia and Moldavia were dominated by these peoples, and it seems that groups of Dacians, perhaps romanised, from north of the Danube found refuge from these raiders to the south of the Balkan range.[37]

Nor were the Ural-Altaic peoples the only invaders of eastern Europe in the period before 1100. It had long been a feature of Byzantine policy to uproot and shift to a different part of the empire any minority which threatened to make trouble. In the late tenth century such a group, the Armenian Manichees, heretical and therefore considered rebellious and antisocial, were settled by the Emperor John I Tzimisces in the Marica basin near Plovdiv.[38] Here they reinforced the northern defences of the empire and were in most respects slavicised. They retained, however, their heterodox religious views which they were later to disseminate westwards through the Dinaric region.[39] They may by 1100 even have spread to Macedonia. Groups of Turks had in a similar fashion been established along the Vardar valley; of Persians at Sinope, and in Constantinople itself were 'Varangians from Thule'.[40]

There is no means of knowing the numbers of the Slav invaders of the Balkan peninsula, nor how densely they were settled through the region. The anthropometric evidence suggests that the scale of movement has been exaggerated. In all probability the Slav penetration of the Balkans resembled the invasion of western Europe by the Germanic peoples. Up to a certain point they were able to impose their language and most aspects of their culture; beyond it, they were themselves assimilated. But the Balkans constitute a region of stronger and more varied relief than western Europe, and groups of pre-Slavic peoples could without great difficulty preserve their identity in such mountainous regions.

It is possible to distinguish two 'levels' of pre-Slav peoples; the one retained in some degree the language and customs of the Illyro-Thracians, and the other those of the Graeco-Roman conquerors. The former had survived, as far as is known, only in the highlands of Epirus and western Macedonia, a region notorious during the Middle Ages for the obstacles which it presented to travel and communication.[41] They had already become known as Albanians,[42] from the name of a clan formerly living in central Albania, and probably inhabited a somewhat more extensive area than today. Slav peoples had moved through the area, but such of them as had remained had become assimilated, though not without influencing the Albanian dialects. At the same time the Albanians had absorbed Vlach pastoralists, and through their service in the Byzantine armies, had been much influenced by the Greeks. Thus the primitive Illyrian culture had come to be overlain by elements derived from Latin, Greek and Slav.

The Vlachs, or Arumãni, though descended from the Illyro-Thracians, rather than Roman colonists, had been more deeply influenced by Rome. Their basic vocabulary is derived from Latin, though terms associated with settled village life are mainly loan-words from Greek and Turkish, suggesting that in Roman times they were exclusively pastoral. They were at this time both numerous and widely distributed. Anna Comnena described Vlach communities in Thessaly.[43] Under the Comneni emperors they paid tithe, and a number of them were subject to the monks of Mount Athos, to whom they contributed pastoral products. Alexius I recruited soldiers for his army from 'the nomadic tribes, called Vlachs in popular parlance'.[44] On the other hand they were restless and difficult to control, and Anna Comnena describes how they led the invading Kumans across the passes of the Zygum, with which they seem to have been very familiar.[45] The Crusaders who took the route along the Dalmatian coast encountered peoples who 'use the Latin idiom'.[46] These appear to have been nomadic pastoralists, and thus Vlachs, and there is evidence that Vlach communities were also to be found throughout the Dinaric region.

The chief importance of the Vlachs lies, however, in the possible relationship to the Romanians. The Romanian view of their own origin is that they derived from the romanised Dacians of the second and third centuries A.D. With this view most western writers have tended to concur. Their argument hinges on passages in the anonymous Hungarian chronicle of the twelfth century. Here *Blachi ac pastores Romanorum* are represented as sharing the grazing (*pascua Romanorum*) with Slavs and Bulgars, and *Blasii et Sclavi*, described as the poorest and most primitive of people, are represented as living in the mountains of Transylvania.[47] There can be no doubt that the pre-Magyar population of the Pannonian plain was ethnically Slav and Bulgar, with an admixture of Avar, and a Celtic and Illyrian substratum.

The Roman occupation of Dacia was relatively short, and left few remains. It is doubtful whether many settlers went to Dacia and unlikely that the polyglot armies of Rome could have had any great civilising power. The alternative view is that the Romanians originated in the Arūmani of the Balkans who, at some time in the later Middle Ages, crossed the Danube into Walachia and continued their pastoral and semi-nomadic life in Transylvania and the Carpathian Mountains. This latter view, which has always been accepted by Hungarian historians, has recently gained some support in the west.[48] It is suggested here that the *Blachi ac pastores Romanorum* may have been such peoples who drifted northwards along with Bulgars, during the period of the First Bulgarian Empire.[49]

The Greeks. The dominant language not only in Greece but also over the eastern half of the Balkan peninsula had long been Greek, and, despite the Slav invasions, Greek it remained in much of Macedonia and Thrace as well as in the peninsula of Greece itself. Slav settlers had been numerous in Thessaly, Epirus and the western parts of the peninsula, but by 1100 most of the Slavs had been assimilated to the Greeks, though communities of Slavs may still have been quite numerous in the rural areas. The narratives of the Crusades, however, mention only Greeks in these regions.

The administrative needs of the Byzantine emperors as well as migration occasioned by the Slav invasions had brought about a wide dispersion of Greeks in Europe. The coastal lowlands of Albania were settled at least in part by Greeks, while the Albanians looked down from their hills. Greeks had also settled in southern Italy and Sicily which retained until the Norman conquest a tenuous link with Constantinople. At the time of the Norman invasion, the Greeks were a very important minority, and their monasteries provided the institutional basis for the preservation of Greek culture. The Normans, however, restored the balance and permitted Latin culture to re-assert itself. By 1100 the Greeks were largely assimilated and only a few colonies remained in eastern Sicily and Calabria; even here Greek lived alongside and intermarried with Latin, and the Greek colonies were evidently declining. Greeks were probably to be found also throughout the Balkans either as merchants or in government service. Edrisi even found them as far afield as Pančevo near Beograd.

The commercial revival of the late eleventh century had taken Italian merchants widely over the central and eastern Mediterranean. In the ports of the Dalmatian coast they reinforced the 'Latins' who had survived from classical times. There were Venetian and Amalfitan colonists in Durazzo in Epirus, and already Italian merchants had established themselves in Constantinople.

The Jews. A large part of the long-distance commerce and in some areas also of the local was in the hands of the Jews, who were by now widely scattered over Europe. They remained, however, most numerous in southern Europe and the Aegean, where Benjamin of Tudela enumerated no less than 50 Jewish communities. There were in addition communities in many of the cities of northern France and the Rhineland, as well as in some of the cities of central Europe. Most Jews were urban, though some were at this time active in rural areas, where they owned land, and engaged in farming, especially in viticulture. Within the cities most appear to have worked at crafts; at Thebes, for example, the Jews were the best silk workers; at Brindisi were 'about ten Jews, who are dyers'. There was already a tendency for the Jewish communities to live in small exclusively Jewish areas, although only Constantinople at this date is said to have had an actual Jewish ghetto. That the institution of the ghetto was in origin a voluntary one on the part of its inhabitants is clear from a question addressed, in the eleventh century, by a Jewish community to its Talmudic scholars: 'One of the home owners in an alley exclusively inhabited by Jews wanted to sell, or lease, his courtyard to a non-Jew. May the other inhabitants of that alley legally restrain him from settling a non-Jew in their midst?'[50]

Benjamin of Tudela mentions the anti-semitism of the Greeks, who 'hate the Jews, and subject them to great oppression',[51] while at the other end of Christendom, the good merchants of Cologne and Mainz participated vicariously in the Crusades by massacring their local Jews.[52] An outburst of anti-semitism is also reported before 1100 in Spain where there had been a furious attack on the large Jewish community of Granada.

It is difficult to form any estimate of the number of Jews in Europe at this time. Benjamin of Tudela tells us the size of the communities which he visited, but his estimates of numbers are in other respects so unreliable that one distrusts even these. In any event, 2,000 appears high for the Jewish community of Thebes. A figure, unfortunately, is not given for Constantinople, and Andréadès, making allowances for this and other omissions, suggests a total of 15,000 for the Byzantine Empire.[53]

SETTLEMENT

The population of Europe throughout the Middle Ages was predominantly rural, and it is very doubtful whether, at any time even in the most highly urbanised areas, the population of the cities ever exceeded 35 or 40 per cent of the whole. Over Europe as a whole urban population probably never exceeded 10 or 15 per cent even in the fourteenth century. During the period under discussion towns were very few, and their population as a general rule was to be measured in hundreds rather than thousands. Even in those areas where urban life was most active, such as the Lombardy plain and Tuscany, less than 10 per cent of the population lived in towns, and over the continent at large, this proportion must have been considerably under 5 per cent.[54]

Rural settlement

It seems probable then that at least 90 per cent and possibly more of the population of Europe at this time lived in villages and hamlets and was employed most of the time in agriculture. Rural settlements were ephemeral. Helmold described in Wagria the evidences of the earlier colonisation of the region under Otto I: 'Traces of the furrows which had separated the plough-lands of former times may be descried among the stoutest trees of the wood. Wall structures indicate the plans of towns and also of cities'.[55] These had been destroyed in the course of the Slav wars of more than a century earlier. Similar campaigns of destruction can be documented from most parts of the continent. Villages were built and rebuilt, and even when there was no whole-sale and deliberate destruction, there was a need for piecemeal rebuilding of the houses. Furthermore there was a tendency, with the growth of population, for building-plots to be divided and subdivided, and for the number of houses to multiply not only within the original village area but also around its periphery. The plan of any rural settlement was thus in process of almost continuous change.

At the same time the number of individual settlements was increasing; as a practical limit to the size of a settlement was reached, villagers hived off to found another. Evidence for this process is to be found not only in the few surveys of this period, but also in the documented appearance of new place-names, and the evidence implicit in the place-names themselves. The total number of discrete settlements had almost everywhere in Europe been in-creasing for a century or more and was to continue to do so at least until the later years of the thirteenth century.

Many settlements which existed about 1100 have since disappeared or have been altered. It is nevertheless possible to see in the plans of modern villages basic patterns which in most instances must derive from the period of their foundation. The main features of these patterns were sketched some seventy years ago by Albrecht Meitzen.[56] They reflected in differing degrees the social organisation, the economy and the technology of their founders, subject always to changes to bring them into line with later requirements.

The basic settlement type in western and central Europe, and also in large parts of eastern and southern was the compact or nucleated village. The houses were clustered, usually irregularly; the area given over to yard or garden was relatively small, and the visible evidence of communal life, as in church and churchyard, was usually conspicuous. On the other hand, the size of the village cluster at all times varied greatly, from a mere half dozen 'hearths' to as many as a hundred.

Mutual protection may have been a factor in the creation and perpetuation of this settlement form, and in the Mediterranean region this motive may have been dominant in the selection of sites which gave protection not only from human marauders, but also from the malaria which infected many of the valley-bottoms. The Life of St Anselm tells how the saint found refuge from the heat of the Roman summer in a village near Caserta, 'which, being on the top of a mountain, has always a healthy and cool air agreeable to those living there'. But the hill-top site had its less agreeable aspects; 'the men of the village suffered daily many inconveniences from the scarcity of water'.[57]

Outside the Mediterranean region wood was more widely used than masonry, and the most sought-after advantages were not natural protection, but a dry site and easy access to both springs of water and the cultivated fields. Very approximately from the Loire to the Elbe the most frequent settlement form was the nucleated village, and outside this area similar forms were to be found in much of England, in parts of southern France and northern Spain, of Bavaria and Bohemia, and in restricted areas elsewhere.

In many instances, the large nucleated village is to be understood as the settlement of a patriarchal group; such an origin is often implicit in its name. Others must have derived from the *villae* or *vici* of the later Roman Empire, but it is noteworthy that, although the village may have been the heir to the villa in that it cultivated the same land, even, in some instances the same fields, it rarely occupied exactly the same site. The nucleated village was without regular plan. It is to be presumed that the settlement was, in most instances, growing during this period, in part by extension but also by subdivision of both fields and housing plots. This is implicit in the number of fractional *mansi* recorded in the polyptyques and surveys.

The nucleated village was closely associated, both geographically and technically, with the open-field system – specifically the three-field system (see below, p. 283). It appears almost as if the nucleated village, which had its origin in the needs of a primitive society for mutual help and protection, was preserved to satisfy the demands of economic feudalism, which imposed a common agricultural regime on the community. But why did it cease at approximately the Loire and Elbe, and why were there *lacunae* within this area where different types both of settlement and of rural economy prevailed?

South of the Loire and west of the Paris basin the large nucleated village was rare, and in its place were smaller clusters or hamlets. Areas characterised by small and scattered settlements existed also within the region of agglomerated villages. The reason may lie in the limitations imposed on the size of a settlement by the physical conditions of relief, climate and soil. In many areas, however, the differences between types of settlements lie in the dates of their foundation or the degree of seigneurial control, rather than in environmental circumstances. Small settlements were often late settlements. They were established after the principal village had attained the maximum practicable size, and they represented its overflow, as population continued to grow. On the other hand, areas where feudal constraints remained strong – as in parts of the Massif Central – the tendency was to restrict the break-up and dispersal of settlements. The lands of the abbey of St Germain were spread across the regions of both nucleated and dispersed settlement, and the contrast between them, implicit in its ninth-century polyptyque, must also have been apparent still in 1100.

To the west of the Paris basin, however, the prevalence of the small village or hamlet is to be attributed in the main to the conditions under which, in the eleventh and twelfth centuries, the area was cleared and colonised. In Maine, for example, where a demesne was unusual, and 'week-work' was not demanded of the peasants, large family units of settlement became the rule.[58] A passage from the Cartulary of Saint-Vincent of Le Mans, quoted by Latouche, describes such a settlement: 'Herbergamentum quoddam ad caput stagni situm, ubi manebat Gausbertus Grandis villanus, et viridarium et olcham adjacentem herbergamento.' A similar pattern of settlement, 'a multitude of tiny agricultural exploitations',[59] developed in the Gâtine of Poitou.

Small, scattered settlements also characterised highland areas of Spain, France, Italy and the Alps, where they doubtless consisted sometimes of isolated farmsteads, but more often of small hamlets, and also the lowland areas of the Netherlands and north-west Germany. The later medieval evidence from the Low Countries suggests that the size of settlements was fairly closely correlated with the physical conditions of soil and terrain. The polyptyques of the ninth century, particularly those of Ghent, Saint-Omer and Prüm, demonstrate the close juxtaposition of large settlements, hamlets and isolated farmsteads. The tendency of the smaller was clearly to grow to the greatest size consistent with their physical conditions and this, in parts of the Flanders plain and in the Ardennes plateau, was often quite small.

Settlement in the mountainous regions of the Jura, Alps, Apennines and Pyrenees was also in the main one of dispersed farmsteads, each standing amid its small patch of cultivated land and surrounded by pasture and forest. The relative importance of social and physical conditions in the formation of this and of other types of settlement has long been discussed. There is no simple answer. The physical environment imposed restrictions in all these regions. The rough terrain and the poor soil prevented the adoption of the relatively intensive three-field system of agriculture; it also made close seigneurial control both more difficult and less rewarding.

East of the Elbe a frequent and, in some areas a predominant form of settlement, was at this time the *Runddorf* or *Rundling*. Its houses were arranged on a round or oval plan, so that they enclosed a green, and presented an almost continuous fence towards the outside. Meitzen regarded the 'ring-fence' village as typically Slav; this however it clearly was not. It seems rather to have characterised the area between the Elbe and the Oder within which the wars between Saxon and Slav were fought in the eleventh and twelfth centuries with the greatest bitterness. The *Rundling* reflected the overriding need for defence for the community and its animals, which could enjoy some degree of protection within the circular wall of cottages and farm buildings.

The settlement forms adopted by the Slavs themselves at this time are obscure. Probably a small village of very loosely grouped houses, each standing within its own garden-plot, was most common. Procopius may have had such a settlement in mind when he wrote of 'the poor huts, standing apart from one another'.[60] If indeed such a settlement type was normal – and its survival in much of the Balkans suggests that it was – a difficult problem is at once raised by its relationship to both the fortified settlements, or *grody*, which were common in the western Slav lands, and to the street and forest-villages which today characterise much of the region. The latter were in all probability the characteristic settlement form of colonists from the west. They were alike in consisting of houses with their ancillary buildings aligned, somewhat irregularly perhaps, along a street or road, with their cultivated land extending in a narrow strip back to the stream, forest margin or other limiting feature.

Such a street, forest or marsh village implies a degree of organisation and control, such as might have been supplied by the *locator* who established it. It was open-ended; it could be added to as and when the need arose, and it offered to each family a separate and self-contained holding – itself no small attraction in Europe that was characterised by bipartite manors and burdened with labour-dues. The street village was not restricted to the areas of German settlement in eastern Europe, though it was here that it was most strongly developed. It was to be found also in parts of France and of western Germany, where it is also to be associated with planned and organised colonisation of forest and waste. It was also at this time beginning to spread into the Slav lands.

Scattered through settled areas of east-central Europe were, however, fortified enclosures, burgwalls or *grody*. In number they seem to have been comparable to the Iron Age forts of southern England. They were very diverse in size, ranging from a hundred acres or more down to less than an acre. In the plain of eastern Germany and Poland, most were constructed on the margin of the damp valley floors and sometimes on islands in rivers and lakes. Some of the largest had, in effect, become proto-urban nuclei, their primitive character obscured by masonry walls and churches, and markets had been established within them (see below, p. 272). But the great majority, in so far as they were still in use, continued to serve essentially rural functions. Too little work has been done on their fragmentary remains for it to be clear whether they were permanently inhabited. Some – notably the proto-urban nuclei – unquestionably were, but others may have served only as refuges in time of war. This is confirmed by Helmold, who wrote of the Nordalbingians that 'the . . . people came out of the strongholds in which they were keeping themselves shut up for fear of the wars. And they returned each one to his own village or holding, and they rebuilt houses and churches long in ruins because of the storms of the wars';[61] and of the repopulation of Wagria: 'The surrounding villages were already coming little by little to be inhabited . . . but with great fear on account of the attacks of robbers, for the stronghold Plön had not yet been rebuilt.'[62] These passages relate to the Nordmark; similar conditions must have obtained elsewhere along the frontier of Saxon and Slav.

The available evidence suggests that such *grody* were fewer in the indubitably Slav lands of central Poland, Bohemia and Moravia. Here, however, one finds a number of large and impressive forts, usually occupying naturally defensible sites and containing evidence of permanent settlement. These were 'tribal' seats, analogous to the *oppida* of western Europe a thousand years earlier. Though resembling the small *grody* in their general nature, they had become in many respects urban, and their form and function are discussed below.

In southern Europe the settlement pattern of classical times had been modified in detail by invading peoples and had been adapted to the prevailing conditions of insecurity. Scattered dwellings had been abandoned in favour of villages, *castelli* or *borgi*, tightly built of masonry and usually sited on the summit of a hill which gave them some physical protection. Occasionally they were surrounded by a wall; more often the outer walls of their houses were linked together to present a continuous rampart to the outside. The Germanic invasions and the inroads of Saracens, Normans and Greeks were the principal causes of this return to an older settlement pattern, but it is likely that in many instances the hill-top site had in fact been continuously occupied since early classical times.

In the eleventh century this process of consolidating settlements was still continuing in some areas whereas in others a new cycle of dispersion was beginning. There is evidence[63] that in Tuscany and near the Lombard cities the greater degree of security was allowing peasants to leave the safety of their *borgi* and to live nearer their fields on the lower ground. But this process was only just beginning; settlement in Italy remained more strongly nucleated than in most other parts of Europe, and only in the Apennines and Alps was the settlement pattern made up largely of hamlets and scattered farmsteads.

Spain, which also inherited a classical tradition of rural settlement, had a not dissimilar history. The mountainous north of the peninsula remained, as in all probability it had always been, a region of scattered and predominantly pastoral settlement. But to the south, the wide areas reoccupied by the Christian kingdoms in the course of the *reconquista* were settled, generally in an organised fashion, by peoples brought in from the north. Reasons of security dictated a nucleated pattern, though it is apparent that the villages were often very small. The large nucleated village was not practicable on the dry steppeland of Castile. In general, however, an open network of castles or cities was established, each with its dependent rural area, and constituted a system of refuges for the villagers whenever they were threatened by a Moorish raid. Beyond the unstable frontier of Christian and Moslem Spain lay a similar system of Moorish castles and forts which provided refuges for the muwallads and mozarabs of the surrounding villages.

There was clearly a relationship between the plan and size of the settlement and its economy. While an agricultural settlement could be of any size, provided only that its most distant fields remained within reach of the village, a pastoral economy, being less intensive, demanded more space and thus restricted the size of the settlement. And the more important the pastoral element in the society, the smaller was the settlement likely to be. In Scandinavia the economy consisted of a varying mixture of crop-farming and pastoralism, and settlements tended to be small. In Denmark they consisted mainly of small villages. A similar settlement pattern was to be found in Sweden, but in the hills and along the Norwegian coast, where pastoralism assumed even greater importance, the settlements became yet smaller, occupying discrete areas of cultivable land and extending their grazing lands into forest clearings and in summer up to the fells. The place-names speak eloquently of the formation of secondary settlements up the valleys, many of them isolated family farms, based more on their flocks and herds than their few small fields. 'There be wide fell-settlements up in the Mark [i.e. Finnmark], some in the dales, and some by the water's side.'[64] The *Egil's Saga* describes how the summer shieling gradually became a permanent but exclusively pastoral settlement.[65] The *Heimskringla* depicts a settlement pattern made up largely of hamlets and scattered farms, and 'in but few places did the people live close together', but rather 'in forest-farms far from other men'.[66]

Generalisation about settlement types is always difficult, and especially so for the earlier Middle Ages. Even in areas as small as a *pays* there is today a variety of settlement types, as there probably has always been. One can speak only of averages and percentages. In west Germany and north-eastern France, the majority of the settlements were nucleated villages but this does not exclude the *Waldhufendorf*, or long, forest-village, the hamlet, and even the isolated settlement in areas newly cleared for settlement. Each type of settlement is a function of soil and terrain, of technology and social structure, and to these should be added a certain random factor in man's decision where and how to live. The pattern of human settlement has at no time been simple, and has always been in process of growth and change. The eleventh and twelfth centuries, marked by strong population growth, saw the filling out of old settlements and the establishment of new. There is no reason whatever to expect the latter to resemble the former, when the social and economic conditions in which they had been created were entirely different.

The castle and the church. Essential features of the landscape of settlement were the church, both parochial and conventual, and the fortified home of the aristocracy. The former is, from its nature, the better documented and the better preserved.

By the year 1100 most of Christian Europe had been divided into units of ecclesiastical organisation, of which the diocese and the parish were the essential components. These corresponded in some degree with the *civitas* and the *vicus* of the later Roman Empire. The older dioceses each centred in a town, and there were few tribal capitals which did not become the seat of a bishop who exercised a spiritual jurisdiction over the tribal area. From the Rhineland cities, a network of dioceses was extended eastward into German and later into Slav lands. By 1100 Europe from Poland, Bohemia and Hungary westward to the Atlantic was divided into dioceses, and episcopal jurisdiction had been extended through Denmark, to Sweden, where the dioceses corresponded roughly with the principal provinces; and Norway, where bishops had taken up their residence at the chief settlements.

The headquarters of the diocese were established in almost every instance within an urban settlement, whose form and function they were to influence greatly. In the few instances where the bishop's seat was established in a rural setting, it seems quickly to have gathered an urban settlement around itself.

The parish was the smallest unit of ecclesiastical organisation. It was an area of land capable both of supporting a church and of being ministered to by its priest. By 1100, a network of parishes spanned most of central and western Europe. In well-populated areas, parishes were small – an average size in southern Europe was about 12 square kilometres, and in the Paris region about 7.5.[67] Adam of Bremen described the Danish islands of Sjoelland and Fyn as having respectively 150 and 100 churches or parishes, whereas Scania had 300. This suggests that their average area may have been about 45 square kilometres. The parochial system had probably not been extended into many of the mountainous areas, and the *Heimskringla* noted that around the heads of the valleys and on the fells 'the greater part of the people were heathen'.[68] The parochial system attempted to keep pace with population changes, subdividing parishes as their wealth appeared to justify it or combining those no longer able to exist independently. The size of the parish was in fact controlled by the value of the tithe which the local region could contribute, and, in theory at least, the tithe of the parish was used to maintain the church and to support its ministers. One of the first acts of the Christian conquerors of the east German Slavs was to build parish churches and impose the payment of tithes.[69]

The church, the focus of the parish as a unit of ecclesiastical organisation, was founded most often in the principal village or hamlet of its parochial territory. In the large nucleated villages of northern France and south-west Germany parish and village community in fact tended to coincide.

Church-building was active at this time, and was subsidised both by the contributions, compulsory and voluntary, of the parishioners and by donations of the local seigneurs. It is demonstrated not only by allusions in the chronicles, but by the fragments of romanesque masonry which have survived the ravages of time and the hand of the restorer, and are today incorporated into the fabric of later churches. These are particularly numerous in England. They are very much less common in rural areas of continental Europe, where warfare has taken a very high toll of religious and other buildings. In the *grody* of the Slav lands of eastern Europe it has been not uncommon for archaeologists to uncover the foundations of a small apsidal church of eleventh or twelfth century date. In southern Europe churches were more numerous and in general older, and in the Balkans, the influence of the Byzantine Empire had led to the building of churches, with cupola instead of tower, and adapted to a ceremonial different from that of the west.

Throughout western, central and southern Europe the architecture and style of the church reflected not only the requirements of a common liturgy, but also the widely differing conditions in which they were erected. Most churches were built of local materials, of stone, brick and wood as occasion served, with freestone, if it was not abundant locally, reserved for the carved mouldings, the tympana of doorways, arches, and the capitals and bases of pillars. Fine Caen limestone was shipped from Normandy for use in churches in southern England; the sites of Roman villas were raided for old bricks; the intractable flint was used where nothing better offered; in Norway and Sweden and probably also in Alpine regions, wood was used, and the king of Norway even sent to Iceland 'timber for building a church'.[70]

The parish church, in its infinite architectural variety, had already become a conspicuous feature of the cultural landscape. But monastic settlements probably exercised the most profound influence on settlement and land use. The foundation of monasteries, slowed by the invasions of the Saracens and the Norse, received a new impetus in the tenth and eleventh century, and by 1100 had made a profound impact on both the urban and rural landscape.

Western monasticism had been dominated by the Rule of St Benedict. Work was encouraged, but retreat to the wilderness had never been strongly advocated. Even in its reformed Cluniac pattern it kept close to settled and well-developed areas. Many of the Benedictine houses had been established within or very close to existing towns. Indeed the suburban situation became a favourite one (see below, p. 265) and exercised a strong influence on the topographical development of many cities. Cluny itself, prototype of the reformed order, was established in a very far from inaccessible valley of the Charollais, in Burgundy, and became in a relatively short time the nucleus of a small town.

In Normandy a considerable number of houses, including Fécamp, Jumièges, Saint-Wandrille and Le Bec, were established within the wooded valleys which trenched the chalk plateau. But they were never far from the cleared and cultivated plateau, and represented a small though significant extension of settlement into the waste. It was no hardship to a landowner who was short of labour, to give some of his uncultivated property to the monks who thereupon established their house and its granges on land they had cleared themselves. Often, too, a diminutive settlement – or *vicus* – passed into monastic hands and was quickly expanded into a large village. The role of monasticism had hitherto been primarily to expand the limits of existing settlement, using customary methods of cultivation, rather than to move *into* the wilderness.

Nevertheless, a movement towards asceticism and complete withdrawal from ordinary life, along the lines of primitive Middle Eastern monasticism, was developing in Italy. A number of hermitages were established in the mountains, where the monks exerted themselves to clear land and provide for their daily wants themselves. Their total impact on the landscape, however, was negligible. Only a few years before 1100, however, two events had occurred which were to have more far reaching consequences. The first was the foundation in 1084, amid the wilds of the Chartreuse (Dauphiné), of the first Carthusian monastery; the second was the establishment fourteen years later at Cîteaux, amid the sandy, forested wastes of the Saône valley in Burgundy, of the first of the Cistercian order. Both orders extolled the virtues of physical labour; both sought remote and unpopulated regions for their settlements, perhaps because in this time of sharply rising population and increasing feudal control of the land, this was all they were likely to get, and both were to make in the next two centuries very significant changes in the cultural landscape of the deserted areas amid which they made their homes.[71]

The building of churches and the foundation of monastic orders, which were features of the eleventh and twelfth centuries, were signs of growing wealth. An increasing surplus from agriculture passed into the possession of the Church, in part through donations of the faithful and the payment of tithe, in part by way of rents and other dues from Church-owned land. Donations included not only money and materials, but also fragments of agricultural land and urban tenements. Tithes, rents, payments in kind and labour-service provided the means for the very active building programme. In general, income of this kind could not easily be transferred or invested. Most had to be used where and when it arose. Building activity may thus be taken as in some degree a measure of local wealth. Despite the loss of much of the church architecture of this period and the occasional uncertainty of the date of what remains, it is nevertheless possible to use it, in conjunction with other evidence, such as coinage and urban development, as a measure at least of the relative wealth of different regions of Europe. Most is found where urbanism and urban pursuits were developing most rapidly: first and foremost, northern Italy, the middle and lower Rhineland and the lower Meuse valley. Lesser clusters of romanesque art occur in Provence, Burgundy and central Italy. The absence of such monuments from the Paris region and their presence in considerable numbers in Cataluña is disconcerting. The former is largely explicable in terms of the flowering of Gothic architecture which in this region physically replaced romanesque. The latter is characterised mainly by monuments of no great size or sophistication, and is a reflection of the immigration from southern France.

If we add the Paris region, in which there is documentary evidence for an ambitious building programme at this time, we shall probably have delimited the most important centres of wealth and of economic progress in western Europe. Each probably represented an area within which there was a relatively high degree of social communication, a fairly rapid diffusion of cultural traits, and, in terms of architecture, the adoption of common building styles and decorative motifs.[72] The unity of the system of communications created by the Seine and its tributaries contributed to the emergence of a local style in this region. Puig i Cadafalch commented that stylistic boundaries tended to coincide with diocesan; this is to be expected, since the dioceses were essentially territorial units, within which communication was well developed. Outside these west European hearths we find lesser areas with some degree of local wealth and creativity in Bohemia and central Poland. In each the growth of towns was matched by the founding of monasteries and the building of churches.

The castle was, in a sense, the converse of the church. In style it was more crude and, as a general rule, it demanded few materials that the local region could not supply. Although it was clearly built by labour which had been taken from the land for the purpose, its existence was less a measure of an agricultural surplus than of political insecurity and the developing and extending feudal organisation of society. The castle – the fortified home of a member of the feudal class – had become by 1100 a feature of the landscape. It is difficult to form any estimate of the number, though attempts to do so for the British Isles appear to have been reasonably successful. Evidence for their construction is usually indirect, unlike that for monasteries which usually preserved the records of their origin. Their foundation, at least under weak rulers, was quite uncontrolled, as the Anglo-Saxon chronicler recorded in Stephen's England: 'every rich man built his castles. They cruelly oppressed the wretched men of the land with castle-works; and when the castles were made they filled them with devils and evil men'. Even under less anarchic conditions the havoc which their construction caused among the peasantry must have been horrifying.

Lambert of Hersfeld described how castles were built on all the 'mountains and hills' of Saxony and Thuringia with compulsory labour and provisioned from the surrounding countryside, 'as if in enemy country'.[73]

In many areas the castle was basically an earthwork – a mound and adjoining courtyard – which could be heaped up in a relatively short period of time by the forced labour of the local population.[74] Its walls were of logs, though these came in the more permanent and important castles to be replaced in time by masonry walls. The large tower or 'keep', round or rectangular in plan, had made its appearance, but was by no means common at this date. It called for a level of wealth and a power of organisation that were beyond the reach of all except the higher members of the aristocracy.

The question is often raised whether castle building at this time accorded with a coherent strategic plan. As far as the lesser nobility were concerned the answer is almost certainly no. They built their castles on their own lands with little reference to those of their neighbours. Kings, princes and members of the higher aristocracy who possessed extensive lands may well have had some concept of strategy even if this consisted of little more than guarding their boundaries from invasion. The German emperors were active in protecting their marches in this way, as were also the rulers of the Spanish kingdoms. The suggestion that castles were built specifically to control routes, more particularly trade routes, is more questionable. There was, in fact, little long-distance trade to control, and routes were too fluid to be easily dominated from a few fixed points. The concept of a network of fortified points, conceived by some military genius and established by an anarchic feudal aristocracy, dies hard.

The establishment of 'castellanies' in the Low Countries was the most orderly development of this kind (see above, p. 234), in part because of the very real military danger from the Norsemen, but primarily because the Counts of Flanders had been able to preserve a great deal of authority in their own hands.

The building of castles was not, however, without its influence on the pattern of settlement. Castles were attracted to settled and populous areas, which could satisfy their need for labour and provisions. It was to the interest of the *châtellain* to encourage rural development in his vicinity. Though tactical considerations may have dictated the site of a castle, it soon became, if it was not from the start, the focus of a village, and sometimes also the site of a market. In many instances, notably in the Low Countries and Germany, the settlement at the castle-gate developed with the aid of a charter and grant of privileges, into a town. In a few instances the town was to grow and devour the castle.

It would be a mistake to assume that in style and purpose there was a gulf between military and ecclesiastical settlement and architecture. They overlapped at many points. Many a monastery was walled and fortified, and employed soldiers for its protection,[75] and the humble parish church often served as a refuge for the villagers, and was occasionally designed for this purpose. On the other hand, the larger castles incorporated churches and the parish church was not infrequently built close to a castle if not actually within its courtyard.

Urban settlement

The late eleventh and twelfth centuries were characterised throughout Europe by the growth – or regrowth – of urban institutions. Within the limits of the former Roman Empire there were considerable and conspicuous remains of the imperial cities, and in southern Europe these had never ceased to be inhabited. This cannot, however, be said of northern France, Britain and the frontier provinces of the Roman Empire. In these urban life, as the Romans understood it, had come to an end; the cities had fallen to ruin, and their sites, if inhabited, served only as villages (see p. 190). In England, of all Roman towns, only Canterbury showed evidence of continuity of settlement through the Dark Ages.

Nevertheless, the city walls had survived in most instances; Roman routeways still converged on the sites, and they could be re-occupied whenever economic conditions justified. Beyond the Rhine and Danube there had been no Roman cities, and an area corresponding roughly with west Germany and Scandinavia was in the early Middle Ages a virgin territory as far as urban development was concerned. Cities developed at this time as small gatherings of merchants and artisans around the nuclei – either military or monastic – provided by those who had acquired and were developing the land. The inspiration of this urban development had come from France and the Rhineland. But east of the Elbe and within the Bohemian and Pannonian basins an analogous development was also taking place. Settlements which we must term urban had grown up around the tribal seats of the Slavs and Magyars, and by the year 1100 many of these had attained a considerable size and had become centres of crafts and trade.

We can, therefore, distinguish four separate but not wholly exclusive zones of urban development in the twelfth century: (1) that which derived its urban life in an unbroken line from the later years of the Roman Empire; (2) that in which city life decayed and may have been abandoned, but where the fabric of the Roman cities survived until a new urban growth could breathe a fresh life into them; (3) the Germanic and Scandinavian zone in which urban development was around nuclei, either castle or religious centre, introduced from the west, and (4) an east European region of autonomous urban growth.

The zonal boundaries were far from distinct; and development was by no means uniform within each zone. There were towns in Mediterranean Europe that were abandoned during the Germanic invasions and never re-occupied, such as Aquileia and Salona, and there were towns in northern Europe, such as Reims, in which there was almost certainly a continuity of settlement on however modest a scale. Nevertheless this fourfold division of the continent constitutes a useful model about which the urban geography of the early Middle Ages can be organised.

The Mediterranean city. It is reasonable to assume that towns of Italy and southern France declined in size and importance during the Dark Ages, without, in the great majority of cases, being abandoned. Long-distance commerce almost disappeared, and most towns were reduced in function to being the centres of small, self-sufficing and largely autonomous city-regions. The renewed growth of commerce in the eleventh century brought with it a revival of urban life, but it does not follow that those cities which had been the most important in the later years of the Roman Empire regained the relative positions which they had previously enjoyed. In fact, the reverse was more often the case. Rome itself, the capital of a spiritual empire over western and central Europe, had little else to recommend it; it was but the ghost of its former self, sitting crowned in a corner of the vast Aurelian city.[76]

In southern Italy the largest cities were probably the ports of Bari, Amalfi, Gaeta and Naples, and, in Sicily, Messina and Palermo, but most cities which had been important under the later Roman Empire continued to be inhabited. Romanesque architecture of the eleventh and twelfth centuries is in itself no measure of population, but it does suggest which cities should be considered among the more important. In the interior Melfi and Benevento; on the coast, Salerno, Taranto, and Brindisi, and, in Sicily, Syracuse and Catania were all populous centres.

Central Italy at this time was a frontier region, where the Norman Duchy of Apulia bordered the principalities which were tributary to the Papacy. Warfare had restricted economic growth; towns were strongly fortified, like Spoleto, but their economic relations were only with their immediate regions. The urban revival was more marked in Tuscany and the plain of Lombardy. Pisa was now at the height of its prosperity as a commercial town, and Lucca was probably the largest and most important in Tuscany. In Liguria Genoa was growing in importance and a generation later was to rebuild its walls and almost to double its urban area. It was in the Lombardy plain, however, that urban life was most vigorous. Proximity to the alpine passes and to navigable rivers were factors in the selective process whereby some of the former Roman cities attained an unparalleled wealth and power. Foremost among them were Pavia, chosen by the Lombards as their capital and now a focus of trade and site of an important fair; Milan, which commanded several of the alpine routes, and had grown during recent years perhaps to the foremost position among Italian cities; Cremona, Piacenza and Mantua, which had the advantage of navigation on the Po; and Parma, Bologna, Reggio and Modena, which succeeded, not without considerable difficulty in some instances, in keeping open their waterborne communications with the Po. Add to these the many smaller towns: Brescia, Verona, Borgo San Donnino and Argenta, and a picture emerges of an urban development more intense and expanding more rapidly than in any other part of Europe.

The cities of the Lombard plain were still contained within their Roman walls. A cathedral and in most instances monastic and other churches had been added, but the layout of the Roman city had in many cases been preserved. It is difficult to form any estimate of their size at this date. Their renewed urban growth had begun only recently, and it is doubtful whether, with the possible exception of Genoa, any had more than five or six thousand inhabitants. The area occupied by a city at this date bore no necessary relationship to its commercial importance. Amalfi covered only 15 hectares, much of which was too steep for habitation, and it is doubtful whether its population could have exceeded two or three thousand. Venice was certainly larger than this, and probably was second only to Genoa. It was already the dominant port of the Adriatic Sea; it had occupied Zara and established a favourable position at Constantinople, and its commercial hegemony was disputed only by Ancona.

Southern France shared in the commercial revival of northern Italy; merchants of Pisa and Genoa visited regularly the river ports of Montpellier, Saint-Gilles and Arles, though by 1100 these were becoming difficult to reach from the sea, and shipping was beginning to use the more accessible port of Marseille. The cities of the interior participated to only a small degree in this prosperity. They remained small, and some among them, Arles and Nîmes for example, occupied only part of the Roman enclosure. In the Iberian peninsula the largest and most prosperous city was Barcelona, which shared prominently at this time in the commerce of the western Mediterranean. The cities of the interior, however, were noteworthy more for their fortifications than for their essentially urban functions. They had been established in the course of the *Reconquista* and served rather as bases for the resettlement of the land than as centres of crafts and commerce.

The city in north-west Europe. Throughout the rest of the former provinces of Gaul the Roman cities had survived as settlement sites, though not, in many instances, without a hiatus of greater or lesser duration. Many became administrative centres and seats of local counts who established their fortified homes within the walls; most became the seats of bishops, whose dioceses corresponded roughly with the sphere of the classical *civitates*.

At the same time however they attracted monastic foundations, and these in turn became the foci of small settlements of artisans and merchants. It might have been supposed that these developments would have been accommodated within the walls of the ancient city. In some instances they were, but examples are numerous of their development outside the line of the Roman walls, and occasionally at a considerable distance from them. The monastery of Saint Rémi of Reims, notable for its ninth-century polyptyque, was founded in the sixth century a thousand yards to the south-south-east of the walls of the Roman *Durocortorum*. It had its own enclosing wall, and was capable of defence independently of the old city. At Arras, which had been a small and relatively unimportant Roman town, the bishop established himself within the Roman enceinte, while the monastery of Saint Vaast, founded about 650, was situated beyond the marshy valley of the tiny river Crinchon. About 1100 Saint Vaast, now the nucleus of a small commercial settlement, was enclosed by a wall, and the two settlements, known respectively as the *Cité* and the *Ville*, continued separate and distinct until modern urban development bridged the gap between them and filled the little Crinchon valley with houses. At Toulouse, also, the *Bourg* grew up around the monastery of Saint-Sernin, some 500 yards to the north of the walls of the Roman *Cité*. Here, however, the gap between the two nuclei had probably been filled by 1100. At Paris, the monastery of Saint-Germain lay on the left bank, well beyond the Roman and early medieval defences. Other examples might be cited: Troyes, Cologne,

Narbonne, Mâcon, Dijon, Auxerre, in whose early history we can detect rival poles of attraction: the Roman and episcopal city and the monastic and commercial settlement before its gates.

It is perhaps easier to explain why the monastery should have been established outside the walls than to understand why it, rather than the walled enclosure, should have attracted a community of merchants and craftsmen. The monastic ideal of at least partial withdrawal from the world, accompanied not improbably by a desire to keep at arm's length from the bishop, may serve to explain the former. It is possible that more space and perhaps a more flexible and sympathetic attitude on the part of the monks may have served in some instances to locate the market at the gates of the monastery.

The second important development of these years was the growth of the city's commerce and crafts. The city was distinguished primarily by its commercial functions. These in the first instance served the needs of the ruling classes, and luxury goods played a proportionately important role, but in the eleventh and twelfth century trade was becoming more broadly based, and the city had itself become a medium of exchange within its region. This development assumed two forms. One was the establishment of artisans and shopkeepers usually within the protection of the town walls; the other was the foundation of a regular market and perhaps also of a fair. These were sometimes located for convenience outside the walls. At Cologne, the markets were situated between the Roman walls and the river; and at Besançon, Orléans, Dijon, Strasbourg, Rouen, Maastricht, Tournai, Troyes and elsewhere the modern topographical evidence suggests that the market place was established outside the line of the earliest walls.

The cathedral was not the only religious building to be erected within the walls of the old city. It was customary for a number of churches, sometimes maintained by canons, but usually parochial in status, to be built. At the same time the urban area was itself divided into parishes to support the churches. Outside the earliest line of walls churches were numerous in addition to those which formed part of the suburban monasteries. The romanesque churches of Cologne, for example, lay outside the contemporary city walls, though enclosed by the subsequent expansion of the city, and this development can be paralleled in other cities of western Europe.

Urban development was most conspicuous in and around the cities of Roman origin, but the process of creating new towns on new sites had already made considerable headway in France. No such city appears at this time to have come into being without a nucleus, around which, as it were, to crystallise. Such nuclei were usually formed by castles or monasteries. They were most numerous in northern France, where the net of Roman towns had been least dense. Of the cities at which the Champagne fairs were to meet, only Troyes was of early origin. Bar-sur-Aube was at this time a small town beside the castle of the Counts of Champagne, and Provins was growing around its twin nuclei, the castle of the counts and the abbey of Saint-Ayoul, 500 yards away in the valley below. The origins of many another city were visible at this time as a small settlement near the gates of a castle. The king of France himself granted privileges to potential burgesses, and the customs and practices of the

Norman town of Breteuil became a model for several in England. Most of these towns, however, were very small, and their range of influence probably did not extend more than a few miles from their market places. It is doubtful whether in many instances they were actually walled.

Monasteries, no less than castles, formed nuclei around which small urban settlements formed. The most important city which had originated in this way was probably Saint-Omer. The abbey of Saint-Bertin had been founded in the marshy valley of the Aa about 648; in the following century it became the nucleus of a small proto-urban settlement, and it is possible that at a later date another such settlement emerged near the later church of Saint-Omer, a half mile to the south. By 1100 these two nuclei had probably joined to create perhaps the largest and industrially the most advanced city in northern France.

Ghent probably originated in a similar way, as a small commercial settlement beside the abbey of Saint-Bavon, and Arras owed everything to the monastery of Saint-Vaast. Nivelles in Brabant, Montreuil-sur-Mer and probably Thourout also grew up on monastic sites.

Flanders. The province of Flanders had been little settled in Roman times, and its towns – Therouanne, Tournai, Bavai – had grown up along the southern edge. Its seaward margin consisted of a belt of recent marine alluvium, lying at most a few feet above the level of the sea, and liable to flooding by the rivers and inundation by the tide. It was dotted with sandy islands, which provided dry settlement sites, and laced by waterways which provided the only practicable routes through this wilderness. To the south lay a gently undulating region, lying generally from 20 to 25 metres above the sea, with relatively dry plateaus, separated by shallow but very damp valleys. It was composed of Tertiary clays and sands, to the south of which the Cretaceous beds came to the surface, giving rise to the dry rolling plateau of Artois and southern Hainaut.

The coast of Flanders, now smooth and fringed by dunes, was in the early Middle Ages interrupted by the broad estuaries of a number of rivers of which the Zwin, IJzer and Aa were the largest, while near Bergues a deep embayment, the Moëres de Ghistelles, was slowly turning itself into salt marsh. Despite the absence of conspicuous physical barriers, this region was in fact divided into a western and an eastern part by an area of forest and marsh which extended northwards from the Scheldt below Tournai towards the mouth of the IJzer. Today this strip of land is more thinly peopled than the areas to east and west; in the early Middle Ages its heavy Flanders clay, and its marshy valleys combined to make it almost impassable – an effective divide between West and East Flanders.

Though both parts of the province were at this date becoming urbanised, development was considerably more advanced in west Flanders. The reason is not hard to find. The west Flanders cities lay on the margin of the chalky upland of Artois, from which they derived much of their food supply, the wool for their growing cloth industries and probably also a large proportion of their townsfolk. The east Flanders towns had no such advantage. The sites of Bruges, Ghent and Antwerp were surrounded by forest, heath and marsh, amid which areas of good farmland were very restricted. At a later date (see below, p. 415) these cities were in part supplied with foodstuffs brought down the Scheldt and its tributaries, but about 1100 this trade had not really been developed.

West Flanders was wealthier than east; it was linked to the road system of northern France and had easier communications with the Paris basin and Channel ports. At least four of its cities were on sites which had been developed by the Romans. Saint-Omer had already grown from its monastic nucleus into a walled town at the limit of navigation on the Aa, and Arras, thanks also to its monastic foundation, had become a centre of cloth-working and commerce. But most other towns were of recent origin. They had been *vici* in Frankish times; they became the centres of castellanies in the tenth or early eleventh centuries, and then small pre-urban settlements. The history of Lens-en-Artois is typical.[77] Its castle was built about 975 at the site of a small settlement near the limit of navigation on the Deule.[78] Between 1030 and 1070, the *vicus* grew into a small town with a group of craftsmen and a modest network of trade. At the same time, similar castles, with small economic and administrative functions, had given rise to other towns along the margin of the Artois upland and the Flanders plain; Aire, Lillers, Bethune, Douai, Valenciennes, and on the fingers of somewhat higher land which extended eastward between the IJzer, the Lys and the Scheldt: Cassel, Ypres, Messines, Bailleul, Hasebrouck, Armentières and Lille. This process of urban growth overflowed the Flanders plain. It spread into Artois at Hesdin, Saint-Pol and Montreuil, and reached Abbeville, and Cambrai in Picardy.

Urban development came later in east Flanders than in west, but in its broad lines it was similar. The origin of the city of Bruges is thus described in the Annals of Saint-Bertin:

> In order to satisfy the needs of the castle folk, there began to throng before his [i.e. the count's] gate near the castle bridge traders and merchants selling costly goods, then innkeepers to feed and house those doing business with the prince, who was often to be seen there; they built houses and set up inns where those who could not be put up at the castle were accommodated. . . . The houses increased to such an extent that there soon grew up a large town which in the common speech of the lower classes is still called 'Bridge'.[79]

This pattern of development was being repeated around innumerable castles in northern Europe. It is possible, however, that the importance of monastic foundations was in fact greater than that of castellanies in stimulating town growth. Their estates were often extensive and widely scattered; the surpluses both of farm products and of the domestic crafts had to be brought to the monastery, where furthermore artisans were employed on the fabric of the church and in providing for the needs of monks and lay-brethren. Inevitably, too, there were surpluses both of foodstuffs and manufactures to be sold to whoever came by, just as there were needs that could be satisfied only by merchants who dealt with distant markets.

The reasons for this urban growth were twofold. In the first place, a growth of population in northern France and the southern Low Countries provided the manpower without which it would have been impossible, and in the second the cloth-making industry itself gravitated from the villages and hamlets towards the small towns. There were unquestionably other predisposing factors; improvements in agricultural tools and technology permitted the creation of a farm surplus, without which the cities could not have been provisioned. At the same time the organisation, if not also the technology, of cloth manufacture improved. It ceased to be primarily an occupation of the women, and weaving became a full-time occupation for male artisans. A class of merchants developed to supply raw materials to the artisans and to market their products. These occupations could at this time be carried on more easily in towns than in rural settlements. Above all, demand was increasing not only for cloth but for other consumers' goods, and this necessitated an increase in the scale of operation in both manufacture and marketing.

It was to assist this small but growing traffic that fairs came to be established in west Flanders before the end of the eleventh century. By 1100 they were being held regularly at Lille, Ypres, Messines and Thourout to which must be added the fair at Bruges in east Flanders. Neither here nor in Champagne did the fairs themselves contribute greatly to the development of the cities, and it is possible that they may have served rather as outlets for what remained of the rural industry.

To the east of the barrier of forest and marsh which separated west Flanders from east, the growth of towns began later and was slower and less prolific. It was not until they were able to capitalise on their most important asset, their proximity to the sea by way of the navigable Zwin and Scheldt, that they began to rival and then overtake the more westerly towns. By 1100 both Bruges and Ghent were small settlements near the castles built by the Count of Flanders; at Ghent the monasteries of St Bavon and St Pierre added to the attractions of the site. In 1100 the Zwin had not been opened for navigation as far as Bruges, and both Bruges and Ghent sent part of their merchandise overland to loading points along the lower Zwin. Ghent was beginning to use the Scheldt, which was soon to provide the town with its chief commercial routeway towards the north-east and the Scandinavian and Baltic market.

It is impossible to form any estimate of the size of the cities of Flanders. They were new; they were small; they were in the throes of organising the supply of foodstuffs and raw materials and the distribution of their products; they had not solved the question of their own autonomy and of their relation to castellan and count; they were at this time growing by migration from the surrounding countryside.

These incipient cities were formerly regarded as essentially groups of merchants, attracted from one knew not whence by the prospect of trade and profit. The possibility that such rootless travellers may have settled in the cities can certainly not be excluded, but the population was derived mainly from the local peasantry.[80] A study of the names of the burgesses of Amiens, which in many instances incorporate evidence of their origin, indicate that almost all came from within a day's walk, no more than about 20 miles, of the city.[81] In those pre-urban settlements which grew up around monasteries it is highly probable that the earliest settlers were in fact servants of the abbey itself, engaged primarily in collecting farm surpluses and artifacts from the scattered *villae* of the convent and in satisfying the needs of the monks by the work of their hands.

The German city. The Romans had built a number of towns in the Rhineland and upper Danube valley, including among them some of the largest in the empire. These lost their military functions with the Germanic invasions, but many – perhaps most – retained some urban functions through the centuries that followed. Several became the seats of bishops before the final collapse of the empire, and this alone would have ensured that the sites were never really abandoned, even though commerce may long have been on a very restricted scale, and the citizens few in number.

At Cologne, Mainz, Speyer, Worms, Basel, Regensburg, and probably also at Strasbourg the Roman walls had survived. Within lay the cathedral, and, since few cathedrals were monastic, the homes of the canons and their servants, and probably of a handful of other persons. Outside lay a *burgus*, *portus*, or *Vorstadt*, a small community of tradesmen and artisans and the site of a market, and commonly one or more monastic foundations. At Cologne, a merchants' quarter had already arisen between the Roman city and the Rhine, where goods could be most easily unloaded from river craft and stored in warehouses built for the purpose. At Metz also the earliest suburb lay *between* the Roman city and the navigable Moselle. At Verdun it lay on the opposite bank of the river.

Not a few Roman settlements, however, had vanished. Carnuntum, abandoned by the legions, seems never to have been re-occupied. Life probably did not continue in Aquincum (near Budapest), nor at Juvavum (Salzburg), and many were the legionary camps which continued to give protection to at most a small and wholly agricultural community.

Beyond the line of the Rhine and Danube were to be found the small nuclei of the later German cities. Only a very small proportion of those which were to gain distinction in the thirteenth and fourteenth centuries can be said to have existed at this date. In Westphalia there were in 1350 about 120 towns; no more than six of these were in existence in 1180, and even fewer in 1100.[82] Much however depends on definition. As an autonomous institution, with its own rules and elected officers, the town clearly did not exist in Germany, as it had begun to do in Italy. But if by proto-urban settlement we may understand a non-agricultural settlement in which business was regularly transacted and crafts practised by professional craftsmen, irrespective of the legal status of themselves and their community, then we find several.

Three categories of German city can be recognised. Clearest is the merchants' settlement, the *Wik*, such as Dorestad, Birka and Hedeby. They were places of exchange, established by the merchants themselves. They owed little or nothing to feudal lords, whether lay or ecclesiastical, and seem in general to have been located without reference to previous *Burgs* or religious houses. But they were well placed for seaborne commerce, which was their primary function.

Equally clearly defined were those which developed around a monastic house. They were fewer than to the west of the Rhine, because monasteries were themselves less numerous. In this category may, however, be placed Sankt Gallen, Höxter and Corvey. The great majority of urban settlements in Germany however developed around small, fortified nuclei. At the heart of the city is generally to be found a *Burg*; either the seat of a local count, a royal *Königspfalz* centrally placed within a complex of royal estates, or a *Fluchtburg*, built as a place of refuge for the local population. Thus Aachen, Nimwegen and Goslar grew up around the seats of the Carolingian or later emperors; Würzburg originated in the Marienberg, the fortress which lies on the hill-top across the river from the later city; Magdeburg was a pre-Carolingian *Burg* or refuge, within which Charlemagne himself established a castle to guard the crossing of the Elbe, and Quedlinburg, Merseburg, Erfurt and many others were in origin early German, or even Slavic forts or *oppida*.

Whatever their mode of origin, it is impossible to dissociate the history of German cities from that of the church. If the pre-urban settlement did not grow up around a church – and most did not – it very early attracted ecclesiastical institutions. Some had become episcopal seats in Carolingian times; monasteries were founded beyond their walls, and in all of them parish churches were established and shared between them the tithes and the spiritual obligations which arose within the city.

If the Rhineland cities were small those which had grown up deeper within Germany were, with few exceptions, very much smaller. Magdeburg, favourite seat of Otto I and frequented by merchants who did business with the Slavs, was one of the largest amongst them, and perhaps the only one to be enclosed by a masonry wall at this date. Merseburg, Erfurt and Würzburg were probably also, by the standards of 1100, large merchants' towns. But most can have been only the small central-places of relatively populous rural areas. The map of pre-urban centres at this time shows some degree of concentration on the west to east loess belt. This has been interpreted in terms of the ancient trade route, the *Hellweg*, which followed this belt of open country.[83] Commerce doubtless added to their size and importance, but such a network of incipient towns is explicable only in terms of the productivity and wealth of the local area.

The Slav town. East approximately of the line of the Elbe, where the Romans had never come and monasteries had not yet been established, the nuclei around which cities were beginning to develop were the fortified enclosures of the Slavic tribes. These forts, *grody* or *hrady*, were protected by earthen banks and wooden palisades,[84] and varied in size from 100 hectares down to less than half a hectare[85] (see p. 256). They dated in some instances from the early Iron Age, but continued to be used, perhaps intermittently, for many centuries. On the other hand, some had been constructed comparatively recently. They appear to have served rather as Fluchtbürgen, as seats of tribal leaders and as the scenes of tribal gatherings and religious ceremonies, than as permanent settlements of large communities. They were very numerous. In Brandenburg alone no less than 90 have been catalogued,[86] and even more have been mapped in Silesia. They appear to have been less numerous in Great and Little Poland, and it is probable that their greater number to the west was related more to the danger of German forays than to the density of Slav population.

It was adjoining or in the vicinity of these *grody* that the pre-urban settlements were growing up, as gatherings of artisans and traders with a few agricultural workers. The forms assumed by this *Vorstadt* or *suburbium* were in general similar to those which prevailed in the west. Sometimes the suburb adjoined the *grod*, as at Płock and Kraków, but this relationship was more often than not ruled out by the terrain. If the *grod* occupied a small island – as at Poznań and Wrocław, a narrow peninsula, as at Kruszwica, or an old meander core, as at Gdańsk, the *suburbium* was developed at any convenient point on firm land nearby. At Wrocław and Poznań the merchants' city spread to the opposite bank of the Odra and Warta respectively. At Łęczyca the constituent elements of the city were half a kilometre apart. The pre-urban settlements of Poland showed a cellular character no less marked than those of contemporary western Europe (see above, p. 265); the difference lay chiefly in the fact that the nucleus around which the cells were grouped was the traditional Slavic tribal *grod*, rather than a feudal castle or a monastery.

The *suburbium* became, as in the west, a centre for crafts and trades. Excavations at Opole and Gdańsk have shown a small but densely settled community. Several of the *grody*, notably Gniezno, Poznań, Wrocław and Kraków, had long been the seats of bishops, whose cathedrals, were built within the *grod* itself, thus repeating the pattern familiar in France and Rhineland. The subsequent partial abandonment of a number of these pre-urban sites has made it possible to excavate them, so that we in fact know more about the Slav cities of about 1100 than about the German. Opole had about 160 dwellings and perhaps 800 inhabitants; Gdańsk may have had a thousand, and Poznań, Kraków and perhaps Gniezno may have been larger.

Polish towns and proto-urban settlements were most numerous in Great Poland. They were fewer in Silesia, Little Poland, and along the Baltic coast. Great Poland was characterised by a relatively large extent of soils of good and intermediate quality, and Little Poland and Silesia were by no means poorly endowed. These areas had all, since at least the Iron Age, been relatively densely populated. A prosperous agriculture had from an early date supported trade and a leisured and aristocratic class. This had in turn encouraged the formation of larger political units than the tribe, and a concentration of political power. There was a growing demand for the products of both domestic crafts and of long-distance trade.

Nevertheless, the earliest Polish kings found it necessary to protect this core-area of their state by building fortresses at points where entry could most easily be effected. Thus originated a line of *grody* along the Noteć, between Great Poland and Pomerania, many of which appear by 1100 to have given rise to some kind of urban settlement. Fortresses, some of them with urban settlements, were established at this time along the eastern boundary of Poland, where their role must have been not unlike that of castles in the Welsh and Breton Marches.

The plains of northern Bohemia and of southern Moravia provided similar physical conditions to those which characterised the more populous parts of Poland. In both areas settlements of an urban character appeared at an early date. Moravia was the earlier to develop, and in the ninth century had formed a state of considerable power and extent. Considerably less is known of its urban development than of the Polish, but it had large walled settlements of an urban character, notably at Starě Město and at Na Valech, near Hodonin. In the former the foundations have been excavated of a stone-built romanesque church.[87]

Similar fortified *hradiště* existed in Bohemia. The most noteworthy are Libice, Levý Hradec and Budeč. Some have been continuously occupied since the early Middle Ages, so that it is impossible to form a clear concept of their earlier shape. Foremost among these is Prague itself, which, even in the tenth century, impressed the Arab traveller, Ibrāhīm Ibn Ja'kub, with its masonry buildings and its commerce.[88] To this group of early urban settlements, in all probability, belonged Brno, Hradec Kralové, Litoměřice and Plzeň in Bohemia, and Olomouc and Znojmo in Moravia.

Similar proto-urban settlements were at this time to be found also in the Hungarian plain, established either by the Slavs and maintained under the jurisdiction of the Magyars, or actually founded by the latter. Among them were Zalazar, to the west of Lake Balaton; Nitra, a former centre of the vanished Moravian Empire, and Székesferhérvár and Esztergom, at this time the seats respectively of the Hungarian king and bishop.

The Byzantine Empire. There is little that can be said of the urban pattern of the Balkan peninsula. Serdica, the modern Sofia, Philippopolis and Odessos, or Varna, probably possessed the distinguishing marks of towns, and the Bulgars had recently established Preslav as their capital, in place of Plisca. Preslav consisted at this time of a *castellum* and palace with storehouses and workshops, contained within a small rectangular enclosure.

Towards the west of the Balkan peninsula the more strategically placed of the imperial cities had retained something of their former importance. The rabble which followed Walter the Penniless was, for instance, refused permission to use the market of Beograd, which appears to have been a small town at this time. Niš, however, was described as 'strongly fortified with a wall and towers', and having enough people to require seven mills.[89] Such ancient cities as Scupi (Skopje) and Stobi continued to be inhabited. Semlin (Zemun) was a Danube crossing for more than one company of Crusaders, and Poetovio (Ptuj) and Emona (Ljubljana) were also probably small towns.

In the areas under Byzantine control urban life was more vigorous, and indeed had suffered no prolonged interruption. Adrianople (Edirne) and Thessaloníki appear to have been cities of size and importance, and later in the twelfth century Edrisi wrote approvingly of Skopje, Ohrid, Edessa and of several towns of Macedonia, Thrace and northern Greece,[90] and others were also listed by Benjamin of Tudela, with the size of their Jewish communities. But in the Greek peninsula itself the glory of the classical city was gone, and Athens was reduced to a small town lying against the northern side of the Acropolis.

Benjamin of Tudela, who was silent on Athens, found Thebes a large and rich city, with an important cloth industry and a Jewish community which he estimated, doubtless with some exaggeration, at two thousand. Despite the raids of the Normans from Sicily, urban life had continued, and even the flourishing silk industry had not been seriously disrupted.

But the largest city of the region, and, indeed, of Europe, was Constantinople itself, with a population that may have been as great as 400,000.[91] Although the jaundiced Liudprand of Cremona had condemned the city, 'once so rich and prosperous and ... now such a starveling, a city full of lies, tricks, perjury and greed, rapacious, avaricious, vainglorious',[92] it remained a focus of trade and the most highly industrialised city in Europe. Benjamin of Tudela pictured the warehouses filled with 'garments of silk, purple, and gold', the wealth of the inhabitants, and the silk-weavers, many of them Jewish, working in the ghetto beyond the Golden Horn. 'Wealth like that of Constantinople,' he wrote, 'is not to be found in the whole world ... and they eat and drink every man under his vine and his fig-tree.'[93] Benjamin undoubtedly exaggerated the wealth and prosperity of Constantinople, just as Liudprand had maligned it. It had passed its peak; the eastern provinces had been taken by Seljuk Turks, the emperor was constantly at war with the Slav peoples of the Balkans, and its commerce was diminishing. Nevertheless, it was a source of wonder to the Crusading horde[94] and of envy to the merchants from the Italian cities who were allowed at this time to establish their 'quarters' within its walls.[95]

As they travelled along the coast of Dalmatia from Aquileia to Dyrrachium, the company of Crusaders who followed the Count of Toulouse passed four large cities: Zara, Salona, which was 'also called Spalato', Ragusa and Antivari, only one of which, Zara, was a survival from classical times. Split (Spalato) consisted at this time essentially of the vast palace built by Diocletian near the water's edge, in which the inhabitants of Salona had, some five centuries earlier, found refuge from the invading Slavs. Ragusa, in origin a small island off the rugged south Dalmatian coast, had similarly been occupied by refugees from the classical Epidaurus on the mainland. Despite the testimony of the Crusaders, these were only very small towns, though, under the fostering hand of the Venetians, their wealth and commerce were growing.

Moorish Spain. Towns were well developed in Moorish Spain, and some of the Arab geographers described them in exaggerated terms. Most, however, represented a direct continuation from the late classical period, when southern Spain was relatively highly urbanised. About 1100 the largest city was probably Seville, the capital of the Abbasides. It covered an area within its ninth-century walls of over 300 hectares, and may have had a population of 50,000. Cordoba and Granada as well as the coastal cities of Algeciras, Malaga, Almeria and Valencia, were all relatively large. Crafts, notably leather and metal working, were carried on in the cities. The surrounding fields as a general rule were irrigated and intensively cultivated. The agricultural lands of Moorish Spain were mainly in the hands of absentee landlords, who swelled the population and added greatly to the purchasing power of the cities. The cities of Moorish Spain, like those of Italy, tended to serve a wider range of functions than those of north-western Europe, where the land-holding classes did not, in general, move to the city.

A description of Seville about 1100 has been left by Ibn 'Abdun.[96] It was a highly regulated city. The streets were kept clean; noxious crafts, such as the burning of bricks and the making of charcoal, were excluded from the city; the washing of clothes was forbidden in those parts of the river from which drinking water was customarily taken, and the city authorities maintained a regular supply of food which was distributed in an orderly fashion to the public. Yet Seville, like most others, was essentially a Middle Eastern city, and Lévi-Provençal rightly says that a picture of the medieval town of southern Spain can best be obtained in Fez in Morocco.[97] The city scene was dominated by its mosques, and near them was the market, either open or covered, in which much of the city's commercial business was transacted. Shops in the European sense were few. The residential quarters were separated from the business, and consisted of a maze of narrow streets bordered by crowded and insanitary houses, which gave place on the margin of the built-up area to the luxurious villas, or *cortijos*, of the landowners and ruling classes.

AGRICULTURE

The Europe of about 1100 was predominantly agricultural in its economy and rural in its habitat. The previous pages have shown how feeble was the development of towns over most of the continent and how insignificant the volume of goods which entered into commerce. Considerably more than 90 per cent of the population, not excluding the Byzantine Empire and southern Spain, must have lived and worked on the land, and the small surplus which they were able to produce sufficed for the provisioning of the urban and quasi-urban settlements. Most of the food was consumed where it was grown. Small quantities were transported to the towns, by water wherever this was practicable, but the extreme difficulties of land transport (see below, p. 300) made it more convenient to take the eater to his food.

Over most of Europe the land-owning classes continued to live in the rural areas where their food was produced. The more richly endowed amongst them passed their lives in almost continuous motion with their servants and retainers, from one villa or estate to another, moving on when the accumulated supply of food had been exhausted, only to return after the next harvest. The king of England, for example, had certain fixed points to which he returned in the course of his annual migration. The German emperors had a similar though less regular pattern of movement between a number of palaces which included Augsburg, Regensburg, the towns of the middle and lower Rhineland, and of Saxony and Thuringia. It is noteworthy that the longest stays and most frequent visits were in areas of the greatest agricultural productivity. Only in Italy and Moorish Spain did the land-owning classes gravitate to the cities, because only here was the movement of food on a sufficient scale practicable at this date. Trade was expanding about 1100, but it was not, with the exception of wine and a few exotic spices demanded by the rich, a trade in foodstuffs.

The extension of cropland

The years around 1100 were a period of growing population and of increasing demand for food. This demand was met by the extension of the acreage under crops and grazing, by the more intensive use of existing cropland and by small improvements in agricultural technique. The increase of cropland took two forms. The more important by far was the extension of agricultural land use around almost every community in western and central Europe, both by pushing back the margin of the forest and the waste and also by establishing new settlements. The other consisted of the eastward migration of settlers from areas which appeared crowded into the less populated lands beyond the Elbe.

To some extent the former movement was the re-occupation of lands which had been settled under the Romans and abandoned during the succeeding times of insecurity and invasion. This was the case in particular in northern France. Of 105 identified possessions of the monastery of Lobbes at this time, no less than 32 have significant Roman remains, suggesting that they were settled in Roman times, and 29 others have yielded evidence of earlier occupation.[98]

In Denmark and the lower Elbe valley contemporaries discovered in the course of their land reclamation traces of earlier cultivation, and realised, not without some surprise, that others had been there before them. The shifting agriculture of the German tribes and the folk migrations of the Celtic and Germanic peoples must have caused the abandonment of so much land that it would be surprising if the colonisation did not in fact lead to the *re*-occupation of large areas that had been cultivated before.

The sources do not allow one to trace the course of land clearance and cultivation during the eleventh and twelfth centuries. One cannot say precisely where it was most active or at what speed it proceeded. It does, however, appear that in southern France and the Mediterranean lands the colonisation of the waste was pursued less actively than in northern France, the Low Countries and Germany. This was probably due to the fact that the early medieval dislocation of life was less marked there and that settlements in general had continued to be occupied through the Dark Ages. Population had continued to grow, and new villages were established, though not on the scale found in northern France. There was progressive division and subdivision of holdings, followed by a consolidation of the scattered parcels which had resulted. At the same time the demesne lands, where they existed, were being broken up and leased, and labour dues commuted for a money payment. By the twelfth century the pattern of land-holding in southern France and much of Italy had come to be dominated by small compact holdings, held by free peasants for a money rent. The resulting tenurial system probably made a very much more effective use of the land and was able to support not only a denser rural population but a growing urban population as well.

In the mountainous areas small and generally self-sufficing communities had continued probably from Roman times, but on the gentler slopes and across the Lombardy plain great changes were taking place. Over the former the cover of scrub and light woodland was giving way to orchards, vineyards and olive-groves. A rudimentary form of terracing was adopted to check soil erosion, and the landscape of central Italy was beginning to assume the features made familiar by the Italian artists of the later Middle Ages.

Wheat cultivation was difficult on the steeper slopes, where the soils were stony and liable to erosion. The largest expansion of cropland at this time took place on the alluvial plains of central and northern Italy. The valleys of the Arno, Tiber and other rivers of peninsular Italy were marshy, and offered little scope for farming, but wheat growing was spreading into coastal lowlands such as the Maremma of Tuscany. The greatest expansion of crop-farming, however, took place in Lombardy. This region of marshy lowland, inadequately drained by the Po and its tributaries, had been somewhat neglected in classical times, though the uncontrolled meanders of the river had produced a plain of great potential fertility. The building of dykes to restrain the rivers and hold back their flood waters had been begun at least by the eleventh century. At the same time canals were being cut primarily for the convenience of navigation, though some must also have served the needs of land drainage. The area under crops – mainly wheat – was greatly extended, and meadows were developed on land too damp to cultivate. Indeed the contemporary expansion of the Lombard cities cannot be understood without a commensurate expansion of food production in the Lombardy plain. But this movement to bring the plain of the Po under cultivation, one of the major land reclamation projects of the Middle Ages, was in 1100 still in its beginning.

France. In southern France there were no such large areas awaiting the peasant's plough, and the slender evidence suggests that cropland did not expand here as rapidly as in northern Italy and in north-west Europe. In part this must have been because the Spanish March and northern Spain, devastated in the course of wars with the Moors, constituted an attractive frontier of settlement. It would be a mistake, however, to underrate the extent to which new settlements were established. There is, for example, good evidence for the creation of villages, especially by the religious orders, near Toulouse,[99] and for the clearing of open woodland in the Alpes Maritimes for the planting of vineyards;[100] but internal colonisation in southern France was on a more restricted scale than in northern.

In many respects the course of cultural development and economic growth in northern France was different from that in southern. This contrast was emphasised almost a century ago by Karl Lamprecht,[101] and nothing has since been found to diminish its significance. The boundary between the two zones is commonly said to follow the Loire. In some respects it lies well to the south of the Loire valley, and in no aspect is it a simple line of division. One can only say that beyond approximately the northern boundary of the Massif Central, land reclamation and colonisation appear to have been pursued more actively; field systems and tenurial conditions were modified, and changes became apparent in such widely separated and apparently unrelated fields as church architecture and public law.

Much of this land reclamation, it has generally been assumed, was directed and controlled by the Church. The role of the Church was unquestionably of great importance, but it is probable that its primacy in this respect tends to be exaggerated because only religious bodies at this time maintained adequate records. The Church at the same time had received extensive gifts of land. The donors, anxious, like all such benefactors, not to pay too highly in material terms for their spiritual blessings, often gave their favourite monastery or chapter mainly forested or waste land. Its value to the recipient was contingent on its development, and the monasteries and cathedral chapters were thus obligated to give more attention to reclamation than lay landlords.

In the record of his administration of the estates of the abbey of Saint-Denis Abbot Suger described how:

> At Vaucresson [in Seine-et-Oise] we have laid out a township, constructed a church and house, and caused land which was uncultivated to be brought under the plough . . . there are already there about 60 tenants. . . The place before . . . had more than two [square?] miles of barren land, and was of no profit at all to our church.[102]

Such activity was common at this time, and the term *burgum* (*bourg*) was often given to these arbitrarily created villages. They became a feature of the early twelfth-century colonisation of the Paris basin, and were established by the king and lay landowners as well as by monasteries.

Not all the new colonists were settled in *bourgs*. The monastic sources contain frequent references to *hospites* who made clearings in the forest and established family-sized farms. The resulting settlements were either scattered or grouped into small hamlets. As a general rule their fields were separate and distinct from one another, and tended to be enclosed by hedges. They gave rise to a landscape which contrasted sharply with that of the compact villages and open fields, of north-eastern France.

In north-western France the spread of settlement and agriculture assumed in large measure the form of small or scattered settlements. At this time peasants were penetrating and settling Maine and western Normandy. The colonisation of the Gâtine, an infertile region to the south of the lower Loire, had hitherto assumed the form of the establishment of new villages, or *bourgs*, or the enlargement of old. Now it changed its character; 'settlers increasingly struck out into the wider, uninhabited countryside to locate their farms', and there came into being 'a multitude of tiny agricultural exploitations, many of which have still preserved the same isolated character today'.[103]

Such scattered settlements were most often held by some form of money rent. They might owe carting obligations to a distant territorial lord, but of the traditional *corveé* they were of necessity completely free. Whatever their initial status, their tenants soon came in fact to be free peasants, in contrast with the degrees of 'unfreedom' which existed on the broad, rich plains of the Paris basin and north-eastern France. The association was clear to Wace in the eleventh century:

> Li paisan et li villain,
> Cil des bocages et cil des plain.[104]

The reasons for this contrast between north and south are obscure. It is sometimes linked with prevailing physical conditions; the *champagne* with the loess soils, the level plateaus and the paucity of wells and springs of much of the Paris basin and northern France; the scattered settlements and enclosed fields, with the more hilly, more densely forested region of the west, with its poorer soils, more abundant water supply and greater reliance upon a pastoral economy. On the other hand the region of large villages and open fields had been more densely settled and intensively developed by the Romans than other areas, and it can be argued that the large village structures derive from the *villae* of the classical period. The latter were rare, but by no means absent in the *bocage* region. Neither explanation alone appears adequate.

The contrasted landscapes were assuming about 1100 the forms they were to retain thereafter. The physical conditions of the Paris basin and north-eastern France permitted easier communication and a tighter seigneurial control than in the *bocage* lands to the west. In the former the integrated village community, subject to *corveé,* practising *vaine pâture* on the stubble, and tied to a regular cropping routine, was both desirable and possible. The increase of the area under cultivation was characterised, not so much by the formation of *hospites* as by the foundation of large villages like Vaucresson (see above, p. 278). There must have been some economic reason for the retention of the demesne and the continued exaction of labour dues. May this be found in the existence of both a market for the agricultural surplus and the means of transporting this surplus to it? The monasteries themselves created a demand for farm produce, and the way in which, about this period, they traded off those of their possessions which were difficult to reach and consolidated those nearest the monastery serves to substantiate this point. At the same time the cities grew in size and created a demand especially for bread-crops. There can, for example, be little doubt that the growth of Paris and the clearing and cultivation of Beauce were closely interconnected. Beauce was 'rich in grain',[105] and it was no accident that the chief grain market in Paris came to be called the *marché de Beauce.* Landowners thus had a vested interest in expanding the agricultural surplus from their lands and in profiting from its distribution to consuming centres. The large, bipartite villa was amenable to this degree of control, while the serf who obtained a *hospitia* in the forest thereby became a peasant in large measure independent of seigneurial control, making little or no contribution to the urban market or monastic kitchen.

Flanders. This region within which colonisation consisted in the main of the further extension of large village communities was bordered on the east and south by others where the creation of hamlet or isolated farm was normal. The boundary on the east was indefinite. Champagne and Burgundy with their large villages passed into hilly and forested regions, such as the Ardennes, the Argonnes, the Jura and the Vosges, within which individual peasant families occupied small holdings.

On the north-east the type of land exploitation met with in the Paris basin ended more abruptly. Large villages ceased roughly where the chalk of Artois dips beneath the Tertiary and Quaternary deposits of the Flanders plain, and Flanders was characterised in general by small settlements and scattered farmsteads.

The land itself was divisible into two regions: the low-lying area of alluvial or marine clays, potentially fertile but in need of drainage and protection from river flood and tide, and, secondly, a somewhat higher area of lower fertility, but drier and less exposed to the danger of flood. Both were at this time being reclaimed and brought under agricultural use. It was clear that the marshlands, which demanded the largest capital investment, were also the most rewarding, and the records of the Counts of Flanders show how high a value was placed on small tracts of reclaimed fen, The cities and industries of Flanders could not have developed without such an agricultural resource in their vicinity.

Land reclamation along the estuaries and lagoons of the Flanders coast may have begun in the ninth century, but was most active after the Norse invasions of the tenth. Dykes were built to exclude the tide; ditches or canals were cut to drain the land thus protected. The process was greatly aided by a slight lowering of sea-level in the eleventh century, but would nevertheless not have been possible without the capital and leadership provided by monastic and lay landowners. The earliest reclamation appears to have been along the shores of the gulf which extended inland between Bergues and Sangatte (fig. 5.4). Lambert of Ardres described the lands around Ardres, which stood near the south-western margin of this gulf as 'waste and desolate'[106] near the beginning of the eleventh century; by 1100 much of the area which extended east to the course of the Aa had been reclaimed, and in the twelfth century the Counts of Flanders built dykes to restrain the Aa along the whole of its course across the maritime plain. By 1180 the bishop was able to consecrate a church constructed in a place which had recently been watery and uninhabitable, but was now drained.

The Acts of the Counts of Flanders contain numerous references to *nova terra*. A grant was made (1107) of 'new land . . . which has grown through the surge of the tide and will grow in the future'.[107] along the IJzer river, and again of land between Watten and Bourbourg (1110) 'which up to now has been recovered from the marsh'.[108] Much of this land was at first not used for crop-farming, but as *berguariae* and *vaccariae*, grazing land for sheep and cattle. It is evident, however, that some crops were grown, and it is probable that as drainage operations became more effective, crop-farming was extended at the expense of pastoral.

Medieval and early modern writers on this area rarely failed to describe its fertility in extravagant terms. Crop yields were high and their variety considerable. It appears furthermore that the practice of fallowing disappeared early, and may in fact never have been adopted on some of the rich peat soils of Flanders. This alone would have justified some of the eulogies of this region. Yet agriculture was carried on under constant threat of flooding by the rivers and inundation by the sea. Dykes were doubtless poorly constructed and easily broken. This was recognised in the regulations drawn up by some of the monasteries.[109] Though the land had been reclaimed by the monastery, the obligation to maintain it and keep the dykes in repair rested on the tenants. They were required to repair the dykes before the feast of St Remigius (1 October) and forfeited their holdings if the land suffered through their neglect.

The less rewarding lands in east Flanders were settled and reclaimed later, and, except near the towns, no great progress had probably been made before 1100; the twelfth and thirteenth centuries were here the chief period of land reclamation and settlement. Water-meadows lined the lower Scheldt and bordered the Zwin and the Honte, but there remained at this time extensive areas of marshland on the lower ground and of heath over the higher.

South of the Scheldt lay a forested and unpromising region which passed eastwards into the sandy Campine. The abbey of St Trond, which owned extensive lands in this area, began about 1100 to clear and cultivate them. The polyptyque of the lands of St Trond, compiled in the thirteenth century, suggests that these poor soils were settled by *hospites* who established scattered farms. To the south lay the region characterised by nucleated settlement and open-fields. The lands of the abbeys of Nivelles and of Lobbes, which mostly lay in this region, were made up largely of bipartite *villae*, like those of northern France.

On the islands of the Rhine mouth and over the sandy heaths and marshes which lay beyond, colonisation was less active than in Flanders, Hainaut and Brabant. Many small settlements, however, were established on artificial mounds (*terpen*) amid the marshy plains, and cultivation was very slowly encroaching on the sandy heaths.

The lands west of the Rhine thus presented a rural economy that had by 1100 probably recolonised the lands that had gone out of cultivation since the Roman period, and had, locally at least, expanded beyond their limits. The formation of new settlements was probably carried on most actively in northern France, particularly in the Paris basin, Flanders and Burgundy, though no quantitative estimate is possible.

Germany. Less progress had been made to the east of the Rhine. Monastic records of the period present a picture of generally small village settlements, fairly widely scattered and each surrounded by its cultivated fields and enclosed by the still almost limitless forest. But, as in areas west of the Rhine, settlements were increasing in size and number. New fields made up of irregular strips were added to the old, and fresh settlements arose in clearings in the forest and waste. To some extent place-names are a measure – qualitative rather than quantitative – of this progress. Certain elements are indicative of forest clearing, but as a general rule they contain no evidence of the date when they were adopted. Jäger has shown[110] how the number of settlements in *Kreis* Geismar, lying to the west of the Fulda river, increased from seventeen at the beginning of the Middle Ages to over a hundred at the end of the thirteenth century, but it is impossible to date this expansion of settlement with precision (fig. 6.16). Much of it probably took place after 1100, and in most of Germany the *Rodungszeit* was the twelfth and thirteenth century, somewhat later than in western Europe.

Early in the twelfth century the Archbishop of Bremen summoned Netherlanders to reclaim and colonise the marshlands along the rivers of Lower Saxony and Holstein, but the more vigorous colonisation of the wastelands along Germany's eastern border, as described by Helmold, had to wait until the middle years of the twelfth century, and the much advertised eastward movement of the German peoples came even later.

Germany suffered at this time from an acute shortage of labour; it is apparent that the appeals for settlers made by both lay and ecclesiastical landowners were inspired, not by the abundance of potential colonists, but by their scarcity. The overall population density in Germany was low; few were available for eastward settlement, and there was generally sufficient potential cropland nearer home. On the other hand, social rather than economic factors stimulated the formation of new settlements along either the eastern frontier or within the forests of western Germany. The peasant in the older villages was in general unfree. A greater degree of personal freedom was a feature of the frontier settlement. When labour was scarce, even the unfree could impose conditions and ameliorate their conditions of servitude, and the principle was accepted that *Rodung macht frei*.

There was a number of relatively large and densely settled regions. These were in general characterised by nucleated settlements, but between them lay forest and waste, receding only slowly before the advance of cultivation. The largest of these cleared lands was along the southern border of the north European plain, widening in the Rhineland, in Lower Saxony and in the Elbe valley. Similar areas were to be found around Frankfurt in the middle valley of the Rhine, and on smaller scale, around Fulda, Speyer, Strasbourg and elsewhere in western Germany and along the valley of the Danube. The larger amongst them constituted the core-areas of the German *Stamm*, or tribal duchies, while the smaller were self-sufficing entities enclosed amid the forest.

Eastern Europe. Between the Elbe and the Oder much of the land was infertile heath and marshy valleys. Though settlements were numerous, good cropland was restricted, and gave little room for expansion. Along the Oder valley, however, they were more extensive, especially in Silesia and on the young moraine soils of Pomerania. East of the Oder, terminal moraine, sandy outwash and the broad, damp river-valleys greatly restricted the area under agriculture. The water table in the valleys was, however, lower than during the later Middle Ages (see p. 13) and the extent of alluvial soil available for cultivation was greater than in later times. The landscape was dotted with small areas of cleared land, smaller on average than those in Germany, but all of them undergoing a similar slow expansion at the expense of the forest.

It is not difficult to relate these centres of agricultural expansion to the better soils – the alluvial terraces along the Vistula, the loess of Silesia and Galicia, and the more fertile boulder clay. The population of Poland is estimated to have approximately doubled between 1000 and 1340; and it had probably been increasing slowly in the centuries before the year 1000. Land reclamation was probably most active in Great Poland and in the region around Kraków. But monasteries were very few at this date; written records scanty, and the progress of land clearance in large measure unchronicled. Conditions in Bohemia, Moravia and the more westerly parts of the Pannonian plain must have been broadly similar, with a slow expansion of agricultural settlements under the impact of a gradually rising population.

Fields and field-systems

It is impossible to study the medieval expansion of agriculture apart, on the one hand, from the growth and mobility of population and, on the other, from the forms assumed by settlements and fields, the agricultural techniques employed, and the crops grown. Nor can it be taken for granted that settlements and field patterns which can be reconstructed for early modern times were identical with or even similar to those of the original settlement. The forms assumed by medieval fields were many and varied. They derived in varying degrees from soil conditions, agricultural techniques and social conditions, all of which were susceptible to change. Two forms, however, can be distinguished: the compact enclosed field and the open-field with strips which usually had no permanent fence between them. Field shape has generally been taken to bear some relationship to the kinds of plough and traction used, with a light plough (*aratrum*) dictating a small and approximately square field, and the heavy, mouldboard plough requiring an elongated strip. Though the heavy plough tended to be used on the long strips, the correlation between plough type and field pattern does not seem to have been particularly close. It has also been argued that, since the strip, especially if ploughed in such a way as to form a high-backed ridge (*Hochäcker*), drains more easily, it would tend to be adopted on damp and heavy soils. The fact is that different types of field were to be found adjoining one another on similar soil-types. While the influence of technology and soil can certainly not be excluded, it appears that social conditions were probably more important in determining both the field-system and the agricultural technique. Verhulst has demonstrated[111] that the oldest of the cultivated lands of St Bavon at Ghent consisted of *kouters,* large open fields in which the land was arranged in strips, and that as cultivation came to be extended it increasingly took the form of small and compact fields. Since the strip needed a large team for its cultivation, open-fields were not likely to be found in association with small hamlets, whose total resources would be unlikely to add up to a whole team. Small settlements are likely to have used a lighter tool in a smaller field. This still leaves unanswered the question how the large village and the open-field came to be associated, and whether the association can be pushed back to the period of Roman occupation of western Europe.

The simplest form of strip cultivation is provided by the *Esch* of north-west Germany and neighbouring regions of the Netherlands. It consisted of a field, irregular in shape, and divided into a number of unequal and often somewhat curving strips, shorter, in general, than those of the true open-fields. Increase in the number and size of such fields was often prohibited by physical conditions, but elsewhere one finds evidence of field laid against field, each divided in the same way. It cannot be said with any certainty that the strip answered to any technical need for the peasant; it was a manner in which he tended to react when faced with the problem of dividing a restricted area of cropland between cultivators.

Elsewhere in north-western Europe, the form of the strips was more complicated than in the *Esch* of north-west Germany. There was more than one field to be divided, and within the individual fields the pattern of strips was fluid; they might be divided between heirs and regrouped; they may have originated in single *Esch*-like fields, to which others were subsequently added. It is certain that north of the Loire and eastwards from Normandy to the Elbe most of the cultivated fields were at this time divided into strips, and each peasant probably possessed several strips scattered through most of the fields. But within this north-west European region of open-fields, of *Gewanneflur,* there unquestionably lay many areas, large and small, in which the opposite type of field, the compact or square *Blockflur* predominated. Foremost among them were the newly enclosed areas of Flanders plain.

West of the Beauce also the open-field tended to give way to small enclosed fields, and strip cultivation was to be found only locally. Large villages were rare; many of the settlements about 1100 were of recent origin; they were small and scattered and each stood amid its enclosed fields. But in these western lands also settlements were to be found, often of hamlet size, surrounded each by a few fields made up of strips. These were the in-field, manured and continuously cultivated. Around it lay the more extensive out-field, portions of which were ploughed and tilled in turn. This in- and out-field constituted an early field-system which may have been very widely distributed through western Europe.

The small enclosed fields of the west and of Flanders were worked in close association with animal farming. Manure was available; thus it may not have been necessary to leave the land fallow every two or three years. Each farm tended to constitute a discrete unit, and, though custom dictated in general the methods of cultivation, there were few social constraints on the peasant, and he was free to experiment within limits. Over most of the open-field region of north-west Europe, manure was not available in quantity, though the polyptyques sometimes note the obligation of the peasants to supply it to the demesne. Fallowing was therefore necessary, and it is probable that a three-field system prevailed, with a third of the cropland under bare fallow in any one year. This system is generally held to have replaced a two-field system, and it is probable that the latter may have lingered on in some areas. The need to feed stock during the winter, coupled with the desire to make the most of their droppings, led to the practice of opening the cropland to *vaine pâture* (common pasture) as soon as the harvest was in. The rural community was thus strongly regimented, and custom ruled supreme.

The village and its fields were bipartite. Part was cultivated by the peasants for their own purposes, and part was in demesne, providing food for the lord with the labour of the peasants. Such is the traditional regime of the open-fields of north-west Europe. The villeins of St Rémi performed ploughing services in autumn and spring on the demesnes of the abbey. Similar obligations were owed on estates in Artois, Hainaut and Brabant. Even in Alsace, the lands of the abbey of Marmoutier were mostly divided into demesne and villeins' land, the latter supplying the labour for the former. Doubtless this was an ideal to which many a landowner tried at this time to approximate, but it was one that was in all probability only rarely achieved. The statistics of the polyptyques can rarely be reconciled with this simple and logical system (see below, p. 204). The number of fields and the variety of crops suggest a pattern more complicated; perhaps the elegant simplicity of the three-field system may have been an end product of centuries of deliberate experiment and enforced change; it does not appear to have come at the start.[112]

Some areas of German settlement appear to have been associated with strip cultivation in open-fields and a bipartite organisation of the land, and it is at least possible that this system evolved, with increasing population and improving technique, from an in- and out-field system of cultivation. East of the Rhine, however, there is little evidence for the bipartite manor. The polyptyque of Prüm showed, at the end of the ninth century, only small settlements, whose inhabitants discharged their obligations in kind (see above, p. 207). The eleventh-century polyptyque of the abbey of St Emmeram,[113] at Regensburg, represented its tenants as discharging their obligations by contributions of oats and rye; chicken and eggs; flax, wool and cloth; nowhere are they recorded to have owed any form of labour on the demesne. Beyond the limits of German settlement, in the Slav regions of Bohemia and Poland, the agricultural system was later to be characterised by open-fields, but it is probable that some form of in- and out-field system prevailed at this time.

In southern Europe, however, the open-field system of agriculture was rare, and in most places unknown. The small, compact fields, inherited from the classical period, continued to be cultivated with a light plough. The absence of heavy clay soils from much of the region and the difficulty of keeping draught animals in a region of Mediterranean climate together tended to discourage the use of the mouldboard plough. The classical alternation of cropping and fallow was preserved. The field-systems of the south, like those of the *bocage* lands of the west, allowed greater initiative to the peasant or cultivator, who was not subject to communal regulations of his cropping and agricultural methods. The peasant could change his pattern of land use without disturbing that of his neighbours. Such flexibility was a feature of southern agriculture, and indeed, it would have been impossible to expand the production of cash cultures, of which the grape-vine and the olive were by far the most important, under any other system of land division.

The crops

Bread-crops continued to provide the staple food of most people, and occupied in every part of the continent the greater part of the cropland. Wheat or spelt and rye were preferred, but all too often the peasant had to content himself with oats. Barley was grown in most of northern and north-western Europe for the purpose of malting and brewing, and malt, or *bracia*, was a common form of rent in kind. But it was an exacting crop, and all too often oats were used instead for making beer. Among the bread-crops, wheat (*frumentum*) was always preferred, but spelt (*spelta*) and rye (*sigillum*) were in fact more extensively grown, and in time of famine oaten bread was eaten. Galbert of Bruges relates that on such an occasion the Count of Flanders prohibited the brewing of beer and 'ordered bread to be made out of the oats so that the poor could at least maintain life on bread and water'.[114]

Wheat was, with barley, the dominant crop in southern Europe, but was grown only on restricted areas in western and central Europe and probably not at all in eastern. It was more demanding than the other cereals in terms both of soil and climate. It was sown in autumn and exposed during the winter months to the hazards of ill-drained fields and severe frost, but, should wheat fail to survive, the land could be ploughed and sown again in the spring with quick-

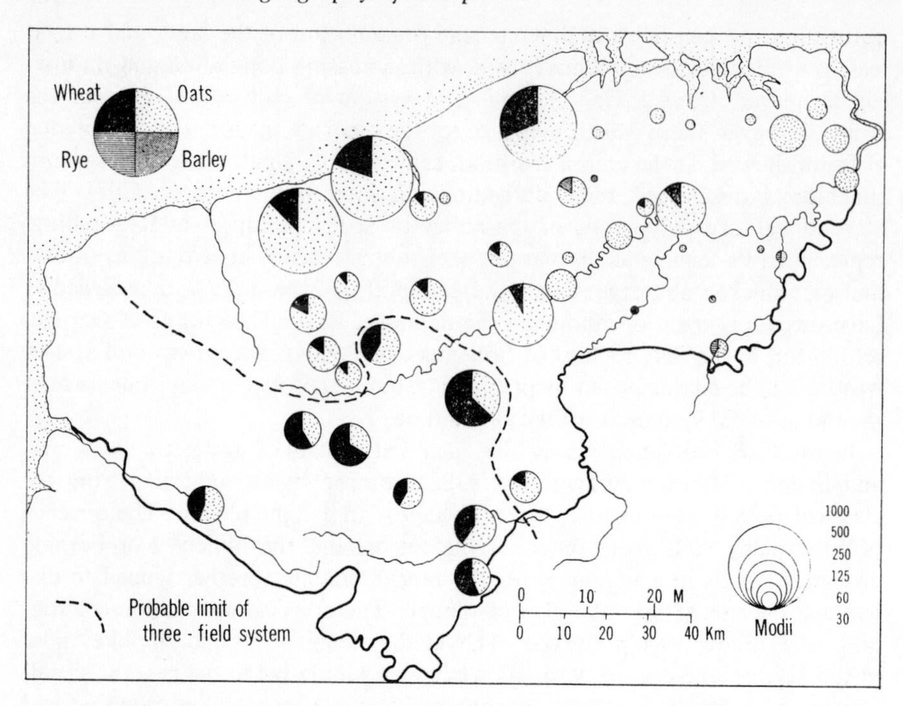

Fig.5.6 Grain crops in Flanders and northern France

growing oats. Barley was an unreliable crop. It was grown on the good soils of Beauce, Lorraine and Burgundy, but elsewhere the peasant grew a little if it formed part of the dues he owed his lord; for himself he was more likely to grow the more dependable and heavier cropping oats and rye.

In the *bocage* regions of the west rye was the chief bread-crop. On the damp, poor soils of such areas as the Gâtine, it was almost the only crop grown; it was predominant in the hills of the Auvergne and Beaujolais, and widespread in the Alps, and in Great Britain it remained the principal bread grain of the working classes until the eighteenth century.

In the hills of eastern France, in the Ardennes and the Eifel, and also in the plain of Flanders and on the newly cleared lands of Brabant, oats were the foremost and in some places the only grain-crop. Apart from an occasional reference to spelt, the oldest register of St Bavon of Ghent records only oats,[115] and it was oats which provided the chief food during the Flanders famine of 1124–5. Oats grew well on light soils, and were also, as is apparent from the Prüm polyptyque, the first crop to be grown on newly cleared forest soils. They were the dominant crop over all of Germany, and must have provided the bread grain for most of the population. Rye came next in importance, and a little wheat was grown for the lord's table in south Germany. The *censier* of St Emmeram at Regensburg[116] of the mid-eleventh century recorded the payment made in kind by the monastic tenants. Out of a total of 895 *modii* of grain contributed in this way no less than 769 – 85 per cent of the total – consisted of oats. Eleven per cent was made up of rye and the rest was wheat. No barley was recorded. Bavarian *censiers* of a somewhat later date continue to show a marked predominance of oats, though the volume of rye contributed as rent payments appears to have been growing during the period under consideration.

Table 5.1 *Grain contributed as rent payments in Bavaria*

			Percentage			
	Date	Total payment (*modii*)	Oats	Rye	Wheat	Barley
St Emmeram	1031	895	85	11	4	..
Salzburg	*c.* 1200	5,600	61	29.6	9	0.4
Tegernsee	*c.* 1250	6,148	79	11	7	3
Ducal lands in Upper Bavaria	*c.* 1280	21,577	55	30	12	3

Source: Data in P. Dollinger, *L'Evolution des Classes rurales en Bavarie* (Paris, 1949), p. 168.

The records of the abbey of Werden-on-Ruhr for this period show oats and rye predominating in north-west Germany, together with a little barley, but no wheat. Indeed the evidence for cropping here suggests proportions very similar to those established for nearby Duisburg in the Prüm polyptyque (see p. 205).

In the Alps of Savoy oats were the most widely cultivated grain. Wheat was also grown in some of the warm and fertile valleys, such as Grésivaudin, but rye and barley were relatively unimportant. Beyond the limits of German settlement oats continued to be extensively grown, together with rye, but the evidence is inconclusive.

Although cereals must have occupied the greater part of the cropland, they were by no means the only crops grown by the peasantry. The diet was varied by a considerable consumption of beans, peas and a number of garden vegetables, and some piquancy was added by extensive use of herbs. Peas and beans were occasionally sown as field crops, but most vegetables were grown as a general rule in small garden plots. They rarely figured in rent-rolls or tithe lists, perhaps because, with the exception of beans and peas which could be dried, they were perishable and thus unsuited to serving as a means of discharging obligations to church and landlord. Most monasteries and, doubtless, many lay landlords at this time possessed gardens from which they were supplied with herbs and fresh vegetables.

Viticulture. The distribution of the grape-vine was of especial importance. Wine was not only necessary in the ritual of the church, but was much sought-after by all who could afford it. It was one of the most prominent items in long-distance trade, and, since its transport was far from easy, the cultivation of the vine was more widely diffused than at any other time in its history. In southern Europe, where the grape vine was climatically suited, it continued to be widely grown, but even here its cultivation appears to have advanced deeper into the mountains than is the case today. Viticulture must have been increased to meet the needs of a growing population, and there was an export of wine from certain favoured areas such as Cyprus, and probably also from Crete and Sicily.

The limit of viticulture was pushed far to the north-west and north during the tenth and eleventh centuries. Wine-growing spread through Normandy, where monasteries, such as Jumièges and Fontenelle, established vineyards on the sunny slopes of the Pays de Caux. Vines were even grown on the coast near Le Havre, but were found not to do too well. On the borders of Brittany and on the poor soils of the Gâtine, landowner and peasant nevertheless often set aside a small parcel of land for vines.

Viticulture spread northwards across the plains of Beauce and the *côtes* of Champagne. Northern monasteries continued to maintain small wine-growing estates in the neighbourhood of Laon and Reims, but the frontier of viticulture had in fact spread northwards into Artois and Picardy. The hill slopes around Nogent-sous-Coucy were covered with vineyards, and the grape-vine was far from uncommon even in Flanders.

But southern England and Flanders were the frontier of wine-growing at this time. In Carolingian times there had been a remote vineyard at Ghent; by 1100 viticulture in the Low Countries had greatly increased; it was practised near Tournai and Antwerp, at Huy and Liège on the Meuse, and doubtless at many other places which have failed to be recorded.

The greatest expansion of wine-growing was however, not in such marginal areas as Normandy, Artois and Flanders, but in those which were both climatically suited to the vine and also endowed with good riverine or maritime communications with the markets of the north.

The chief exporting regions at this date were Burgundy, Lorraine, the Moselle valley and the Rhineland, and of these the most important were without question the Moselle valley and the Rhineland. The rivers provided the usual means of transporting the wine, and much of it must have made long and circuitous journeys to the northern monasteries to which it belonged. The monastery of Saint-Trond, to the north-west of Liège, owned several vineyards along the Rhine above Cologne and at Pommeren and other sites on the Moselle. The Abbey of Nivelles, to the south of Brussels, also owned vineyards which lay scattered on the Moselle and Rhine. One presumes that the wine barrels were carried by boat down the Rhine and up the Meuse or Scheldt.

Vineyards in Lorraine probably sent wine down the Meuse and Moselle. Burgundy, which extended from the Jura Mountains to the headwaters of the Paris rivers, possessed, at least in its more westerly parts, advantages similar to those of the Rhineland. The Seine and its tributaries flowed northwards and had probably begun to carry wine to the Paris region which was to become one of the largest markets for wine in medieval Europe. There is evidence that much of the Burgundy wine-production trade had been in the hands of Jewish entrepreneurs, but that at the end of the eleventh century they were moving away. This may afford at least a partial explanation of the apparently low level of this wine trade as compared with that of the Moselle and Rhine.

Viticulture was established in south Germany, where it may have survived from the Roman period, but the evidence is that production was small, and that this region received wine from lower down the Danube valley. Nevertheless the grape-vine had been carried far into Germany by colonists from the Rhineland, and although it had not by 1100 reached its greatest areal extent, it was already being grown on a small scale in areas from which it has since vanished, including Lower Saxony and Thuringia.

In those wine-growing areas which lay close to the climatic limit of the vine, one must distinguish between the vineyards which made wine for local consumption, and those which produced for distant markets. Most of the latter, it would appear, distributed their wine by boat, and the vineyards lay almost exclusively close to navigable rivers. This was most apparent along the Moselle and Rhine, but even in Normandy, where the vintage can hardly have been of high quality, the vineyards lay close to the Seine or to its navigable tributaries, and were unknown in valleys which did not have the advantage of a navigable waterway. Cities clearly provided the largest market for wine; it was wine which emboldened the men of Cologne to revolt against their archbishop. Vineyards were extensive around the larger cities by 1100. William of Poitiers described the destruction of those around Le Mans,[117] and Edrisi mentioned those at a number of cities of northern France, including Angers, Tours, Chartres, Reims, Laon and Paris itself.[118]

Industrial and fodder crops. Agriculture contributed primarily to the nourishment of man, and the area under industrial and fodder crops made up only a very small proportion of the whole. Nor was the variety of such crops great. Most of the oil-bearing plants which have since become important were either unknown or little grown. The chief exception was, of course, the olive, but its climatic requirements restricted it to the Mediterranean region. The monastic records relating to southern France and Italy frequently mention the cultivation of olives, and olive trees were of such value that the records mention the ownership of a single tree or even of part of a tree. Olives were extensively grown in Spain, where the earlier *Lex Visigothorum* had protected trees by providing severe penalties for those who cut them down; olives were also grown in Greece and along parts of the Balkan coast. But olive oil seems rarely, if ever, to have entered into long-distance commerce, presumably because there were adequate substitutes in non-Mediterranean Europe.

There is evidence later for the cultivation of rape- and poppy-seed for the sake of their oil, and flax was sometimes allowed to ripen and its seed was then crushed. These sources were of trifling importance about 1100; vegetable oil at this time was derived almost wholly from nuts which were gathered in the woods and crushed. Oils and fats, including those burned in lamps and made into candles, were of animal origin.

There was on the other hand a widespread and important cultivation of vegetable fibres. Hemp, the material universally used for rope and coarse fabric, was the more extensively grown. Evidence is scanty or lacking for central and western Europe, but in western, there were in all probability few communities which did not grow hemp, at least occasionally, for their own use. Flax was a more valuable, a more demanding, and thus a more restricted crop. It yielded a smoother and stronger fibre, and was more prominent in tithe and rent payments than hemp. It was conspicuous among the payments demanded by monasteries, and some communities, both rural and urban, appear already to have achieved some importance for their manufacture of linen. Flax appears to have been grown most intensively in north-western Europe, in approximately the area where it was recorded in the ninth century. Already we find in this area the beginnings of a linen industry.

The dyestuffs used in cloth-making were almost wholly of vegetable origin; they were few in number and they gave a very restricted range of colour. They were relatively valuable in relation to their weight and bulk, and their cultivation tended to be highly localised and thus gave rise to long-distance trade. The vegetable dyes in most common use were woad (*gauda*) which yielded a blue dye, and was grown at this time in Artois, Flanders and the Limousin, as well as in other parts of north-western Europe; weld or pastel, which gave a yellow dye, and was widely grown and very much used in France, and madder, or *garantia*, which continued until modern times to be an important source of red pigments. Saffron was used at a later date, but no evidence has been found for the cultivation of the crocuses from which it was prepared as early as 1100.

Mineral dyes do not appear to have been used, but two of animal origin were locally important. The *purpura* or *murex* had been known in the Mediterranean since classical times, and continued to be employed in the workshops of the Byzantine Empire. From here it was carried to Sicily in the twelfth century. There is no evidence of long-distance trade in purple, and in north-western Europe a reddish dye was obtained from the kermes. The collection of the insect (*vermiculi*) and the preparation of the dye were occasionally cited as obligations of the peasants. But this was not a colourful time; much of the cloth was undyed; the splendour of medieval glass and wall-painting had not begun to appear, and there was little in 1100 to suggest the later importance of the production and trade in dyestuffs and pigments.

Animal farming

Meat and milk products formed only a very small part of the human diet in most parts of Europe, and with the exception of the pig, animals primarily served other ends than the provision of food. The ox was used mainly as a draught animal, and the paucity of references to the dairy cow, or *vacca*, suggests how unimportant was cow's milk in most areas. The sheep was bred more for its wool than for its food value, though ewe's milk, and the cheese made from it, were locally important. The horse was the principal beast of burden, carrying not only the knight to battle, but also the bale of wool and the roll of cloth to the market. Only the pig served no other end than the provision of food, and bacon and pork were, of all forms of meat, those consumed most widely and in the greatest amounts.

An important reason for the relatively small number of farm animals was the lack of fodder crops, the scarcity of meadow and improved grazing, and the consequent difficulty, especially in northern Europe, of keeping animals through the winter. In Mediterranean regions the difficulty was rather the lack of summer grazing which restricted animal husbandry and contributed most to the widespread practice of transhumance.

The ox was the principal draught animal in most of northern Europe, though the horse, with its greater speed, was preferred in some areas. In southern Europe the donkey and the mule remained the chief beasts of burden. For this there were good reasons in their greater ability to withstand the summer heat and to thrive on the poorer fodder that grew in a Mediterranean climate.

Over much of humid, north-western Europe, swine were the most numerous of farm animals. Their natural habitat was oak and beech woodland. Here they needed and received little attention. Their numbers were reduced each autumn, and their carcasses salted for winter consumption. Woodland continued to be valued as much – if not more – for its ability to support swine than for the provision of lumber. Nowhere, however, were cattle and sheep excluded; their number depended on the supply of winter feed, much of which went to the support of draught animals. Sheep were, for example, most often mentioned in Flanders, where the comparative scarcity of forest restricted the rearing of pigs. That sheep were kept on a very large scale is shown by the rent of 450 measures of land (*mensuris terrae*) at Eecloo, which consisted yearly of 50 poisses (*pensas*) of cheese, 8 of wool, and 200 lambs.[119] The same records frequently mention the bercaria, or sheep-runs; but *vaccariae*, or dairy-farms, very much less frequently. The Count of Flanders at this time had his own sheep-runs, which were maintained by overseers and earned a money income for him. Sheep appeared to have been important also in the marshlands which fringed the Danish peninsula, and were doubtless also kept in the coastal marshes of Frisia. Local wool at this time probably was a very significant factor in the cloth industry which was now becoming established in the Low Countries. But sheep-rearing and wool production were also important over the plateau of the Ardennes, in the Alps, and the Massif Central of France, and on the damp and poorly drained grazing lands of the Lombardy plain. But it was in the dry summer conditions of the Mediterranean region that sheep were, relative to other animals, the most important. Large flocks were to be found in southern France, throughout peninsular Italy, and above all in Spain.

Extensive sheep-runs were not new in southern Europe. They had covered large areas during the later years of the Roman Empire, and the devastation which accompanied the invasions of the early Middle Ages served only to increase the area that could be used in this way. In the Spanish peninsula, the *Reconquista* and its accompanying depopulation added greatly to the extent of wasted and empty land, and when settled agriculturalists again moved into parts of the Meseta, it was into a land in which the pastoralists had already established a prior claim and were able to impose restrictions on the extension of cropland. At this date, however, neither the merino sheep nor the *Mesta*, the organisation of sheep owners, had yet made their appearance (see p. 384).

In a Mediterranean climatic setting any extensive animal husbandry was dependent upon a seasonal movement between summer and winter grazing lands. In much of Italy, southern France and perhaps also the Balkans, this migration was over only short distances, from coastal lowlands and valley floors to nearby highlands. On the Spanish plateau, where contrasts in relief and climate were less marked, migrations were of necessity longer. Visigothic law had protected the transhumant sheep, and had given their shepherds wide privileges to conserve and protect the grazing, and these rights were later confirmed and even strengthened.

Cattle were numerous and important only in north-western Europe. This was to a very large degree a matter of climate which restricted the cultivation of bread-crops and encouraged the growth of grass. William of Poitiers presents a picture of conditions in Brittany, where milk was more important than bread and over large areas the flocks and herds had not been displaced by arable farming.[120]

In Scandinavia also pastoralism was dominant. The Norse sagas portray a society in which crop-farming was little more than an adjunct to pastoralism. Animals were generally grazed near the coast and on the valley floors during winter, where they helped manure the fields. In spring they migrated to the upland grazing, returning to the lowlands when early frosts and the first snows made the former no longer tenable.[121]

Cattle appear to have participated most in this transhumant movement, but sheep also took part, and, if Egil's Saga is to be believed, sometimes remained high in the mountain valleys through the winter.

> When Skallagrim's livestock was much increased, then went the cattle all up into the fells in the summer. He found there was great odds in this, that those beasts became better and fatter which went on the heaths, and this too, that the sheep throve a-winters in the mountain dales, even though they could not be driven down. So now Skallagrim let make a farmstead up by the fell and had a dwelling there: let there tend his sheep.[122]

MINING AND MANUFACTURING

Society was mainly rural, and social relationships were predicated upon labour obligations on the land and upon payment made in agricultural produce. It was these that were recorded. Manufacturing, on the other hand, was primarily a domestic occupation; society was more or less self-sufficing, and only rarely were rents paid or obligations satisfied in handicrafts. Yet crafts were practised everywhere; cloth was woven, ores smelted, ironware forged, glass made in the most unlikely places. It is only by the chance discovery of a furnace that we know that window glass was made in central Poland and iron smelted on the Norwegian coast at this time. The literary record is almost silent.

Mining

Minerals must have been mined and smelted on no inconsiderable scale. Iron was used in weapons and armour, in the tools of the mason, woodsman and farmer. Yet one is at a loss to know where it was made and how it was distributed. Most of the Roman mines appear to have been abandoned during the early Middle Ages, to be re-opened gradually as the Middle Ages progressed. The mines of Styria were probably in use, and their ores were smelted in the neighbourhood of Eisenerz. It is certain that iron was being smelted at this time in the Siegen district of north-west Germany and possibly also in Lorraine and Burgundy, in the Alps of Savoy and northern Italy, in northern and southern Spain and in Sweden, and primitive smelting furnaces have been excavated at several sites in southern Poland. About this time the newer monastic orders, notably the Carthusians and Cistercians, began smelting iron, since their isolated communities could obtain it in no other way.

Smelting was carried on with small hearths, roughly built of stone and clay; the metal obtained was poor in quality and full of slag. It appears to have entered into long-distance trade to only the smallest degree, and is scarcely ever mentioned in inventories of goods and schedules of tolls. The account in Egil's Saga of iron-working along a Norwegian fjord could doubtless have been paralleled from much of Europe if only there had been someone to record it:

> Skallagrim was a great iron-smith and had great smelting of ore in winter time. He let make a smithy beside the sea ... the woods lay not over far away there. But when he found there no stone that was so hard and so smooth as might seem to him good to beat iron on ... [he found one] ... and bare the stone to his smithy and laid it down before the smithy door, and thenceforward beat his iron on it. That stone lieth there yet, and much burnt slag nigh.[123]

Iron-ore was the most important mineral mined and worked at this date, but it was far from being the only one. The increasing volume of trade necessitated an ever larger circulating currency. The precious metals were in demand, and wherever they were found, they were vigorously exploited. Perhaps the foremost centre of silver-mining in Europe at this time was the Rammelsberg, the north-eastern bastion of the Harz Mountains, which overlooks Goslar. It was silver from this source, it is claimed, which circulated in England and was used to pay the Danegeld. If other mining centres were active and important, they are not known, with the exception of southern Spain, where the mining of the precious metals, as well as of copper and other non-ferrous minerals, had perhaps been practised continuously since Roman times.

Quarrying was carried on wherever building materials were to be found. Most domestic building was of wood, but there was a large and growing demand for building stone for churches and monasteries, castles and urban defences. Architecture was generous in the safety margins which it allowed; walls and pillars were thick and window openings small. Though much of the vast bulk of romanesque building was of rubble masonry, this was usually faced with prepared stone and often carved and decorated. The local stone could often be used for the former, but superior qualities of free-cutting stone were needed for the latter. The fabric of most buildings came from nearby quarries and was extracted, carted and set in place by the unfree labour of peasants. Every town, monastery or parish was likely to have its quarry, from which much of its building material came, but good-quality stone had often to come from a distance and was used sparingly.

The transport of stone was difficult and costly. Overland movement was reduced to a minimum, and the greatest possible reliance was placed on waterborne traffic. Limestone from the quarries at Caen was, for instance, widely used along the south coast of England. The precocious architecture of the Paris basin owed much to the fact that its network of navigable rivers facilitated the movement of high-quality building stone, and there was a significant movement of stone along the Rhine and its navigable tributaries. Much of the building stone of Venice was brought by ship from the opposite shore of the Adriatic.

The Romans had made great use of brick, but its manufacture appears to have died out in most of Europe. Here and there, however, Roman sites were raided, and the ancient brick re-used, often in rubble masonry, occasionally in the construction of the decorative features of pillars and windows.

Manufacturing industry

The simple, unrecorded crafts were carried on in all parts of Europe. Wool and flax were spun and woven; cloth was fulled and dyed; garments made and leather tanned and fabricated into footwear, harness and domestic utensils. There were masons and carpenters everywhere. Lime must have been burned to make mortar. Metals were smelted and fabricated by armourers and jewellers; in a few places a coarse glass was made. But of all these activities we know almost nothing for the years around 1100. The craftsmen were few and probably unorganised. If they kept records these have not survived. The labour of many of them was part of their obligations to their territorial lords, prescribed by custom and unrecorded.

The tenants of the abbey of Stavelot-Malmédy were required, early in the twelfth century, to gather limestone and burn it to make quicklime or, if the monastery should already have enough lime, to deliver building timber instead.[124] In the ninth century, the obligation to make rough woollen or linen cloth for the monks was imposed widely on the peasants; there is no reason to suppose that conditions had greatly changed by 1100. Other industrial processes were the smelting and fabricating of metals, the making of pottery and glass, and the manufacture of jewellery and of other articles of personal and architectural adornment. These required at least supervision by professionals. The larger monasteries must have had workshops, at least during their periods of construction, but these crafts, with the exception of smelting, were chiefly carried on in the larger towns.

Glass for the windows of the churches was being made in central Poland, and, since much of the artistic inspiration of this region derived from eastern France, it is probable that glass manufacture was more widely practised in Burgundy and Lorraine. At the same time there were other specialised craftsmen at Kruszwica, and doubtless also in other towns of central Poland, who worked in bronze and amber. It is probable that crafts technically as advanced and specialised were also to be found in Bohemia, the lower Rhineland, eastern France, the Paris region and, above all, in northern Italy, and the cities of the Byzantine Empire and of Moslem Spain.

Among the more localised products which entered into long-distance commerce was salt. Salt was relatively more important during the Middle Ages than in modern times, owing to the shortage of winter feed for animals and the need to salt down meat in the autumn. It was obtained from brine springs, of which those of Lorraine, Upper Austria and Poland were amongst the more important (see below, p. 399). But most came from coastal salt-pans. These were to be found even on the cool, humid coasts of Flanders and eastern England, where the natural evaporation of the water must have been a slow process. They were more important on the Bay of Biscay coast, from which the salt was taken by boat up the Loire for distribution over central France, and above all along the shore of the Mediterranean. From Narbonne to the *étangs* of the Rhône there were salt-pans, whence the salt was shipped up the Rhône until it met the competition of the brine-springs of Lorraine and Burgundy, and up the alpine valleys to Piedmont until it encountered the salt brought up the Po by boat. The marshes which fringed the head of the Adriatic Sea had long been important for their salt, trade in which had contributed to the rise of Venice. It now supplied much of the Lombardy plain and was carried into the Alps and the valley of the Danube. Around the coasts of Spain and Italy and probably also of Greece were salt-pans which supplied local needs and those of their hinterlands.

In Europe at this time it is possible to distinguish only three areas which, by reason of the concentration and specialisation of their industrial structure and of the long-distance trade to which this specialisation gave rise, deserved to be called industrial regions. They were Flanders, northern Italy and the more dispersed manufacturing centres of the Byzantine Empire.

The more southerly parts of the Low Countries had already in the Carolingian period developed a cloth industry of more than local importance. By 1100 a significant revolution was taking place: the manufacture of cloth was moving from the rural areas to the towns, and at the same time was becoming a predominantly male occupation. Why this revolution should have taken place first in Flanders and Artois is far from clear. This was an area where sheep were kept in large numbers. It was also one where feudal control was less complete than in most other areas of western Europe. It was a frontier of settlement (see p. 267) where the peasant enjoyed a greater personal freedom and there was more scope for the entrepreneur than elsewhere. This does not, however, altogether explain the rise of long-distance trade, which alone could justify the kind of specialisation which we find by 1100. It is possible that this can be associated with the Vikings, who performed the positive role of redistributing goods within north-western Europe, thus developing a knowledge of and taste for distant products. Small commercial settlements, such as Bruges, had developed early in the eleventh century; by 1036 it was worth the trouble of the abbey of Saint-Vaast at Arras to establish a very detailed table of tolls on a wide variety of goods, including wool and cloth, passing through the town. A cycle of fairs developed within Flanders, of which those at Messines, Thourout, Saint-Omer and Douai were already in existence by 1100, and those at Ypres and Lille were active a few years later.

The chief cities of Flanders – Valenciennes, Saint-Omer, Ypres, Bruges – together with the small towns and villages which had become economically dependent on them, were active in the manufacture and sale of cloth, both woollen and linen. The chronicle of Galbert of Bruges and the Acts of the Counts of Flanders show that there were settlements in which manufacturing and commerce had assumed a leading role. In most of them the merchants had already grouped themselves into gilds or *Hanses*, and the creation around the margin of the Flanders region of stations, as at Arras and Bapaume, where duties were levied on commerce, shows that a regional was giving place to an interregional trade.

The second centre of manufacturing at this time lay in northern Italy, where also specialised industrial centres were emerging and a local exchange of goods was being gradually transformed into an interregional trade. The Flemish and the north Italian industrial regions were already in commercial contact. Italian merchants had appeared in the cities of Flanders, and the beginnings of the Champagne fairs, the meeting place of Flemish and Italian merchants, were already apparent. We may be sure, however, that commerce between the two was at this date still restricted to goods of small volume and high value.

Though the manufacture of cloth and leather, of metal and ceramic goods was important in the north Italian cities, their economic foundation lay rather in their commercial functions. Venice, Genoa and Pisa were the principal intermediaries in the commerce between western Europe and the Byzantine Empire, the Levant and North Africa. From these ports – great ports by the standards of the early Middle Ages – the luxury goods which made up much of the commerce of the time were transported northwards to central and western Europe. Apart from the ports, the chief intermediaries in this commerce were the inland cities of Pavia, possibly the largest at this time, and of Turin and Piacenza, at the effective limit of Po navigation; of Asti, Chieri and Novara, all of them centres for the monetary transactions on which the trade was based, and Milan, Brescia, Borgo San Donnino, Ferrara, and Lucca, the foremost city of Tuscany. There were also many others, since urban life, though enfeebled, had never completely disappeared from northern Italy. About the year 1100 a rapid growth was taking place in these cities, but it was a selective growth, responding to political forces and commercial needs, which were different from those of the Roman period. The growing cities were in general those which could profit from the trade which now followed the Po and its tributaries towards the alpine passes and northern markets.

The only other part of Europe where manufacturing was of more than local importance was in the Byzantine Empire. Constantinople, despite the economic decline of the empire, was still by far the largest city in Europe. It remained the policy of the emperors to keep its population employed, to maintain the supply of foodstuffs and to facilitate the export of manufactures. Despite the conquests of the Seljuk Turks, which had greatly reduced the commercial orbit of the city, its commerce remained prosperous, and its manufactures, primarily silks and brocades, continued to command a large western market. There was also a number of smaller towns with highly specialised craft industries in both the European and the Asiatic parts of the empire. In Europe, the more important were Thessaloníki, Thebes and Corinth, where there were communities of Jewish craftsmen. According to Benjamin of Tudela the artificers in silk and purple cloth at Thebes were only excelled by those of Constantinople itself.[125]

Moorish Spain also had a number of cities in which commerce and crafts were well developed. Among them were Seville and Cordoba, the largest cities of Andalusia. Silk cloth was made at Jaen from raw silk prepared in the surrounding villages, and at Almeria, where Edrisi found 800 silk weavers.[126]

TRADE AND TRANSPORT

These centres of manufacturing could not have been developed without a parallel expansion of commerce. It is impossible to measure either the volume of trade or its rate of growth during this period, and we know only in very general terms how it was composed. It is convenient to divide this trade into, first, local or short-distance trade and, secondly, long-distance commerce. The former consisted essentially of the products of agriculture and forestry, which were transported to the nearest monastery, market or town either to discharge feudal obligations or in exchange for fabricated goods. The latter consisted in general of luxury goods, which were distributed from a relatively small number of trading or manufacturing centres. In terms of volume the local trade was much the greater. The monastic records give, in a few instances, a rough measure of the volume of grain, flax, wool and other products which made their slow way in ox-drawn carts to the monastery gates. The roads which converged on the *caput* of an important estate must indeed have been busy after harvest, when the wine was made, or the sheep shorn. But the aggregate *value* of goods entering long-distance trade was considerable. Their price, when they reached their destination, included not only the production cost, but also the value added by transport by sea-going ship, by river boat and pack animal, together with the cost of tolls and doubtless a high insurance against all risks which the merchants ran.

The commodities of trade

Local trade was predominantly in foodstuffs and raw materials of low value, such as wool, flax and lumber. Long-distance trade, by contrast, is generally assumed to have been made up of high-value luxury goods, such as silks, spices and perfumes. This however, was only partially true. There was also a long-distance movement of certain low-value goods, such as salt, furs, raw cotton, and wine, which could not be widely produced, as well as of wool and even grain. The larger cities had already outgrown the capacity of their local regions to supply them with food, and some, such as Venice, Genoa and several of the trading cities of southern Italy and Flanders, lay in areas geographically unsuited to the production of grain-crops.

Part of the food supply of the towns of Flanders and the lower Rhineland was brought in by boat from the plains of Hainaut, the upper Rhineland and even northern Germany. The rivers which converged on the Paris region carried grain and wine to Paris; the south of Italy, particularly Apulia, and Sicily, supplied grain to the commercial cities, like Amalfi and Bari, and to Rome itself, and Constantinople was heavily dependent on the import of foodstuffs, not only from the Aegean region and the coastlands of the Black Sea, but also from Asia Minor.

Clothing was important in an age as heavily overlaid with ritual as the Middle Ages. Rank, both spiritual and lay, was expressed in dress, and there was a large and continuous demand for silks, furs and fine cloth. Furs entered largely into the commerce of northern Europe, and were shipped from the Baltic region. Wool was perhaps still transported in only small amounts. Flanders could possibly produce from its marshy grazing lands much of what it needed, but there was already some import from England. Cloth from Flanders sold widely in north-west Europe, but in 1100 Flemish cloth had not yet begun to find a market among the cloth finishers of Italy. In many respects the Mediterranean and north-western Europe constituted two separate markets, between which travel and transport formed a tenuous and difficult connection.

The alpine passes were crossed only by goods of high value and small bulk, which were carried on pack animals from the limits of navigation in the Italian plain to the navigable Rhine, Moselle, Meuse or Seine. Prominent among the goods transported in this way were silk, in great demand among the rich laity and the dignitaries of the church; spices and perfumes, parchment and manuscripts, ivory, alum, and dyestuffs. Most of these goods had been handled or even processed in the cities of the Byzantine Empire. Much of the silk fabric originated in Constantinople, though the raw silk from which it was made had possibly come from Persia or the Middle East. Other exotic products were obtained in the markets of Alexandria and the Levant. Most were transported to the Italian ports, and were paid for by the export of manufactured goods, especially woollen cloth and linen.

As Lopez has shown,[127] western Europe was beginning to lose its 'colonial' status and to look for markets in the Levant and North Africa for its manufactured products. The Genoese were beginning to ship European cloth to African markets from Tunis to Morocco, and one may look on the Crusades as, in part, an attempt to extend this market to the whole Mediterranean world. There can be no question that, with the Crusades, the actual conduct of this trade passed into the hands of the Italian merchants, who acquired and extended their commercial privileges in Constantinople and in other ports of the Byzantine Empire and the Levant.

The commerce which flowed to the Italian ports and doubtless also to those of southern Spain and Provence, was not made up entirely of manufactured and luxury goods. It embraced also skins and hides, wool and grain, to supply the manufacturers and the populace of the cities along the northern Mediterranean shore. These were requited by the export of woollen cloth; western Europe was very slowly reversing its earlier commercial role of exporter of primary goods.

It is easy to exaggerate the scale of commercial operations at this time. It remained very small; how small can only be guessed from the size of the minute harbour of a port, such as Amalfi, which was reckoned to be amongst the richest and most prosperous. Yet the years around 1100 were a period of growth and change, the magnitude of which can only be grasped by comparing

Fig.5.7 Mediterranean ship at the time of the First Crusade (relief from the Campanile, Pisa)

commercial conditions in the early eleventh with those in the late twelfth century and thirteenth.

Markets and fairs

Much of the commerce of the time was conducted at periodic markets and less frequent fairs. The distinction between them was still far from clear, and some fairs were in process of evolving from markets. The latter tended to serve only local needs and to provide a medium of exchange for the products of local agriculture and crafts. Fairs were periodic gatherings of merchants engaged in long-distance trade, and tended thus to concentrate on goods of higher value than were to be found at the markets.

The location of fairs was related to the great commercial and manufacturing regions of medieval Europe, and most were held at seasons of the year when the roads were open and the difficulties of travel reduced to a minimum. Many of the fairs had been established by the monasteries, and were often held on the name-day of the local saint; the growth of fairs was inextricably bound up with the urge to go on pilgrimage. One of the most important fairs at this time was the Lendit fair, at Paris, which was 'owned' by the abbey of Saint-Denis. Other monastic fairs at this early date were at Saint-Omer and Messines in Flanders, and there was also an important fair at Visé, on the banks of the Meuse below Liège.

The essence of the fair was its regularity and its freedom from interruption by any kind of civil commotion. Groups of interrelated fairs began to emerge, of which the most well known at this date were the fairs of Flanders. Eventually a cycle of five fairs came to be established, but by 1100 it is probable that only those of Thourout and Messines were well established, though the Ypres and Lille fairs were recorded a year or two later and the Douai and Saint-Omer fairs were beginning to be active.

The most famous and important of all medieval fairs, those of Champagne, had already begun by 1100, but it is doubtful whether at this date they did much more than serve local needs for primarily agricultural goods. It is uncertain whether Italian merchants had yet begun to visit them, but we may be sure that Flemish cloth was not, as it later became, an important article of commerce.

There were fairs in northern Italy, notably at Pavia, where oriental goods were brought by merchants from the north, but they seem to have been much less important than those of northern Europe. This arose probably from the fact that urban institutions were more widespread and more strongly developed and that, with merchants doing business throughout the year within the security of city walls, there was really little need for periodic fairs. Thus a network of markets spread over much of Europe. Most towns and many villages had markets all the way from Spain to Poland, and from southern Italy to Scandinavia, but of their actual number and distribution almost nothing is known.

The merchants. Even less is known of the merchants, the social classes from which they derived, and the origin of their trading capital. In general they do not appear to have been of noble origin, though many of the more successful among them came to be ennobled. A considerable number of them were Jewish, especially in the Rhineland and central Europe. The records of their transactions and of their disputes with one another at the end of the eleventh century throw some light on the great distances travelled and the fairs which they frequented.[128] A majority of the merchants, however, was non-Jewish, and in all probability sprang from the ranks of the lesser landowners. From Italy to Poland evidence is found of merchants who lived in towns but derived at least their initial capital from their land possessions. The ranks of the merchant also included the self-made man like Godric of Finchale, who obtained his trading capital by robbing a church, but used his gains so lucratively that he became wealthy enough to endow a monastery and so attain sainthood.[129]

Not all merchants enjoyed so meteoric a rise as Godric, but in northern Europe most probably came from the same social class and probably had at first almost as little working capital. In Italy, by contrast, the gulf between the noble and common people was narrower and more easily crossed. The former never despised trade; many of them moved to cities, where they built their tall 'towers', and at times fought one another in the streets, while their younger sons, with no lands to inherit, were often absorbed into the merchant class. This circumstance may explain the apparently greater amount of capital available to the Italian merchants than to others; it also suggests a reason why the Italians participated in the Crusades to a lesser degree than the noble families of northern Europe. The 'commoners' formed the mass of petty traders in the Italian cities and greatly outnumbered the 'nobles' in all known cases. It was the latter, however, who more often controlled the city government and dominated long-distance trade.

Means of transport. The increase in the volume of trade was not matched by any improvement in the means of transport. As much use as possible was made of navigable waterways for the movement of bulky and heavy goods, but most were transported overland on the backs of pack animals. The few itineraries suggest that the traveller, whether noble or ecclesiastic, went by horse, and that the humbler merchant and pilgrim had often to walk. Wherever possible the Roman roads were still followed, but these had never been built north of the Danube and east of the Rhine, except in the *Agri Decumates,* and elsewhere bridges had in many places collapsed and the roads had fallen into disrepair. Furthermore, the cities on which the Roman roads had focussed were not necessarily those which the early medieval traveller wished to visit. Thus new segments of road came to be established both to make connections which the Romans had not needed, and also to avoid obstacles created along the Roman roads themselves. Joris, for example, has shown how the route from the Low Countries to the alpine passes was shortened by the opening up of a new route to avoid Trier.[130]

Some attempt appears to have been made to keep the Roman road surface in repair, but most roads were merely tracks. Bridges were very few, and rivers were generally forded. Many roads, especially in central and eastern Europe, were not clearly defined, consisting of multiple paths through the forest and waste. The medieval traveller does not appear to have been greatly deterred by mountains and areas of sparse population, and seems to have made little effort to avoid them. There was, for instance, a steady flow of traffic across the Ardennes and Eifel, and travellers to the south of France crossed the Central Massif as readily as they followed the Rhône valley to its mouth.

The Alps, on the other hand, aroused considerable fears, and the roads across their passes were avoided whenever possible. Few passes were available (see below, p. 413) and the snowline may have been lower than to day. Bernard of Clairvaux and Peter the Venerable, neither of whom was given to complaining about the hardships of the road, nevertheless listed the 'snows of the Alps' (*Alpes ipsae gelidae*) along with the dangers of the sea and the 'great expense', as the difficulties in the way of travel to Rome.[131] The journey to and from Italy of Anselm of Aosta was 'beset with many dangers'.[132]

Crossing the passes was strenuous and dangerous, but the most hazardous part of the journey was usually the approach to the high mountains. Here, where the roads ran alternately through defiles and wide valleys, they could be more easily blocked and the traveller harassed by robbers and burdened with tolls. The 'closing' of the passes always meant the fortification of castles on their approach roads.[133] For this reason the crossing of the Jura was often more hazardous than that of the Alps themselves.

Travel was conditioned by the availability of hospices and inns, and, at a time when 20 miles constituted a good day's travel, these needed to be closely spaced. Once provided, they tended to stabilise routes. The surviving itineraries, which increase in number as the Middle Ages wear on, consist in fact of little more than the names of the places along the routes where overnight accommodation was available. The building and endowment of a hospice, like the construction of a bridge, was a humanitarian act for which many a traveller must have been grateful.

The alpine crossings played a vital role in the development of commerce between northern Europe and southern. In spite of the great number of possible routes, a small number of passes was in fact used, largely perhaps because others lacked the amenities required by the traveller. At this time transalpine

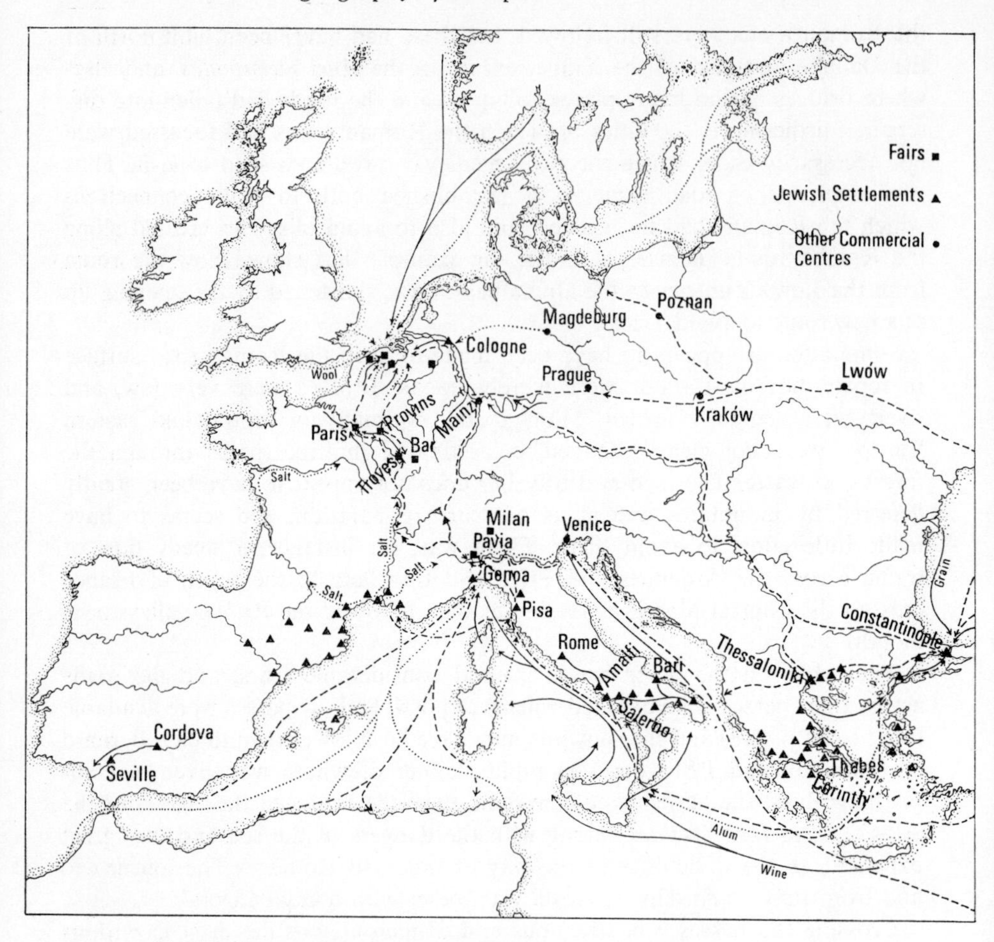

Fig.5.8 Commercial routes about 1100

traffic rarely used any pass in the western Alps other than the Mont-Cenis and Great St Bernard, and it was the foundation of a hospice on the Great St Bernard pass that confirmed this as the major transalpine route. The relatively low and easy Simplon pass probably was not used, owing to the difficult gorges along its southern approaches. Of the central passes only the San Bernardino appears to have been much used, and traffic had not yet begun to use the more easterly passes of the Graubünden on any significant scale.

By contrast, the evidence for the use of navigable waterways is considerable. Boats were small and of shallow draught, and were used on many rivers not now considered navigable. The Po, with the lower courses of its tributaries, and the Rhine and Meuse appear to have been regularly used. The chronicler of Conrad II described Cologne as very well provisioned thanks to the Rhine shipping.[134] The local needs of the Flemish towns were partially served by the small boats which plied their narrow waterways; Paris drew much of its wine supply by boat from Burgundy, and there is good evidence for the regular use of the Rhône, Meuse and Loire and of the waterways of Poland.

The availability of water transport was, in fact, a condition of the large-scale production of certain bulky commodities. The production of salt from the lagoons of Comacchio and Bourgneuf is inconceivable, at least on the scale to which it was developed, without river transport. The vineyards of the Moselle and Rhineland, some of them owned by distant monasteries, distributed much of their bulky produce by boat.

The medieval road system was fluid. Neighbouring towns were often linked by several routes, between which the traveller could choose according to the physical and human hazards of each. Though one can determine the fixed points at which merchants gathered, it is not easy to trace the routes by which they got there; any map of roads in the Middle Ages must necessarily be very highly generalised.

The chief foci were the commercial cities of northern and southern Italy, from which the principal axes of medieval trade ran by way of the Rhône valley and alpine passes to the Paris basin, the Rhineland and the Low Countries. The second most important focus was Flanders and neighbouring areas of Artois, Hainaut, and south-eastern England. During the following century the Champagne fairs emerged as the chief intermediary in the commerce between these two regions. By 1100 the fairs were not yet regularly frequented by Italian merchants, but Paris was already a commercial centre of more than local importance, and its Lendit fair was probably more used than the fairs of Champagne.

Ibn 'Abdun's description shows a busy traffic in Seville, and this was probably the chief port of Moslem Spain, and carried on its commerce with ports of Italy and North Africa.[135] Between northern and southern Spain, between Christian and Moor, there was little exchange of goods, but north-eastern Spain at least, with its ports of Barcelona, Valencia and Tarragona, and its inland cities of Zaragoza and Tortosa, was in touch with the merchants of the southern France, notably those of Narbonne, Montpellier, Arles and Marseille, and of the Italian cities of Genoa and Pisa.

Most of the luxury goods imported from the Byzantine Empire and the Levant were handled by the Italian merchants. The geographical pattern of their ports was at this time changing rather rapidly. In the later years of the eleventh century, the more important ports had been those of southern Italy, especially Amalfi and Bari. The importance of Amalfi is not easy to explain. Its inhabitants, wrote Benjamin of Tudela, 'are merchants ... who do not sow or reap, because they dwell upon high hills and lofty crags, but buy everything for money'.[136] They lay within the limits of the Byzantine Empire, which probably gave them some commercial advantage, but within reach of Rome, which constituted one of their markets. Of their wealth and prosperity there was no question, and even after the capture of Amalfi by Robert Guiscard and the Normans in 1077, they retained some importance.[137] The import of foodstuffs to supply Amalfi raised no difficulties once trading connections had been established; the problem is rather to explain how the colony of merchants had come to establish itself in the first instance in this rocky cleft in the Sorrento peninsula.

In 1100 Amalfitan merchants were of diminishing importance in the commerce of the eastern Mediterranean, but their role remained an important one in the local commerce of southern Italy and the Tyrrhenian islands. Control of long-distance trade was passing to the northern ports: Pisa, Genoa and Venice. Amalfi, Bari and the lesser ports of southern Italy were ill-placed for trans-alpine trade; the use of Genoa or Venice reduced and simplified the land journey at the expense of increasing the length of the sea voyage. They introduced such economies in the commerce between central Europe and the eastern Mediterranean that their growth was almost inevitable. At this time it was probably the Pisan merchants who played the leading role in the commerce of the western Mediterranean, but Genoa was growing in importance, lay nearer to central Europe and had yet easier connections with its northern hinterland. It seems likely that both Pisa and Genoa had previously obtained much of their goods from south Italian middlemen, but were now beginning to deal directly with Constantinople and the Levantine ports, where gradually they displaced the Amalfitans.

Venice was the most favourably placed of all the Italian cities to serve as an intermediary between central Europe and the east. Its merchants had already, in 1082, obtained a preferential position in Constantinople. Its commerce suffered from the campaigns of Robert Guiscard into the Balkans, but was able to profit soon afterwards from the success of the First Crusade, to strengthen its position in Constantinople, and establish bases in the Levant. The small ports of the Dalmatian coast, from Zara (Zadar) to Durrës, were in some measure under the protection and control of Venice, and served as staging points on the long voyage to the Bosporus and the Syrian coast.

In the Aegean itself were a number of ports from which shipments were made to Italy. Among them 'Armylo [on the Gulf of Volos, in Thessaly], a large city ... inhabited by Venetians, Pisans, Genoese, and all the merchants who came there',[138] and Thessaloníki itself. But foremost in the commerce of Europe at this time was Constantinople. Its empire no longer extended from Venice and Sicily to the Middle East, but it remained the focus of Mediterranean trade, 'a busy city and merchants come to it from every country by sea or land, and there is none like it in the world except Baghdad ... they fill strongholds with garments of silk, purple and gold. Like unto these store-houses and this wealth there is nothing in the whole world to be found.'[139] Raw silk and silk fabric, spices, gemstones and exotic wares from Asia came by caravan not only to Constantinople but also to the ports of Asia Minor, where they were picked up by Italian merchants. The effect of the Crusades was to open up trading stations yet farther to the east. Part of the commerce was now beginning to bypass Constantinople, but it is doubtful whether in fact the city suffered any absolute decline.

That the overland route from central Europe to Constantinople was used is apparent from the routes taken by some of the Crusaders. Several groups crossed the Hungarian plain to Beograd, and thence through the hills of Sumadija to Niš and by the old Roman route to Sofia, Philippopolis (Plovdiv). and Adrianople (Edirne). Another followed the coast road of Dalmatia until at Durrazzo it reached the western end of the Roman *Via Egnatia*, which it followed through Macedonia and Thrace. Another group reached Durrazzo by sea from Apulia. Branches from the road through Macedonia gave access to Skopje and Niš on the one hand and Thessaloníki on the other. These roads were direct and probably well-marked, but they were notoriously unsafe. Though about 1100 there was outward peace between the Byzantine Empire and its Serb and Bulgar neighbours, the Tatar Pechenegs and Polovtsi from beyond the Danube threatened the traveller, and the predatory nature of the Dalmatian, Albanian and Vlach tribesmen were the source of continuing disorder. Edrisi described[140] this network of routes in the Balkan peninsula, but it is very doubtful whether the pilgrim and the merchant took them when the easier and safer journey by sea was available.

There was, however, no real alternative to the land route eastward from the Rhineland to Bohemia, Poland and Kievan Russia, and travellers and merchants were obliged to follow the tracks through the widespread forest. Very little is known of the routes themselves, but the Jewish *Responsa* and the evidence of coins show that they must have been well used. The most frequent starting point of such eastward journeys was Mainz, from which Jewish merchants frequently made the journey to Hungary, and, on occasions, to Constantinople by way of the Balkan routes already described. They were also to be found in Prague, and in southern Poland as far as Przemysł, where they probably were working the route from Kraków to Kiev. A branch from this route followed the outer curve of the Carpathians to the lower Danube and thus to Constantinople.

Coin finds in central Poland suggest that trade with the west was especially active in the eleventh and twelfth centuries. The cities of the Low Countries and Rhineland were most frequently represented in the hoards, which were very abundant near the commercial centres of Poznań, Gniezno, and Kruszwica. From Mazovsze a trade route ran up the Bug into Ruthenia and on towards Smolensk. Furs and skins were probably prominent among the exports of eastern, as they were of northern Europe. Imports are likely to have been made up of cloth and luxury goods.

CONCLUSION

The economy of Europe about 1100 was still primitive, even by medieval standards, but it was at a stage of rapid growth. So quickly was it changing with the colonisation of new land, the foundation of cities and the redevelopment of trade by sea and land, that it is difficult to represent the stage of development in each sphere at this time. About two and a quarter centuries later this period of growth had run its course, and, locally at least, there were signs of recession as much as a generation before the Black Death (1347–50) struck its fatal blow to medieval society.

As in all periods of rapid growth, there was a sharp contrast between the most developed and the least developed parts of Europe. The pre-eminence of northern France, the Low Countries and the lower Rhineland was apparent in both urban and commercial development and in building and cultural activity. Second only to north-western Europe was northern Italy, including Pisa and Lucca. Between these two lay the chief commercial axis of this period.

If a line be drawn from Rouen to the Rhône mouth, and another from the Zuider Zee to Venice, they will contain between them the largest towns, the densest population and the most developed trade in Europe at this time. It is no accident that most of the monastic and church reforms were conceived and implemented in this area, nor that the great international fairs of the 'high' Middle Ages were located here. Both German emperor and French king lived most of their lives in this area, and from it they derived their wealth and power. England was an appendix to this axial belt, the terminus of its routes and source of some of its principal raw materials.

West and east of this axial belt lay the less precocious regions of central and southern France, northern Spain, and Germany. In the former, the cities of the Roman Empire were being slowly redeveloped, and its cropland brought again under the plough. In the latter, towns were being created *ab initio*, and the virgin forest cleared for human use. Beyond these 'developing' regions lay 'undeveloped' Europe: the Celtic north-west, the Scandinavian north, and the Slavic east, in each of which, however, lay the nuclei of future growth.

In southern Europe, from Moslem Spain, through the south of Italy to Greece and Constantinople, lay the remains of classical Mediterranean civilisation. Many of its cities were still inhabited; much of its road system was still in use, and its ships still sailed the sea, linking its port-cities with one another and with North Africa and the Middle East. During the later Middle Ages, the urban, commercial civilisation of the axial belt and of the Mediterranean was diffused slowly through the west, the north and the east of the continent.

This period from the eleventh century to the early fourteenth was one of creative effort rarely equalled in European history. It was one in which cities expanded in number and in size; in which artistic creativity and metaphysical and political speculation reached new heights, and wealth and ostentation were greater than at any previous time. It was also a time when the aggregate population of Europe doubled; when human pressure built up and the level of welfare of the masses was depressed as never before. The gulf between the very rich and the very poor was probably wider than at any time before the recent past.

The wealth which was accumulating supported the apparatus of state-government. Throughout these centuries administrative procedures were being refined, and the functions of government widened. The more politically developed states – the Duchy of Normandy and the County of Flanders at first, and then the kingdoms of England and France – each built up a highly sophisticated bureaucratic structure.

At the head of the administrative structure was the king, prince, duke or count. The medieval ruler was a busy person; he was judge and law-giver, tax-collector and administrator, the source from whom privileges and immunities derived, and, in his private and personal capacity, a seigneur, interested in the work of his tenants and the income from his lands. The kings who discharged their manifold duties conscientiously – and there were many – found themselves gradually building up a bureaucracy, staffed by men of ability who made a career in the royal service. The activity, both judicial and administrative, in the royal courts, grew steadily during these centuries, and at the same time intensified the central control over all parts of the state.

It was nevertheless an uneven process. In this pre-constitutional era much depended on the character and personal attainments of the king and his closest advisers. In most parts of Europe even the rule of succession was not clearly defined; kingship was too important to be left entirely to primogeniture. There was a presumption that son would succeed father, but he had in some indefinable way to be suitable to bear the rule that devolved upon him. This process may have spared a country the disasters that would have resulted from a succession of weak princes, but it nevertheless placed contraints on kingship. The king made new law, but his powers were tempered by the weight of tradition and by the freedom which remained to the barons to rebel if their rights were thereby infringed. The customary law of the feudal kingdoms had failed to satisfy the problems of growing centralisation. Strong kings were able to make new laws; the weak succumbed to the pressures of their own feudal baronage. These facts have to be remembered when comparing the political map of Europe in the fourteenth century with that of about 1100.

By about 1100 certain trends which were to characterise the later history of Europe, were already apparent. On the one hand there was the disintegration of the German Empire; on the other, the strengthening of what one may begin to call the national monarchies.

In prestige, if not also in power, the German emperors stood first among the secular rulers of Europe. If Henry IV had lost his struggle with the Papacy, the Concordat (1122) made by his son with the Papacy showed that the latter had achieved no unequivocal victory. The imperial power was nevertheless weakened, and the attempts of Frederick I of Hohenstaufen to rebuild a territorial base for an empire of the Christian west – this time in Italy – failed, and in the following century, with the collapse of the schemes of Frederick II, all semblance of unity disappeared from the empire.

The Papacy, which had done so much to destroy the medieval empire, performed a dual function. It exercised a spiritual rule over all of Christendom that did not accept the rival pretensions of the eastern Patriarch. It defended the rights and protected the doctrines of the Church against layman and heretic alike, and claimed and exercised the right to excommunicate and depose even kings and emperors. At the same time the popes were the secular rulers of the city of Rome and of central Italy. The lands under papal rule were not immune to the movements which characterised the cities of northern Italy. The towns of the Patrimony of St Peter were often as hostile to the popes as those of northern Italy were to the emperor. The politics of Italy thus came to be polarised, as those of Germany had been, Guelf against Ghibelline, with, in the thirteenth century, The Guelf's representing the papal interests, and the Ghibellines those of the Hohenstaufen.

The final victory of the Guelfs in Italy was achieved only by introducing a power stronger than either of the contending factions and by humbling the Papacy before the throne of France. Charles of Anjou, brother of Louis IX of France, accepted a papal invitation to rid southern Italy of the Hohenstaufen. He succeeded only too well and established a state in Italy which gravely impaired the pope's freedom of action. A succession of weak popes, punctuated by interregna, when the cardinals, under pressure from outside, could agree on no one, ended with the election of Boniface VIII. He was the last medieval pope, as Frederick II had been the last medieval emperor. He tried to assert the universal rights of the church, as if he had been Innocent III. He failed, and in 1303 was arrested at Anagni by a group made up of French soldiers and Ghibelline conspirators. The humiliation of Boniface was as great as that of Henry IV had been at Canossa, and it proved to be more lasting. In 1305 a French pope, Clement V, was elected, and took up his residence at Avignon, across the Rhône from the kingdom of France. The imperial city was abandoned to the feuds of its noble families, and southern Italy to the undisputed control of the Angevins.

France thus assumed a dominant role in Europe. It was the most powerful and the best governed of the national monarchies, and its kings pursued a policy as consistent and constructive as that of the German emperors had been ill-considered and reckless. The boundaries of France underwent no significant change, but within these limits the area controlled by the king increased many times. In 1100 the French king could control directly only a small area extending from the Paris region to Orléans. By 1328, the royal domain covered at least two-thirds of the country. This extension of royal authority was the achievement of Capetian policy.

During this period of almost two and a half centuries fiefs of the crown were brought into direct dependence upon it by war and conquest, as happened in Normandy; by marriage and inheritance, as was the case in Champagne, and by the extinction of their own ruling families, as in Poitou and Toulouse. But, despite the occasional big haul, it was a slow and piecemeal process. Nor was it without its reverses. The kings recognised the obligation to provide for members of their own family, and alienated duchies and counties as *apanages* to their brothers and younger sons. Important regions within the body of France, including Burgundy, Artois and Anjou, lay outside the royal domain, though their rulers were none the less vassals of the king of France.

On the other hand, the immense holdings of the English kings had been broken up by the policies, at once astute and aggressive, of Philip Augustus (1180–1223) and his successors. By 1328 it had been reduced to a tract of land bordering the Bay of Biscay and centring in Bordeaux.

For two centuries the Capets had been thus eroding the independence of the greater fiefs of the French crown. On two of them, however, they had made very little impression: Flanders and Brittany. The latter, insulated by distance and protected from French influence by its own Celtic language, stood obstinately aloof. Flanders, by contrast, lay at the centre of economic life in north-west Europe. Throughout these centuries its manufactures and its commerce had been growing, and, with them, its wealth and political power. Even by 1100 it was importing England's most important raw material, wool, and the volume of its trade steadily increased. Though its markets lay partly in France, the most important of them, the Champagne fairs, lay, until the fourteenth century, beyond the direct jurisdiction of the French kings. At the same time, trade was growing with the Rhineland, and the Count of Flanders, as a general rule, was firmly opposed to any close dependence on the French king.

The progress of the Christian states of Spain was slow. Toledo had fallen to the arms of Castile before 1100 and the rudiments of constitutional government appeared during the following century. Then the expansion of economic and cultural life were slowed. The Moors were reinforced first by the Almoravides and then the Almohades. Both were fanatical and uncultured Berbers from North Africa. They posed a greater obstacle to the advance of the Christian states, without at the same time offering them any kind of intellectual challenge. The *Reconquista* was long drawn-out, and its conclusion, 'a triumph too long deferred'. It was not in fact, until well into the thirteenth century that much progress was made. Then, under the leadership of James I of Aragon (1213–1276) and of Ferdinand III (1217–52), who finally joined the crowns of Leon and Castile, a more rapid advance was made into Andalusia, and by the end of the century all that remained to the Moors was the small kingdom of Granada.

The several Spanish kingdoms were in the process reduced to three: Portugal, Castile and Aragon, in addition to the ambivalent Pyrenean state of Navarre, unable to decide whether its fortunes lay with France or the Spanish peninsula. The long war with the Moors had a profound influence on Spanish character and institutions. Feudalism, as a political-economic system, was intensified at a time when elsewhere in Europe it was breaking down. The war itself became a crusade, with its fighting orders of Calatrava, Alcantara, Evora and Santiago, closely linked with the monastic order of Cîteaux. Commercial developments in Lisbon and along the coast of Aragon were inadequate to offset the reactionary trend in the rest of Spain.

The heroic age in northern and east-central Europe had ended. The Vikings no longer set out on raiding expeditions across the seas, and in Denmark, Sweden and Norway centralised states were gradually emerging. In east-central Europe also boundaries were stabilised. The three states of Poland, Bohemia and Hungary, each under its historic dynasty, extended from the Baltic Sea to the head of the Adriatic. All were in some degree under pressure from German expansion, and Poland had lost substantial areas – Pomerania and the Oder valley – to German rulers. In the course of the fourteenth century, all three national dynasties came to an end, first the Hungarian house of Arpád (1301), then the Bohemian Přemyslids (1306), and finally the Piasts of Poland.

The peoples of these eastern kingdoms thus lost a link with their tribal past. Kingship was deprived of the sanction of tradition and became a pawn in European politics. It was inevitably weakened in the process. At the same time, the aristocracy became more independent and more feudalised in the western sense, and the eastern kingdoms were drawn into the dynastic rivalries which marked the rise of the nation-state, as well as into the western cultural tradition. In these centuries east-central Europe became European.

The same, however, cannot be said of the states of the Balkan peninsula. The short-lived state of Croatia was absorbed into Hungary. Over the rest of the peninsula, the Byzantine Empire claimed a purely nominal jurisdiction. Political control in fact rested, after 1186, with the revived Bulgar state, which spanned the peninsula from the mountains of Albania to the Black Sea, and with the Serb state which developed at about the same time. Both derived their religion and much of their artistic inspiration from the Byzantine Empire. They owed little or nothing to the west. Thus, while east-central Europe was being assimilated, politically, culturally and economically, to western, the barrier was intensifying between the Slav world of the Balkans and the rest of the continent.

But the Byzantine Empire, most tenacious and long-lived of all the political divisions of Europe, lived on. It declined in area and authority under the Angeli emperors; its capital city was pillaged by the soldiers of the Fourth Crusade (1204); a Frank sat upon the throne of Justinian, and those parts of the empire which could still be controlled from Constantinople were distributed as fiefs among the Frankish lords, but in Epirus, north-western Anatolia and in even parts of Thrace and Peloponnese Greeks held out against the Franks, until the Latin Empire disintegrated and disappeared. In 1261, Michael Palaeologus, who had retained control in Asia Minor, recaptured Constantinople, and the Byzantine Empire was restored.

Economic growth. The economic expansion which was in full flood about 1100 continued until the fourteenth century. Population was increasing everywhere. The most reasonable estimates suggest that the overall growth was of the order of 60 to 65 per cent, but locally, as for example in northern Italy and the Low Countries, it must have been considerably higher. The increase in population was accompanied by the multiplication of settlements and the continued erosion of the forest and waste.

The growth of urban population was more marked than that of rural. It is difficult to form an estimate of its size during the later and better documented years of the period; impossible for the earlier. The increase in the *number* of towns can be calculated more readily. Haase has shown[141] how they multiplied in Westphalia:

Founded	Number
Before 1180	6
1180–1240	27
1240–1290	30
1290–1350	22
1350–1520	20
	105

The second half of the thirteenth century seems throughout central Europe to have been a period of the most intense urban growth, characterised not only by the greatest number of new foundations, but also by the rapid expansion of the older towns.

In western and southern Europe, however, the period of most rapid urban growth had come earlier. The larger towns had redeveloped around their Roman nuclei. The foundation of 'new towns' had been most active in the late eleventh and twelfth centuries, and the *bastides* of the thirteenth century marked the end of the process. It is doubtful whether in Italy any new town was established after the foundation of Alessandria in 1168. On the other hand, the eastward diffusion of the town continued with scarcely diminished vigour into the fifteenth century.

The growth of at least the larger towns was closely related to the expansion of commerce. They were most numerous near the commercial arteries which ran from central Italy to the Low Countries. This urban net already existed in embryo by 1100. During the next two centuries the emergence of regions of specialised manufacturing in Italy and north-western Europe greatly intensified the volume of trade. From early in the twelfth century the chief intermediary in this trade was the fairs of Champagne, where Italian merchants met those of Flanders. This commerce reached its peak in the thirteenth century. Then the trading pattern slowly changed. The Italian merchants, now grown to positions of wealth and importance, stayed at home, while their agents bought cloth and wool in the northern cities and shipped it direct to Italy. Trade deserted the Champagne fairs and used the Rhine and the alpine passes. The distribution of manufacturing also showed an eastward shift. The older centres, especially of cloth manufacture, ceased to expand, while the industry grew in Brabant, the Rhine valley, Swabia and Bavaria.

Though this urban and commercial growth was most conspicuous along the trading axis between Italy and Flanders, it was present no less in France and in Germany, and eastwards it died out only on the borders of the steppe and the forest of Russia.

Rural population was increasing steadily, though less sharply than urban. The status of the peasant changed, but not his agricultural techniques. The tendency was for the landowner to commute the services of his tenants; to accept a money rent or a rent in kind in lieu of labour, and to lease his demesne. The rigidity – such as it was – of the tenurial system was breaking down, and with it the mutuality of the earlier village community. The serf was gradually transforming himself into a peasant. With the gradual disappearance of villeinage, a new tenurial basis for peasant land-holding had to be devised. This consisted most often of a fixed contribution of agricultural produce, together with, in many cases, a money payment. When the demesne itself came to be divided and leased, this was often done on a share-cropping basis. The servile obligations of the villein were gradually eroded, and a peasantry, free though bound by contractual relationships with its territorial lord and dominated by the traditions of its community, began slowly to appear.

The disintegration of manorialism and villeinage occurred unevenly. Much depended on the landlord's evaluation of the possibilities open to him; on his desire for a cash income, and on the availability of currency. In the long run, it was the peasants who benefited from the commutation of services and the payment of fixed rents, since the latter tended to diminish in real terms with the price inflation of the twelfth and thirteenth centuries. On the other hand peasant labour was a cheap and increasingly abundant commodity. Landowners, blessed with a surplus of tenants, could, and did, exact increasingly onerous conditions. Peasant holdings were divided, rents and other obligations were increased, and in this period of rising production and ever greater ostentation, the peasant class was more and more depressed. One's admiration for the thirteenth, 'greatest of centuries', must be tempered by the knowledge that wealth was increasingly concentrated, and that the gulf between rich and poor was growing larger. It is chastening to remember that just as the Erechtheon had probably been built with slave labour, so the beauty of Chartres was, in the last resort, paid for by serfs scarcely living at a subsistence level.

6

Europe in the early fourteenth century

By the end of the first quarter of the fourteenth century the period of medieval economic growth was over, and with it the cultural flowering to which it had given rise. The late medieval population decline had not yet begun, or, at least, had not reached measurable proportions. But there had been a run of bad harvests a few years before, and mortality had been heavy. Most medieval cities had already achieved their maximum size, as measured by the circuit of their walls. The great cathedrals and churches were built, and the few that still lay unfinished were to remain so into modern times. It is the Europe of this late summer of the Middle Ages that this chapter seeks to describe.

POLITICAL GEOGRAPHY

By the fourteenth century the political map of Europe had begun to assume the form which it was to retain for several centuries. In the west and north the national monarchies had taken shape. In the east, the tribal kingdoms of Poland, Bohemia, Hungary, Serbia and Bulgaria had ripened into states with well-defined boundaries and established administrative structures. Only within the limits of the German Empire was the trend towards the centralisation of political power reversed, with the fragmentation of the German Empire into an immense number of quasi-sovereign territories and self-governing cities. In Italy, too, where the imperial authority had rarely been more than nominal, the land was breaking up into independent duchies and *contadi*, while the Balkan peninsula had already become the prey of plundering bands of Crusaders from the west and of Turkish armies from the east.

Iberian peninsula

Spain provides a partial exception to the generalisation that kingship was in western Europe becoming more absolute and political control more centralised. The Moorish kingdom of Granada had long been reduced to an area which corresponded roughly with the Sierra Nevada and the neighbouring coastal region of southern Spain. The kingdom of Castile had finally incorporated Leon, and Spain from the Pyrenees southwards to the Mediterranean coast, was ruled from Burgos. But the map is misleading. A succession of weak kings had permitted too great a licence to the feudal nobility, whose feuds and wars had produced conditions of near anarchy. The chief force for order and stability in Castile was not the king but the league, or *Hermandad*, which Castilian cities had formed at the end of the thirteenth century to protect their interests against the feudal class.

Other states of the Iberian peninsula were but enclaves around its coast. The kingdom of Aragon had been formed from the union of Aragon with the

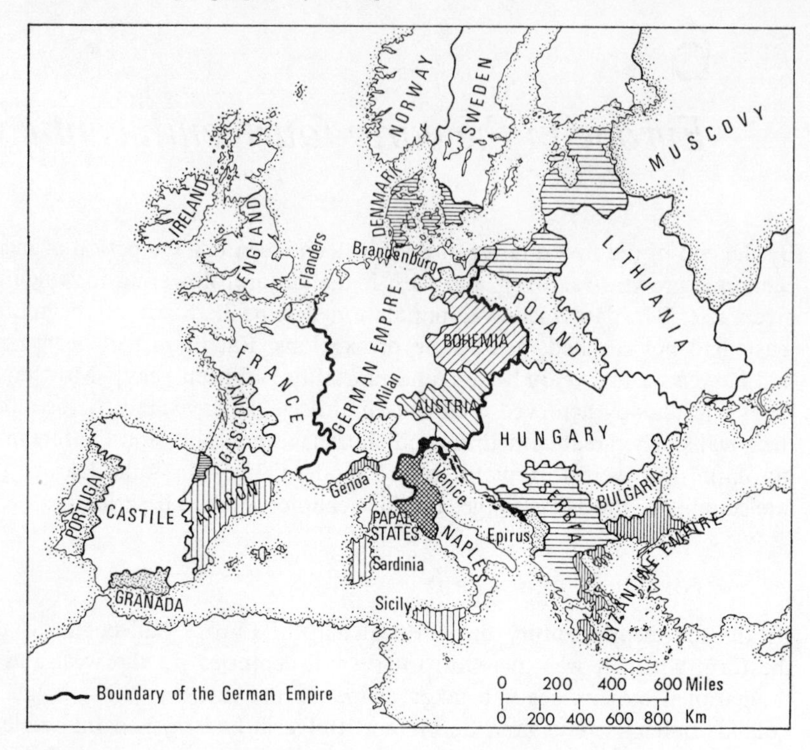

Fig.6.1 A political map of Europe in the mid-fourteenth century

counties of Cataluña and Valencia. But the union had been more legal than real; each retained its own identity, and the power of their common ruler was hedged around by the established privileges of the nobles and the cities. Cataluña, with its port of Barcelona, and Valencia had long been interested in Mediterranean trade, and this had contributed both to the annexation of the Balearic Islands and to intervention in Sicily and control of Sardinia. Aragon had indeed created a maritime empire in the western Mediterranean, and was more interested in consolidating its control over the islands than in extending its rule within the Iberian peninsula itself.

The petty kingdom of Navarre lay across the western Pyrenees, extending from the Ebro river to the plains of Aquitaine, hemmed in on the south by Castile and Aragon, and on the north by the Duchy of Gascony. It had little claim to importance beyond its control of the mountain passes, but had escaped incorporation into either of the Spanish kingdoms. Its rulers at this time were more interested in the affairs of France than of Spain, but succeeded in playing an important role in neither.

Portugal alone of the Iberian kingdoms showed any unity and cohesion. It had broken away from Galicia in the mid-twelfth century, and in 1267 had completed what has since been its national territory by annexing the Algarve from the Moors. Without either the distraction of Moorish neighbours or of an unruly aristocracy, Portugal was quietly developing its resources and building up a maritime commerce along the sea-lanes of western Europe.

France

In 1328 the house of Capet, which had provided the kings of France since the tenth century, ended in the direct line, and the crown passed to Philip VI of Valois, a cousin of the last Capet. This remarkable dynasty had succeeded in converting its nominal kingship over France into an effective control of much of it by eliminating the intermediate dukes and counts. The most recent triumph had been the occupation of the counties of Champagne and Brie on the death (1274) of their last count, and the marriage of his heiress to the future Philip IV of France. At about the same time control was won over Poitou, the Limousin and western Languedoc. The acquisition (1312) of jurisdiction over the territory of the Archbishop of Lyon extended the eastern boundary of France almost consistently to the line of the four rivers, the Scheldt, Meuse, Saône and Rhône, and locally, as at Chalon, Avignon and in Provence, even beyond it.

It must not be assumed, however, that the French kings had pursued with singleness of purpose the policy of bringing the great territorial lordships under their direct control. They alienated duchies and counties, chiefly to members of their own family, as frequently as they acquired them. In time most returned to the possession of the king, but the map of the royal domain underwent rapid changes. Fig. 6.2, which is based on the survey of parishes and hearths of the royal domaine in 1328, would have been significantly different if the survey had been made a few months later because the County of Chartres returned to the crown on the death of the count, the younger brother of Philippe VI, in November 1328.

The extension of royal power has been regarded, if only because it proved in the end to be successful, as in some way inevitable. In its most naive expression this view regards the history of France as written in its rivers. The ultimate goal of unity was, however, achieved only through a series of fortuitous incidents; the course of events could so easily have moved in the opposite direction, and France could have broken up, like Germany and Italy, into a number of independent principalities. For a policy, similar to that pursued by the kings of France, was practised by many of their leading seigneurs. These were also engaged in turning a title into the reality of power; in organising their exchequer, building up a system of courts, asserting their feudal rights, subjecting their own counts, and excluding the influence of the king.

If, in 1328, over half the territory of France was under the direct rule of the French king, there were nevertheless four major principalities in which his power and influence were small and where rival ducal bureaucracies were being created, modelled in part on that of Paris, and in some ways more effective than the king's. The most important of these, because the most urbanised and industrialised was Flanders and Artois (see below, p. 316). From the western outskirts of Paris to the peninsulas of Brittany lay a second such area, made up largely of the County of Chartres and the Duchies of Anjou and Brittany. Across the centre of France, from the Limousin to the Saône valley stretched another extensive bloc of alienated territory, which included the counties of Marche, Bourbon and Nevers and the Duchy of Burgundy, nucleus of the Burgundian state of the following century. In the south-west, lastly, the king had no control over the counties of Béarn, Armagnac and Foix; and the English held the Duchy of Guyenne, or Gascony. A few smaller fiefs, such as Rethel, Bar and Blois, had escaped the Royal net, but some 24,000 parishes and about 250,000 square kilometres were under royal control; and the king was able to conduct a survey and count households within most of this area.

To the east the boundary between France and the Empire had gradually been gaining precision as disputes regarding its course were settled one by one. It was thought of as the line of the 'four rivers', though in important respects it departed from them. There was a tendency to adopt the line of rivers in settling boundary disputes, if only because it was an obvious point of reference; precision rather than expansion was for long the objective of the French kings. Feudal boundaries tended to be personal rather than territorial; they separated the loyalties and obligations of people rather than sovereignty over their land, and in case of doubt or dispute it was a practice to consult the local people and to ask them whose they were, to whom they belonged.

Nowhere between the North Sea and the Mediterranean were boundaries of less consequence than in Flanders. Here an economic region, perhaps the most developed in Europe at this time, had taken shape, spanning the Scheldt, and embracing the lands of the royal domaine around Lille and Tournai, as well as the semi-independent counties of Flanders and Artois and the imperial duchies of Hainaut and Brabant. The boundary between France and the empire followed the Scheldt, except in the north-east where the territory lying to the south of the Honte channel (fig. 5.4) lay within the empire though forming part of Flanders. Conversely Ostrevant and the Cambrésis, parts of the imperial fief of Hainaut lying west of the Scheldt, were regarded as part of France, and the Duke of Hainaut unwillingly did fealty for them.

Nevertheless, the king exercised little power in this northern corner of his realm. The Flanders cities were dependent mainly on England for their wool supply and on Germany and southern Europe for their markets. In the past the Champagne fairs had provided an important instrument of Flemish commerce, but these had declined in importance. Flanders responded sharply to the twists of English commercial policy, but was not in any major respect dependent upon the rest of France.

Within Flanders, the authority of the count had been much reduced by the almost independent posture of the patrician rulers of the Flemish cities. There thus developed a four-cornered conflict between the king, the count, the patricians and the humble workers, in which for a time count and proletariat had united to resist the pretensions of both king and bourgeoisie. England's interests clearly coincided with those of the latter. The wool trade was manipulated in order to put pressure on the count, and refugee Flemish artisans were welcomed and settled in England.

Of the other semi-independent duchies and counties within France, only Guyenne calls for comment because it alone had important links beyond France. It had been part of the English crown since the mid-twelfth century. Its extent had fluctuated, but its prosperity had grown with the increasing demand in England for Gascon wine. It was the most independent of the duchies, and far from doing fealty to the French king for it, the king of England even claimed that he was himself the rightful king of France. In terms of political geography, we have thus two provinces of France, Flanders and Guyenne, where the authority of the king was openly defied; and we have a number of minor boundary problems along the landward limits of France, and, within France, a number of semi-independent duchies and *apanages*. On the other hand, the royal domain was scattered over most parts of France, thus providing bases for the extension of royal power. An able and energetic ruler could from this foundation have gone far towards creating a political unity within France. The early kings of the house of Valois showed neither ability nor energy.

The empire

The political fragmentation of Germany had continued without interruption during the past two centuries. Since the end of the Hohenstaufen dynasty, the elected emperors had been feeble and ineffective. They made no attempt to control the hundreds of fiefs and free cities; and Rudolf I, the first of the Habsburgs to assume the dignity, had (1279)) renounced all imperial authority over central and southern Italy. He nevertheless used what authority he had to endow his family with the Austrian lands stripped from the king of Bohemia. It was not that Rudolf and his successors lacked ability, but the elections of each was disputed, and divided the German princes into factions; the emperors pursued conflicting policies, and made grants and promises inconsistent with the maintenance of the imperial dignity. Above all the future of Germany and of the empire was at stake. The idea of a strong emperor was opposed, not only by the leading German princes, but also by the king of France, who was having some success in attracting to himself the allegiance of the peripheral areas of the German Empire, and by the popes, who remembered only too well how hard it had been to overcome the Hohenstaufen.

The division of Germany into a small number of coherent states was proposed. This recommended itself to the Papacy and above all to the Habsburgs, who had in fact carved out such a realm for themselves in south-eastern Germany. It was opposed by the Rhineland principalities – the greater number of them ecclesiastical – which saw in such a step a drastic diminution of their own not inconsiderable power to influence imperial elections and control imperial policy. And so the German anarchy continued, with real power passing to an oligarchy of princes, the electors.

The empire embraced at this time the provinces of northern Italy, Provence, Arles and Dauphiné, the Burgundian Franche Comté and the Low Countries from Hainaut to Holland. In none of these was imperial authority more than nominal. They played little or no part in the selection of their ruler, and were more susceptible to the influence of the king of France than to that of the emperor. All western Germany had fragmented into a mosaic of duchies and counties, margraviates and burggraviates, many of which were themselves fragmented. Prominent among them were the church lands of the archbishoprics and of a few well-endowed monastic houses, such as Freising and Fulda.

The Rhine, bordered by the lands of Strasbourg, Speyer, Worms, Mainz, Trier, Cologne and Utrecht, had already become the *Priesterallee.* To the east, the bishops of Bremen, Münster, Paderborn, Hildesheim, Halberstadt, Würzburg and Bamberg controlled lands which were vastly more extensive if less valuable. The church lands had the advantage that they were handed on intact by each incumbent to his successor. They thus tended to grow as bequests were made to them.

The lay fiefs, on the other hand, became smaller, as they were divided between heirs or split up to make dowries and gifts. Against the background of petty states, however, were a few of greater size and significance, such as Hainaut, Brabant, Luxembourg and the Palatinate (Pfalz) to the west of the Rhine; Braunschweig-Lüneburg, Thuringia, Meissen, Württemberg and Bavaria to the east.

The states which made up the eastern half of the empire were larger and very much fewer than those in the west. The lands of the Teutonic Order, which bordered the Baltic Sea from Pomerania to the border of Lithuania, made up one of its largest territorial units. To the west lay the duchies of Pomerania and Mecklenburg, and south of these, the Mark of Brandenburg. But the strongest of these eastern fiefs of the empire at this time were the kingdom of Bohemia with its dependency, the margraviate of Moravia, and the complex of Habsburg lands developed around the duchy of Austria. None of these was wholly German. Bohemia was a Slav state, into which Germans and German culture had begun to penetrate (see p. 341). The last of its native Přemyslid dynasty, Vaclav III, had died as recently as 1306, to be succeeded four years later by the German, John of Luxemburg.

The Hapsburg family had established itself in Austria which it had taken from Bohemia (1282) in a dispute which had overtones of a German–Slav struggle. To this nucleus had been added Styria (Steiermark), Carinthia (Kärnten) and Carniola (Krain). The duchy of Silesia, which had formerly been part of Poland, had fragmented into a number of small counties which in 1335 accepted the sovereignty of Bohemia.

Scattered throughout the empire, and in some degree an exception to its prevailing feudal structure, were the cities. The majority were small and dependent upon a feudal suzerain who had created them and whose needs they served. These were part of the feudal framework of society; but among them was a smaller number of cities of higher status and generally larger size, the *Reichsstädte.* They were conceived as feudal entities directly dependent upon the emperor; in fact, they were exceptions to the prevailing way of life, and were interested primarily in the profits to be obtained from crafts and commerce, rather than from land. That their interests were often in conflict with those of the feudal class is apparent from the leagues which they formed to protect themselves and from the wavering policy of the emperors, favouring now the rights of the cities, now the archaic pretensions of the landed nobility.

One alliance, not so much of cities as of peasant communities, deserves notice. The mountain valleys around the headwaters of the Vorder Rhine and Reuss had remained of little importance until, some time about 1230, the St Gotthard pass became available for traffic between Germany and Italy. The small communities of free peasants which lay along the alpine routes began to acquire both political sophistication and economic importance; and the Habsbergs, who had inherited feudal rights over much of this area, found it profitable to enforce them more vigorously. At this time the family, whose ancestral home lay at Habsburg in northern Switzerland, was creating its immense family holding in Austria and the eastern Alps. In 1291, when Rudolf of Habsburg, architect of the family's fortunes, died, the leaders in the three mountain valleys of Uri, Schwyz and Nidwalden formed a league, or *confederatio*, to protect their interests. The Habsburgs failed to reconquer the mountain cantons, which, joined by Unterwalden, became the nucleus of the Swiss Confederation.

Italy

The fragmentation which characterised Germany seemed to have been carried to extreme in Italy. The territory which lay south of the curving line of the Alps was divisible into northern Italy, still in name a part of the German Empire; central Italy, where the Papacy by a combination of ruthless imperialism and judicious forgery, had carved out a state of its own; southern Italy, where the French Angevins had replaced the Normans, and the islands which were falling under the commercial domination of Aragon.

Northern Italy. Two centuries earlier northern Italy had witnessed a remarkable development of urban life and institutions (see p. 236). A number of self-governing communes had appeared. By the fourteenth century their number had been greatly reduced and in almost all of them the reigns of power had passed into the hands of a small local oligarchy, or *signoria*. The process by which the larger and more powerful city-states had absorbed the smaller had been assisted by the interminable feuds which destroyed the harmony within each of them. A legacy of the wars of the Guelfs and Ghibellines had been the armies of mercenary *condottieri*, who sold their services to any city willing to pay and whose leaders often rose to be despots of the cities they protected. The Visconti had thus come to control Milan, and the Gonzagas, Mantua. Northern Italy was dominated by a small number of powerful and aggressive city-states. If Venice and Genoa were content at this time to make their conquests beyond the seas, Milan, Verona and Florence were interested in expanding the boundaries of their *contadi* and in absorbing smaller and less fortunate cities such as Pavia, Cremona, Piacenza, Bergamo, Vicenza, Prato, Pistoia and Volterra. Florence had not yet occupied Pisa and Siena, but was exerting a growing pressure on them. The Visconti of Milan were creating a large and rich domain across the middle of the Po basin from the Alps to the Apennines, a partial justification for which lay in the extensive land-drainage projects then being undertaken. A few other city-states, though smaller and weaker, nevertheless succeeded in playing the political game and insuring their own future; Mantua under the Gonzagas, Modena under the Estes, Padua under the Carraras, Parma under the Correggio who subsequently sold it to the Visconti, and Lucca under a succession of despots and dynasties.

Venice had not yet encroached significantly on the eastern margin of the

Lombardy plain, but in the west the Duchy of Savoy, originating to the west of the Alps, had extended its political control across the passes and down into the plain of Lombardy, where it already controlled the future capital of Piedmont-Savoy, the city of Turin.

Central Italy. Obliquely across central Italy, from the marshes of the Po delta to those of Terracina lay the lands of the Holy See. Their prince was elected by a conclave of cardinals, in which the noble families of Rome, the Orsini, Colonna and Savelli in particular, fought for position and power. Boniface VIII (1294-1303) had tried in vain to revive the older papal claims to universal rule; he had been defeated by the growing sense of nationalism of France and England. His successors made no such mistake; and in 1309 Clement V, a French pope who never ventured into Italy, took up his residence at Avignon, a small town in the county of Provence, across the Rhône from France, where he and his successors obediently followed the will of the French kings. The lands of which he was the temporal head were rich and varied. They were made up of many small city-states and lay fiefs all of which owed loyalty and usually more to the Papacy itself. Among them the popes were accustomed to carve out *apanages* for their families, while the towns strove to minimise this spiritual control and to extend their own commercial privileges. Foremost amongst the towns were Perugia, Orvieto, Spoleto, Todi, Viterbo, Narni, Terni and, of course, Rome itself. Orvieto was typical, a city of perhaps 20,000 and a *contado* of about 800 square miles,[1] to which a limit of growth had been set, if not by papal vigilance, at least by that of neighbouring Perugia, Todi and Narni.

Beyond the Apennines lay the Romagna, less urbanised than the rest of the Papal States, but an important source of grain and salt to much of Italy. Rome itself did not share the prosperity enjoyed by the city-states within the papal lands. It enjoyed no self-government, and had no merchant and artisan class of significance. It was ruled arbitrarily by the pope, and in his absence was a prey to the factions which disrupted the life of the city as much as they did the business of the *curia*.

Southern Italy. The last of the Hohenstaufen had succeeded to the Norman kingdom of Naples and Sicily, and had thus enclosed the Papal States on both north and south. In his anxiety to free himself from this danger to his territorial aspirations the pope (Urban IV) had not hesitated to call the French to his aid and had offered the crown to Charles of Anjou, brother of Louis IX of France. Charles made himself master of the whole of southern Italy from the borders of the Papal States to the western promontories of Sicily, until the French were driven from the latter island (1282) in the rising popularly known as the Sicilian Vespers. The kingdom of Naples, after the loss of Sicily, remained in the hands of the Angevin dynasty, which continued to be the overzealous supporter of the Papacy and implacable opponent of all imperial pretensions in Italy.

The islands. The short-lived Angevin control of Sicily was replaced by that of Aragon. In the last resort it was Aragonese sea power, to which the king of Naples had no answer, that secured the victory for Spain. The other islands of the western Mediterranean were in dispute among the sea-powers which alone could reach and occupy them. A long struggle between Genoa and Pisa was fought in part for the possession of Corsica and its wealth of ships' timber. In the early fourteenth century Genoa was victorious, in so far as she was able to control the island's commerce, but did little to change the state of anarchy to which the interior of the island had been reduced.

In Sardinia the dispute between Pisa and Genoa for control ended in the triumph of neither. The pope in 1297 conferred the island on the king of Aragon in an attempt primarily to weaken the Ghibelline city of Pisa. In 1326 the island was conquered by the Aragonese though parts of the interior continued in practice to be independent of any outside authority. The Aragonese had already made good their claim to the Balearic Islands, and thus ruled a maritime empire which spanned the whole western Mediterranean.

Eastern and northern Europe

Beyond the eastern limits of the empire lay states – Poland, Lithuania, Hungary, Serbia, Bulgaria and the principalities of Romania – which derived from the tribal kingdoms of the earlier Middle Ages. Northward, beyond the Baltic, lay the Scandinavian kingdoms; their heroic age had ended, and their role in the politics and commerce of Europe was in general a passive one.

Scandinavia. The ports of Scandinavia still played an important role in the commerce of northern Europe, but their trade was controlled by the German merchants of the Hanse. The kingdom of Denmark had challenged the Hanseatic League, with no great success and a generation later was to be humiliated by it (see p. 409). The kingdoms of Norway and Sweden, which at this time included Finland, were temporarily united, but this brought them no great power or influence. The Swedish advance towards Novgorod had been halted many years earlier, and the Swedish role in the affairs of the Hanseatic League was one of minor importance.

Poland. During the previous two centuries the Germans had encroached upon the Polish state, and both Silesia and Pomerania had fallen effectively if not formally under foreign domination. To the north the German knights had taken possession of both west and east Prussia, and had even occupied Toruń. Hemmed in on the east by the still barbaric state of Lithuania and constantly threatened from the south-east by inroads of the Tatars and Russians, Poland had been unable to recoup in the east for its losses in the west. Nor had internal unity been preserved. More than once in its history had the Polish state been broken up into a number of feuding duchies, but each time a 'restorer' had arisen to re-unite the fragments, only to have them again divided under his successors (see p. 240). The early years of the fourteenth century were a period of fragmentation, made worse by the fact that the Přemyslids of Bohemia not only laid claim to the Polish throne but also invaded the country to make it good. This proved to be the turning point. Some element of nationalism, together with loyalty to the family of Piast and the opportune death of the Bohemian king, led to the gradual re-unification of the semi-independent duchies; and in 1320 Ladislas Łokietek was crowned king of a re-united Poland in the capital city of Kraków.

The East Prussia of the German knights had been called into being by the Poles in their struggle to hold back the fierce Prussians and Lithuanians. As the latter became more civilised, the balance was changed; and Poles and Lithuanians found themselves in alliance to restrain the aggression of the knights. The grand-duchy of Lithuania extended from the borders of Prussia and Mazowsze eastwards through White Russia and into the Ukraine. It was a vast inchoate state, in which primitive tribalism was being modified by incipient feudalism. Polish cultural influences were strong in Lithuania, and force of circumstances – their common exposure to dangers from both east and west – was driving the two states towards a close alliance, though it was not until 1386 that they were joined in a personal union under Jagiełło.

North of Lithuania, over an area corresponding roughly with the later republics of Latvia and Estonia, the knights of the German Order of the Sword had established their rule. They held the land lightly; very few German settlers came, and these lived mainly in Riga, the headquarters of the order and in the very few other towns (see below, p. 348). Nevertheless, the knights succeeded in reducing the native Baltic peoples to serfdom and in holding the land against the attacks of both the Russians and the Lithuanians. About a century before our period the Order of the Sword was merged with that of the Teutonic Knights.

Hungary. The boundaries of Hungary had changed little. They still encompassed the Hungarian plain together with the bordering highlands to the north and east and included an extension to the Dalmatian coast. A common boundary with Poland and common exposure to pressure from the Germans tended towards a close alliance of the two countries. To the east, the weak Romanian principalities gave little protection from the nomads of the Russian steppe, and to the south both Serbia and Bulgaria proved awkward neighbours, while the Republic of Venice had gained control over the more important ports of Hungary's Dalmatian coastline. The Hungarian leaders at this time tended to think of themselves as the guardians of the Christian west against the pagan Tatar, the schismatic Slav and the heretical Bogumil. But only against the last and weakest of these – the Bogumils of Bosnia – did they have much success. Bosnia, conquered and ruled by its local *ban*, was loosely attached to Hungary. He had indeed overrun the neighbouring Zahumlje, corresponding roughly with the later Hercegovina, and had even acquired an outlet to the Adriatic Sea, near the mouth of the Neretva river, but still needed Hungarian support against the Serbs.

The Arpád dynasty, which had ruled Hungary from the foundation of the state, died out in the male line at the beginning of the fourteenth century. The crown became elective, but for a century kings were chosen from among more distant relatives of the last Arpád. As in Piast Poland, the family represented the country, and mediated the incipient sense of nationalism.

Romania. The Romanian principalities were beginning to take shape in the early years of the fourteenth century. They had been a frontier zone of Europe, alternately ravaged by the Tatars and tributary to the Hungarians. Romanian tradition claims the formation of a state in Walachia in the last decade of the thirteenth century under Vlach leadership, and in 1330 the Hungarian king was defeated in his attempt to reassert his authority. The history of Moldavia took a similar course to that of Walachia. Here also a state was formed under Vlach leadership about the middle of the fourteenth century. Both Walachia and Moldavia were able to hold off the Hungarians and Tatars, but both succumbed eventually to the attacks of the Turks.

Balkan peninsula. The Balkans, between the Danube and the attenuated remains of the Byzantine Empire, were occupied by the states of Bulgaria and Serbia. The second Bulgarian Empire had been formed at the end of the twelfth century, and it is reasonable to assume that its leadership was also Vlach. Its capital was established at Trnovo, where the ruins of Asen's fortress can still be seen. Profiting from the weakness of the Byzantine Empire and the capture of Constantinople by the Crusaders in 1204, the Bulgarian rulers extended their rule as far as Macedonia and Thessaly. But much in the fortunes of the east European and Balkan states depended upon the character and initiative of an individual. In Bulgaria the House of Asen was shortlived. The Byzantine Empire revived and reoccupied Thrace and the lower Maritza valley, while the Serbs transformed themselves into a powerful and aggressive force. By about 1330, Bulgaria was again reduced to a relatively small area lying mainly between the Stara Planina and the Danube. Serbia, on the other hand, was at the height of its medieval glory. Territorially it had grown from the province of Raška amid the mountains of Old Serbia. With the occupation of Dioclea (Zeta, or approximately Montenegro) the Adriatic coast was reached. Serb rule was extended northwards to the Sava and Danube, though not without opposition from Hungary. Though Bosnia and Zahumlje were lost, significant gains were made in the south. Under Stefan Dušan, Albania, Epirus and Thessaly were added to the Serb empire, and lands were taken from the Bulgars. This was a period of great prosperity; the mines of Serbia were active; commerce flowed through the port of the Republic of Dubrovnik, and in Serbia itself monasteries, which have remained among the architectural gems of the Balkans, were built to commemorate Serb victories.

Byzantine Empire. In 1261 the emperor had been restored to Constantinople, from which his predecessor had been driven by the Crusaders more than half a century earlier. At most, he was able to occupy effectively only Thrace. His European lands were poor and limited in extent, and the Turks were constantly encroaching on those of Asia. The emperor was ringed with enemies; the miracle is that this feeble and ineffective empire proved to be so durable. The Genoese dominated the Black Sea, and in Constantinople itself had fortified themselves in the Galata suburb. The Venetians could hold the emperor to ransom, while a band of Catalan mercenaries wrecked havoc in Thrace and Asia Minor.

The Latin conquest of more than a century earlier had left as its legacy a swarm of petty states in Greece and the Aegean, ruled by Franks and in a few instances by Greeks. A feudal order of society was introduced, as both the *Assizes of Romania* and the ruined castles of Greece still testify. In time the Frankish and Lombard conquerors became hellenised; some of their conquests were re-absorbed into the Byzantine Empire; others continued in practice to be independent. By about 1330 there survived only the feudal states of the Morea and Attica, the latter now in the possession of the notorious Catalan Company of mercenaries.[2] The Angevin kings of Naples showed great interest in reviving the political connection between southern Italy and the lands of Albania and Epirus, and succeeded, indeed, in establishing bridgeheads on the opposite coast. The Sicilian débâcle put an end to such dreams, but the Naples kingdom continued to hold some scraps of territory, including Corfu and the Albanian ports of Valona and Dürres. In so far as the hinterland was held by anyone it was by Uroš III (Dečanski) of Serbia. In fact with the extinction of the Despotate of Epirus, most of the area had reverted to the primitive isolation, autonomy and self-sufficiency that had characterised much of its history.

Empire of Venice. Foremost amongst those who had helped themselves to the lands of the Byzantine Empire was the Republic of Venice, which had successfully manipulated the Crusades to its own ends. Venice already enjoyed a kind of protectorate over the Dalmatian cities and held valuable commercial privileges in Constantinople and the Levant. To these it was able to add the island of Euboea, or Negroponte; the south-westerly headland of the Peloponnese (Morea), with its ports of Modon and Coron, the city of Thessaloníki, and the island of Crete. In addition, an enterprising Venetian had seized the island of Naxos, together with most of the other islands in the Cyclades group. These were organised as the duchy of Naxos, or of the Archipelago, in which the lesser islands became fiefs of Naxos. Venice was able to retain all these territories and dependencies until finally driven out by the Turks.

The Genoese, throughout this period the mortal rivals of the Venetians, controlled the Black Sea, on whose shores they had about a dozen commercial stations, of which the most important were at Soldaia and Caffa in the Crimea. In the fourteenth century the Genoese also occupied Chios and other islands in the north-eastern Aegean.

Already by 1330 the shores of Europe had been attacked by the Turks. They had by this time gained almost complete control over Anatolia, and, if they had not yet set foot permanently in Europe, they had nevertheless raided the eastern shores of Greece and carried off the subjects of the Venetians from the islands of the Aegean. Their common danger led the Italians and the Catalans of Athens to discuss an alliance against the Turks, but little came of it. Latin princes, Byzantine emperors and the Slav rulers showed a complete incapacity for joint action against the common foe, and all succumbed in turn to the Turkish invasion.

Conclusion

One can recognise in the political geography of Europe at the end of the first quarter of the fourteenth century two conflicting trends. The first was one towards centralised rule and the creation of nation-states. A true sense of nationalism had not yet emerged, but there was a broad coincidence between kingship in Europe at this time and later national organisation. Nationalism probably owed much to the administrative organisation of medieval kings, and there was an incipient sense of cultural unity in the future nation-states.

The second trend was in the opposite direction, towards the creation of coherent states within the compass of the medieval feudal units. These were necessarily smaller in terms of total resources, but often far better organised, and in very many instances they were founded upon a firm basis in strong local sentiment. This trend towards political fragmentation triumphed in Germany and Italy; in France it was a powerful force, and it was strongly marked in Spain. Not until the fifteenth and sixteenth centuries do the centralising forces gain the upper hand, and the nation-states begin to dominate the European scene.

Another feature of the political geography of this period was the rise of the European city-state, which, since it constituted in many ways an exception to the feudal order, had, of necessity, to protect itself and its interests against the latter. The consequence was, in those countries in which central authority was weak or non-existent, such as Germany and Italy, a proliferation of city-states. In others, where political control of an area continued to be exercised, there was a continuing struggle to control the cities, as in France, or to enlist them on the side of the monarchy, as in England.

The political map, lastly, was gaining in precision as the boundaries of territorial lordship and sovereignty came gradually to be defined. This process can be traced in the greatest detail along the boundary between France and the empire, but it was active also on the border with Spain, between the states of eastern Europe and the principalities of Germany and Italy.

POPULATION

The population of Europe, which had been increasing since at least the tenth century, had ceased to grow by the early fourteenth century, and the later Middle Ages were marked by a undoubted decline in total population.[3] It has been generally assumed that this decline was the direct consequence of the Black Death of 1348. This was probably not the case. Not only was the Great Plague the culmination of a series of natural disasters but the evidence of wages and prices suggests that population had fallen somewhat in some areas quite early in the century, if not in the later years of the thirteenth. Pestilence, though not necessarily the bubonic plague, had already attacked one area after another.[4] Commonly these visitations followed a sequence of bad harvests, such as those which north-western Europe experienced in 1315–17. During these years there were prolonged periods of cool, wet and cloudy weather, when crops failed either to germinate or to ripen. Meteorological conditions may, in turn, have been related to a cyclical deterioration of climate, which had begun in the previous century.

The climate did not itself cause the epidemics; it merely provided the conditions in which infections could spread. Even so, food shortages would not have been so disastrous if much of Europe's population had not been living close to the limit of subsistence and if much of the land had not been fully cultivated within the limits of existing technology. For an understanding of conditions in the early fourteenth century, it is necessary to know just how great this population pressure was, and how close to the starvation level most people lived.

The data for such a study is inadequate, and comes largely from the years after the Black Death. Until we know the extent of the losses caused by the Plague and the speed of recovery from it, it is impossible to extrapolate backwards from post-plague data. Furthermore, the latter are usually couched in terms that do not permit direct use. They appear as the number of births and deaths, of 'hearths' and householders, of craftsmen and soldiers. In each case it should be possible, by the use of a multiplier, to obtain an approximate figure for the total population. The most used unit of measurement is the 'hearth'. If this means merely a family group, then a multiplier of between 4 and 5 would appear reasonable. It seems however that the term was beginning to acquire a purely notional quality, as the *mansus* and *hide* had done earlier, and that it could sometimes represent something very much larger than the family.[5]

Two categories were regularly excluded from medieval enumerations of hearths and population: the clergy and other members of the monastic and mendicant orders, and the beggars whose lack of property excluded them from the tax-rolls. This probably made no great difference in rural areas, but in the larger towns, the ecclesiastical component of the population was probably considerable, and an addition should probably be made to the computed populations of 5 to 15 per cent to cover all the excluded categories.[6] If the average hearth was at this time about 4.5, a multiplier of 5 would seem reasonable.

In the Middle Ages the average expectation of life was relatively short, and infant mortality was high. The birth-rate must have been unusually high during the previous centuries – particularly in view of the celibacy of the 'religious' – in order to achieve the very considerable increase that actually took place between the tenth and the fourteenth centuries. The sex ratio, furthermore, appears to have been somewhat in favour of males, and it has been suggested that the number of females was deliberately restricted.[7] The net reproduction rate was greater in rural than in urban areas. This alone would have necessitated a migration from the country to the town. In fact, urban population had been growing far faster than rural, and in a few towns – notably the Hanseatic ports – this rapid growth was still continuing about 1330.

A number of estimates, all of them based upon inadequate samples, have been made of the total European population before the Black Death. The data upon which such estimates are based are scanty in the extreme. For France there exists a table of parishes and hearths of 1328,[8] covering about half the country, together with a few local registers and hearth rolls, and some population data in the *pouillés*, or parochial records of the bishops.

The density of rural population in much of France, of the Low Countries and of Italy was high before the Great Plague. The survey of Beaumont-le-Roger,

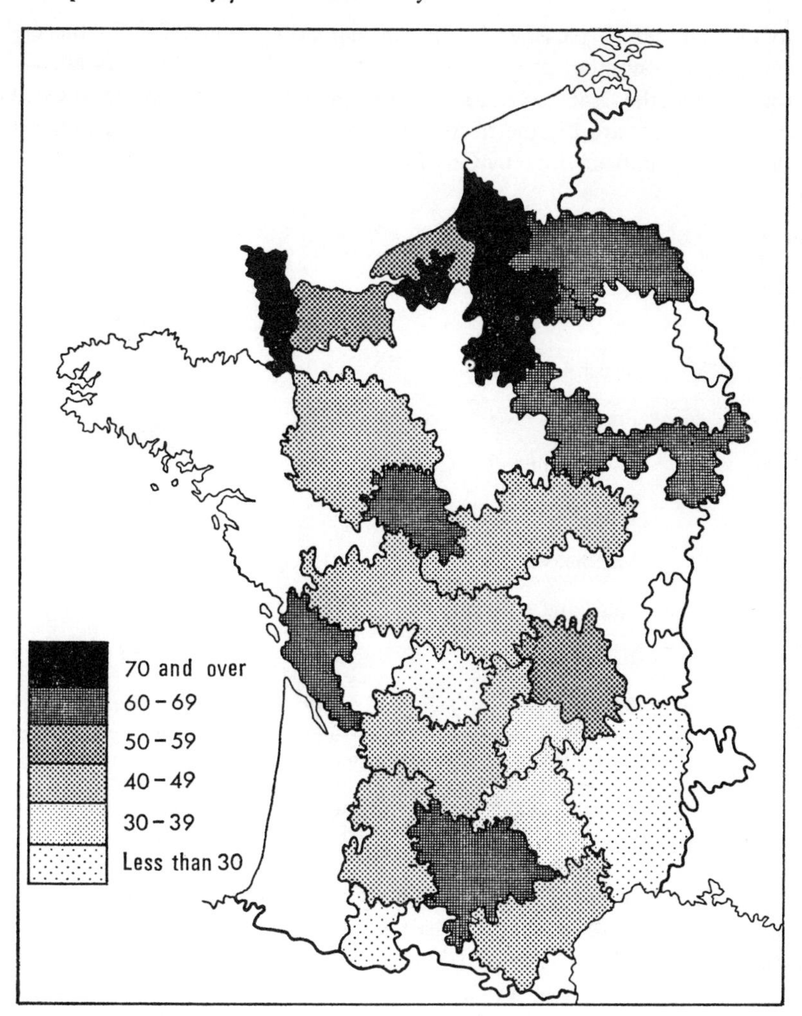

Fig.6.2 The population of France, 1328

described as 'one of the most detailed descriptions of the domains of a great lord ever made',[9] described 23 rural communes in Normandy. In 1313 they had between them 6,093 hearths, but, since the communes were not contiguous, it is difficult to estimate their area and thus the density of population. Nevertheless, their average size was very large. One is 'left with the impression', wrote Strayer, that 'the population . . . was not greatly inferior to that of today, and that it may actually have been denser in some of the less fertile agricultural areas'. A similar conclusion has been drawn from the registers of Givry in Burgundy,[10] where there were 310 hearths in 1360 *after* a severe outbreak of plague, and there are about 2,000 inhabitants today.

The French parish and hearth list gives totals only for large territorial units, such as *bailliages* and *sénéchaussées*. Only the vicomté of Paris is presented in any detail. The number of hearths is given for about 23,000 parishes, very close to the number of parishes in the corresponding *généralités* in the eighteenth century. The number of gaps in the survey must have been very small, and the total of hearths, 2,469,987, not far from the truth. It is possible to calculate the density per square kilometre for some of the territorial units enumerated, and this has in fact been done by Guy Fourquin for the Paris region.[11]

Table 6.1 *Total European population before the Black Death* (in millions)

<table>
<tr><th></th><th>Beloch*</th><th>Russell†</th></tr>
<tr><td>British Isles</td><td>4.0</td><td>5.3</td></tr>
<tr><td>France</td><td>14.0</td><td rowspan="2">19.0</td></tr>
<tr><td>Low Countries</td><td rowspan="2">15.0</td></tr>
<tr><td>Germany</td><td>11.0</td></tr>
<tr><td>Scandinavia</td><td>1.9</td><td>0.6</td></tr>
<tr><td>Spain and Portugal</td><td>6.0</td><td>9.5</td></tr>
<tr><td>Italy</td><td>11.0</td><td>9.3</td></tr>
<tr><td>Totals</td><td>51.9</td><td>54.7</td></tr>
</table>

* J. Beloch, 'Die Bevölkerung Europas im Mittelalter', *Zeitschr. Soz.*, 3 (1900), 405–23.
† J. C. Russell, 'Late Ancient and Medieval Population', *Trans. Am. Phil. Soc.*, n.s., 48, Pt. 3 (1958), 14–16.

Confirmation of the general accuracy of the survey is provided by a separate enumeration for the Sénéchaussée of Rouergue. In 1328 this province was credited with 577 parishes and 52,823 hearths. Another list.[12] attributed to 1341, listed by name 578 parishes and gave the number of hearths in each. The total hearth count at the later date was 50,125. It is probable that the two lists derive from the same source, despite small discrepancies in the number of parishes and hearths.[13]

There were marked inequalities in the density of population. The city of Paris is itself credited with 61,098 hearths. Philippe Dollinger has argued[14] that this figure is so contrary to the facts and so out of keeping with the rest of the document, that it must be regarded as an error. Historians have in general tended to accept a lower estimate, even though this involves rejecting, at least in this respect, a document that has gained general acceptance. More recently, however, the figures given for Paris in the 1328 survey have received some support.[15] The high estimate of a quarter of a million or more does, however, seem, in the light of data for other cities (see below, p. 349) to be excessive. Nevertheless, if one can postulate a small average hearth size for the city and include within it extensive suburbs, a total of 61,098 hearth may not have been impossible.

The city was surrounded by a zone within which there was a moderate density of from 15 to 17 hearths per square kilometre. Beyond this the density fell off sharply, and was less than 9 in the areas of poor soil to the west of the city and in parts of Brie. To the north of the city, in a belt of territory extending west to east, from Pontoise to Meaux, the density was over 18 hearths (90 persons) per square kilometre, which may well prove to have been one of the highest for a predominantly rural area in north-west Europe. It was from this area that a large part of Paris's food supply was obtained.

The Paris region, with an overall area of about 4,225 square kilometres, had a hearth density, even if the figure for Paris is adjusted downwards, of about 19. The record does not allow the rest of France to be examined in as great a detail, and here, even more than in the Paris region, one is faced with the uncertainty of feudal boundaries and of the areas which they embraced. The calculated figures may thus be subject to a small range of error. It is, nevertheless, apparent that the density of hearths was high (figs. 6.2 and 6.6) in a wide belt of territory, which extended from Flanders into Normandy and Beauce, including the Paris region, and was continued by a region of somewhat lower density to the Loire. Fairly high densities are also met with in Bar, Champagne, Brie, and Burgundy. The County of Toulouse also shows up as one of the most densely populated, and if similar densities are not recorded for the coastal plain of Languedoc, this is probably because the average was reduced by its inclusion with the thinly peopled Cevennes. A problem arises with regard to the small territory of Mâcon, which is recorded as having no less than 1,029 parishes and 111,912 hearths. The area covered must have been far more extensive than the Mâconnais, which had only 130 parishes, and may have included, in addition to the Lyonnais, which does not appear in the list, a considerable area of southern Burgundy.

The hearth list of Rouergue for 1341, which gives the number of hearths in each parish, shows a gradation in population density from about 40 persons to the square kilometre in the lower, more westerly parts of the province to less than 10 in the Causses and on the flanks of the mountains of Aubrac (fig. 6.3).

Between 1293 and 1322 a series of surveys was made of lands in Languedoc over which the French king had recently acquired suzerainty. The lands extended from the densely peopled coastal plain, where the seigneurie of Lunel had almost 70 people to the square kilometre, to less than 5 in the rugged terrain of the Cevennes. The relatively low overall density suggested by the 1328 hearth list for Languedoc must be viewed against the very great variations which are indicated by the earlier and more detailed surveys of parts of the region.

The County of Provence, east of the river Rhône, was ruled at this time by Charles II of Anjou as a fief of the empire. It was excluded from the *énumération* of the French king, but was subjected to a number of partial hearth counts by its own feudal lords. These have been exhaustively studied by Baratier, who has tabulated the hearths, either recorded or estimated, for the years about 1315.[16] The total, 63,000, is quite close to Lot's estimate, based on the number of parishes, of 69,500.

Fortunately these figures are broken down by administrative districts. A relatively dense population of over 20 per square kilometre is recorded for parts of

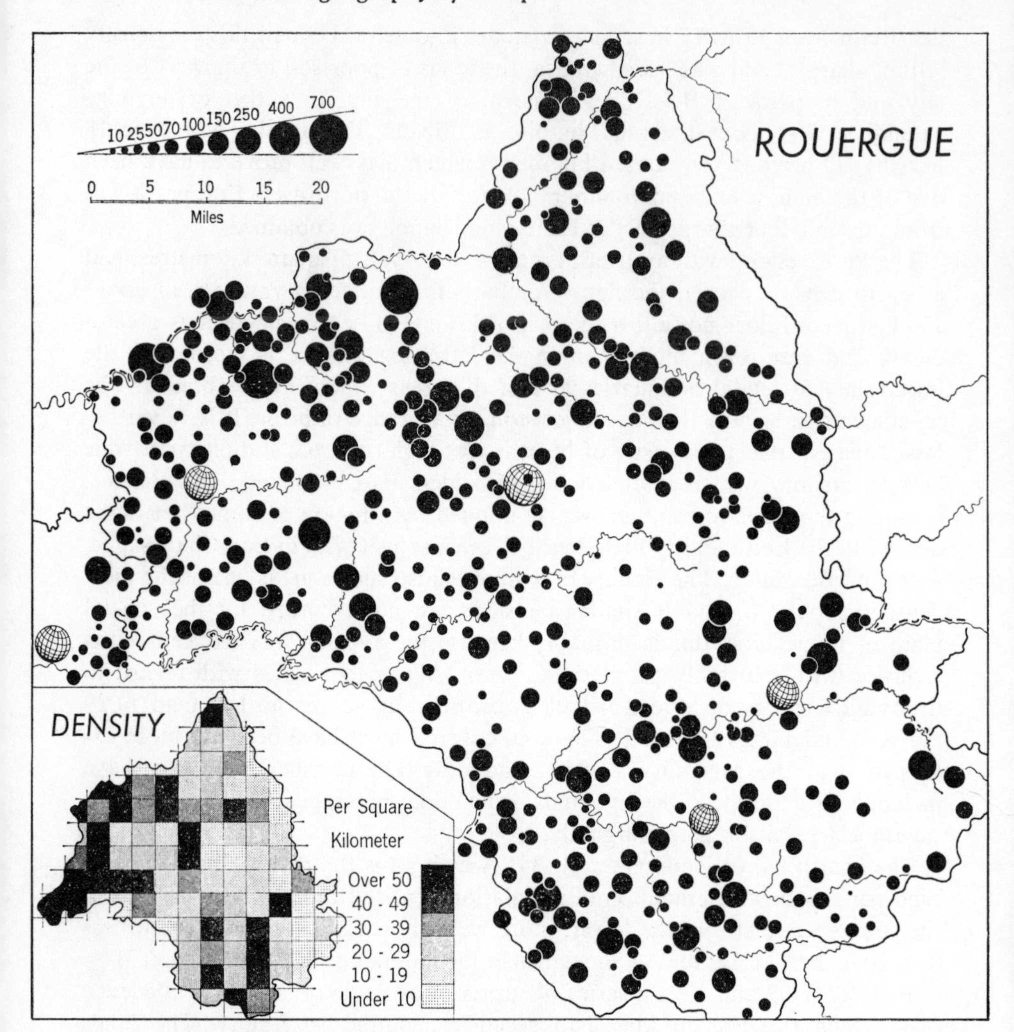

Fig.6.3 Population in the Rouergue, 1341

the coastal region and for much of the Durance and Verdon valleys. Low densities – less than 10 and locally less than 5 – are recorded for the rugged region that lies between the coast and the Durance valley, while intermediate densities characterise the gentler hills to the north of the Durance.

Lot made an estimate of the population of those part of France not included in the *énumération* of 1328 by finding the number of their parishes and assuming that each represented about a hundred households. This is a very rough and approximate method of computation, and it is likely that his totals may have been somewhat high (see table 6.2).

This hearth count for the whole of the royal domain can be supplemented by a few more detailed surveys of more restricted areas. Foremost amongst these is the so-called polyptychum of the diocese of Rouen, ascribed to Archbishop Eudes Rigaud.[17] This was a compilation made in the mid-thirteenth

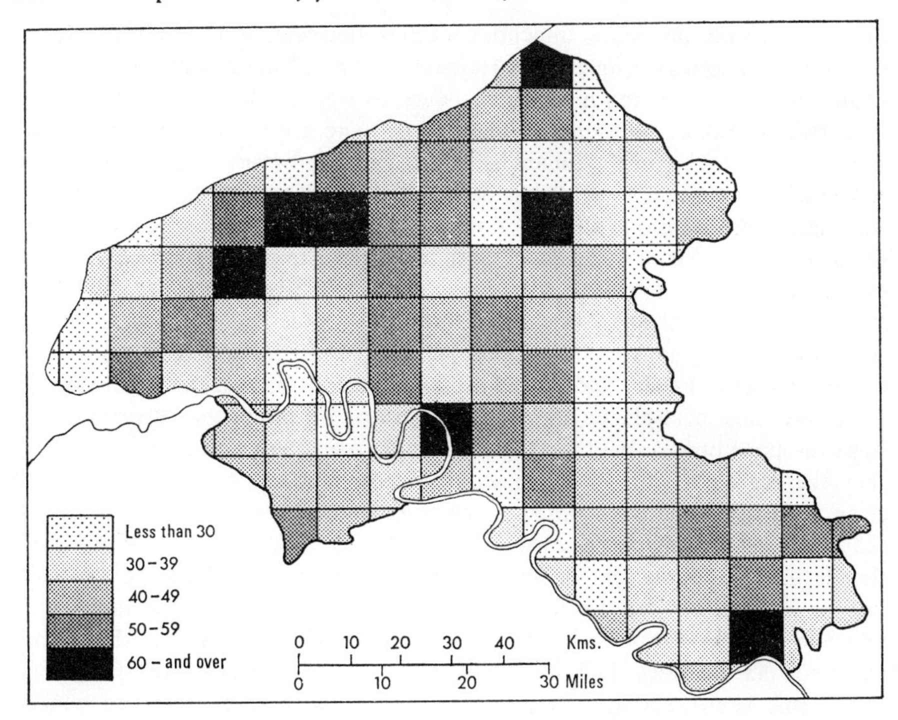

Fig.6.4 Population in the diocese of Rouen

century from the *pouillé* or survey made of each parish when a new priest was inducted. Densities varied with the quality of the soil. Population was, however, growing rapidly, and the overall density of about 57 to the square kilometre was less than that suggested by the 1328 hearth list.

Table 6.2 *Ferdinand Lot's estimate of the population of those parts of France not included in the énumérations of 1328*

	Parishes	Hearths	Estimated population (hearths × 5)	Area (km²)	Per km²
Domaine	24,500	2,469,987	12,213,500	313,663	38.9
Fiefs and *apanages*	7,500	893,763	4,468,800	110,721	40.3
Others not accounted for	500			424,384	39.3

A survey was made about 1300 of the smaller Sergeantry of Porcien,[18] a district of northern Champagne. Hearths were enumerated for 101 settlements, and suggest, with 5 persons to the hearth, a density of about 33 to the square kilometre.

Low Countries. For no other part of Europe, except a few small areas in Italy, are there comparable data for the pre-Black Death period. For the later years of the fourteenth or for the fifteenth centuries, however, there are hearth lists, or closely analogous documents for Hainaut, Brabant, Namur, Luxembourg, the principality of Liège, and much of Flanders, as well as for Artois and Picardy in northern France. For some of these areas there are several such lists, and most appear to be fairly reliable. The Liège list, however, gives only the monetary assessment of each settlement and not its size, and the Namur list is incomplete. In addition, there is a series of hearth surveys – the earliest of them of 1375-6, for most of ducal Burgundy, and similar lists for the diocese of Lausanne, the cantons of Zürich and Basel, and parts of Normandy. This body of statistical data is distributed through a whole century, during which there were fluctuations in the total population, as its natural tendency to increase was offset locally and irregularly by the effects of war and pestilence.

To use this post-plague data as a measure of pre-plague distributions, however, presupposes that the earlier geographical pattern was perpetuated and that the regrowth of population was not accompanied by migration. It is clear from the numerous studies of the *Wüstungen* of the later Middle Ages that some settlements were consolidated and others abandoned as a result of the decline in population. A further difficulty in the use of the hearth lists lies in the fact that the 'hearth' tended during this period to become a fiscal unit and that the number of actual hearths may have differed from that computed for tax purposes. The number of urban hearths, furthermore, is omitted from some hearth surveys and in all provinces one finds enclaves for which no data are available.

This group of territories, nevertheless, spans a great deal of eastern and northern France, the Low Countries and Switzerland, and, provided no great precision is expected of it, it can be used to provide a figure – perhaps a minimum figure – for the density of population. The pre-Black Death figures would have been higher in most areas – perhaps 25 per cent higher – but for this there is no real evidence.

In North Brabant, a region of poor, sandy soil, with narrow tracts of alluvium along the valleys of its rivers, there was a low overall density: about 5 hearths, or 25 persons, to the square kilometre. South Brabant, along with much of Liège, Hainaut and the French provinces of Artois and Picardy, constitutes an undulating and generally well-drained lowland with areas of very good soil. Densities in this region vary greatly, from under 5 to over 20 hearths – 25 to 100 persons – to the square kilometre.

Hainaut and Luxembourg together embraced much of the Ardennes and part of the Eifel. Densities throughout this area were low, often no more than two hearths, or ten people, to the square kilometre.

Southern Luxembourg and Lorraine are part of a hilly region where limestone plateaus and escarpments alternate with broad, flat areas of lowland. The resulting population densities were lower than those of the Hainaut–Brabant–Artois plateau, but very much higher than in the Ardennes and Eifel. They ranged from 8 to 14 hearths – 40 to 70 persons – to the square kilometre.

To the south of Lorraine lay the Duchy of Burgundy, for which a series of hearth lists has survived from the late fourteenth to the mid-fifteenth century. This region, in relief and soil, resembles Lorraine but areas of unproductive

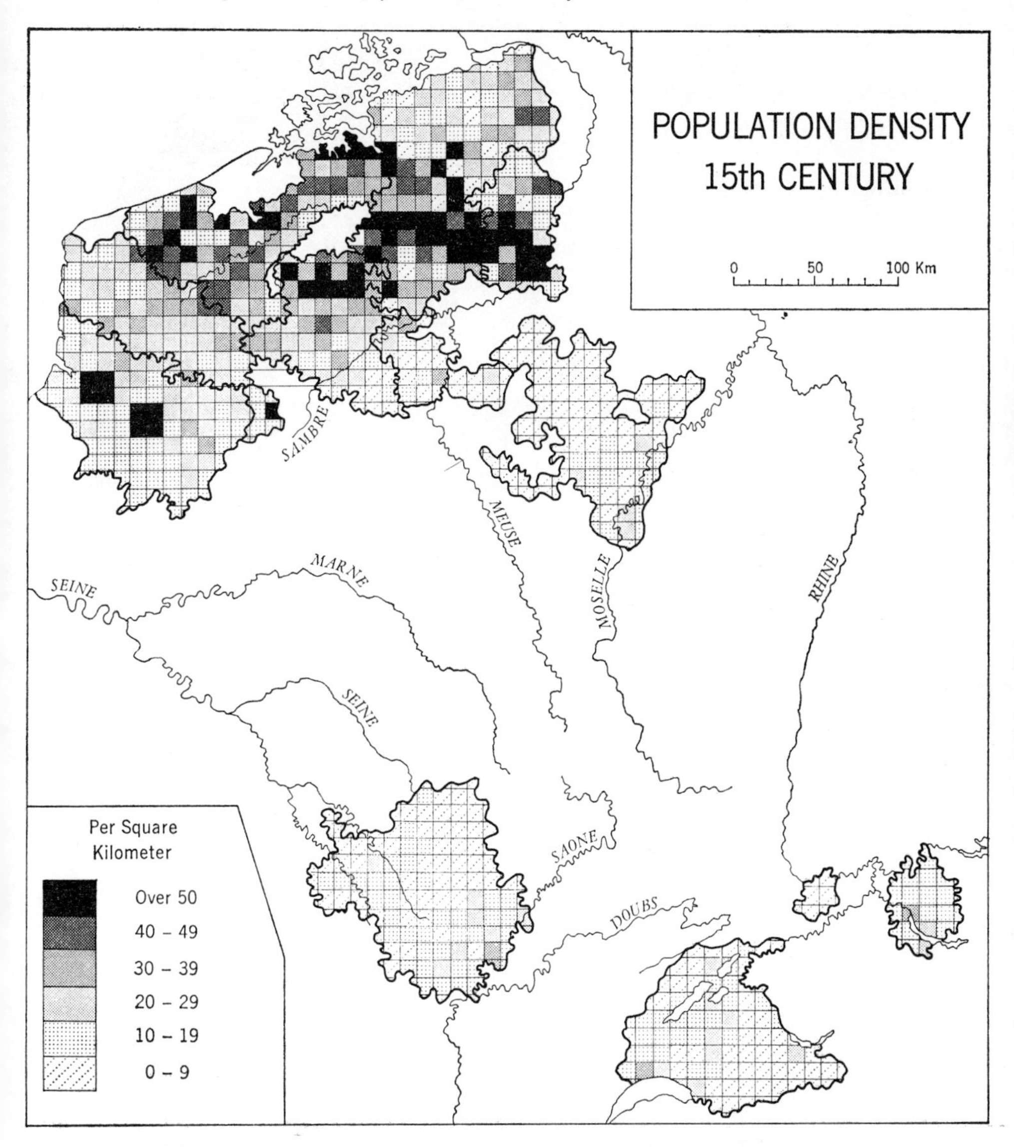

Fig.6.5 Population of the southern Low Countries and parts of Burgundy

limestone upland are more extensive. It was only lightly urbanised, and population densities ranged from 5 to 15 persons to the square kilometre in the hill country, to 20 or more in the plain of the Saône. A similar contrast is apparent in the diocese of Lausanne, where they range from less than 5 in the Jura – confirmed by the figures for Basel – to more than 20 in the more fertile and productive areas of the Swiss plateau, and drop again to well under 5 in the Alps.

Population densities were very much lower in the Netherlands than in the southern Low Countries. In the polderlands of Friesland and Holland the number was probably over 20, but in the *Geest* regions of Utrecht (the Veluve)

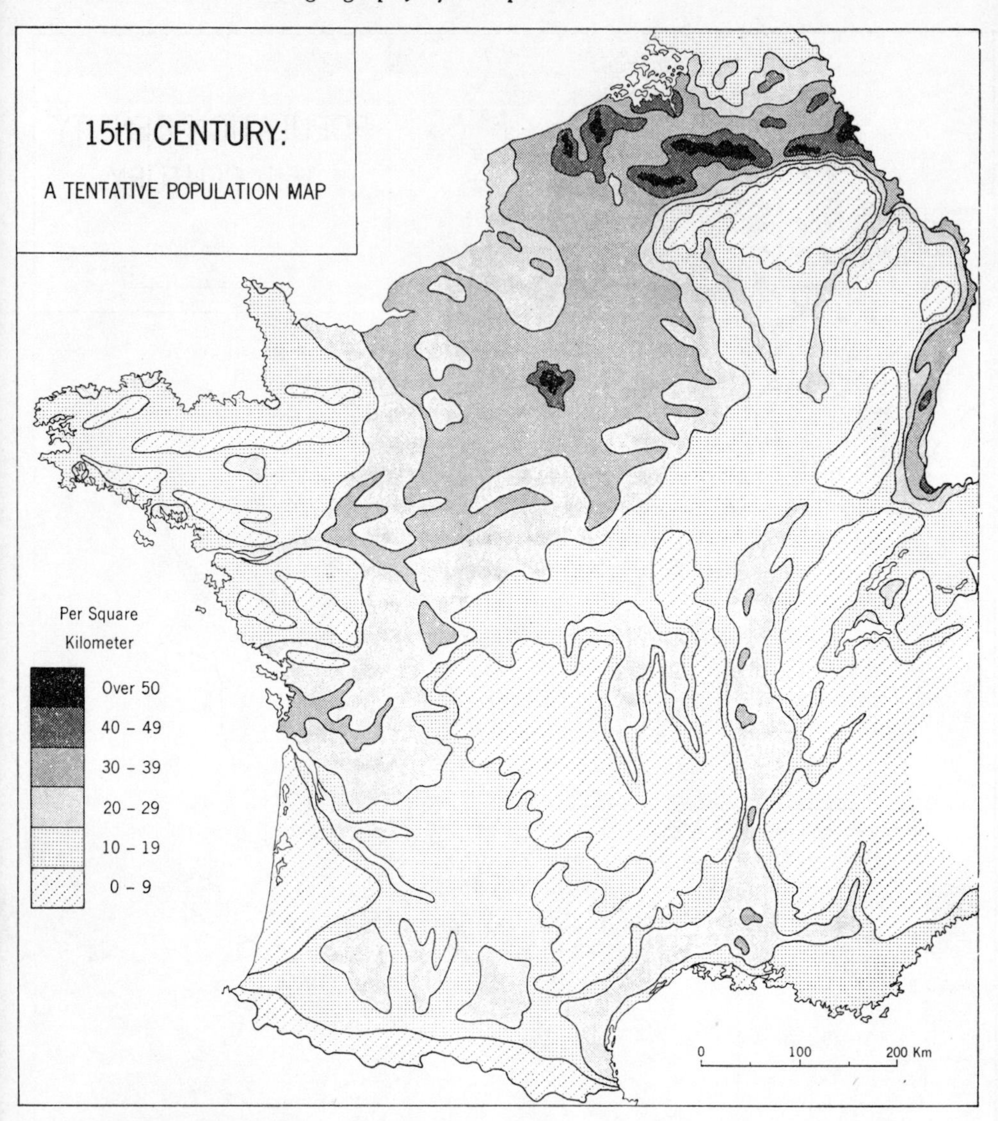

Fig.6.6 A tentative population map of western Europe in the mid-fifteenth century

and Overijssel – as unrewarding as the sands of North Brabant – it cannot have been much more than 10 to the square kilometre.

Switzerland. Ammann[19] has estimated the population in *western* Switzerland – approximately the French-speaking area – at 140,000 to 145,000 about 1400, with population densities ranging from 9 to 65. This is entirely in keeping with the findings of Schnyder from his examination of the Zürich tax rolls for 1467. The overall density of the canton was estimated to have been about 22 per square kilometre, with a range from 40 to 80, along the shores of Lake

Zürich, where an intensive viticulture was practised, to less than 10 in the more hilly parts of the canton. Generalising from these fairly well authenticated figures, Bickel[20] estimated that the population of Switzerland, within its present boundaries, was in 1400 about 600,000 to 650,000, an average density of about 15.

No other data for western and central Europe are at once so complete and so reliable as those already discussed. For Germany and eastern Europe such sources are almost completely lacking, though, as will be examined later (p. 349), there are materials for the study of urban population. A hearth list of the sixteenth century for Saxony has recently been used by K. Blaschke,[21] who projected the figures back to the fourteenth, making allowances for the *Wüstungen* of the later Middle Ages. The method is suspect both because of the imprecision of the term 'hearth' in the sixteenth century and of the uncertainty regarding the course of population history during the intervening centuries. Blaschke suggests for 1300 an overall density of 25 per square kilometre, with a range from 9 in parts of the Erzgebirge to 41 on the limon soils around Leipzig.

A survey of settlements and farms was made in 1337 in the Brandenburg Neumark.[22] It shows a total of 15,843 *Hufen*, or farm settlements in 365 of the 442 villages. This suggests a total of about 19,200 farms in an area of about 10,600 square kilometres, a density of a little under 10 persons. A register of the Herrschaft of Sorau (Żary) in the Oder valley gives details for 55 contiguous villages with altogether about 1,350 households and a population of 6,000.[23] This yields an overall density of about 10 to the square kilometre. A similar estimate for the Mittelmark gives a density of 9.3.

A number of attempts have been made to determine the population of Poland on the basis of Peter's Pence assessment of 1340.[24] Estimates vary greatly, from an overall 3.1 per square kilometre to 8.1. The tendency today is to accept the higher figures.[25]

A population map for the diocese of Kraków, corresponding approximately with the province of Little Poland, has recently been compiled by Ładogorski on the basis of ecclesiastical records.[26] It shows, not only a very great range of density from over 20 along the upper Vistula to less than 3 in much of the Carpathian region, but also a remarkably close adjustment to soil quality. Relatively dense population – more than 15 per square kilometre – occurs, with very few exceptions, only on areas of loess soil, and populations of intermediate density are found on brown forest soils. The correlation, however, is not perfect. Mountain basins, where there is no expanse of good soil, nevertheless, show densities considerably above the average, while the highly fertile Lublin plateau has a density which is scarcely greater than that of the sterile sands of the Sandomierz Basin.

It may perhaps be concluded that population pressure was relatively weak in southern Poland; that, far from cultivating sub-marginal land, as had happened in western Europe, the population had not even developed fully the existing areas of high-quality land. It is possible however that an eastward movement of settlement from the populous Kraków region to the equally promising but thinly peopled Lublin region, had been inhibited by political insecurity, and that the Vistula gave some protection to its western bank from the steppe peoples beyond.

Comparable data does not exist for other parts of eastern and south-eastern Europe. 'There is,' wrote Charanis of the Byzantine Empire, 'absolutely no statistical evidence for this period.'[27] The unsatisfactory data for Hungary have been used by Györffy,[28] who estimated a total population of two million or less before the Tatar invasions of 1241. The losses from the latter were severe, especially in the Great Plain to the east of the Danube, and it is not known when population regained its earlier level. By 1500 it is said to have been between 3.5 and 4 million,[29] and in 1330 about 2 million in the whole Hungarian state, a density of about 8 to the square kilometre. Density was certainly greatest in Transdanubia and Transylvania; least in the Great Alföld. Densities were very much lower in the Romanian principalities.[30]

In the Balkan peninsula the rugged terrain and the frequency of war and invasion combined to reduce population. The very few narrative accounts emphasise the emptiness of much of the region, and it is very doubtful whether the overall density could have exceeded 5 and it may not have been more than 2 or 3 to the square kilometre. On the other hand, there were small areas of denser population, notably the valley of the Serbian Morava, the Raška region, the Kosovo, Metohija and Sofia basins, the Marica valley and the plain of Macedonia, near Thessaloníki. Here the density may have been comparable with that in some of the well-settled parts of central Europe, perhaps from 15 to 25 to the square kilometre.

Greece, Macedonia and Thrace had suffered greatly from the aftermath of the Fourth Crusade, when the territory was fought over by Frankish and Catalan bands. Urban life almost ceased to exist, and population densities cannot have been higher than in the Serbian and Bulgarian Empires to the north. Constantinople remained, however, a large, if decadent, city, and this must certainly have been reflected in the greater rural population of eastern Thrace.

Italy. At this time Italy was the most highly urbanised part of Europe, and political control of much of it was vested in a highly literate bourgeoisie. Records were better kept than in most of feudal Europe.[31] Lists of hearths and taxpayers, of gild members and men of military age have survived, together with well-informed estimates, like those which Giovanni Villani made of Florence. Most of this data is for the period following the Black Death, which was particularly virulent in Italy. Much of it, furthermore, relates to the towns rather than the countryside. While one can compile a satisfactory map of cities for the period about 1330, one is less certain of rural densities. Italian population had been increasing for several centuries, but in the late thirteenth century rural population at least seems to have reached a peak and to have begun slowly to decline in northern Italy and Tuscany.

For three only of the 'states' of Italy has it so far been possible to arrive at even an approximate figure for the population of both city and its *contado*. Best documented of these estimates is that made for Pistoia by Herlihy.[32] The *contado* of Pistoia lay in Tuscany, to the west of Florence, and extended from the marshes along the Arno to the summit of the Apennines. It was made up of a segment of the alluvial plain, at this time poorly drained and not intensively cultivated; a belt of hills, rising to some 1,600 feet, in which the larger part of the rural population lived; and the thinly peopled mountains. Table 6.3 shows the distribution of population in 1344.

Table 6.3 *Distribution of population in the* contado *of Pistoia in 1344*

	Area (km^2)	'Mouths' in 1344	Estimated population	Density per km^2
Plain and low hills	195	7,593	9,263	47.4
Intermediate hills	140	7,278	8,879	63.4
Mountains	565	4,772	5,822	10.3
City	1.2	..	11,000	..
Totals	901.2		34,964	38.8

Contemporary Florence was even more densely populated. Giovanni Villani stated that there were in the city about 25,000 men fit to bear arms, and altogether 'some 90,000 mouths',[33] while in the *contado* there were a further 80,000 men. Herlihy has estimated a population of 255,790, and an overall density of about 65 to the square kilometre.[34] The density in the 'states' of Siena and San Gimignano may have been even higher. If such densities could be generalised for the whole of Tuscany, the total would be about 1.2 million.

A Malthusian situation appears to have been reached in Tuscany, with over-intensive cultivation, fragmentation of holdings and very large numbers of poor and indigent. Poverty, malnutrition and exploitation of the peasantry by the bourgeoisie were widespread, and provided a fruitful soil for the plague.

While it is difficult to establish similar densities for the plain of Lombardy, the region was, nevertheless, very populous. What is considered the best crop-land today was at this date so poorly drained that it was little cultivated. Padua must have had an overall density within its *contado* of between 35 and 40 to the square kilometre in 1320.[35] Pavia, much of it marshy land along the Po valley, may not have had more than 15 to the square kilometre, but densities over the somewhat higher and better drained lands of Piedmont rose to over 25. Since part of the state is mountainous, densities in its more fertile valleys and basins may have risen as high as those in Tuscany. The only other north Italian states for which any sort of estimate is possible are Bologna, which may have had a density of about 40 at the end of the fourteenth century, and Mantua, which appears to have had about 45 part way through the next century.

The Papal States were a region at once more mountainous, poorer in agricultural resources and less developed commercially than Tuscany and the plain of the Po. Rome itself was reduced to a fraction of its former size, and appears to have been capable of drawing most of its food from the surrounding Campagna. The population of the latter, reduced by war and disease, was in no way comparable in density with that of Tuscany. The Patrimony of St Peter had in the seventeenth century an overall density of less than 25, at a time when the population had risen elsewhere to twice its pre-plague level. This suggests that the density about 1300 could not have been more than 15, and might have been as low as 10.

Umbria was more populous, even though, with the exception of Perugia, it contained no large towns. The city and *contado* of Orvieto may have had a density of over 60,[36] and one assumes that comparable densities were to be met with in the upper Tiber valley, near Assisi and Perugia. Within the Apennines, in the Duchy of Spoleto and the March of Ancona, population densities fell off sharply. Romagna, east of the mountains, was a richer and more developed province; it had a number of towns, and its agriculture supported a significant export trade in wheat. In 1371, according to a description of the province, there were 34,644 hearths, suggesting a population of about 150,000 and a density of about 36.[37]

The data for the kingdom of Naples and for Sicily are even less satisfactory than that for the Papal States. Though a number of tax rolls survive, the basis of the taxation is unknown, and at best they suggest which areas were the more prosperous, and, by implication, the more populous. The pre-plague population of the kingdom of Naples may have been of the order of two million, and that of Sicily, about 600,000, densities of about 27 and 24 respectively.

Iberian peninsula. Data for the Iberian peninsula are very scanty, and only for Cataluña are specific figures available. There were here in 1359 about 84,000 hearths, suggesting a population of at least 350,000.[38] Before the plague it would have been larger, probably over 400,000, and the density about 13. Castile remained more pastoral than the states bordering the Mediterranean, and there was, furthermore, a continual flow of population southwards as the Reconquest advanced. There is no satisfactory evidence, but densities must have been appreciably below those of Cataluña and Valencia. Russell's estimate of about 9.5 millions for the whole peninsula before the Plague would give an overall density of about 16, a not unreasonable figure in comparison with estimates obtained for France.

Scandinavia. Direct evidence is almost entirely lacking for Scandinavia, though use has been made of the somewhat uncertain evidence of the collection of Peter's Pence. This suggests that Denmark may have had 5 persons to the square kilometre within its fourteenth-century boundaries, while the rest of Scandinavia had an average density of less than 0.5 persons.

Conclusion. The demographic evidence for the pre-plague era is fragmentary, uneven and ambiguous, and to supplement it with the generally more satisfactory data for the late fourteenth and the fifteenth centuries may raise more problems than it solves. It has been argued,[39] in the opinion of the writer convincingly, that a quasi-Malthusian situation had arisen in much of Europe in the early years of the fourteenth century, and that the Great Plague, together with the epidemics of lesser proportions which preceded and followed it, were of the nature of 'Malthusian checks'. There were, of course, social restrictions on land use, such as 'forest' reservations, in which cutting and settlement were forbidden, but, with these exceptions, one may assume that land, at least in Italy and most of north-west and probably much of central Europe, was fully cultivated within the limits set by current technology. The evidence for the abandonment of cropland during the later Middle Ages suggests very strongly that at this time the extent of arable land was greater than it has been at any subsequent date.

Migration

By about 1330 the period of migration was over. There was increasing movement along the roads and navigable waterways, but it was the movement of traders and pilgrims, of bands of mercenaries and Crusaders, not of migrant peoples. The last invasion of southern Spain by Berber peoples had taken place more than two centuries earlier. The Viking expansion had long since ended, and increasing storminess in the north Atlantic was restricting the contacts between the Scandinavians and their outlying possessions in Iceland and Greenland. Tatar inroads has ended almost a century earlier, and though the Empire continued over the Russian steppe, its influence on peninsular Europe was slight. Turkish bands had landed on the shore of the Balkans, but their invasion of Europe, the only large-scale migration of the later Middle Ages, lay in the future. The eastward movement of the German peoples had in effect ended. Though the area of Germanic language was to continue to expand, this was rather through the Germanisation of the Slavs than the physical movement of the Germans themselves.

Yet there was migration, chiefly the short-distance movement of small family groups and of individuals within Europe. The towns had been growing in size despite a birth-rate which was in general lower than that of the rural areas. This was made possible only by a steady trickle of peasants who defied the legal and social restrictions on movement and migrated from the villages to the towns. Town records, especially those of the Hanseatic cities, include lists of new burgesses, and show that the annual increase of population by migration was high. At Lübeck it averaged 179 a year in the first half of the fourteenth century; at Stralsund, 123, and at Wismar, Hamburg and Bremen the average was over 40. Much of this migration, which reached its peak around 1350, was from west Germany, and represents all that was left of the German eastward thrust.

For a number of towns in western Europe also there is evidence in the personal names of new citizens of their places of origin. As a general rule they came from within a few miles of the city. Those from more remote settlements commonly used a well travelled highway, as if learning from the merchants who frequented it, of the opportunities offered by the distant town. Table 6.4 shows the distances which migrants to four west European cities are presumed, on the evidence of their personal names, to have come.

In all, 50 per cent of the total migrants had travelled from settlements which were less than 25 kilometres from the city, and the homes of a further 30 per cent were less than 50 kilometres distant. The larger towns as might be

Table 6.4 *Places of departure and distances travelled of migrants to four west European cities*

Distance in km	Toulouse*	Beauvais†	Douai‡	Provins§ (data incomplete)
Over 125	7	3	..	
101–125	3	1	..	
76–100	12	3	6	5
51–75	20	7	17	34
26–50	36	46	27	61
1–25	21	93	91	81
Totals	99	153	141	181

* C. Higounet, 'Le peuplement de Toulouse au XIIe siècle', *Ann. Midi*, 54/5 (1942–3), 489.

† Marie-Thérèse Morlet, 'Les noms de personne à Beauvais au XIVe siècle', *Bull. Phil. Hist.* (Paris, 1955–7), 295–309.

‡ Georges Espinas, *La Vie urbaine de Douai au Moyen Age*, vol. 4 (Paris, 1913), plate 2.

§ Marie-Thérèse Morlet, 'L'origine des habitants de Provins aux XIIIe et XIVe siècle d'après les noms de personne', *Bull. Phil. Hist.* (1961), 95–114.

expected, generally drew their population from a wider area than the smaller. Only in east-central Europe does the evidence in general suggest that migration to towns was over really great distances, and the reason is probably to be found in the lower density of the local rural population and its reduced pressure on the local agricultural resources. It appears then that the larger part of the urban population derived from what might be called the market area of the city region, and that much of the remainder came from distances not much greater. Acute rural overpopulation would be expected to lead ultimately to urban overpopulation, which would be reflected in a large urban proletariat, and this would give rise to a large body of mendicants. Such a view is entirely consistent with labour relations in the Flemish cities in the fourteenth century. The existence of large groups of mendicants in the cities is probable, but particularly difficult to prove. Villani noted that there were over 17,000 paupers in Florence in 1330; if this statement is even approximately correct, they must have made up a third of the population. The hearth lists, particularly those of Brabant and Burgundy, give the numbers of the poor, but this was presumably those unable to pay the levy, rather than actual beggars. In Brussels, for example, there were 670 'maisons pauvres' out of a total of 6,376 in 1437; clearly these were not all mendicant families; just the poorest members of the urban proletariat.

The language map

The ethnic map had, at least in its broad outline, assumed its modern form. Language boundaries were becoming sharper, and, at least in western Europe, were showing little tendency to shift. The Romance–Germanic boundary was well established from the North Sea coast to the Alps. To the west of it French was slowly acquiring the character of a unified language, though Romance dialects continued to be used among the less literate and travelled. The prestige of standard French enabled it to make some headway against the Germanic dialects spoken to the east of the language boundary. St Omer was a bilingual town in which French was gradually replacing the Low German dialect of west Flanders. In Artois also Flemish was in retreat and the Middle High German of Luxembourg and Lorraine was yielding slowly to French.

The Germanic languages had barely begun that process of unification which was making French the *lingua franca* of western Europe. Within the Low Countries the Flemish dialect of Brabant had assumed a literary pre-eminence, but it is doubtful whether it could have been understood in west Flanders or Holland. There is no evidence that the Flemish-Dutch peoples were at this time conscious of their ethno-linguistic ties.

In Germany the situation was similar. The many dialects were roughly grouped into Low, Middle and High, with none of them acquiring any primacy before the sixteenth century. The eastern boundaries of German speech were vague and fluid. The line was more social than geographical, and separated classes rather than areas. Poles and Czechs were becoming Germanised owing to the attraction of German culture and to the powers of compulsion of German clergy and landowners. If Lower Silesia was largely German, Upper Silesia remained at least partially Polish and eastern Pomerania was largely so. Danzig was in part at least a German city, and Germanic settlement and language had pushed up the Vistula from the ports of Elbing (Elbląg) and Marienburg (Malbork) to Thorn (Toruń).

There has been a tendency to exaggerate the extent of German settlement in the east. There is little evidence for the expulsion of the Slav population, communities of whom were prominent in Brandenburg and Mecklenburg in the fourteenth century. Labour was scarce on Europe's eastern frontier, and the German settlers wanted the Slavs, provided that these could be converted to the Germans' faith and technology. It is doubtful whether German settlers ever constituted a majority in more than a few areas east of the Oder. Among these was the Lordship of Sorau (see above, p. 335) where there were 764 'deutsche' households and 514 presumably Slav.

The plains of Bohemia and Moravia continued to be inhabited by Czechs, though groups of Germans, many of them encouraged by the Luxembourg kings, had settled there. In the eastern Alps, the valleys of the Mur, Drava and Sava were settled by Slavs, but Germans, entering the valleys from the west, were advancing the frontier of the German realm eastwards towards the Hungarian plain.

In the Alps Romance-speaking communities, living between the areas of German and Italian speech, were larger and more numerous than today. In Italy itself the Tuscan dialect, now being developed by Dante, Petrarch and Boccaccio into a literary instrument, was becoming standard Italian and displacing the Italian dialects from at least northern and central Italy.

The Spanish peninsula, more united politically than the Italian, had, nevertheless, made very much less progress towards cultural unification. The media of social communication were less developed than in Italy, and if Castilian Spanish was beginning to dominate the northern Meseta, it showed no tendency to displace Catalan, Galician and Portuguese, or even to threaten the Valencian and Andalusian dialects of the south.

Slav dialects were spoken over most of the Balkan peninsula. In the west the Slovenes and Serbo-Croats, in the east the Bulgars were clearly distinguishable from one another, even if it is not possible to delimit precisely the areas within which they lived. The pre-Slav peoples were more numerous and geographically more extensive than in subsequent centuries, and included not only the Illyrians of Albania, Epirus, and Montenegro, but also the numerous Romance-speaking peoples who had survived from the period of Roman rule.

The latter belonged to two groups. One was the Vlachs, or Maurovlachs. who practised a pastoral economy within the more rugged mountains. Their habitat at this time extended from northern Greece, where the 'Walachians' had threatened the Crusaders on their way to Constantinople, to the Velebit. the *Montagna della Murlacca,* or 'Mountain of the Maurovlachs' of the Venetians.[40] No estimate is possible of their numbers, though it was evidently greater than in later times, since they tended to be assimilated by the Slavs. They were, however, a numerous and politically powerful group, and it is generally assumed that the leadership of the Second Bulgarian Empire had, at least in its earlier phases, been Vlach and that it originated in Macedonia. The other group comprised the settled peoples of the Dalmatian cities, who also spoke a language derived from Latin and were known in the documents as Latini. Throughout the later Middle Ages they were subjected to pressure by the Slavs and, along the southern part of the Dalmatian coast, by the Albanians. Nevertheless, Slavs appear to have comprised only a very small proportion of their citizens in the fourteenth century, and the Albanians were held back from the coast until the end of the Middle Ages.

In Greece the Slav invaders had by the fourteenth century been completely assimilated, except perhaps in the north, where they were in close touch with the Serbs. But other ethnic groups had been added: Normans from Sicily; Franks after the capture of Constantinople in 1204, the Catalan Company, and Venetian and Genoese merchants. Individually they were small, and would have been quickly absorbed by the Greeks if they had not tended to form the ruling aristocracy in the areas which they had conquered and settled. There is, in fact, good reason to believe that the small towns which they occupied and used as fortresses to command the surrounding countryside, were at this time more Frankish or Catalan than they were Greek.

Despite frequent pogroms and occasional wholesale expulsions, as from England in 1290, the Jewish population had increased and flourished. The Jews succeeded in the main in retaining their identity, and their settlements about 1330 were larger and more numerous than they had been in 1100.

Their largest communities were urban, and there was a tendency for them to occupy distinct quarters in the larger cities. In Oxford this consisted of a large part of a single street;[41] in Cologne it was a corner of the old city, up against the relics of the Roman wall, [42] and in Nürnberg 'they occupied the best houses on one entire side of the market square, where their synagogue also stood'.[43] There were at about this time Jewish quarters in no less than forty German towns. That some were of considerable size is shown by the fact that at

Cologne it consisted of 70 houses and about 140 families, with a total population of some 700. At Nürnberg in the mid-fourteenth century, even after the pogrom of 1298, there were still 212 Jewish citizens.[44] But this was one of the largest communities. Most would have been a good deal smaller. Roth has estimated the Oxford community at about 200. Jewish communities were by the fourteenth century widely scattered over all of western Europe. They continued to be numerous in southern France; they played an important role in the commerce of Spain and were found in much of eastern Europe and the Balkan peninsula. There was a large Jewish quarter in Constantinople, and a community of 300 occupied a 'long street' outside the Venetian fortress of Modon.[45]

It would be a mistake, however, to assume that the Jews were all urban. In England they were distributed, apparently in small family groups, through the countryside. They were engaged in the main in commerce (see below, p. 252) and their role as middlemen in the trade in furs and skins, wool and salt, required that some live more or less permanently in the areas where these were produced. A Jewish community, doubtless very small, was based on Dürres, in Albania, and was engaged chiefly in buying and exporting salt.

Jewish communities were fewer, or at least less conspicuous, in Italy, and were rare or non-existent in the Celtic fringe of the British Isles and in Scandinavia. Their heaviest concentration appears to have been in western Germany, southern France, and Greece, where they had been noted two centuries earlier (see p. 252). It is difficult to estimate their total number; Starr put their number at about 12,000 in the Byzantine Empire.[46]

SETTLEMENT

The expansion of European settlement reached its peak during the years preceding the Black Death, and in the later years of the fourteenth century settlements began to contract in both number and size. The decline, like the growth of the settlement pattern, was uneven, and varied from region to region. From most parts of Europe there is evidence for the abandonment of fields and the decay of villages during the later Middle Ages. The larger cities, with the exception only of a few of those of the Hanseatic League, ceased to grow. Many declined in size, and some of the smallest even disappeared entirely.

Urban settlement

The most conspicuous changes that had overtaken the landscape of Europe during the previous centuries had been the extension of the cultivated area and the growth of cities. The two were of course related, in so far as urban growth would have been impossible without the creation of an adequate food base. But the expansion in the size and number of cities was related also to the growth of long-distance trade and to the rise of manufacturing industries which, in some instances, fabricated materials of distant origin.

The fourteenth-century city, at least in continental Europe, was a compact and well-defined settlement. There were towns without a defensive curtain of walls, especially in England, but they were few. Some only had towers and gates to safeguard the entrance, while the outer walls of the houses were interconnected to protect the town (see chapter 2, p. 60). A very small number still made do with wooden palisades which lack of resources had prevented them from replacing with masonry. Walls were not only a necessity in this rough age, if the town's shops and warehouses were not to be pillaged by every marauding band; they were also a mark of status, cutting off the free burgess from the generally unfree peasant. For this reason the town gates were often designed to impress the visitor with the might and wealth of the town. In northern Europe the walls also separated the small but growing middle class from the feudal gentry. Not so, however, in Italy and a few other areas of southern Europe. Here the landowning class moved to the towns, where they built their 'towers' from the profits of their rural estates. There they carried on their feuds, fighting from their high 'towers' and skirmishing in the streets, in a manner immortalised in *Romeo and Juliet*. Living amongst merchants and craftsmen, the feudal class here came to participate in commerce in a manner unknown in most of northern Europe, and the profits of the land were used directly to fit out trading ventures.

The majority of the towns in Europe had grown up around a pre-existing nucleus or within an older framework, such as the circuit of Roman walls. At first subject to the castle or monastery which had brought them into being, most towns had by 1330 established, at least *de facto*, their independence. Most Roman towns had revived and had again become centres of urban life, either smaller, as at Avenches and Tongres, or very much larger, as at Cologne, than during the period of the Roman Empire. More recently towns had been founded without a pre-existing nucleus around which to crystallise. These included the *bastides* of southern France, built mainly in the later years of the thirteenth century and early years of the fourteenth by the king and nobles, and also many of the Hanseatic towns of north Germany and the new towns of Britain and central Europe.

The city plan. The existence of city walls imposed certain restrictions on the growth of the medieval town. It would be wrong to assume that the area within the walls was always fully built up; there were, for example, considerable areas devoted to gardens within most towns, and the size of market places and church properties was often generous. The walled was, nevertheless, more compact than the unwalled town, such as was to be found in England. It is convenient to divide fourteenth-century towns into those which conformed with or derived from a preconceived plan, and those which did not. The former was largely made up of rectangular blocks. Sometimes, as in many of the north Italian cities and also in northern towns such as Cologne and Winchester, the block pattern was a distorted version of that established by the classical town planners (see p. 123). In others it was the product of the medieval planner, and survives today very littled altered in many central and east European towns and in the French *bastides*.

Most other medieval towns lacked a regular plan; some had grown from villages in which the roads were first marked out by the farm carts as they journeyed to the fields. And always the plan, whether regular or not, was being

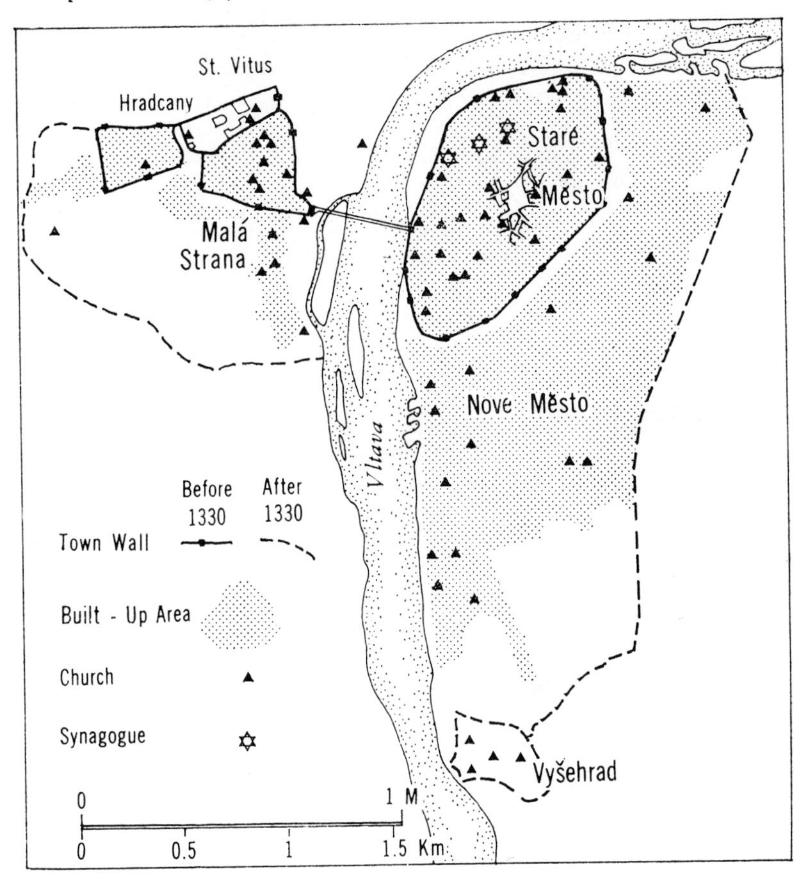

Fig.6.7 Prague, Bohemia

modified as the privileged encroached upon the streets and religious bodies acquired properties, absorbed them within a common fence, and diverted the thoroughfares.

The more compact the town, the shorter was the curtain wall which enclosed it. Ideally, the town should have been circular in plan, but few were permitted by the terrain to approximate this shape. An irregular oval or trapezium was a more usual plan. Sometimes a highly irregular shape, with a great length of curtain wall in relation to the area enclosed, as at Laon, Perugia, and Rothenburg ob den Tauber, was adopted in order to secure the maximum tactical advantage from the terrain.

A large number of European cities existed in intimate relationship with a castle or monastery. In a very few instances the castle was imposed upon an existing town – most often one of Roman origin. More often the town had grown up beside and under the protection of the castle, with which it was linked by a continuous curtain of walls. Only rarely, as at Goslar, Würzburg, Brno and Prague, was the town physically distinct from the castle–palace complex. In central Europe we have the peculiar phenomenon of the *Doppelstadt*. It has been pointed out above (p. 270) that many German cities

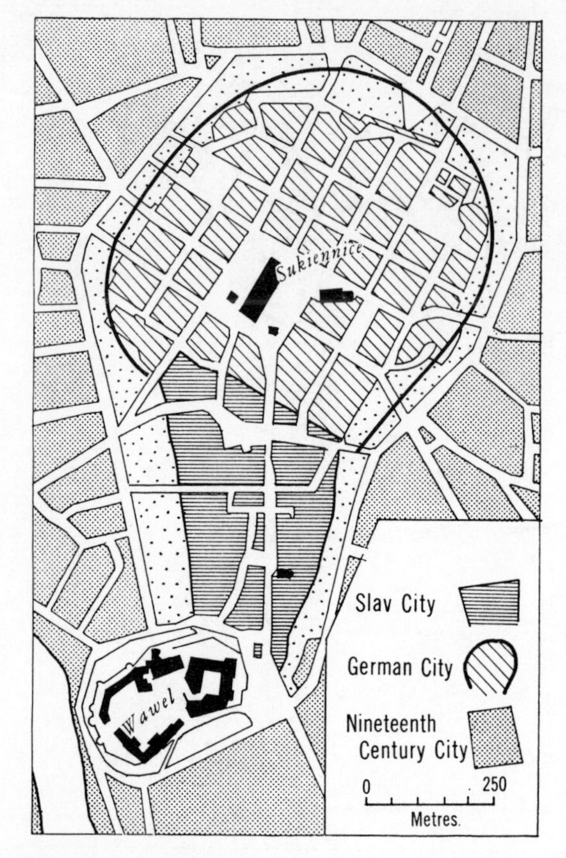

Fig.6.8 Kraków, Poland

developed from two or more nuclei. In most instances, however, the built-up areas merged, and became functionally if not morphologically uniform. In a few instances this process was never completed, and we have *Doppelstädte*, in which the separate entities remained capable of independent defence, as well as retaining their own distinctive functions. In Prague the Malá Strana, with its primarily governmental functions, contrasted with the commercial Staré Město on the other side of the Vltava. A similar dichotomy existed between Buda and Pest, which were not formally merged until the nineteenth century.

Most often, however, this 'double' quality arose from the addition of a later medieval planned town to an earlier town that was small and oriented more towards defence than commerce. The *bastide* of Carcassonne, for example, was established below the embattled walls of the older city and on the other side of the river Aude. The phenomenon may have arisen from the small space and difficult access of the older town, but it seems in most instances to have been due to a contrast in urban 'law' and institutions. The cases of Poznań (Posen) and Wrocław (Breslau) are in this respect typical. The original town in both cases was an island in, respectively, the Warta and Oder rivers, and consisted of the fortified *gród* and suburbium. It was here that the cathedral was located.

During the thirteenth century a planned city with German institutions and some German settlers was established in both instances on the west bank of the river. In time the original island settlements came to be distinguished primarily by their ecclesiastical functions, in contrast with the commercial preoccupations of the new settlement.

As a general rule, fourteenth-century towns had a market place, which was their commercial focus, a town hall, which was the centre of local administration, and religious buildings. The relative importance of these three in the landscape and the life of the medieval town varied greatly from one part of Europe to another. As a general rule, in much of France and in southern Europe the cathedral was the focal point within the town. This arose from the fact that in the declining years of the Roman Empire the bishop had generally been the most powerful local figure, and the town itself had become largely an episcopal preserve. In northern Europe, the bishop was not infrequently a late-comer, and his cathedral had often to take a less conspicuous location. In Troyes or Orléans the cathedral lay at the centre of town life; in Arras, Ghent or Bruges it lay back behind the civil buildings by which it was overshadowed and almost hidden.

Italy, however, was somewhat of an exception. The course of political events, from the eleventh century to the fourteenth, placed most of the leading towns under the control, first of their local communes, and later of a *podestà*. In any event, the ecclesiastical power was in most of them eclipsed by the lay, and this fact was reflected in the town's building programme. Communal buildings, like the *Bargello* at Florence and *Palazzo del Popolo* or *Palazzo Communale* of numerous cities of central and northern Italy, antedated the town halls of north-western Europe by up to a century.

Not all, even among the larger towns, were episcopal, but most had parish churches, and in these, since they belonged to the city, the burgesses were apt to take great pride and make a not inconsiderable investment. The larger towns comprised several parishes and had a corresponding number of churches. Only in the newest urban foundations might the church be lacking, since the hierarchy was apt to be slow in recognising secular changes by the establishment of new parishes. The principal parish church, that is the church which represented the oldest parish, was commonly centrally placed and often elaborately built.

All except the smaller towns are likely to have had at least one house of monks or friars. Where an abbey served as the nucleus of the town, it is likely to have been centrally and conspicuously placed, as in such monastic towns as Fulda and Prüm, Bury St Edmunds and Glastonbury. Elsewhere monastic institutions, like the houses of the friars, were later intruders into the town, insinuating themselves wherever they were able.

The market place was always present, and was as much a definite feature of the town as its walls. It might only have been a simple widening of a main street. In the planned towns and many of the unplanned, like Bruges and Arras, it was a large open square. In some the 'squares' assumed shapes that were very far from rectangular, and in the largest towns there may have been several squares which, as their names indicate, specialised in specific kinds of goods.

The town hall or guild-hall was the meeting place of the town's *élite*. Its more

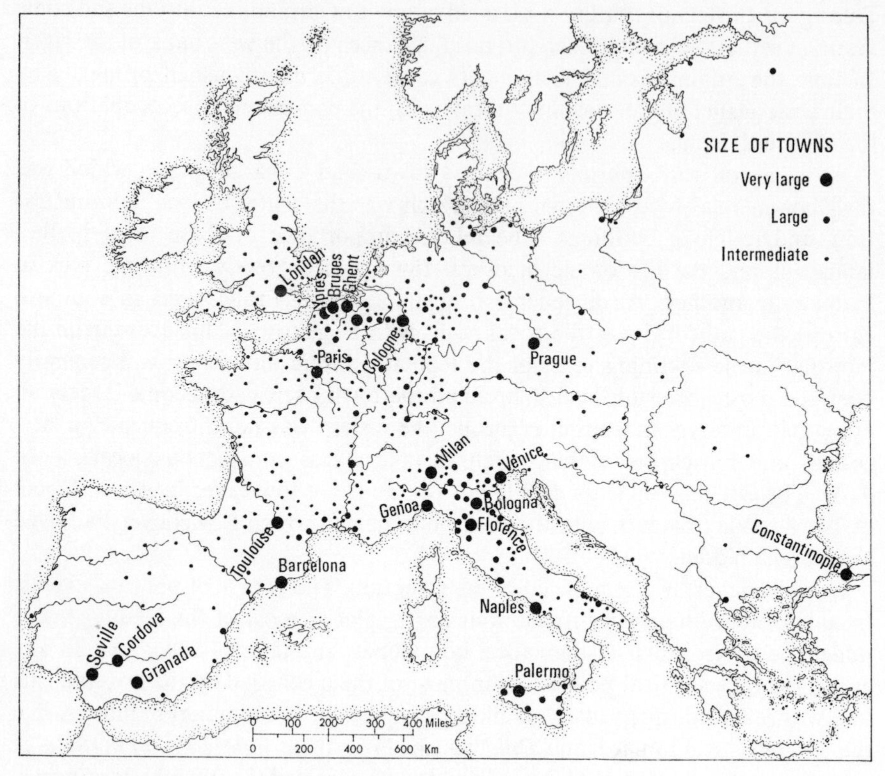

Fig.6.9 Distribution of the larger towns of medieval Europe

elaborate forms were a late medieval development, and about 1330 its existed as a large, independent building in only a minority of towns. Not infrequently it served also as a permanent market for the town's chief manufactures, and a number of the towns of the Low Countries had acquired 'cloth-halls' by the fourteenth century. But most of the elaborate city halls which today witness to the civic pride of the burgesses were built during the fifteenth century.

City size. The medieval city, even at the peak of its prosperity remained small. Unwalled suburbs were rare, and most burgesses lived within the shelter of their fortifications. Very few embraced more than 100 hectares, and the great majority covered less than 20. Within this restricted area were usually extensive open spaces, occupied by market places and churchyards and not infrequently extensive areas of garden. The maximum population that could have lived within them, even when allowance is made for a far greater congestion than is normal today, exceeded 10,000 in only a very few instances.

The data for assessing the population of medieval towns are indirect (see above, p. 326), and subject to a considerable margin of error. They usually exclude the mendicants, clergy and religious orders, which were as a general rule far more numerous in the towns than in the countryside. Furthermore very little of this statistical data is from the period before the Black Death, and estimates for the year 1330 must be obtained by extrapolating from later data.

A second method of estimating population lies in measuring the area of the towns themselves. This is defined by the circuit of the city's defences. The density of settlement within the walls varied from one town to another. It was usually densest near the centre and in close proximity to the markets, while areas close to the walls were sometimes not built up at all. On the other hand, there was little tall building in the medieval town, except the masonry construction of the churches and public buildings. During the previous centuries of urban growth, the tendency had been for population to fill out the area originally enclosed by walls and then to overflow and establish suburbs, which were then incorporated by extending the walls. All cities which were large in 1330 had grown in this way, and in some instances – Cologne, Paris, Antwerp, Ghent – the city had grown by several stages. In most of these cases the final phase of wall-building, generally in the early or middle years of the fourteenth century, enclosed a larger area than was justified by subsequent growth, and they remained only partially built up at the time when the sixteenth-century town plans were made.

The smaller and medium-sized towns appear in general to have been more intensively settled than the large. Their walls, built usually in the twelfth and thirteenth centuries, enclosed an area that was not fully adequate for subsequent growth, while their wealth and expectations did not seem to justify rebuilding the wall and extending the city. A considerable variation in the density of urban settlement is to be expected, and to this difference between cities must be added the extraordinary fluctuations in the population of most of them. These may have been due in part to epidemics, to which urban areas were more prone than rural, but also in part to changes in the economic fortunes of towns. In general, however, an index of between 100 and 150 persons per hectare fits the evidence, especially if it can be assumed that fluctuations beyond these extremes may be explicable in terms of temporary deviations.

On the basis of all the data available, one may group the towns of medieval Europe into:

1. *Very large cities.* Very large cities formed only a small group; their population ranged from 25,000 to 50,000, and in a few instances, Paris, Milan, Venice and Florence, for example, actually exceeded this latter figure at the peak of their medieval prosperity. They owed their great size to their ability to combine a number of distinct urban functions. They were manufacturing as well as commercial cities; seats of government and, in the case of the Italian cities, places of residence of an aristocracy which derived its wealth from landed possessions in the surrounding region. Their food supply had necessarily to be obtained from a considerable distance, and their growth was heavily dependent upon the existence of an effective transport net, in which rivers played a very important role. It is difficult to assess the importance of the several factors in the expansion of the giant cities of the Middle Ages. In the case of Florence and probably also in most others, the city crafts were the chief source of employment and wealth.

2. *Large cities.* These were more numerous. They are taken here to have been those with a population of more than about 10,000. They included the

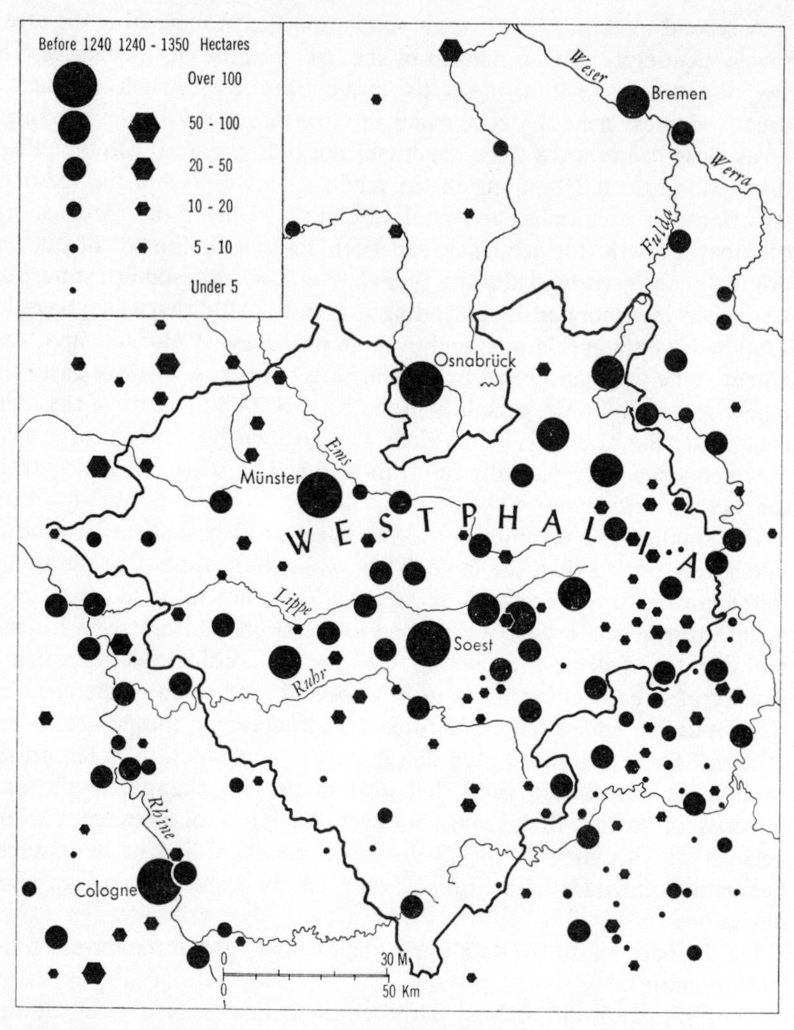

Fig.6.10 Towns in Westphalia and north-west Germany

larger of the cloth-making towns of Flanders and Brabant. Ghent, Bruges and Brussels greatly exceeded this figure, as also in all probability did Louvain, Ypres, Aachen and Antwerp. Hamburg approached 20,000, but Lübeck in 1330 was probably considerably smaller. Nürnberg, one of the best documented of German cities, had at least 20,000, and Frankfurt, Strasbourg, Basel, Bordeaux and perhaps Dijon, more than 10,000. Several of the Italian cities were of comparable size: Pistoia, Siena, Pisa, Lucca, Rome, and, in the northern plain Parma, Piacenza, Padua and Verona.

Eastern Europe could show few cities of this size, though Prague and Vienna had probably more than 20,000 and Brno perhaps over 10,000. France also had relatively few large cities apart from Paris and the clothing towns of French Flanders. Reims, Rouen, Bourges, Toulouse, Béziers, Narbonne and probably Troyes had populations exceeding 10,000. The removal of the Papacy to

Avignon had brought about a rapid growth of its population, which probably also exceeded 10,000 at this date.

In the Spanish peninsula, southern Italy and the Balkans there were few large cities. Barcelona, Valencia and the Andalusian cities of Seville, Cordoba and perhaps Granada exceeded 10,000. In southern Italy, Naples may have approached 25,000 and Palermo already had over 10,000, but in the whole Balkan peninsula there were only two large cities, Constantinople and Thessaloníki, the former with over 200,000, and the latter with over 20,000.

3. *Intermediate cities.* Towns of intermediate size had populations ranging from 2,000 to 10,000, and contained among their number most places that were recognisably urban at this date. At the one extreme they merged into the category of large cities, and it is impossible to know whether such towns as Valenciennes, Mons and Frankfurt lay on the higher or the lower side of this arbitrary dividing line. In all probability their fluctuations placed them in turn in both categories. At their lower extreme the intermediate towns contained many whose functions were only marginally urban, and which were, in effect, market centres for small regions of a few square miles.

These three categories embraced almost all towns which in some way – by their manufactures, commerce, banking or administration – served the needs of distant areas, and undoubtedly also some whose functions were restricted to their immediate vicinity. It is to be expected that, if travel and transport were difficult, a relatively large proportion of the urban services would be performed at the local level, and correspondingly few at regional or national levels. This is borne out by the size pattern of thirteenth-century towns. The number of large and intermediate towns was few, and their spacing wide. Bechtel has estimated that at most 35 German cities out of an estimated total of about 3,000 could have belonged to these categories.[47] The author's estimates for the whole of Europe at this time are given below (p. 358–9).

4. *Small towns.* These made up the great majority of all urban settlements. In France perhaps three-quarters belonged to this category and in Germany the percentage rose to over 90 per cent. They are here defined as towns having a population of less than about 2,000. Small though towns were, they must, nevertheless, be further subdivided into those which had a fully independent status, together with craftsmen and shops which served the needs of a distinct region, and those which were in fact nothing more than agricultural villages of somewhat higher status. Typical of the former was the Swiss city of Rheinfelden in Canton Aargau. It lay on the Rhine 10 miles above Basel. It had a walled area of only 10 hectares, and on this basis its population is put at about 1,000 to 1,200. An important part of its population was engaged in agriculture in the surrounding fields, but it obtained part of its food supply from as far away as the Bernese Alps and Alsace. A majority of the families of Rheinfelden, however, probably drew their main support from handicrafts, and, though few of their products entered into long-distance trade, they, nevertheless, supplied the needs of the town's market area which extended at least 12 miles into the northern Jura (fig. 6.11). The barrier nature of the Rhine clearly had the effect of restricting its service area in this direction, so that it extended a much shorter distance into the foothills of the Black Forest than into the Jura.

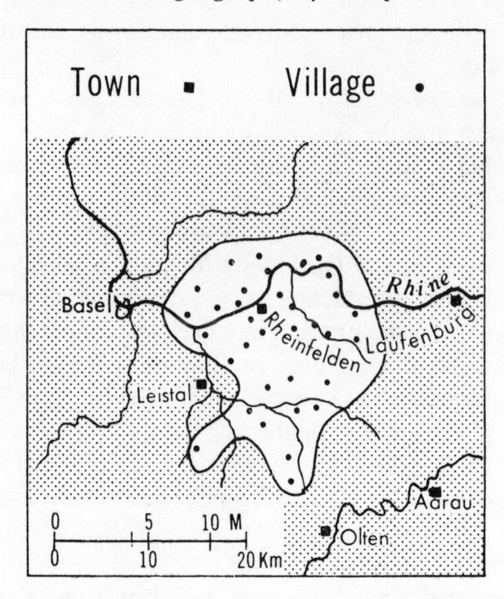

Fig.6.11 The service area of the town of Rheinfelden, Switzerland

At least as numerous as these small towns which actively served as central-places for their local regions were those in which it is very difficult to detect any functions that do not belong to a village. Not even in status were they always clearly differentiated from rural settlements. Most owed their origin to the deliberate act of a seigneur, who doubtless hoped, when he laid out the town and tried to attract settlers by the offer of certain privileges, that he was laying the foundations of a prosperous city. But not all could succeed. Many remained *Agrarstädte*; they had markets, but attracted little business, and agriculture, instead of being an important subordinate activity, assumed the dominant role in the total income of the town.

Such was the Swiss town of Lenzburg, to the west of Zürich, which covered an area of less than 3 hectares and probably never had a population of much more than 400. It is doubtful whether the market area of the small town covered more than 18 square kilometres. To this category of towns belonged also most of the *bastides* of south-western France, whether founded by the English king or the French seigneurs. A great deal more than a hundred were established in Gascony alone.

Other areas notorious for the large number of such small towns were the Swiss plateau and Swabia. In Switzerland as a whole, only 13 towns out of about 200 could qualify even as medium-sized. All the others are estimated to have had less than 2,000 inhabitants; some could not maintain the status even of agricultural villages, and disappeared before the end of the Middle Ages. Many covered less than 5, and one only 0.5 hectares. Bavois, at the foot of the Jura scarp, had only 26 households at one time in the fourteenth century, and the population of Bourjod, in Vaud, may never have exceeded a hundred.

Swabia was also a region of small and closely spaced towns. Along one valley near Schwabish Hall 5 towns were to be found within a distance of less

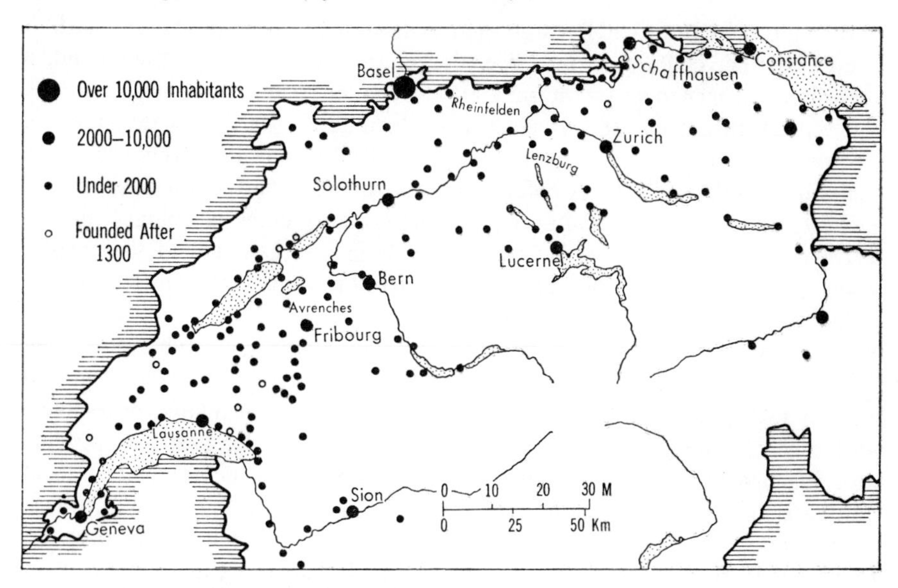

Fig.6.12 Towns of Switzerland

than 12 miles. Württemberg as a whole had only one large town and eight medium-sized, as against 140 small.

West of the Rhine and Jura the average size of towns was larger and their pattern less dense. This is likely to have been due to the possibilities for urban growth presented by the remains of Roman towns and their surviving road nets. With very few exceptions the Gallic cities had continued through the early Middle Ages as centres of ecclesiastical administration and as foci of some form of economic activity. As urban life revived during the eleventh and twelfth centuries, these towns found themselves in a position to dominate their regions. If France had more medium- and large-sized towns than Germany, it was not necessarily an indication that her economic life was more advanced. Her urban functions were more concentrated than in Germany because she had inherited their structures from the Roman Empire.

The function of the city. It is customary to regard the medieval town as a centre of commerce and of crafts, a market for its local region and in certain respects also for areas much farther afield. By and large this is true, but every town, including the largest, participated directly in agriculture. The fields which adjoined the town not only contributed to the urban food supply, but were often owned and cultivated by its citizens. In some towns, especially the Italian, there was an urban class of land-owning rentiers who lived by the rents of their rural holdings.[48] But in all of them a class of urban peasants was to be found. There is no direct evidence for their numbers and importance. If, however, it can be assumed that such urban peasants could not have effectively cultivated lands lying more than 3 kilometres from the town walls, we have a factor which clearly limits their numbers. When allowance is made for land not used agriculturally, it seems unlikely that much more than 1,000 in any town could have been wholly supported by farming, and the pure *Agrarstadt*, if it ever existed, could not have been larger than this. On the other hand, a very much larger population may have retained some small stake in the countryside, insufficient to provide either a livelihood or a significant source of income, but enough to give them some link with the land. For Toulouse, Wolff has estimated that half the population in some quarters had retained a parcel of agricultural land. The agricultural component of such cities as Cologne, Frankfurt and Basel cannot have been significantly larger than that of the smaller towns, because of the barrier nature of the river on which the towns lay. Nevertheless, in some towns of medium size, such as Mons, rents were being paid in 'oats, capons and pence', and in the smaller towns such dues were often rendered exclusively in agricultural produce. Around many towns was a narrow zone of intensive agriculture. It may have been under fruit and vegetables; where the climate permitted it was not infrequently under grape-vines. This land was clearly worked by citizens, though it is possible these were also local craftsmen and traders. It seems certain that, with the increasing size of towns, the percentage of the population engaged in agriculture must have dropped from considerably over 50 per cent in the smallest to less than 5 per cent in the largest.

Most of the remainder were engaged in crafts and in trade. Among the former, those engaged in the processing of food and of locally produced materials would, as a general rule, have predominated. One is constantly surprised at the number of butchers, bakers, millers and practitioners of other such simple crafts in the medieval town. Toulouse, for example, had 177 butchers in 1322,[49] in a population that cannot have exceeded 30,000 at this date, and Frankfurt had 101 bakers and 88 butchers for a population of no more than 10,000, as well as an immense number of smiths, carpenters, shoemakers and other craftsmen who served directly the needs of the town and its immediate region.[50] Florence, at the time of Giovanni Villani, had 146 bakers.

In all towns, except perhaps the very smallest, for which data is lacking, there were many who worked only as porters, waiters and servants. Frankfurt had 25 *Sackträger* and a similar number of *Weinknechte*. At Zutphen in the Netherlands, a small town in which agriculture played a prominent role, there was a large number of wage earners who were employed for at least part of the year in maintaining the roads, dykes and sluices and for the rest on the farm holdings of the richer citizens.

Only in a few towns, generally the larger, was there a dominant or basic craft industry, which produced not only for local but also for distant markets. In Florence there were 200 or more workshops engaged in the *arte della lana*, 'and more than 30,000 persons lived by it'.[51] After allowing for exaggeration in Villani's figures, this, nevertheless, remains a quite remarkable concentration of industry, probably unequalled in any other medieval city. In the Flanders towns the number engaged in the various cloth-working processes was, nevertheless, considerable. In the fourteenth century most of the small towns of Hainaut and Brabant must have been heavily engaged in cloth weaving. The clothing industry is said, late in the fourteenth century, to have supported the greater part of the population of Beauvais. Spinning and weaving provided the mainstay of many of the towns of intermediate size in northern France and the Rhineland. In Swabia the linen industry played a similar, though somewhat smaller, role (see p. 391).

The metal industries were of much less significance in providing a basic occupation for the medieval town. Dinant, with its copper and bronze working, was perhaps the only one in which it was dominant, though clearly it was practised, if we may judge from the number of smiths and the volume of trade in lead and iron, in most towns of intermediate or large size. Other crafts which produced goods for long-distance trade were either rural – like charcoal-burning and salt-making – or were practised outside the limits of the towns, like the manufacture of bricks and tiles and the burning of lime.

Only in the intermediate and large towns were the crafts organised on an individual basis, and craft guilds were rarely to be found in the smaller. In the former the crafts were closely controlled and the masters in each usually played a corporate role in the administration of the town. In the largest there was an immense proliferation of craft and other guilds, with an increasingly narrow specialisation in their activities. Etienne Boileau's *Livre des Métiers*[52] listed no less than a hundred crafts that were practised in Paris about 1300 and described the ordinances regulating each. His enumeration included even those who pressed oil from nuts and olives and those who weighed and carried salt. No such specialisation was possible in the smaller towns, to which the words of Xenophon, describing the smaller Greek city-states (see above, p. 80) would still be applicable.

Merchants were to be found in most towns, and must occasionally have been present even in the smallest. One can distinguish those who travelled from town to town and from fair to fair, like the Bonis brothers of Montauban. They dealt in wool and cloth, bought and sold spices and candles, armour and jewellery; they undertook commissions to acquire wedding and baptismal gifts and served as bankers and rent collectors;[53] they shipped the local products to the distant markets and fairs, and kept the local markets supplied with such goods as salt and spices, wheat and wool. Most had their homes in the intermediate and larger towns, and visited the smaller in turn.

Beside these was the more numerous class of petty traders who retailed in the local markets the goods they had obtained from the travelling merchants. These were assisted by an immense number of people whose labour in plying the barges, driving the caravans, unloading and carrying, made possible any form of long-distance trade in this unsophisticated age when almost all productive activity derived ultimately from the physical strength and endurance of countless common workers.

To summarise, it may be suggested that all towns of intermediate size had some form of basic activity, by which they supplied goods and services to distant markets, and that the large and very large towns were based far more on long-distance than on local trade. Only the lowest category of towns would have had no regular and continuing basic function, though even these must have bought their salt and spices, metals and perhaps weapons from travelling merchants in return for wool or flax or charcoal.

Urban food supply. It is implicit in this that the smaller town, with its region, formed an almost self-sufficing entity, and that it was, in general, able to satisfy its own requirements in food and the basic raw materials from within only a few miles. The intermediate and large towns were in some degree dependent on the supply of grain and meat from areas lying beyond what may be termed the local region. Most of these towns lay in areas which were themselves of above average agricultural productivity, and many also had the advantage of waterborne transport for bulky foodstuffs. Indeed, it may be suggested that no large town could have grown up in this age without one or both of these advantages. Paris, for example, was very heavily dependent upon boat traffic on the Seine and its tributaries for the supply, not only of wheat and wine, but also of fish, salt, lumber, charcoal and other commodities. The royal accounts[54] for this period provide evidence for the purchase for the court of consignments of wheat in the small towns of the Paris basin – Melun, Gonesse, Le Vandreuil, Breteuil – all of them on or close to a navigable river by which it could be shipped to Paris. Indeed, the food supply of the city was throughout this period a major, sometimes even a dominant, preoccupation of royal policy. The Flanders towns derived much of their grain from Artois, and the Lys and Scheldt were in general used for transport (see below, p. 426). But already the Flemish towns were beginning to draw upon the immense reserves of grain that were beginning to accumulate in the Baltic lands and were brought to the Flemish ports by the merchants of the Hanse.

Food had long been important in the waterborne traffic of the Rhineland towns. Though grain and other foodstuffs were brought by farm carts to the markets, even more was transported – and over much greater distances – by the *Marktschiff*, which linked the riverine villages with the towns. Food supply to the Italian towns was in general more difficult, because the towns were themselves larger and the rivers fewer and less navigable. Genoa, Venice and others which lay close to the sea relied heavily on grain from Apulia and Sicily, but inland towns, such as Florence, experienced greater difficulty and the supply of the city from the Romagna or the plains of Tuscany presented political as well as logistic problems. The cities of the Dalmatian coast, with their overwhelming dependence upon seaborne commerce and their scanty agricultural land, had also to import much of their food supply from southern Italy or Albania.

The distribution of towns. By the mid-fourteenth century a network of towns had been spread over the whole of Europe, from Poland to Spain. It was a very uneven network. Some areas, notably western Germany and the Low Countries, had a dense pattern of small towns; in others, including parts of the Spanish peninsula and western France as well as much of eastern Europe and the Balkans, this network was very much more open. A distinction has already been drawn between the functions of the small towns and of the intermediate and large; the former were closely linked with their surrounding regions, whereas towns of large size served regional needs, for which developed means of transport and communication were essential.

Fig. 6.9 is an attempt to show the geographical pattern of urban development during the years preceding the Black Death. In some instances – perhaps about fifty – there is evidence from tax or hearth rolls for the size of the population. In all other instances this has been derived from the area of the walled town, on the assumption that the average density of settlement was about 125 persons to the hectare. This method can at best give only an approximate figure for the urban population. Not all towns were walled and not all citizens, in any event, lived within the walls. The density of building within the city varied greatly, and no allowance can be made, in measuring the urban areas, for military works, such as castles, and the relatively lightly occupied area of monasteries and cathedrals. There were fluctuations in the population of medieval towns. Avignon, for example, was growing very rapidly, and no measurement can recognise its large floating population, attracted by the papal court, and the extreme congestion that resulted. The map does a great deal less than justice to the Slav lands of eastern Europe, in which many towns never had walls at all. Nor can it be demonstrated in every case that the walls were actually in existence in the first half of the fourteenth century. These reservations apply particularly to the small towns, for which the documentation is either inadequate or non-existent. With the intermediate and other large towns we are on safer ground, because there may be documentary evidence and the date of the walls and of the enclosures of suburbs was generally recorded.

It is apparent from the map that there were in Europe two urbanised zones. The smaller extended over northern Italy, from the Alps as far south as Orvieto or perhaps Rome. The map shows a preponderance of intermediate and large towns and a relative paucity of small. This may, however, be deceptive, because the legal status of the smaller, town-like settlements in Italy was different from that in northern Europe. There were many settlements, regarded as villages in Italy, which would have been considered towns beyond the Alps. In some degree this ambiguity is due to their lack of formal walls and gates. Furthermore, by about 1330 many of these settlements had been reduced to dependence upon the larger towns. They had ceased to be self-governing, and had, as it were, disappeared from the list of towns.

The second urbanised region extended from the Paris basin across north-eastern France and the Low Countries to the Rhineland, and thence along the *Hellweg* towards Saxony, up the valleys of the Main and Neckar and through the Swiss plateau. The pattern of towns grew thinner in Brandenburg, Saxony, Bohemia and Bavaria, and they were rare phenomena in Poland, Hungary and the Balkans. In the opposite direction towns were less dense in central and southern France and in the Iberian peninsula.

Three factors appear to have determined the urban pattern shown on this map. The first was the urban inheritance from the Roman Empire. Roman towns were rarely abandoned. In general they continued to be occupied, even if they were structurally too large and geographically poorly sited for medieval needs. This factor would help to account for the relatively high proportion of medium-sized and large towns west of the Rhine. A second factor was the development and productivity of the rural areas, for which the net of small towns provided market and service centres. The third factor was the concentration in relatively few areas of specialised crafts with long-distance trading connections. This was pre-eminently the case in northern Italy and in northern France, the Low Countries and the Rhineland. The rise of such industrial and commercial cities as Florence and Milan, Bruges and Ghent, Cologne and Frankfurt was accompanied by the growth of smaller towns along the commercial roots which supported them. These provided lodging and food for the merchants; their citizens manned the boats and led the pack animals, levied tolls, and received payment for their services which was probably out of all proportion to the services rendered. The line of small towns along the IJssel and lower Rhine, which were at this time being opened up to the merchants of the Hanse, offers an example of this secondary urban growth.

In western, central and southern Europe the pattern of towns, established by the time of the Black Death, underwent no significant addition or change until modern times. In eastern Europe, however, towns continued to grow in number and size. The expansion of the Hanseatic cities along the Baltic coast was the reflection of continued economic growth in their hinterlands. Here the growing surplus from the land was being channelled by kings and seigneurs towards the creation of towns, and the urban net continued to spread through Bohemia, Silesia and Moravia; it extended from the Oder to the Vistula, the Bug and the upper Dnestr and into valleys of the Carpathians.

In Hungary and the Balkans, towns were unquestionably fewer and probably smaller even than in the western Slav lands. Buda had been established and fortified by the end of the thirteenth century and may well have had two or three thousand inhabitants by 1330, but other towns in Transdanubia and neighbouring Croatia were all small, and most were dependent upon a feudal nucleus.

Table 6.5 *Size of cities and towns in northern Europe*

Towns	Number	Estimated total population
Giant (over 50,000)	1	*c.* 90,000
Very large (25–50,000)	8	240,000
Large (10–25,000)	38	570,000
Intermediate (2–10,000)	220	1,100,000
Small (less than 2,000)	3,000	2,250,000
Totals	3,267	4,250,000

Table 6.5, which can only be regarded as approximate, gives the number of towns in each size category for the area lying north of the Alps and the Danube. If the population of the small and predominantly agricultural small towns is excluded, this region of Europe had an urban population of only 2,000,000 probably no more than 4 per cent of the total population. If the agricultural population is screened out from each of the size categories, the total population engaged in essentially urban occupations could not have been much in excess of 3.5 per cent of the population. The urban population ranges from perhaps 30 per cent in northern France, Flanders, Hainaut and Brabant,

down to 2 per cent in the rest of France and Germany, and to even less in eastern Germany, Poland and the remainder of eastern Europe.

Urban population formed a larger proportion of the total in Italy. In addition to the 'giant' cities of Florence and Milan, were several 'very large' and 'large' towns. The smaller towns in Italy passed into large villages. Table 6.6 shows the approximate urban population of Italy at this time. Reasons for this heavier density of urban population in Italy are to be found, in part, in the tradition of urban living which dated from the Roman period. A consequence of this well-established tradition was that the feudal classes themselves moved to the towns, where they built tall defensive towers, such as still distinguish the skyline of San Gimignano and Bologna. Each was, in effect, a castle, inhabited by a large family unit and its retainers, all of them supported not only by urban pursuits, but also by the surplus from their rural estates.

Table 6.6 *Size of cities and towns in Italy*

Towns	Number	Estimated total population
Giant	3	150–200,000
Very large	5	125–175,000
Large	24	300–400,000
Intermediate	*c.* 160	600–1,000,000

This urban tradition does not alone account for the large size of Italian cities. They were also, by the standards of the fourteenth century, very highly industrialised. Most were engaged in the cloth industry. It seems probable that in Italy more of the cloth-working processes were carried on in workshops in the cities than was the case in northern Europe. This may have been due to the fact that in southern Europe agriculture provided more nearly a year-round occupation than in northern Europe, leaving less time for the practice of domestic crafts in the villages.

The conduct of long-distance trade was more concentrated in Italy than in most other parts of Europe, and merchants with their servants and retainers must have made up a large part of the population of Venice and Genoa, and must have been numerous also in Milan, Florence and other towns.

The few large towns of southern Italy and Sicily – Naples, Messina, Palermo – had no well-developed industrial and commercial base. One must think of their large populations as made up largely of an underemployed and undernourished proletariat, as indeed they have since remained.

The evidence for Spain is scanty. Along the Mediterranean coast lay a number of commercial towns of relatively large size. Barcelona was itself a very large town at this time; Valencia was at least a large town. Andalusia probably continued to be relatively highly urbanised, but over the Meseta it is likely that towns were few and in general small.

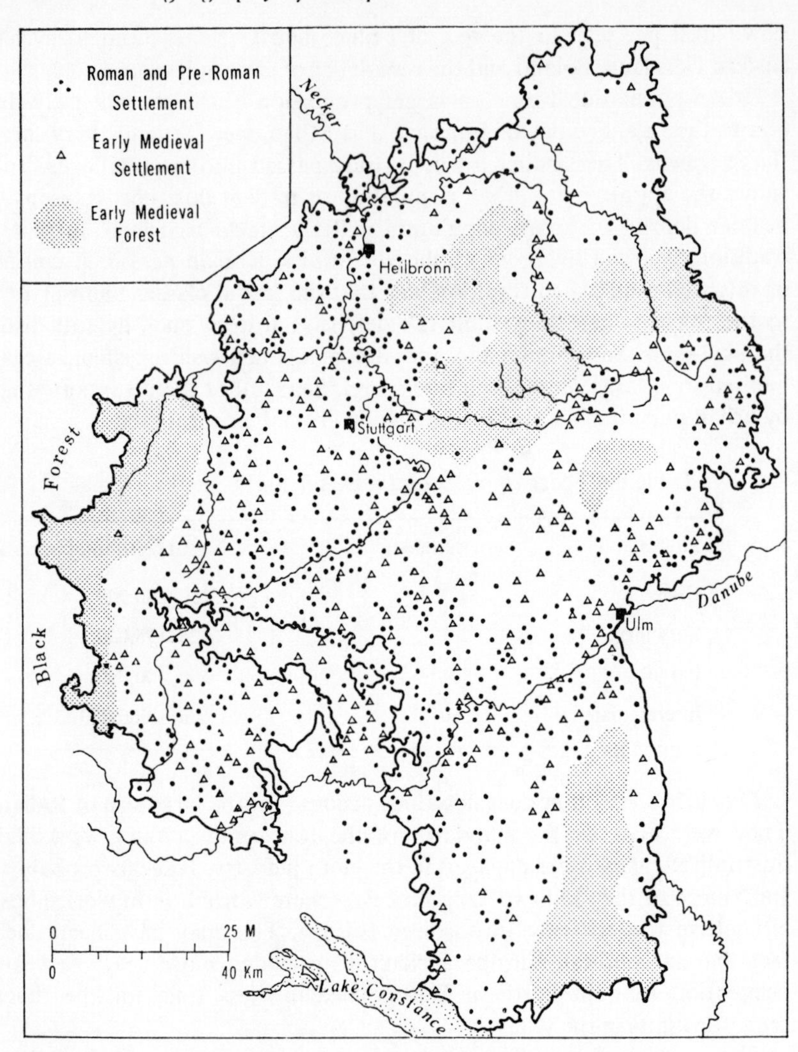

Fig.6.13 Settlement in Württemburg (after Gradmann)

Rural settlement

Careful examination of one area after another in central and western Europe shows that its rural population in the period preceding the Black Death was larger than at any other time during the Middle Ages, and suggests also that this level of density was not regained until the nineteenth century. It follows that the pattern of rural settlement had reached its maximum development; that almost all the villages and hamlets known today were in existence then and, furthermore, that some, inhabited about 1330, have even disappeared.[55] Except in such peripheral areas as Scandinavia, eastern Europe and Spain, the medieval period of the expansion of settlement was over and the medieval frontier was closed.[56] Villages were more crowded; the cultivated fields more fragmented than at any previous time.[57]

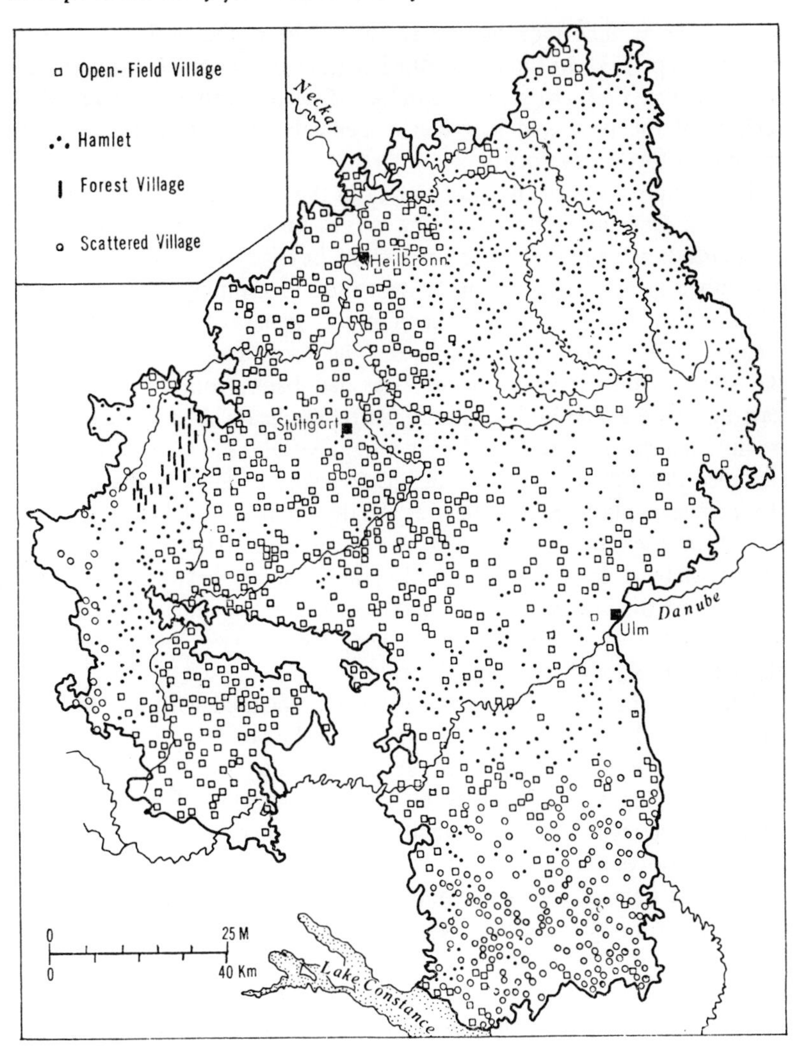

Fig.6.14 Village types in Württemberg (after Gradmann)

This was technically an unsophisticated age, and all communities lived within a narrow margin of subsistence. They had neither cropland nor man-hours to spare. The journey to the fields was minimised, and this would suggest small settlements. At the same time, it was a lawless age, and the larger community provided mutual security which was denied to the hamlet and isolated farm. Furthermore the practice of three-field agriculture and the use of the heavy plough necessitated communities large enough to furnish the equipment. It was also a conservative age; custom ruled the community, and made any significant departure from established forms difficult and improbable, the rural settlement tended to keep its form long after the reasons for it had ceased to exist, and it is the retention of early patterns of settlement into the era of modern survey and cartography that makes it possible to study the landscape of the medieval rural village and farm.

Village forms. The morphology of the village in the twelfth century has been described (p. 253). In the long-settled regions of western Europe there had been little change. Only where the frontier of settlement had advanced – in hilly regions of the west and in much of central and eastern Europe – were new settlement forms developed.

The eastern limit of the *Haufendorf* in the fourteenth century was very close to that of German settlement some two to three centuries earlier. By 1330 the Germans had extended their settlement beyond the Oder, and had adopted a number of settlement forms more suited to the exigencies of their environment and the demands of their economy. In general some variant of the 'street' village (*Strassendorf*) predominated. The farmsteads lay along each side of the street, while their cropland stretched back from the road in relatively narrow strips to the limit of the forest or the heath. There were many variants of this general pattern: the forest village (*Waldhufendörfer*) of Bohemia and southern Poland; the marsh village (*Marschhufendorf*) of the damp valleys of the northern plain. Sometimes, especially in the land between the Elbe and the Oder, the houses bordering the street drew back to enclose an oval or spindle-shaped area, a compromise, as it were, between the simple street village and the *Rundling*. This was the *Angerdorf* or *Owalnica,* and, like the former, probably reflected both the need for a closed space to corral the farm animals at night and the desire for security.

The forest or street village is generally associated with late medieval settlement. In the Herrschaft of Sorau (Zary) in Lower Silesia, where German and Slav villages were distinguished, we find that the former are today represented by linear villages, and the Slav by loosely nucleated. A similar correlation of street villages with late settlement has been noted by Gradmann in Württemberg (see above, p. 256).

These settlement forms extended into the Danubian lands and the Balkans. German settlers introduced compact villages into Transylvania, where they had built strong fortified churches as further protection against Tatar and Magyar. In more easterly parts of the Hungarian plain the Kumans and probably some of the Magyars, still practising a predominantly pastoral economy, had adopted for their permanent settlements a variant of the *Rundling* form.

Southern Europe was a region of predominantly nucleated settlements. Large, compact settlements, defended sometimes by walls, more often by the steep slopes above which they were sited, were in large measure a human response to the insecurity of the centuries following the collapse of the Roman Empire. In the mountains, notably the central Apennines and those of northern Spain and the southern Balkans, the demands of a predominantly pastoral economy triumphed over the desire for protection, and settlements consisted mainly of isolated farmsteads or small hamlets. In parts of the western Balkans the need for mutual help contributed to the formation of the *Zadruge,* or patriarchal villages. In the mountains of Albania and Epirus similar conditions were leading to the fortification of the scattered homesteads and the construction of *külle*, broadly similar in form and function to the peel towers of the Scottish Border.

In most parts of Europe the growth of settlements had been restricted by the fact that with increasing size the length of the journey to the more distant fields became too great. It was easier to establish secondary settlements at a distance from the primary, each set within its own field system. Such a pattern is sometimes implicit in the place-names, the later and subordinate settlement being characterised as 'little', 'new', 'upper', or 'lower', or by some other such distinctive and even pejorative term. The establishment of new settlements had, however, rarely brought about any change in ecclesiastical organisation. Parish boundaries remained where they had been; the church, usually in the primary settlement, continued to serve the needs of the whole parish, and only occasionally was a church of subordinate status, a 'chapel-of-ease', as it was sometimes called, erected in the secondary settlements.

Villages in general attained no great size, except perhaps in southern Europe. It has been estimated that in Germany, a country noted for its large and clustered settlements, the average size in the fourteenth century was no more than about fifteen households, or, at most, 90 inhabitants. The fertile and well-settled region along the middle course of the Weser, in northern Hesse, is said to have had in general only small hamlets of less than ten houses, and similar conditions were to be found in Bavaria and Westphalia. It has been assumed by some German scholars that, with the abandonment of settlements – perhaps as much as a fifth of the total – during the later Middle Ages, there was a tendency for population to concentrate again in the central village settlement, which in consequence grew larger.[58]

The hearth lists are essentially catalogues of settlements, with the number of households in each. Their accuracy is in some instances open to question owing to the tendency to subsume a number of small, discrete settlements under one name. Hearth lists of the Low Countries, eastern France and Switzerland have already been used for their evidence for population. They are here used for the light which they can throw on the size and distribution of settlements.

The plains of Central Belgium were characterised by a great range in settlement size, but the average remained fairly low, between 20 and 30 households, and the pattern dense. There was a rather higher proportion of small settlements towards the west, where soil conditions were more varied and fertility generally lower. It is noteworthy that the rich limon region between Louvain and Liège had a dense population, but a relatively small settlement size.

In Lorraine and southern Luxembourg small nucleated settlements predominated, with their most usual size between 10 and 20 hearths – say 50 to 100 inhabitants. The Ardennes plateau, by contrast, had on average settlements of less than five hearths and a great many settlements consisted of one or at most two or three households.

The late fourteenth-century hearth list for Burgundy suggests that settlements were in general fairly small. In the plain around Beaune and Dijon they rarely numbered less than 5 or more than 40 households. In the hilly country of north-western Burgundy, the average size was lower; few exceeded 25 and many had less than 5 households. The average size of settlement was larger in the diocese of Lausanne but here the figures given are for parishes, not for discrete settlements, and it is thus impossible to say how many villages, hamlets

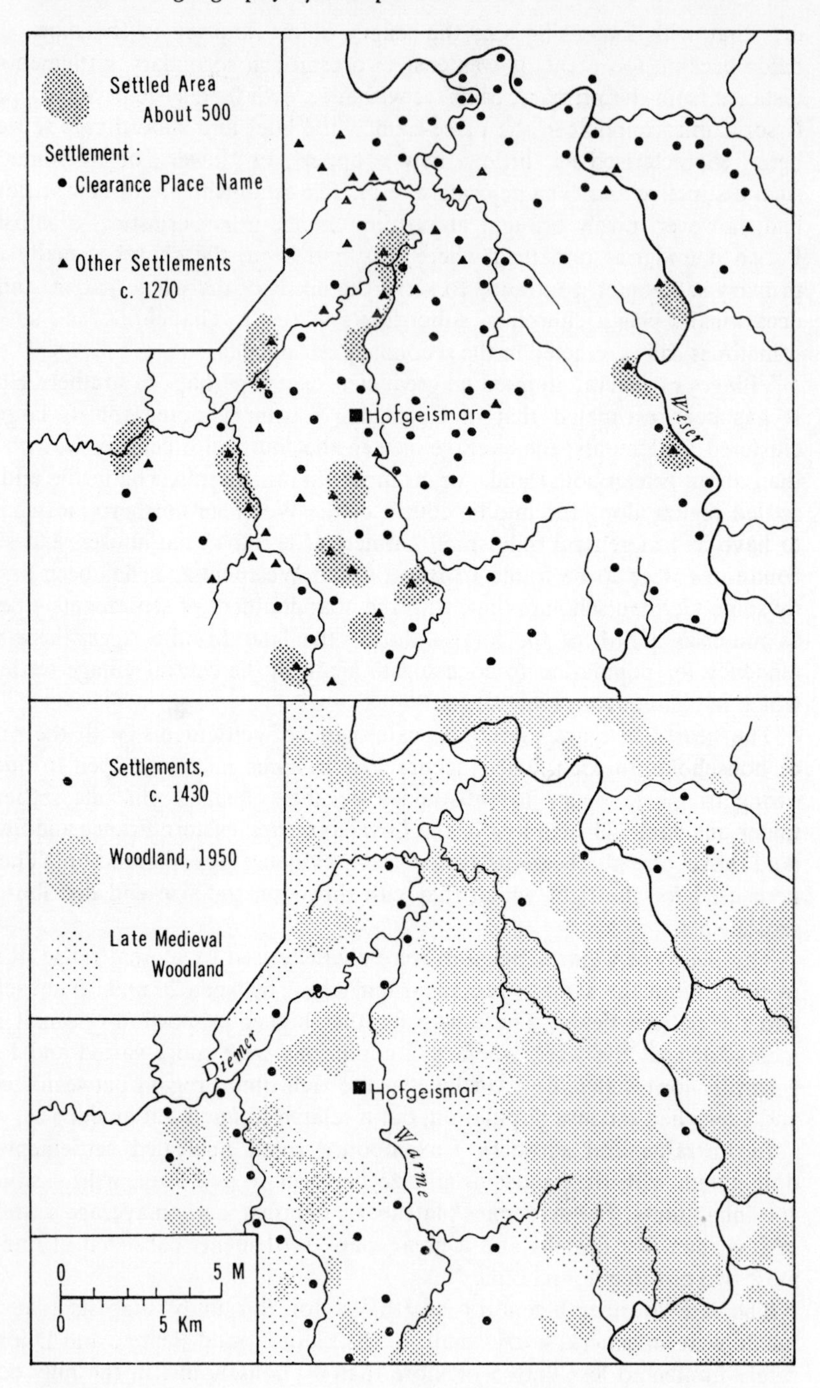

Fig.6.15 The expansion of settlement in the earlier Middle Ages and its contraction in the later

and isolated farmsteads are covered by each figure in the list. Overall population density was broadly similar to that of Burgundy, and it would be surprising if the pattern of settlement differed greatly. The same can be said of rural settlement in the Canton of Zürich.

It is more difficult to establish the pattern of urban settlement in Germany and eastern Europe, because few hearth lists exist. A very late survey of hearths in Mecklenburg, compiled to assist the collection of a subsidy exacted by Maximilian I in 1495, gives the number of adults.[59] If these totals are converted into households, the settlements are found to be relatively small, with few of less than 5 or more than 20 households.

If one may generalise for Europe north of the Alps, most people are seen to live in villages of an average size of from 20 to 30 households. Much larger villages were to be found, but many of the very large units were certainly composite and consisted of several discrete settlements. On the other hand, settlements ranged in size down to hamlets of less than 5 households in hilly and other less productive areas. The isolated farmstead appears from the evidence of the hearth lists to have been something of a rarity.

Very little indeed is known about the architecture of the medieval village. The oldest surviving cottages belong to a very much later date and the oldest informative drawings are of the sixteenth century. Excavations have revealed the ground plans of very simple, sod-built structures, whose architectural style can have changed but little since the Neolithic.

The Middle Ages were a period of great insecurity, and the danger and exposure of small and isolated settlements increased in many parts of Europe with the breakdown of central authority in the fourteenth and fifteenth centuries. The fortified village was not unknown, but the anarchy of the times was reflected more in the use of the church for protection and in the tendency for the population to abandon small settlements and to concentrate in the larger. The former practice extended from Scotland to Transylvania, but the following passage from the Chronicle of Jean de Venette shows it within a few miles of Paris: 'the peasants dwelling in open villages, with no fortifications of their own, made fortresses of their churches by surrounding them with good ditches, protecting the towers and belfries with planks as one does castles, and stocking them with stones and crossbows ... At night the peasants ... slept in these strongholds in comparative safety. By day they kept lookout on top of the church towers.'[60] The Annals of Ghent describes in no less certain terms the digging of protective ditches around villages and the fortification and defence of church towers.[61]

AGRICULTURE

In fourteenth-century Europe agriculture provided employment in some areas for perhaps as much as 90 per cent of the population and also the food supply for the remainder. There was no inhabited region of Europe where agriculture of some kind was not practised, and difficulties of transport made it necessary to grow food crops in areas that have now been given over to grazing or entirely abandoned. The varieties of environment in which farming was conducted were matched only by the variations in the structure of the farming economy. In the popular view the medieval village community was a self-sufficing and tightly organised group of peasants. It cultivated in common both its own individual holdings and at the same time discharged its obligations to its 'lord' by tilling the demesne. It practised a three-field system of agriculture, used a heavy plough and large team, and divided its cropland into long, narrow strips for the convenience of ploughing. In parts of north-western Europe, agricultural conditions approximated this model, but it was rarely realised in practice. The variety of crops, the structure of the fields, the varying relationships between arable and pastoral farming, and changing personal and legal relationships together made for a situation that was at once more complex and less stable than the popular view would admit.

In this section we are concerned primarily with the structures and organisation of medieval agriculture: the field systems and crop rotations used; the arable-pastoral relationships; the crops grown, and the commerce in agricultural products. It is difficult to trace changes in agricultural activity through the Middle Ages; impossible to formulate any clear and coherent picture of their predominant types and geographical distributions at any particular period of time. The rural economy of the Middle Ages is very much less adequately documented than the urban, and the literary sources that are available for its study relate primarily to the lands of the Church. Though we have limited means for analysing the structure and economy of the royal estates of France and those of a number of the leading fiefs such as Champagne and Luxembourg, we have virtually no means of finding out how the lesser landowners ran their estates.

The field systems. The contrast between the landscape of small, compact, enclosed fields and that of open-fields, with no more boundary between the individual parcels of land than a grassy balk, is fundamental to the study of the rural geography of the Middle Ages. The boundary between the two is none the less hard to draw, in part because it is difficult to isolate the essential features of each. The medieval poet Wace associated some degree of individual freedom with the one, and subjection to the will of the lord or the custom of the community with the other (see p. 278). In this he was, by and large, correct. The 'champagne' – the 'champion' of English writers – was made up in the main of large fields, each divided into strips or furlongs, but all subjected to the same routine of ploughing, cropping and fallowing. This does not imply that the peasant of the *bocage* was completely free, but he lived in a smaller community within which the rights of the seigneur were less clearly defined and the force of custom less powerful. It is sometimes said that he was free to experiment. The medieval peasant had neither the instinct nor the opportunity to innovate, but it is possible that he worked harder and perhaps achieved more because his efforts were directed towards his personal rather than communal good.

Champion farming characterised the English Midlands together with much of northern and north-eastern France, and extended eastwards across central Belgium to the Rhineland and beyond. Its distribution in continental Europe has been related both to the areas of light, fertile and easily cultivated soils, and also to those of Frankish settlements. These were also regions of fairly intensive settlement and cultivation during the centuries before the Germanic invasions, and here too the heavy plough appears first to have been widely used. It is not impossible that these circumstances are interconnected. At least, other field systems than champion appear in general to have characterised the areas of medieval forest clearance and settlement.

The fact that, as a general rule, the champion village practised a three-course rotation has led to the supposition that it normally had three large open-fields, of approximately equal size. This is not borne out by the known facts. There was usually a considerable number of fields, each probably resulting from a single act of clearance. There might be a number of ways in which they could be combined into a three-field system; at the village of Westerham in Kent, there were 'about forty fields, ranging in size from 1 to 100 acres', but the areas under bread and fodder grains and under fallow in successive years appear to have been approximately equal.[62]

It is rare to find evidence for the existence of three fields of closely comparable size, analogous to those of Saint-Amand, which were noted earlier (see above, p. 204). A significant exception was Vaulerent, a grange belonging to the Cistercian house of Chaâlis. A survey of the mid-thirteenth-century showed its lands divided into three *aristae*, containing respectively 367, 323 and 333 arpents and under respectively rye, spring grain and fallow.[63] Higounet observed that he knew of no other instance of an estate 'réparti rigoureusement en trois soles' at this date. The explanation lies probably in the fact that it was Cistercian, and that the land had probably been cleared and was now operated directly by the monks and their lay brethren, who were thus able to impose their own organisation upon it. This simple arrangement did not, however, last long. In the early fourteenth century the monks abandoned the direct exploitation of the lands, and leased them at a rent.[64] A Cistercian grange at Ouges in Burgundy also showed a rigorous three-field system.[65]

A polyptyque of the lands of Sainte-Waudru of Mons, of the late thirteenth century, showed the lands which the house possessed at Hal[66] divided into four 'coutures', containing respectively 15½, 9, 15 and 15 *bonniers*, of which all except the last yielded 4 stiers per *bonnier* when sown with wheat and a single *muid* (*modius*) when sown with oats. The three-course rotation appears here to have been restricted to three unequal fields, while the fourth field, together with two other small parcels of land, seem to have been excluded from the regular rotation.

The Hospital of St John in Bruges,[67] on the other hand, held lands which, in the total at least, show proportions as mathematically correct as those at Vaulerent. In general, however, the figures which one can derive from the extents and surveys, both monastic and royal, by no means suggest a simple system of three fields of approximately equal extent, and the cases of Vaulerent, Ouges and St John of Bruges must be regarded as the result of highly successful estate management (see table 6.7).

If, as appears in general to have been the case, the open-fields were given over to grain cultivation, the numerous other plots and enclosures of the medieval village must have been used, continuously or in a rotation, with or without manure, to produce legumes and vegetables and to even out as far as practicable the inequalities in food production that must have resulted from the rotation cultivation of fields of unequal size.

Table 6.7 *Division of the lands of the Hospital of St John in Bruges*

	Total measures	Measures sown	Measures of pasture and fallow
Zuyenkerke	561	375	186
Trenton	98	65	33

The south-western limit of champion farming in the eighteenth century, probably did not differ greatly from that of the fourteenth century. It ran from near the Seine estuary in Normandy southwards to the Orléanais, and thence south-eastwards to the Saône valley near Mâcon. Its eastern boundary was very much less distinct. It was widely practised in the plain of Alsace, and doubtless also in the corresponding plain of Baden. Gradmann found abundant evidence for its wide distribution during the Middle Ages in the Neckar valley, and it would have been surprising if it did not predominate in the loess-belt from Cologne eastwards to Saxony.

But the champion landscape was the product of both seigneurial control and also of certain physical conditions which eased the heavy burden of ploughing and favoured the cultivation of cereal crops. A passage in the late thirteenth-century account book of Guillaume de Ryckel, abbot of the Belgian monastery of Saint-Trond, described the obligations of the monastic tenants on newly cleared land;[68] for six years they might grow whatever they pleased and sow as much as they chose. Thereafter 'they will observe the common custom of sowing, so that in one year they sow wheat or rye; in the second, barley or oats or whatever spring crop was usually grown; and in the third year they will sow nothing'. Even here, on the rich loess soil of central Belgium, exceptions were made, if only for a few years, to the customary routine. Presumably, if the heavy hand of Guillaume de Ryckel had been removed, even more peasants would have sown 'quiquid eis placuerit et quociens voluerint'.

East of the Rhine, both physical and seigneurial conditions were less conducive to champion agriculture, and we must assume that other systems of organising cropland and pasture gradually took over and dominated the rural landscape. Even within the region of champion farming there were cultivated fields lying outside the open-fields themselves, as, for example, at Schoten, where the abbey of Villiers had the respectable area of 27 *bonniers* 'extra culturas hinc et inde'.[69] At the same time the College of St John of Liège had a large number of *curtes nove*, all of them enclosed and very small.[70]

To both west and east of the regions of predominantly champion agriculture, settlements were in general smaller. These areas had physical conditions less suited to the rotation of grain crops which characterised champion farming. Soils were poorer and rainfall heavier. Without the addition of manure, the land could be cropped even less intensively than in champion country. Pastoral farming was necessarily more important than in the open-field system, and may locally have displaced arable farming entirely. The tendency nevertheless was for fields immediately adjacent to the settlement to be manured and perhaps also to be cropped continuously. Farther afield, where manure was less readily available, the land might be cleared and cultivated only intermittently. The weight of local custom and of seigneurial control weighed less heavily on the peasantry. Their greater freedom to cultivate their lands as they pleased and to sow what crops they chose led inevitably to a much greater variety in the agricultural 'structures'. The 'in-field', as the regularly or continuously cultivated area close to the settlement was called, might consist of small irregular or compact parcels of land, cultivated in 'several', or of miniature open-fields, consisting each of a small bundle of misshapen strips. Once established, the structure of the fields was modified only very slowly in response to technical changes.

This in- and out-field system was widespread in 'Atlantic' Europe, but its predominance did not preclude the existence of open-fields which might in size and organisation even approximate those of champion farming. Within the region of champion farming, variants of in- and out-field farming were also to be found, as a general rule, in areas of poorer soil and rougher terrain, such as the Sologne, parts of the Paris basin, and on the Geest of north-western Germany. This association with marginal soil again suggests small settlements and a relatively late foundation.

The in-field–out-field system, with its continuous cultivation of an in-field, was clearly dependent upon a large and continuing supply of manure. This might, locally at least, include seaweed and calcareous sea-sand, but as a general rule it was made up of animal manure obtained from stall-feeding of the farm animals. Every rural community produced a supply of manure, but it was relatively scarce in champion regions, and it was there that seigneurial demands for the tribute of manure or of straw for bedding the animals were most burdensome. Amounts available did not, as a general rule, permit the continuous cultivation of more than small garden plots, though the land destined to receive winter grains was used as *vaine pâture* during the previous year and may have received a small dressing of manure. Animal husbandry was more important, in general, in areas of in-field–out-field cultivation than in champion. Manure was commonly available for the continuous cropping of the in-field, whatever the rotation adopted. In such farming, which tended to emphasise the autonomy of the peasant and had little use for fallow grazing, there was less need for field and rotation 'systems', and one must conceive of year-to-year changes in both in response to natural conditions and human needs. It was in such conditions, and under the pressure of rising population, that continuous cropping under a more or less regular cropping system was evolved.

The remaining group of field-systems, those associated with the ribbon-like street, forest and marsh villages, cannot be as easily related to the traditional systems of north-west Europe. They were created at a later date, in areas reclaimed from forest and marsh, by someone who had, as it were, a kind of planning authority over the area. They were inhabited by peasants whose degree of freedom was in all respects greater than that of the 'serf of the champagne'. There was no question of peasant services on the demesne, which, as a general rule, did not exist, nor of a uniform system of cultivation. Apart from common-rights to meadow, pasture and woodland, the peasant holdings were individual and consisted, in general, of elongated fields which extended back from the road frontage of the village. There must have been, as in all simple, agrarian societies, some sharing of equipment and co-operation in the major tasks of the farming year, but in the main the 'colonist' had some degree of freedom in determining what to sow and when to do it.

But this type of settlement did not preclude the appearance of others, if only because some lands had been cleared and cultivated long before the period of German colonisation. Central and eastern Europe also had its two- and three-field systems of agriculture, with open-field strips and communal right to graze the stubble. A two- or three-field system was widespread in Poland in modern times, and was in existence in some areas by the end of the Middle Ages. It is not clear, however, whether this system evolved in eastern, as it did in western Europe, during the earlier Middle Ages. It is at least possible that it was the result of the introduction of better farming equipment and of the intensification of seigneurial control, which were a feature of the fifteenth and later centuries. The earlier system, from which the east European field rotation was developed, consisted probably of some form of in-field–out-field method of cropping.

Most writers have distinguished between the agricultural structures just discussed and those of southern Europe. The latter, it is held, owed much to classical tradition; they were adapted to the use of the light plough; they used crops and rotations suited to the Mediterranean climate, and the paucity of fodder crops led both to a reduced emphasis on animal farming and also to the practice of transhumance. This probably exaggerates the contrast. Some form of open-field cultivation was practised in many areas, notably in the south of France; the heavy, or mouldboard plough was not unknown, and the animals were more numerous and their manure more readily available than has been sometimes assumed.

The climate, with its relatively dry and hot summers, did nevertheless impose restraints on the system of farming. Spring-sown grains would in general not do well and this in turn would tend to restrict the supply of fodder, the number of animals, and the amount of manure. For these reasons, in part at least, the light plough was preferred to the heavy, and fields tended to be small and compact and cropped in alternate years. Physical conditions as well as social organisation militated against demesne farming, and there was a strong tendency in the fourteenth century to lease demesne land on a share-cropping basis. By and large, farming in southern Europe was more individual and the peasant had a greater freedom of choice on his plots, which were usually enclosed. This is not to say that they were better off or more progressive. Herlihy commented on the 'tiny plots often scattered widely over the infertile slopes' of the *contado* of Pistoia, making 'little use of cattle, or of other aids in tools, fertilizers, seeds or hired labor which high capital investment might have secured for them'.[71]

The crops. The pattern of cropping in medieval Europe was a response to a number of factors, some only of which can be isolated and measured. Foremost, of course, were the physical demands of the crops themselves, and the conditions of soil and climate under which they were grown. The agricultural equipment used – particularly the type of plough – indirectly influenced the choice of crops. There was, lastly, a kind of institutional control over the pattern of cropping. Change came slowly, resisted by the innate conservatism of the peasant and by the structures he had erected. The area under any particular crop could be increased only by sowing more and consuming less for a time, and this under medieval conditions of low yield was by no means easy.

The grain-crops dominated the farming system in all parts of Europe and, either as bread or as a kind of gruel, provided the greater part of the calorie intake of the human body. The division into autumn- and spring-sown corn corresponded with that between the bread-grains and the others. Wheat, spelt and rye could be used, either separately or mixed, to produce a flour for baking purposes. They were the grain-crops in greatest demand, and they alone, it is safe to say, under normal conditions entered into long-distance trade. It is clear also that all these grains were used for human consumption. They were sown in autumn – usually October or November – and reaped in late summer. They were exhausting as well as essential crops, and were, for this reason, commonly sown after a period of fallow, when in perhaps most instances, the animals had been allowed to graze the stubble. There is reason to suppose that this may have been supplemented by the direct application of manure from stall-fed animals.

The spring-sown crops – the *tremois* or three-month crops, as they were sometimes called to distinguish them from the *bladum hyemale*, *ivernagii*, or winter crops, were most often oats, but occasionally barley and, locally, millet. These demanded less and gave less. Many are the occasions when one reads in the monastic farm accounts that the winter crop had been lost through flood or other catastrophe, and that the peasants had hastily to prepare the ground in March for a crop of oats. Barley was more tolerant of a dry soil and summer drought, and oats of poor and acid soils, than the winter crops. In general, however, they were fodder crops, though barley and even oats were used for malting and brewing. But in the Middle Ages, the distinction between bread- and fodder-crops was neither precise nor generally observed. As in Dr Johnson's Scotland, oats entered human diet as a kind of gruel, and in emergency made up the greater part of it.

The two-field system, which has commonly been regarded as the forerunner of the three, was essentially an alternation of bread grain and fallow. The adoption of the three-field system introduced a fodder crop, but clearly had the effect of diminishing the production of bread-crops. The reason for the change, it has been argued, was to provide feed for the plough teams, which, in turn, permitted the cultivation of the intractable but potentially fertile clay soils. The equation, however, may not have been so simple. It has been argued that in Basse Alsace and Hesse the two-field rotation was re-introduced because the overwhelming demand in the Rhineland towns was for winter-sown bread-crops, rather than for spring-sown coarse grains.[72] It is very doubtful whether the adoption of the three-field system thus led indirectly to the greater production of bread-crops, and, with population increasing up to the late thirteenth or early fourteenth centuries, the human animal had to dip heavily into those grain-crops which, we assume, were generally reserved for the horse and ox. Walter of Henley[73] was thus able to show how advantageous it was to increase total grain production by the use of the three-field system.

It is apparent from the monastic rolls that by the early fourteenth century oats had entered to a very significant degree into the regular human diet. Wheat, spelt and rye were, in that order, the preferred cereals, but even in the most fertile and productive parts of Europe the poor had usually to make do with a diet in which oats predominated. At many monasteries rye or wheaten bread was the principal food of the abbot and monks, while oats were given to the more humble servants and the deserving poor. At St Pantaleon at Cologne, the monastery received by way of rents and dues 466 measures of oats, 'de qua summa dantur domino abbati ad pabulum equorum suorum et hospitibus suis supervenientibus'.[74]

The listing of tithe payments should be a fairly accurate measure of the nature and volume of agricultural production, though it was sometimes customary not to exact the tithe of certain crops. The grain-crops, however, were always tithed. For manors and even individual tenants it sometimes happens that both the tithe payments and rents in kind are listed. The discrepancy between the two sets of figures is sometimes so marked that one is forced to assume that oats with perhaps a little rye formed the bread-crop of the peasantry, while the bread grains proper went to feed the monastery.

In many parts of northern Europe conditions of soil and climate made it impossible to cultivate the better quality grains, and oats of necessity formed the peasant's staple. In these regions (see below, p. 382), however, the supply of animal protein was probably greater than in those of champion farming. Nowhere, however, was food abundant. The crop yields were in general low, but varied sharply with both the soil and the season. In Flanders the ratio of crop to seed sown was, in the fourteenth century, on average about 4.5 to one for wheat and 6.5 for oats.[75] These yields were high by medieval standards, and indicated a very proficient standard of farming. They were exceeded possibly only by those of northern France. On his rich lands in Artois, Thierry d'Hireçon was getting unusually high yields in the early fourteenth century, an average return of over eightfold for wheat,[76] but his account books show a very considerable use of manure on the land destined to receive the bread-crop. The yields of oats were less generous: on average 6.6 times the amount of seed at one manor and 4.5 at another. A treatise on agriculture, cited by Richard, regarded a fivefold return as normal for wheat, a sevenfold for rye and a fourfold for oats. The long run of figures prepared by Beveridge from the records of the manors of the Bishopric of Winchester, suggest much lower returns for even fertile lands in southern England: 3.9 for wheat, 3.8 for barley and 2.4 for oats.[77] Slicher van Bath has assembled a very large number of estimates of crop yields from all parts of Europe.[78] They were consistently higher in England and the Low Countries than in other parts of the continent, but even here the average was less than a fourfold return in the thirteenth century and only about

4.5 in the fourteenth and fifteenth. Yields were lower in France and central and eastern Europe, but were everywhere tending to rise.

The returns were also erratic and irregular. They responded sharply to fluctuations in weather, and it was far from easy to guard against such natural disasters as floods and pests. On the Roquestor estate of Thierry d'Hireçon wheat yields varied from 7 to 11.6 times the volume of seed, and the yields of oats, from 2.6 to 6.4. Variations of similar magnitude can be demonstrated for a great many other estates and communities.

This wide year-to-year variation in crop yields meant that between the rare years of abundance there were years of hardship or of real starvation. The Register of Eudes,[79] the hard-working and conscientious Archbishop of Rouen in the late thirteenth century, shows how near even the convents of his diocese often were to the margin of subsistence. Of the community of Bellencombre he wrote, 'they have little wheat, which will not last until the feast of St John (June 24), but they have enough oats and barley to last out the year'. At the Priory of Saint-Saëns, 'they had only two muids of wheat to last them until August'. Many of the monastic institutions visited in the course of the archbishop's perambulation of his diocese had a shortage of one essential foodstuff or another, and if there was hardship in the monasteries there must have been destitution in the cottages of the peasants.

The probability of a disastrously bad harvest might perhaps have been of the order of about 15 per cent. The chances of three, or even two, such years in succession would have been relatively remote, yet this is what happened in the disastrous years, 1315 to 1317. The general scarcity was intensified by the hoarding and speculation which it prompted, leading to the most disastrous famine perhaps of the century. On such matters of probability hangs oftentimes the course of human history.

The pressure of a growing population on agricultural resources had reached a very serious level long before 1300. Not only was a class of landless labourers emerging in western and parts of central Europe, but the size of farm units had in many areas reached uneconomically small proportions. This process is best documented for England, but was common to much of continental Europe. Towards the end of the thirteenth century, in the County of Namur, peasant holdings had by a process of subdivision been reduced to sizes that were clearly too small to support a family. In the châtellenie of Furnes half of the peasant tenancies were of under two hectares, and 43 per cent of the tenants of Saint-Bertin at Beuvrequem were no better endowed. Such conditions probably existed over much of the southern Low Countries and in parts of northern France.

That the medieval peasant was obliged to extract the maximum return from his soil is apparent, and it has often been suggested that he did so at the cost of permanently lowering its fertility. Lennard's verdict is that the evidence for the Middle Ages is inconclusive.[80] It is true that runs of statistics from before the Black Death to the end of the fifteenth century do not show any decline, and that in some places even a small increase can be detected. This does not, however, contradict the thesis, because there was during the period a contraction in the extent of arable land. If there had been soil deterioration on any measurable scale, one would expect such wasted land to have been abandoned, and cropping to have been concentrated on the better soil, It would have been surprising if medieval methods of agriculture, which included ploughing up and down the slopes, and leaving large areas of bare fallow, did not lead to the physical erosion of the soil, if not also to its chemical destruction. Herlihy indeed suggests that soil deterioration was significant on the lands of Pistoia.[81] No one today maintains that 'soil exhaustion' was the reason for the late medieval grassing down of cropland, but it is possible that turning marginal land over to grazing may in fact have disguised its deterioration and prevented its condition from becoming worse.

The references to grain-crops make up about 90 per cent of all the allusions to vegetable foodstuffs found in the monastic records. It is certain, however, that they were less predominant in the human diet. In addition to animal products, which were of varying and locally of quite considerable importance (see below, p. 382), there was also a large consumption of vegetables and legumes. The former were grown in garden plots, and included the *brassicae*, hops (*umblones*), herbs, and such plants as onions and leeks. The legumes consisted of peas (*pisae*) and beans (*fabae*), which were certainly grown in the small garden plots and at least occasionally as field crops, sharing the ground with the spring-sown grains. The legumes formed an essential part of the human diet, and were the principal source of protein. They appear infrequently in the records and only rarely were they listed amongst rents and tithes, probably because, being grown on the small, private plots and gardens of the peasantry, they were not subject to such demands.

Fallow in the cereal rotation was the only alternative to manuring, and was necessarily practised where, as in the champion country, animal manure was scarce. The regularity of fallowing varied also with the crops grown. Wheat was a very much more demanding crop than either rye or oats. One not infrequently finds references to these coarser grains on the same land in successive years, but, as a general rule, wheat and spelt could not be sown successfully more than one year in three, or one in two, if the land was fallowed in the intervening year. Wherever the tyranny of the champion system was relaxed one begins to find evidence that the rotations, deemed to be theoretically desirable, were in practice not regularly used. Not only did rye follow rye, but fallow sometimes followed fallow, and a three-phase rotation could be seen to pass gradually into a system of ley-farming. On the other hand, there were places where the practice of fallowing itself came to be abandoned. It is highly improbable that it was used on the small and probably well-manured garden plots. In parts of Flanders also continuous cultivation was beginning to be used in the early fourteenth century, with the cultivation of legumes on land which had customarily lain fallow. It is noteworthy that this revolutionary change was made in an area recently cleared from the forest or reclaimed from the marsh, where the restraints of the bipartite manor were, by and large, absent; where the peasant was relatively free, and the Flanders towns consituted a cash market for farm produce. It was long before social conditions permitted this revolution to take place elsewhere, even in areas physically suited to it.

These changes were facilitated by the practice of leasing the demesne. It was proving uneconomic to cultivate the 'reserve' with the unwilling labour of unfree tenants; it was more profitable to lease the demesne, and to buy foodstuffs in the market. This change was made possible by the rise of a class of merchants who were willing to buy up the rural surplus and ship it to the towns for sale, and was encouraged by the growth of the towns themselves. The Register of Eudes of Rouen shows monastic institutions supplementing the production of their own lands by purchasing in the market. This practice grew. The Cistercian house of Chaâlis, which, in the thirteenth century had practised a three-field system of text-book simplicity (see above, p. 367), by the fourteenth had leased much of its land for a small amount of grain and a large annual rent in money. Many other examples could be cited. Rents, when paid in kind, show a ratio of spring to winter grains which was adapted to the needs of the monastery but ill-suited to the restraints of the three-course rotation. It is probable that a factor leading the monasteries to lease their demesnes was the over-production of spring-grain, particularly of oats.

Monastic estates were, especially in the case of the richer houses, very extensive and widely scattered. They consisted in the main of bequests from the laity. The general practice had been to organise them where possible as a number of bipartite manors; then to sell or exchange the more remote and less valuable for more conveniently located lands, and to lease the demesne of the others. The transfer of farm products from the monastic granges to the monastery involved a complicated problem in logistics, and it is not surprising that the monasteries were content to leave it to professional merchants.

In southern Europe also the tendency was to divide up and lease the demesne, but here, especially in Italy, the leases tended to be on a share-cropping basis. In the fourteenth century *mezzadria* was widely practised in Tuscany, leading not only to an increase in the production of grain but also to an improvement in the status and welfare of the peasant. The trend towards the leasing of large estates was present, though very much less well documented, in the Byzantine Empire. Some of the lands of the Mount Athos monasteries, for which records have survived, had very small areas of demesne, and a large number of small to medium-sized peasant holdings.[82]

Despite the irregular rotations which had come to characterise the in-fields of Atlantic Europe and the progressive agriculture of Flanders, agriculture in most of Europe continued to be dominated by the restraints imposed by the three-field system of grain production. Winter and spring grains were in joint production, and, though their yields commonly differed (see above, p. 372). they might be expected to be produced in approximately equal quantities. This relationship might be expressed as:

$$\text{Wheat}+\text{Spelt}+\text{Rye} \geqslant \text{Oats}+\text{Barley}+\text{Millet}$$

Leaving millet, which was of very restricted importance, out of consideration, and ignoring the fact that barley was occasionally grown as a winter crop, we have almost twenty possible combinations and most of them can be demonstrated from the monastic accounts and registers. A count shows that some combination of wheat and rye as winter crop with oats was the most widespread, though spelt commonly appeared in place of wheat in parts of eastern France and the Low Countries.

The combination adopted in any particular area appears to have been a result of the local, physical conditions, though it must not be forgotten that the initial combination, dictated by the seed grains available, could be modified only with difficulty. The evidence for such systems consists of rents and leases, and, except in the case of tithe payments, is not closely related to the cropping pattern. They show a preponderance of bread-crops, and if coarse grains, such as oats, predominate, this presumably is because the area in question could produce nothing better. Legumes and vegetables were mentioned infrequently. The coarse grains, such as oats, may have been under-represented in the records. Nevertheless, the figures appearing in the monastic rent rolls satisfy the equation given above (p. 375) sufficiently for one to be fairly sure that they represent roughly the proportions and rotations actually used. In table 6.8, the totals, which are taken from the *summae* which often conclude the subsections of the rent rolls, illustrate this recurring pattern.

Where figures depart radically from the formula given it must be inferred either that payments in kind were highly selective or that the cropping pattern departed radically from the winter-grain–spring-grain–fallow rotation which has been suggested as the norm. It is assumed that if, in any particular area, the departures from the normal ratios do not exceed a quarter of the total occurrences, they are not statistically significant and can be understood in terms of the peculiar demands of the monastery. On the other hand, if they are more frequent they must be taken to suggest a crop-rotation other than the normal.

On this basis, an 'abnormal' cropping pattern is to be found at some place or other in most parts of Europe. Indeed, an approximate equality of winter- and spring-sown grains, and thus a three-field rotation, can be demonstrated with a very high degree of probability only for areas clearly associated with champion farming. Elsewhere there was an irregular rotation, with in many areas a predominance of spring-sown grain.

The Paris basin and northern France, approximately the area of champion farming (see above, p. 283) was characterised by a wheat–oats–fallow rotation. Rye was rarely grown, and barley made its appearance only in a few areas of particularly dry soil. The southern margins of this region are obscure. It appears that there was in Normandy a gradual transition from wheat–oats to rye–oats. On the south the wheat–oats rotation appears to have yielded more abruptly to the somewhat irregular cultivation of rye and oats on the sandy soils of the Gâtinais and Sologne. The boundary of the region towards the north-east is made very clear by a somewhat earlier source, the accounts of the Count of Flanders for 1187.[83] The payments in kind, listed for each of over thirty *bureau,* show a gradual transition from an approximate equality of wheat and oats in Artois to the almost exclusive cultivation of oats on the poorer soils of eastern Flanders (fig. 5.6). Nothing demonstrates more clearly than this distribution the basis for the flow of wheat from Picardy and Artois north-eastwards towards the towns of Flanders and Hainaut (see below, p. 426).

A line drawn across Flanders and Hainaut to the Meuse serves to separate

Table 6.8 *The proportions of grain-crops shown in Monastic and other rent rolls*

Place	Wheat	Rye	Total winter-sown grain (w)	Oats	Barley	Total spring-sown grain (s)	Ratio W : S	Source
Clermont (Oise)	2		2	2		2	1	*Cartulaire de l'Abbaye de Notre Dame de Ours-Camp*, No. VXI, p. 12.
Fontainebleau	2	2	4	2	2	4	1	*Cartulaire de l'Eglise Notre Dame de Paris*, Liber 4, No. XXXVI, p. 345.
Penne (Tarn)	255	66	321	284	92	376	0.85	*Comptes Royaux* (1285–1314), vol. I, No. 9263.
Süchtelen (Köln)	..	100	100	100	..	100	1	*Die Urbare von S. Pantaleon in Köln.* LXXVIII p. 340.
Raaût (Werden-Ruhr)	..	175	175	136	96	232	0.75	*Die Urbare der Abtei Werden a.d. Ruhr*, IX, p. 295.
Luxembourg	11	10	21	19	..	19	1.11	*Dénombrements des Feux des Duché de Luxembourg et Comté de Chiny*, vol. I, pp. 487–566.
Maraie (Champagne)	3	..	3	..	3	3	1	*Documents relatifs au Comté de Champagne et Brie, 1172–1361*, vol. 2, II, 474.
Bohemia	18	20	38	32	..	32	1.19	*Urbář zboží rožmberského z roku 1379, passim.*

the region of champion farming with wheat and oats, from one in which the cropping pattern is far more complex and, in all probability, far less regular. The transition from wheat–oats to rye–oats is not abrupt, and is in some degree masked by the widespread cultivation of spelt, a bread-crop which was generally cheaper in price and commonly regarded as poorer in quality than wheat. The ecclesiastical and other administrative records show spelt as the predominant winter grain along the middle section of the Meuse valley and over the northern margin of the Ardennes. In the vicinity of Liège it is unusual to find any winter-grain payments except in spelt. This intensive use of spelt does not extend as far as the Rhine, and east of the Rhine spelt is very rarely met with.

In the Ardennes and Eifel wheat was virtually unknown and spelt was rare. The dominant crops were rye and oats, and contributions were so often made in oats alone that one is tempted to assume that, in many parts of this infertile and inhospitable plateau, there was no winter crop, and that oats were grown as often as soil conditions allowed.

The wheat–oats rotation reappears in the Rhine plain and in parts of the lower Rhineland near and to the north of Cologne, but east of the Rhine wheat is uncommon. Wheat bread was important in the human diet in Cologne, but at Essen, to the east of the Rhine, wheat was so scarce that the Archbishop of Cologne arranged for the canons to have possession of a parish that could remedy the deficiency.

Germany as a whole is poorly documented in this respect. The scanty evidence of cropping, however, confirms that rye and oats dominated the agricultural practice in northern Germany, with oats alone being cultivated in areas of poor soil and adverse climate. Towards the east, particularly in Brandenburg, rye was grown for export on a considerable scale. Towards the south of Germany and in areas of better soil wheat achieved some importance, but in Hesse and the Rhine plain above Frankfurt, the dominant crops were still rye and oats. An *Urbar*[84] of 1379, relating to estates in southern Bohemia, shows rye and wheat making up the winter crop in proportions of about four to one, while the spring crop consisted of oats with a very little barley. Totals for places for which the amounts of grain are given are: winter (wheat, 26; rye, 106); spring (oats, 136; barley, 5). Polish evidence suggests a similar combination with a tendency for rye to grow in relative importance as a winter crop to the exclusion of wheat.

An early polyptyque of St Emmeram of Regensburg[85] indicates an increasing ratio of wheat to rye, in Bavaria, and this is supported by the *Liber censualis* of St Ulric[86] and other monastic records of South Germany. Soil quality varied greatly in southern Germany, and it is to be presumed that oats, which exceed as a general rule all other cereals in the volume produced, came in large measure from the poor, sandy soils which characterise much of this region. Over part of south Germany the now extinct species of grain, *dinkel* (*Tr. spelta*), was cultivated, as a spring-sown crop. Its distribution was as restricted as that of spelt, to which it was related, and has even been regarded as a feature of the Alemannic settlement area. The data are unusually abundant for Austria, thanks in large measure to the publication of the *Urbare* of the Austrian monasteries.

These relate to lands extending along the Danube from near the confluence of the Inn to the Marchfeld, together with possessions scattered through the valleys of the Enns, Traun, Mur, Drau and Sava. This was a rye–oats region. Wheat was a secondary winter crop along the Danube valley and in sheltered valleys within the mountains, and oats were the only grain grown in regions of poor soil and severe winter climate. Many a small settlement in the Niedere Tauern contributed only a few cheeses and a very little oats, and some would appear from the rent rolls to have been entirely pastoral. The lands surveyed in the Austrian *Urbare* extend southwards to the Drau and even the Sava valley in Krain (Carniola). Here wheat began to displace rye as the predominant winter crop.

The western and south-western margins of Europe are less adequately documented than those already examined. From the Paris basin westwards to Brittany there was a gradual transition from wheat–oats to rye–oats, but in the humid climate of Brittany wheat was nevertheless grown. At Fongeray (*dépt.* Ille-et-Vilaine) the abbey of Redon received, in addition to live animals, 18 *muids* of wheat, 9 of rye and 24 of oats.[87] The formula for the three-field system is satisfied, here, as indeed it is by several monastic estates in Brittany. But it was rye which predominated in rent payments, and probably oats in the overall cropping pattern.

Towards the south however, as summers lengthen and become warmer, and the winters progressively less inclement, the cropping pattern again changed. The alteration is masked by the occurrence of highland areas, such as the Massif Central and sterile regions like Sologne, which imposed their own conditions on farming. The wheat–rye ratio again tipped in favour of wheat; with the lengthening summer drought, spring-sown grains became increasingly hazardous, and the three-course rotation of north-western Europe ceased to be the rule in Gascony. Rye–oats predominated in sandy areas such as those of Sologne and in those which, like the Massif Central and the Alps, combined poor soil with harsh climate, but elsewhere barley, tolerant of summer drought, especially when winter-sown, increased at the expense of both rye and oats.

Wheat and barley were the prevailing grain crops in areas of true Mediterranean climate, as they had been since classical times, but millet (*milium*) was also grown in the warm valleys of southern Austria and Italy. When a group of crusaders in 1147 seized and looted the small town of Lisbon, they found a typically classical store of crops: wheat, barley and olive oil.[88]

Viticulture. The grape-vine continued to be one of the most valued plants of medieval Europe, and a plot of land capable of producing a good vintage was valued above all others. The import of French and Mediterranean wines had, however, driven the grape-vine from the fields of England, and had made inroads upon those of northern France and Flanders. The Flemish cities were, by and large, supplied by sea. But only short distances inland from the Flemish ports a local wine was being produced for consumption by those who could not afford the highly priced clarets and malmseys.

In the north-western region of Europe the vine grew with difficulty, produced little sugar, and yielded only a sour wine. The region was, however, linked by two river systems, the Seine and the Rhine, with more southerly areas where the grape grew more abundantly and yielded a better vintage. The Paris region, the Oise valley, and even Normandy, produced wine on no inconsiderable scale in the fourteenth century, and much of it went to supply Paris, but wines of a better quality were obtained from the more southerly parts of the Paris basin, where 'burgundies' could be made and sent down the rivers to Paris.

The valleys of the Rhine and Moselle supplied wine to the lower Rhineland and the Low Countries. Vineyards were maintained as far north as Kettwig and Werden on the Ruhr, and Louvain in Brabant, but the wine produced was of indifferent quality, and even the monasteries which owned them preferred to obtain their wine from more sunny and southerly areas. There was thus in the later Middle Ages some expansion of the vineyards along the Upper Rhine, just as there was at the same time in Burgundy and Aquitaine, and wine provided one of the more important downstream cargoes on the Rhine. Nevertheless the vine continued to be cultivated, and wine to be made in areas that were far from suited to it. The Abbey of Saint-Martin at Tournai still had vineyards near Valenciennes, though much of its wine came from the Oise valley, and that of Saint-Hubert even produced wine within the Ardennes massif. Elsewhere in north-western Europe, a poor quality wine was produced, in Brittany, Picardy, Artois, for example, wherever imported wines proved too costly.

The developing wine trade (see below, p. 428) was encouraging the concentration of wine-growing in areas, such as Aquitaine and Burgundy, which combined a suitable climate with adequate means of transport. In Germany, however, this specialisation was less advanced. Apart from what passed along the Rhine and its navigable tributaries, there was little long-distance trade in wine. South Germany could satisfy much of its needs, supplementing its own production where necessary with shipments brought up the Danube from Austria or across the passes from Italy, where Tyrol was a centre of a wine trade with Germany.

If the area under viticulture was contracting in north-western Europe, it was still expanding in central and eastern. Throughout central Germany and even in part of northern, small vineyards provided their local regions with a vintage of indifferent quality. In Bohemia and even in Poland a poor quality wine was made in a few well-favoured areas, but in the northern ports the merchants of the Hanse were already bringing in the wines of southern France and the Mediterranean in return for the export of rye and the forest products of northern Europe.

Viticulture had spread down the Danube valley from Bavaria. Vines were at this date still rare in Upper Austria, but were becoming increasingly numerous along the Danube valley in Lower Austria. Small vineyards were scattered through the Alpine valleys, but, like those in north-western Europe, were tending to contract, as the vine spread over the sunny plains below Graz, on the margin of the Hungarian plain.

Olive oil, which constituted, with grain and wine, the basis of the classical rural economy, was of less importance during the Middle Ages. This is probably to be explained in terms of the increased significance of animal fats and legumes. Oil remained important in those Mediterranean regions where the olive grows well, but there was no longer an important trade in oil. It is not impossible that the wars and disturbances of southern Europe during the Middle Ages were more injurious to the slow-maturing olive than to the vine. Nevertheless there was an export of olive oil from southern Italy, Crete and Greece to northern Italy, but the amounts were small.

Industrial crops. Medieval society had to produce its own textile and other industrial materials no less than its food. With the exception of wool, these materials entered little into long-distance trade, and the local region was as a general rule self-sufficing. A little cotton was grown in southern Spain, in Italy and in Greece, but the chief vegetable fibres were flax and hemp. These could be grown almost everywhere, and were tolerant of cool and moist climates and poor soil. The preparation of the fibres was arduous and exacting, and the spiritual and lay seigneurs could not but profit from dues paid in flax and hemp, retted, combed and ready for use. Flax and, to a smaller extent, hemp were grown in the damp polders of the Low Countries and valleys of the Alps, in southern Italy and the Balkans, and in Brandenburg and Bohemia. Sometimes their manufacture into fabric was carried on in the cottages of the peasants whose dues took the form of coarse linen cloth. Switzerland and southern Germany produced large quantities of flax, which was used in the important linen and barchant (fustian) manufacture of this region (see below, p. 391).

Pastoral farming.

A feature which distinguished medieval from classical agriculture was the greater importance attached to farm animals. This arose in part from the somewhat different diet pattern, but also from the necessity to make a fuller use of marginal land. Crop farming was interlinked with animal farming in all parts of Europe, and a system with many local variations was evolved in which each was dependent on the other. Champion farming, with its heavy emphasis on grain production, needed ox-teams for ploughing and carting, and manure for its enclosed plots. In areas where grain-crops grew less well, animal farming was an essential supplement to arable farming in providing enrichment for the soil and food for men.

The scarcity of meadow and of fodder crops limited the scale of animal husbandry. In general there was enough feed during the growing season; the problem was to maintain the animals during the winter. Hay was the chief fodder, but meadows were poor in quality and very limited in extent. They were nevertheless amongst the most valuable land in the community. They lay along the valley bottoms, where winter floods and occasionally irrigation in summer encouraged the growth of grass. Some field crops were shared between man and beast: oats, barley, peas and beans, and a species of vetch, which appears not infrequently in the records, but it is more than probable that an increasing proportion had been going for the sustenance of man himself. Cattle and pigs were all too often turned loose in the woods, the former to feed on leaves and twigs, the latter to root for mast and acorns. It was customary, however, each autumn to slaughter the animals that could not be fed during the ensuing winter, and to salt down their meat.

There was in medieval Europe a gradation from the arable farming community with little more than draught animals to the pastoral community whose arable was restricted to a few garden plots. These extremes were to be found in northern France, on the one hand, and in the Alps on the other. One can, however, divide this spectrum into at least four parts, each corresponding with a distinctive physical environment. The first was the champion farming of north-western Europe, in which animal husbandry was subordinated to arable. It is not easy to form an estimate of the numbers of animals in relation to cropland. This clearly depended in part on the ratio of sheep to other animals, and sheep were particularly numerous in parts of the region of champion farming; and in part on the extent of meadowland, which was greater, for example, near Paris than in Artois.

In 'Atlantic' Europe and also in many hilly areas of north-western and central Europe the ratio of animals to cropland was very much greater. In the west, the growth of grass continued for much of the year, thus minimising the difficulties of winter feeding. At the same time the area of rough and hill grazing was greater. Cattle were numerous, and, if the wool was coarser than in drier areas, the greater number of the sheep was some compensation. Sheep were numerous on the poor, dry soil of Sologne and Gâtinais. On the Ardennes and Eifel, animal farming increased in relative importance. Locally cereal crops disappeared from rent payments which were made entirely in cheeses. In Flanders, also, where grazing, much of it recently reclaimed from the sea, was abundant, there was also a high ratio of animals to cropland. The Hospital of Saint-Jean at Bruges appears to have had on only two farms, in addition to extensive cropland, no less than 266 head of cattle and over 1,100 sheep and lambs. In the more northerly parts of the Low Countries and in northern Germany the ratio of animals to cropland was probably even greater.

Pigs tended to be numerous, and were the chief source of meat in much of Europe. Their numbers were roughly proportionate to the extent of the woodland, which provided their chief source of food. A rough indication of the relative proportions in which animals were kept in 'forest' Europe is provided by a table of the losses sustained by a group of estates in Lower Silesia as a result of military action:

Sheep and lambs	690
Swine	541
Cattle (cows, 116; oxen, 102; calves, 30)	248
Horses	56
Goats	40

The losses on this occasion included, in addition to the livestock, no less than 210 hives of bees.[89]

In the third region, crop-farming was clearly subordinate to animal-farming; rents were paid mainly in wool and cheese, and bread-crops were probably imported into the region. Among such areas were the Pyrenean and Cantabrian mountains and the Alps. In the mountains of Savoy and Dauphiné immense numbers of sheep and goats were kept and a fifteenth-century source suggests that there may have been in some areas as many as a hundred sheep to each household.[90] Cheese and a poor quality wool were the chief products. The Austrian *Urbare* show an immense number of small *Schwaigen* or *Schwaighöfen*, which were probably exclusively pastoral and paid rent only in cheeses. The number of cheeses credited to some of these small settlements was in many cases so large that one is left wondering how they were marketed. The cheeses seem here to have been made largely from ewes' milk, though it is probable that goats were also kept.

It is suggested that in the Alpine region pastoralism had been slowly increasing at the expense of crop-farming.[91] Lands in the Bavarian Alps which at the beginning of the thirteenth century had been growing rye and oats were, before its end, producing only dairy products. Such lands lay at the margin of crop-farming; they would have experienced more acutely than others the consequences of any slight deterioration of climate, but the change in their economy taking place during these years is more likely to be related to the improving means of exchange and the increasing degree of local specialisation in agricultural production.

The fourth region of pastoral farming at this time was Mediterranean Europe, where the summer drought imposed restraints on pastoral no less than on crop-farming. Cattle were relatively few, and sheep predominated. Even so they could live through the dry summers only by migrating to the mountains. Pigs, which could not migrate as easily as sheep, were rare in most parts of the Mediterranean basin. The quality of the sheep had probably not yet been improved by the widespread introduction of the merino from North Africa, though merino wool was already being imported for fabrication in Italy.

Transhumance. The seasonal movement of flocks and herds had been known in classical times. It was practised on a greatly increased scale during the Middle Ages in all parts of Europe, from the British Isles and Scandinavia to Spain and southern Italy, but, similar as the pattern of movement might have been throughout this area, its relationship to the farming regime differed between the north and the south. In Mediterranean Europe there was, for climatic reasons, a severe shortage of feed in lowland areas in summer. The animals were thus moved to the mountains or to areas where a somewhat different climatic regime prevailed. North of the Alps, the movement away from the lowlands in summer was for the purpose of leaving the land free for cropping or growing hay for winter.

Much of the medieval transhumance was vertical, in the sense that the vertical component of the annual migration was the more important. Horizontal, or long-distance migration was restricted by feudal jurisdictions, which were apt to burden such movements with heavy tolls.

In Italy sheep moved in large numbers between their winter grazing on the coastal plains, such as the Tuscan Maremma, Romagna and the Campagna, and the summer grazing in the Apennines. The mountain pastures of Camaldoli were open from April until October, and were rented by the monastery to landowners from plains. From the northern margin of the Lombardy plain, the flocks and herds were driven into the Alps in summer.

The Alps were the scene of the most intensive development of transhumance. Though crops were grown along the floors of all except the upper Alpine valleys, the higher pastures could be grazed only in summer, but were then often of the highest quality. Spring-sown crops in the valleys and summer grazing and haymaking in the 'alpine' meadows constituted a practicable combination which was used throughout the alpine system. This narrowly vertical transhumance was sometimes supplemented where political conditions permitted by a seasonal migration over greater distances. The western Alps border the dry-summer regions of southern France, and a transhumant movement developed between Provence and the Rhône valley on the one hand, and the Alps of Dauphiné and Savoy on the other. The monastery of St Julia at Brescia in the Lombardy Plain, held *alpae* in the distant mountains. No doubt cattle were important in the short, vertical transhumance *within* the Alps, but it was mainly sheep and goats which made the longer journeys from the Mediterranean plains to the alpine pastures. In the case of the long-distance movements, it seems probable that the lowland owners of the flocks paid a rent or a tax for the use of mountain grazing. The monastery of Camaldoli in the northern Apennines profited greatly from the rents received in this way, and in Savoy the *anciège*, a tax payable in cheese, was levied on transhumant sheep. The practice of transhumance extended through the Balkans, where political conditions imposed a variety of distinctive forms. It was carried on, not only by the Slavs, but also by the Romance-speaking Vlachs who have perpetuated their transhumant system into modern times.

It was in Spain, however, that long-distance transhumance was most fully developed. By the fourteenth century the reconquest of Spain from the Moors was almost complete, and the valley of Andalusia and the wide spaces of the Meseta were slowly being settled and developed. Sheep-rearing was at this time a profitable way of using large and thinly peopled areas; wool was in demand in the markets of western Europe, and the merino breed, now being introduced from North Africa, was beginning to yield a fleece of high quality. At the same time, however, the climatic variations between northern and southern Spain made the seasonal migration of the flocks desirable, and the unified political control of much of the area rendered it possible. Sheep had been numerous and seasonal migration probable even before the Moorish period. During the earlier Middle Ages, the shepherds had begun to acquire an organisation to settle the problems of ownership that naturally arose amongst them and to protect their interests against settled cultivators. In 1272, Alfonso X of Castile brought these associations together into the 'Assembly of the Mesta of the Shepherds', which continued to exist into the nineteenth century to protect the interests and later the entrenched privileges of the sheepowners.

The grant of privileges by Alfonso X not only affirmed the right of the shepherds to use migration routes with their flocks, but also defined their width when they ran through settled territory. They linked the winter grazing in Estremadura and La Mancha, where vast areas had passed into the hands of the religious and military orders, with summer pastures in Leon and Old Castile. These *cañadas* were protected from encroachment by neighbouring landowners, not only by royal decree, but also by the ever watchful and often bellicose shepherds. By about 1330 the system was far from the peak of its development. In the mid-fifteenth century taxes were paid on about 2.5 million sheep; numbers are not known a century earlier, but it would be surprising if it did not exceed 1.5 million.

The forest in the rural economy

The medieval peasant lived surrounded by the forest. It gave him protection; provided him with fuel, with building material for his house and with feed for his pigs. It harboured the deer, which the noble hunted, as well as the outlaws who lived beyond the jurisdiction of his courts. But for a period of at least three centuries inroads had been made on the forest by the land-hungry peasant. In western Europe it is doubtful whether any potentially good agricultural land remained to be cleared by the early fourteenth century. Over large areas the extent of the forest had been cut back to the limits which we see today, and in the more populous and intensively cultivated regions there must have been a growing shortage of timber and other forest products. On one great estate after another inventories were prepared of the woodlands. The Count of Champagne ordered a survey to be made of the *Etat des Bois* in the region of Troyes, in the course of which some two hundred areas of woodland were listed with the area of each.[92] Even in the vicinity of the well-wooded Ardennes some anxiety appears to have been felt, and about 1300 the Count of Namur also had a record made of his woodlands.[93] Landowners had always been jealous of their game; they now began to guard and husband their pannage and rough grazing. In 1343, John of Luxembourg undertook not to confiscate any pigs belonging to the abbey of Saint-Hubert-en-Ardenne that might stray into his own forest, but to return them to the monks. He made it clear, however, that this unaccustomed generosity would not be extended to others who might offend.[94]

In central and eastern Europe forest clearance had not proceeded as far as in western, and it continued through the later centuries of the Middle Ages and into modern times. The competition for woodland was less intense, and forest products, such as wax, furs and honey, continued to be exported.

Mediterranean Europe had less woodland than the rest of the continent, and stands of timber were proportionately more valuable. Forests of fir and beech might still cover parts of the Apennines, but they were growing smaller rapidly. The monastery of Camaldoli, for example, contracted with a group of Florentine merchants in 1316 to sell 3,000 fir trees. Charcoal burners were active in the forests of Tuscany, and boat-builders of Pisa, Genoa and Venice were turning to the forests of Sardinia and Corsica, Albania and Greece for supplies.

Timber was everywhere used for building. Churches, castles and public buildings were usually of stone, but even they made some use of wood, and many a small town had fortifications and gate-towers of timber. Domestic architecture made extensive use of stone only where it was abundant locally. Elsewhere building was in wood. Monastic tenants supplied building lumber and rafters and shingles for roofing (see above, p. 206). Agricultural tools and equipment were made mainly of wood, and in vine-growing areas immense numbers of vine-poles were needed. In the *Amt* of Sankt-Georgen, in Lower Austria, no less than 20,000 vine-poles were cut and prepared in one year.[95]

The forest provided all the fuel used in most parts of Europe, and was supplemented by peat, turf and charcoal only in a very few areas. All smelting and metal-working operation used charcoal, and for this reason were generally situated within the forests. Many a landowner was beginning to derive an income from the sale of lumber and the concession of forest rights, and if the forest lay close to a navigable waterway, as the Black Forest did, it acquired an even greater value.

Add to these products of the forest, the beeswax and the honey, the oak bark for tanning, the acorns and the beech mast which provided the chief feed for the millions of pigs, the game and the skins of the wild animals, and we have a very significant component of the medieval economy.

MANUFACTURING

The medieval community was as self-sufficing in manufactures as it was in agricultural goods. It made within its borders a wide range of coarse fabricated goods, from home-spun cloth to bricks and roofing tiles, which were rarely, if ever, transported beyond its boundaries. Architectural styles were adjusted to the supply of local materials. As a general rule, there were very few necessities, apart from salt and iron tools, which need ever have been brought in from outside the community. Long-distance trade was thus almost exclusively in goods which might be regarded as luxuries. These would include the more costly fabrics, ornamental and building stones for church or castle, which could not be obtained locally, window glass, the precious metals, and such items as lead which were used in the larger structures.

Superimposed as it were, upon the mass of non-specialised manufacturing were the few specialised centres of industrial production. These produced goods of higher quality and greater price; they had advantages in the supply of raw materials, in technical skill or in transport, and they catered for the demands of a luxury market which was spread thinly over the whole of Europe and extended even beyond its limits. The existence of this higher level of manufacturing did not preclude the existence of the humbler crafts even in its own vicinity. The two existed side by side, and it was often difficult to draw a line between them. Bruges produced quality cloth for export to Italy or the Champagne fairs, and its surrounding villages a poorer quality fabric for local use. Nor did the artisans of Florence wear the artistic fabrics which they were so skilled at making.

Clothing industries

Next to food production, the manufacture of clothing was the most important occupation of medieval man. Spinning and weaving, the dressing of skins, and the tanning of leather were universal. Every cottage had its handloom, and the stench of the tanyard was one of the facts of medieval life.

The medieval cloth industry was based on wool and flax. Cotton, grown in the south of Spain and imported from the Levant, played only a minor role, and silk was used only in luxury fabrics. Hemp, coarser and less durable than flax, was used mainly in the manufacture of rope. Wool and flax could be produced in most parts of Europe, though their quality varied greatly from one area to another. In the early fourteenth century, English wool was still the most highly prized, though that of Spain was rising steadily in repute. But wool was bulky and awkward to transport, and most was spun and woven near where it was clipped from the sheep. Only the best quality wool could bear the cost of long-distance transport, and even so, it was generally marketed by ship. The Flemish and northern French cloth industry had developed on the basis of a local wool supply, but this was easily supplemented by imports from the British Isles.

Flax was relatively more important during the Middle Ages than subsequently. It grew well in the damp regions of 'Atlantic' Europe and the north German plain, which were not congenial to the fine-woolled sheep, and the preparation of the material, spinning and weaving were technically simpler than the corresponding processes in woollen manufacture. The advantages of flax were, however, offset by its lack of felting qualities and its inability to produce a thick and warm fabric. In consequence a cloth was produced, especially in the Rhineland and Low Countries made of part wool, part flax. Cotton, produced in southern Europe or imported from the Middle East, was of only local importance. It was used in Italy, but north of the Alps was significant only in parts of Switzerland, south Germany and the Rhineland. It was used here, as a general rule, together with either wool or flax (see below, p. 391) and was almost certainly brought across the more easterly alpine passes from Venice. Raw silk was also a product of southern Europe, where silk cloth was chiefly made. It was, however, carried northwards to Germany, giving rise to small-scale silk-fabricating industries in the Rhineland, especially at Cologne, and in the Low Countries.

The widespread manufacture of poor quality cloth for local use is very poorly documented, because it concerned few besides the peasants who carried it on. By contrast there is a wealth of data regarding the making of better quality fabric. This was carried on in many towns of Italy and north-west Europe, but was chiefly found in two industrial regions: Flanders, Brabant, and neighbouring areas of Hainaut and northern France, and in Tuscany and the Lombardy Plain.

The cloth industry of the Low Countries. This, the most important clothing industry of medieval Europe, extended from Picardy to Zeeland, and from the coastal towns to the Meuse valley. It possessed the primary advantages of wool supply and a network of internal communications. Its coastal position gave it easy access to England, the source of its better quality wool, and to its north European markets. The Rhine opened up a vast hinterland, but the chief markets for its more valuable fabrics had previously been overland to the south, in the Champagne fairs.

The range of kinds and qualities of cloth produced was immense, but in the early fourteenth century the balance was tipping away from the production of

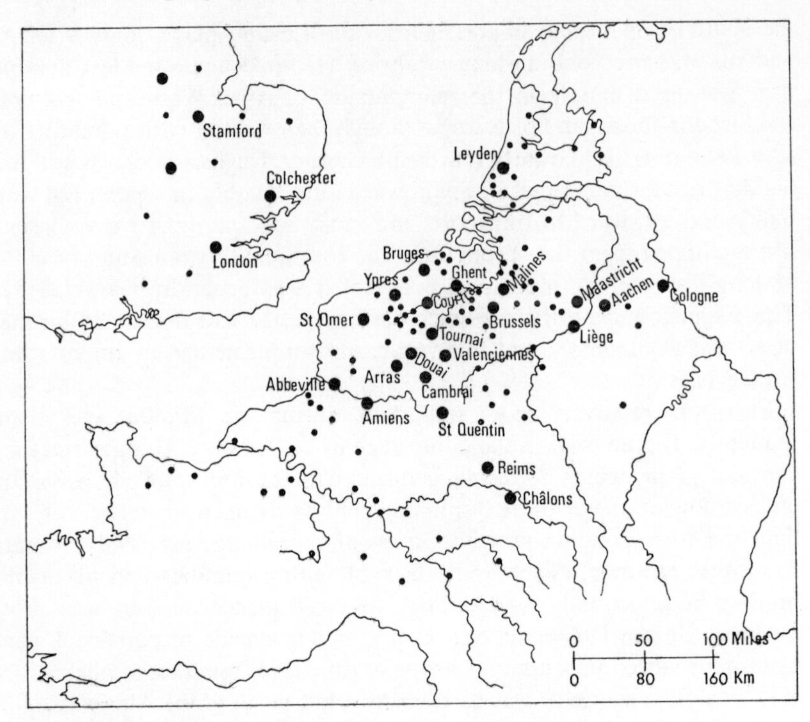

Fig.6.16 The north-west European cloth-making region in the fourteenth and fifteenth centuries

quality cloth for distant and predominantly Italian markets, and in favour of a somewhat cheaper cloth. The latter demanded less costly means of transport than the pack animals that took the road through Bapaume to the fair towns. These it found by water up the Rhine and along the German coast towards the Baltic.

The slight shift in emphasis had been accompanied by a similar movement between the centres of cloth manufacture. The older cloth towns, Ypres, Bruges and Ghent, had reached the peak of their production during the previous century. It is hard for manufacturers whose primacy had long been recognised to realise that the pattern of demand could change to their disadvantage and difficult for them to adapt their methods to new market conditions. Their tendency was to resist change, and the newer developments in consequence, took place elsewhere, in Brabant, Hainaut and in the rural areas of south-western Flanders.

Nevertheless, the older centres of the cloth industry remained the more important, and, even when the volume of cloth produced began to decline, they continued to market and export the cloth made in the smaller towns. The cloth manufacture of the older clothing towns in the main used English wool. It was a complex process and in all probability needlessly expensive. But it did produce a fabric of even texture and great strength. The 'new draperies', on the other hand, simplified the processes of preparing the wool, and of spinning, weaving, and finishing and, furthermore, used a less expensive wool obtained, in part at least, either locally or from Scotland or Ireland. The result was the production of says, fustians and serges.

In the fourteenth century, the market for these varieties of cloth was increasing, especially in the Baltic countries, and areas marginal to the traditional seats of the industry were quick to profit from it. Western Flanders had long been 'saturé de draperie';[96] in all the villages and small towns of the region there was a manufacture of coarse cloth. The industry was growing. Villages and small towns, like Cassel, Bourbourg, Furnes, Hasebrouck and Hondschoote expanded their production. Some of the larger towns, such as Arras, Douai and Valenciennes, whose traditions were less deeply rooted in the past, were able to make the transition to the newer types of manufacture. In all, about thirty towns and villages of western Flanders and neighbouring Artois enjoyed some measure of importance in the clothing industry.

A second region of industrial expansion at this time lay in south Brabant. Unlike western Flanders, this area had not been a significant centre of the industry before the thirteenth century, but in its closing years it increased rapidly in importance. This development was encouraged by the sympathetic government of the Dukes of Brabant, by the expanding Rhineland market to which it had access, and by the opening up of the port of Antwerp, which was tending to replace the ports of Flanders. Brussels, Louvain and Malines quickly grew to be major centres of textile manufacturing, with Diest, Vilvorde, Enghien and Hal of secondary importance. The industry was established in small towns like Ath, and it expanded in the towns along the Meuse: Namur, Huy, Liège and Maastricht. The *Dit du Lendit*, a rhyming poem describing the business done at the Lendit fair at Paris, shows the changed distribution of textile production.[97] Brabant comes first in the number of clothing towns represented and is followed by western Flanders. Bruges is not mentioned at all.

The Italian cloth industry. The second cloth-manufacturing region of medieval Europe was central and northern Italy. Its industrial structure was, like that of the Low Countries, highly complex, consisting of the manufacture from local materials of a relatively poor quality cloth for a predominantly local market, as well as the production of luxury fabrics. But the Italians established themselves as cloth merchants before they gained prominence in the manufacture of cloth. They had long been active in the markets of Flanders, and the fairs of Champagne could not have achieved their international role without them. They had for a couple of centuries been distributing the cloth of north-western Europe to markets around the Mediterranean. They understood these markets; they handled – or even controlled – the market in mordants and dyestuffs used in cloth finishing. It is not surprising that this led them into the business of cloth-finishing.

Though the finishing industry was associated at this time primarily with Florence, foreign cloth was imported into many other Italian cities; it was dyed and stretched; a nap raised and shorn, and the fabric prepared for export. At Florence, this finishing of high quality cloth was the monopoly of the *Arte di Calimala*, so named from a small street in which it was carried on. There were about twenty workshops, which according to Villani, each year worked up about 10,000 pieces of cloth, imported mainly from the Low Countries, for export to all parts of the Mediterranean world.

Important though it was, the *Arte di Calimala* was exceeded in all respects by the *Arte della Lana*, which used wool from Italian flocks, from Spain and North Africa, and later from England, to produce a cheaper cloth that was in wider demand. In Villani's words, 'the workshops of the *Arte della Lana* were 200 or more, and they made from 70,000 to 80,000 pieces of cloth . . . and more than 30,000 persons lived by it'.[98]

In neighbouring Prato, Pistoia and Pisa the cloth industry had been introduced, in most cases by craftsmen from the plain of Lombardy. It spread southwards into Umbria. Perugia and Orvieto became centres of manufacture, but in southern Italy it never rose above the status of a domestic craft.

In volume, if not also in quality of its product, the Tuscan cloth industry was exceeded by that of the plain of Lombardy. Here the cloth industry was carried on in every town, and in some of them it was the dominant industry, and claimed an unbroken history from the period of the Roman Empire. The towns competed with one another for skilled artisans, for the raw wool from the Alps and Apennines, and for access to the ports and overseas markets.

Specialised manufacture of cloth for distant markets was not limited to the Low Countries and northern Italy. It was to be found in many of the towns of northern France and the Rhineland. The *Dit du Lendit* (see above, p. 389) mentions the cloth of Abbeville and Amiens, Beauvais and Reims, as well as of the 'fair' towns of Troyes and Provins. Along the lower Seine valley and in Normandy were clothing towns, of which the most important were probably Rouen, Montivilliers and Louviers. Paris itself was an important source of cloth, and its merchants were careful to keep their products separate from those of nearby Lagny and Saint-Denis. In all the towns of central and southern France cloth was made, for sale, in most instances, only to their local regions, but occasionally a local fabric, such as that of Bourges, by reason of its colour or texture, gained a wider repute. In addition to the fine cloth of Malines, Ypres, Brussels, Beauvais and Montivilliers, the Bonis brothers, merchants of Montauban in Gascony, dealt in the local 'rosets', 'greys' and 'burels', noted for their firm weave and fast dyes.[99] The brown cloth from Rouergue and the 'whites' from Aveyron and Rodez competed with those of north-west Europe in the markets of the South. In Spain also the local market was largely satisfied by the local product, for which there was no shortage of wool, and the towns of Cataluña – particularly Barcelona, Sabadell, Tarrasa and Valls – achieved something more than local importance, but cloth of better quality was regularly imported from north-western Europe and from Italy.

In the cities of the Rhineland, the cloth industry was well established, and profited not only from expanding markets to the east but also from the opening up of the more easterly of the alpine passes and a developing trade with Italy. It would be a mistake, however, to regard the cloth trade of central Europe as a one-way movement. While the flow of the better quality fabrics was from Brabant and the Rhineland eastwards towards the Baltic provinces, Poland and Bohemia, there was a manufacture of coarser fabrics in all these areas. Here labour was cheap and raw materials abundant if not of the highest quality. There thus developed a reverse flow of cheap cloth, particularly the 'grauen Laken', towards the Rhineland and Low Countries, where it competed with the less sophisticated of the local products. At the same time German and Polish cloth sold in Russia, Hungary and the Balkans, where in relation to the local products, it doubtless seemed to be of luxury quality.

Flax and cotton fabrics. In many parts of Europe, the manufacture of woollen cloth faced the active competition of vegetable fibres, primarily flax and cotton. Flax was widely grown. It was cheap; spinning and weaving were easy, and the manufacture of linen had long been a cottage industry (see above, p. 213). Linen cloth was woven very extensively, and seems in some areas to have been even more important than woollen. The best of it entered into long-distance commerce, and the not unimportant cloth trade of south-western Germany and eastern Switzerland was in fact based on linen. Even in Flanders and neighbouring parts of the Low Countries, the linen industry was carried on in competition with the woollen, and in northern France, where the manufacture of woollens was less firmly entrenched, that of linen became proportionately more important.

The linen industry was widespread through the plain of north Germany. Westphalia sent linen cloth to Scandinavia, to other parts of Germany and even to the Netherlands and Italy. In south Germany and neighbouring parts of the Swiss plateau, the linen industry was more fully developed, and produced a good quality cloth which it exported to Italy. Its most important centres lay around Lake Constance, especially Zurich, Constance and Sankt-Gallen, and in Swabia, where the chief manufacturing towns were Ulm, Augsburg, and Ravensburg. The industry also extended down the Rhine, and Frankfurt made linen along with woollen cloth. In many of the towns of this region the weaving of the flax prepared in the surrounding villages was the leading industry; flax was widely cultivated in the upper Rhine and Danube valleys, and the retting and preparation of the thread was a leading domestic industry.

South German linen was sold in Italy, and it was probably the Italian example that led to the addition of cotton thread to the flax to make the barchant (fustian) for which south Germany became widely noted. The spinning and weaving of imported cotton was already established in northern Italy, particularly in Milan. No real attempt appears to have been made at this time to introduce it into north-western Europe, where the traditional woollen industries were entrenched. Raw cotton was, however, carried across the passes to south Germany by the early years of the fourteenth century, and cotton cloth was probably being made by 1320. The high cost and limited supply of cotton no doubt prevented its wide adoption in Germany, but it quickly came to be used as weft in the manufacture of mixed fabrics, for which many of the south German and Swiss cities quickly became noted. The industry had probably been introduced into Germany from Constance only a few years earlier, but soon spread to Ulm, Augsburg, Memmingen and Biberach, which were to become its chief centres.

There is evidence that such mixed fabrics, or fustians, were being sold at the

Champagne fairs, and even that it was being manufactured at Provins and other fair towns.[100] But in northern France and the Low Countries it made slower progress. Fustian had been imported, but there is no clear evidence by the early fourteenth century that it was being manufactured there. In Italy, on the other hand, cotton had long been used to make fustians and bombazines, not only in Genoa and Venice, ports through which the raw cotton had been important, but also in the towns of Lombardy to which it was sold, particularly Milan and those of western Lombardy. Flax was also mixed with wool to make a true fustian, known also as *Tirtei*. The Rhineland, particularly the city of Cologne, and the Netherlands achieved a certain importance for the manufacture of this particular kind of mixed fabric.

Silk manufacture. The best of the woollen fabrics produced by the *Arte di Calimala* were of a very high quality, but the most esteemed and the most expensive fabric of the Middle Ages was silk. Much of the silk marketed in western Europe was imported by Italian merchants from the eastern Mediterranean. Constantinople remained one of the chief sources of quality silks (see above, p. 296) and had, in fact, recently improved the quality and the design of its fabrics.

The rearing of silk-worms was introduced into Sicily, perhaps by the Arabs, and the art of preparing and weaving the silk was carried northwards, perhaps by the Jews, who had practised it in the south Italian cities of Amalfi, Salerno and Gaeta. It was thus that the silk industry came to be established in Lucca, then the capital of Tuscany, by the mid-thirteenth century.

The silk industry differed from the woollen and linen and even from the cotton industry in the high value of its raw material, in the great skill required in its manipulation, and the restricted market for its product. Only the wealthy could afford to buy it and it was fabricated only where rich merchants and the leaders in Church and State were likely to congregate. Lucca, situated on the Via Francigena which led to Rome, and also the seat of the Dukes of Tuscany, was such a place. Only part of its raw material was produced either locally or in southern Italy and Sicily. The rest was imported by the Genoese from Syria, Asia Minor, and the eastern and south-eastern shores of the Black Sea; some came from as far as Persia and Turkestan.

The Lucchese silks were sold in the markets of northern Italy; were sent to the Champagne fairs by Genoese merchants, and, as these began to decline in importance, were sold in Paris and other large towns of north-western Europe. It became the practice for the wealthy to commission the making of silks emblazoned with their personal insignia, and immense numbers of such fabrics were woven and embroidered for cardinals and kings.

Lucca was the most illustrious centre of the silk industry, but it could never exercise a monopoly. Not only did the craft spread up the Arno valley to Florence, but another group of silk manufactures was based on the import of raw silk through the port of Venice. The silks of Lucchese had been heavy, richly embroidered fabrics. Those produced in Venice and the towns of Venezia were lighter and cheaper. Early in the fourteenth century the Venetian silk industry had been greatly assisted by the influx of refugees, driven from Lucca by the wars of the Guelphs and Ghibellines. The success of Venetian silks, like that of the 'new draperies' in the Low Countries, was closely related also to the simpler quality and lower price of the fabric produced and to its resulting appeal to a wider market.

The European silk industry at this time was not restricted to the manufactures in the environs of Lucca and Venice, though it had not achieved any great importance beyond these areas. Lucchese craftsmen had migrated not only to Venice. Some had gone to France, and it is just possible that a small silk manufacture had sprung up at Avignon, the seat of the papal court, and that the industry had already taken root at Lyon.[101] A silk manufacture was also beginning to develop to the north of the Alps, supplied with raw silk brought across the passes from Venice.

Other materials of the textile industry. Only the most rudimentary textile industry could be carried on without other materials, both vegetable and mineral. Foremost amongst these were the mordants, lyes and dyestuffs, the clays used in fulling, teasels and other devices for raising a nap on the cloth before shearing, and the gold and silver thread used in embroidering it.

The dyestuffs familiar two centuries earlier continued to provide much of the colour in medieval clothing and decoration. Woad, the most versatile and widely used had by the fourteenth century become a specialised crop in the vicinity of Amiens and Abbeville, in northern France. It was in the fourteenth century beginning to give way to indigo, which was now being imported in small quantities from Asia. Madder (*garantia*) was much grown in west Flanders, and saffron was also used. Production was somewhat localised, and there was thus an important trade in dyestuffs.[102] Some were produced more easily and cheaply beyond the boundaries of Europe, and were imported and marketed by the Genoese and Venetians; brazil-wood from India was in this way gaining in importance. In addition to the vegetable dyes, a scarlet dye was obtained from the kermes insect.

No less important than the dyestuffs themselves were the mordants, with which the fabric was prepared to receive them. Several mineral substances were used for this purpose, but the most favoured and the most important was alum. The alums formed a complex group of salts, found most often in areas of volcanic activity, and usually obtained from the shallow pits in the volcanic rocks. During the fourteenth century, much of the alum used by the textile industry of western Europe came from the Lipari Islands, particularly from volcanoes; from near Cordoba, in Andalusia, where it was important in the preparation of the soft Cordovan leather, and from Castile whence it was exported through Valencia.

Alum from the western Mediterranean enjoyed no high reputation, and was supplemented by a better quality product from the Middle East. In the later years of the thirteenth century the situation was revolutionised by the opening up of the Phocaean deposits, near the western coast of Asia Minor. The Genoese, who controlled this trade, were able to import this superior alum at, apparently, a lower price than was charged for the western.

Bright colours added very greatly to the value of a medieval fabric. The dyers were among the aristocrats of medieval labour, and were commonly organised independently of other cloth workers. It is very doubtful, however, whether the use of alum and of any but the commonest dyestuffs was known outside the textile manufacturing centres. The grey and brown fabrics, coarse in texture and dull in colour, were all that the smaller centres were capable of producing, or the peasants able to buy.

Tanning and leather industries

The tanning of leather and the dressing of skins and pelts was no less widespread than the weaving of cloth. Wherever cattle were reared there was tanning of leather, and if cattle were reared in large numbers, as they were in the alpine foothills and the Jura, the leather industry was proportionately greater. It required few materials. Oak bark to supply the tannin was widespread; alum was desirable, but not always used. Every small town had its tanyard, and much of the leather, like the greater part of the cloth, was used where it was produced. A few places, by reason of their greater volume or higher quality of production, sold into distant markets. But tanned hides were difficult to handle and were not prominent in long-distance trade. It was the thin, soft leather, commonly made from goat-skins, at once more valuable and more easily transported, that entered into commerce as Cordovan leather, or cordwain. This was a more localised product, and southern Spain, which had given its name to it, was still an important source.

The metal and mineral industries

The extractive and mineral industries were poorly developed during the Middle Ages, and little technical advance was made during the whole period. Amongst the difficulties faced by mining and quarrying were the lack of rock-cutting tools, the inability to pump water from the deeper mines, and the prevailing ignorance of the chemical and physical processes involved in smelting and refining the minerals.

The precious metals. Gold and silver were in universal demand, not only for the minting of currency, but also for the fabrication of art-work and the decoration of fabrics. The medieval miner was able to work deposits too small in size and too low in grade to attract the modern prospector, but he was restricted by his inability to penetrate more than a short distance below the surface, and by the inefficiency of his crushing, separating and smelting devices.

The deposits of the precious metals were few and small, and it is probable that the medieval miner had located and opened up most of them. Much of the gold obtained within Europe appears to have come from alluvial deposits, but Europe was never well endowed with gold reserves and the amounts obtained were probably quite small. Foremost amongst the alluvial workings were those along the headwaters of the Vltava, in the southern part of the Bohemian massif, and in the Ješenik mountains of northern Moravia. Other alluvial gold workings lay in the Black Forest, around the headwaters of the Rhine and in the alpine valleys of northern Italy. Silver reserves, on the other hand, were more abundant. They commonly occurred in association with lead, itself in great demand, and also, though the medieval miner was unaware of the fact, with zinc. Several large ore bodies had been developed, but there is unfortunately no means of estimating their production. The chief centres of production were the Harz Mountains, where the Rammelsberg mines had been active for several centuries but were now beginning to decline in production owing to the technical difficulties of extracting water from the pits; the Black Forest, the Erzgebirge of Saxony and the rich deposits in central and southern Bohemia, where the Kutná Hora region may have been one of the foremost silver producers in Europe at this time. Silver was being obtained from several small deposits in Austria, particularly in the Niedere Tauern, and the exploitation of the far richer deposits in the Slovak Rudohorie had already been begun.

Mining was well developed in the Balkan peninsula, and the argentiferous lead ores of Bosnia were worked at places with such strongly suggestive names as Olovo, Srebrnica and Rudnik.[103] The Kopaonik Mountains of Serbia, known to the Italian merchants as *Montagna della Argentino,* also yielded silver, much of which was transported across the rugged country which separated the mines from the Dalmatian ports, and lodes were opened up near Trepča and at Novo Brdo in Macedonia.

The ferrous metals. The mining and smelting of iron must have been during the Middle Ages the most widely practised and the least adequately documented of all branches of metallurgy. For this there were good reasons. Iron was both perishable and in universal demand. The nature of its raw materials, charcoal and ore, tended to isolate it from human settlements and to keep it mobile, and the iron-worker in general did not, like the peasant, hold lands in return for a fixed payment in kind; nor did he form guilds and fix prices. He tended rather to possess the right to cut lumber and dig for ore over the uncultivated lands of his lord, with whom his relations were often of a more intimate and personal kind than those of most peasants. Very few of the monastic rentals contain any reference to iron-working, not even those of monasteries which we know to have owned forges and refineries, and there is little evidence of long-distance trade in ironware.

Yet, the consumption of iron in arms and armour, hand-tools and agricultural implements must have been large. Neither Gothic sculpture nor the open-fields would have been possible without it. It was produced almost everywhere, but only the better qualities ever entered into long-distance trade.

The blast furnace had not come into use. The ore was smelted on a hearth and was hammered to expel as much slag as possible. Temperatures were too low to yield cast-iron, and control of the processes too uncertain to produce steel. The latter was either imported from the east, or fabricated in urban workshops by methods that were tedious, uncertain, and in some respects unnecessary. Ores varied as greatly in quality as did the techniques that were used to smelt them. Some areas acquired a high reputation for the quality of the metal they produced; the hills lying east of the Rhine, the eastern Alps, the mountains of northern Spain. These supplied iron to the more famous fabricating centres, such as Milan and Toledo. As a general rule, the excellence of the material was due to the quality of the ore rather than to the technical superiority of the iron-workers. In the early fourteenth century both the few centres of quality production and the many local forges were increasingly productive.

Iron-working was to be found over much of the Paris basin, where a low quality but easily extracted ore occurred in association with the limestone rocks. In Lorraine, the *fer fort*, a residual deposit on the plateau surfaces, was smelted. In Normandy, Berry, Poitou and even in Brittany iron was worked. The Carthusian monks had developed iron-working in the forests and mountains of Dauphiné; they had opened up the rich deposits of Allevard, established forges in the Bauges and Vercors massifs, and had made Grenoble the centre of an iron trade. In the Pyrenees of Béarn, Navarre and Ariège, iron-working had again developed, and not only supplied the towns of southern France with bar-iron, but also shipped it to the coastal regions of north-west Europe.

Iron-working had probably never ceased in the Rhineland since it was first established by the La Tène Celts. It was supported by ore-bodies of high quality, and assisted by the Rhine and its tributaries which permitted the bar-iron to be marketed at a relatively low cost. There were forges throughout the hilly areas which bordered the river, but the most important were in the region of Siegen, where the accumulations of slag from the medieval forges are numerous and widespread. Bog-ores were dredged from the northern marshes, and iron was dug from the harder rocks of the Harz Mountains, the Bohemian massif and the Swiss Alps.

The Austrian Alps had a long tradition of iron-working. The industry had been revived, if indeed it had ever lapsed, and was now being carried on by the *Eisenbauer* who had founded small villages along the valleys of the Niedere Tauern. The chief centres at this time were Eisenerz, with its large and easily worked deposits of high quality ore, and the towns of Leoben and Judenburg, but smelting had already spread to the Hüttenberg of Carinthia and even across the Karwanken into Krain.

At the same time iron-working was widespread in the hills of southern Bohemia, to the south-west and south of Prague, as well as in Moravia and Slovakia, in the Odra valley of Lower Silesia, and the hills of Upper Silesia, the Kraków Jura and Holy Cross Mountains. These simple iron-works have left a legacy of place-names, most of which can be assigned to the later Middle Ages.

At the same time the iron industry was developing in southern Sweden, under the influence, in all probability, of German entrepreneurs. Merchants of Lübeck were investing in Swedish enterprises at the beginning of the fourteenth century, and iron was being smelted at many places in central Sweden and shipped from Kalmar by the merchants of the Hanse.

Italy was also an important producer of iron and ironware. The ores came mainly from the island of Elba, which, in the words of Biringuccio, 'is so overflowing and rich in this ore that it surpasses every other place where it is found',[104] and from the Alps of Bergamo, Brescia and Tyrol. The Elban ore was shipped to the mainland and smelted with charcoal from the forests of Tuscany to produce the highest quality iron. But ore was also extracted from the Massa Marittima and from the mountains of Calabria and Basilicata. Italy was also supplied with iron from the ports of Dalmatia, whose merchants in turn had purchased it from the metal workers of Bosnia and Serbia.

The majority of the small and primitive *forgiae errantes* produced only such simple ironware as ploughshares and coulters, spades and domestic hardware. The few areas of more concentrated iron production, such as Tuscany, the Italian and Austrian Alps, and the Siegerland made bar-iron for the urban crafts. No town, in all probability, was entirely without some form of iron-fabricating industry. Often it was no more than blacksmithing. In some towns the iron-processing industries were more highly developed, and those who practised them were organised into specialised guilds, and the lists of tolls payable on river and road traffic rarely omitted bar-iron. Paris had no less than eight iron-working guilds, ranging from blacksmiths to armourers. In all the larger towns of the Low Countries and the Rhineland we find the manufacture of tools, weapons and armour. One town might be noted for its swords, another for its knives, yet another for the quality of its helmets or breastplates. Cologne was outstanding in Germany for its metal work, but the most noted centre for the manufacture of weapons and armour was Milan. The city derived its iron from the forges of the nearby Alps, but 'owed its excellence to the personal skill of the craftsmen', who worked up the metal in over a hundred workshops.[105]

The non-ferrous metals. Lead, copper and tin played an important role in medieval life. They were easily smelted, and lent themselves to a variety of purposes for which cast-iron, mild steel, glass and ceramics are now used. Lead was used as a roofing material, and also for pipes and water conduits. Copper was beaten into sheets or alloyed with tin to make bronze or with calamine for brass. These metals did not occur widely, like iron-ore, nor was any landholder free to exploit them. They were, with the exception of zinc, regalian metals, belonging to the monarch, who alone authorised their extraction in return for a 'royalty' for himself.

Lead was obtained from a restricted number of mining areas. Apart from those of England, the most important in western Europe were the Ardennes, to the south-east of Liège; the Moselle valley, near Bernkastel; the hills of Nassau and the northern Black Forest; the Harz, the Erzgebirge of Saxony and Bohemia, and southern Spain. It was also mined in the mountains of Bosnia, along with silver, and shipped from the mouth of the Neretva.

The evidence for medieval copper mining is sparse. Copper was obtained from most of those mineralised regions which yielded lead. It was important in the Harz Mountains and at Mansfeld, in Saxony, and it had been recently discovered at Falun, in Sweden. It was found in the Ore Mountains and in Slovakia, and was of secondary importance in Bosnia, southern Italy and the Alps, but southern Spain was possibly the largest producer in Europe.

Tin was chiefly used in making bronze and pewter. It is one of the most narrowly localised of minerals, and during the Middle Ages production was in practice restricted to Cornwall and Bohemia. Zinc in the form of calamine ($ZnCO_3$) was a widely distributed ore that does not appear to have been smelted before the sixteenth century. It was often found in association with lead, and as an 'earth' was added to the copper which absorbed sufficient of the metal to form the alloy, brass.

Smelting of the non-ferrous metals, like that of iron, was a rural occupation but the making of copper alloys and the fabrication of pots and pans was carried on in many of the larger towns. There was, however, one centre of outstanding importance, Dinant, on the Meuse, in the bishopric of Liège. Its beaten copperware, often known as *dinanderie*, was fabricated from copper obtained generally from the Harz, with tin from Cornwall, and local zinc ores from the Ardennes, and sold in the towns of the Low Countries.

Building materials. About 1300 a period of intense building activity was drawing to a close. The cathedrals, monasteries and castles had been built. An immense number of towns, perhaps 3,000 of them, had been surrounded with walls, and most had been further adorned with gatehouses, towers and public buildings. Even if most domestic building was in wood the quarrying of stone must nevertheless have been, at least in terms of employment, the most important extractive industry. The unskilled work both in the quarry and at the building site was probably in all cases done by such local labour as could be impressed for the purpose. But there were groups of skilled architects and masons, who moved from one building operation to another, carrying with them tricks of style which help to delimit regional styles of building.

Most of the stone used in medieval building can have travelled at most only a mile or two from the quarry to the building site, and only the wealthy could afford the high cost of transporting a particular building stone over any considerable distance. Gothic architecture made excessive demands. Window tracery required a free-cutting stone, and tall, thin walls could not safely be built of rubble masonry. It is no accident that the finest examples of gothic architecture had been built in areas well endowed with good limestone. Paris, for example, was able to obtain its preferred stone, the *calcaire de Brie*, from the plateau to the east, and to transport it by river to the city. Not infrequently a monastery was granted the right to use a certain quarry for its building operations, and no doubt was faced in many cases with the difficult problem of deciding whether it could afford the transport costs.

Even when building accounts survive, it is not always easy to discern the actual source of the stone. A glance at any surviving medieval building is likely to reveal a predominance of local materials, however unsuited they might be for building construction. All too often a building style was adapted to the medium, as in the round, flint-built church towers, and the simplified gothic tracery executed in intractable granite. The humbler medieval builders of town walls and parish churches rejected nothing. Masonry was re-used as regularly as scrap metal is today. Boulders from the moraine, soft chalk and the irregular nodules of flint were all material for the medieval mason.

Public buildings demanded something better than coarse materials such as these, and the building accounts of the Cloth Hall at Ypres show that large quantities were obtained from Bray-en-Hainaut, a distance of some 60 miles over which waterborne transport could have been of little help. Ornamental stones were carried even greater distances. The Caen stone of Normandy was used in southern England, and Venice and towns in the plain of Lombardy actually obtained much of their building material from quarries along the Dalmatian coast. The volcanic rock from near Andernach, on the Rhine, was distributed not only over much of the Rhineland for use as millstones, but even reached northern England, and the so-called marble of Tournai was widely used for decorative columns and carved fonts.

In northern Europe, from Flanders to Poland, the problem of obtaining a suitable building material was greatly increased by the thick cover of moraine and diluvial deposits under which the older structures were buried. Deposits which yielded no freestone were at least likely to supply clay, and across the northern plain of Europe bricks and clay tiles tended to replace ashlar and rubble masonry, as in the brick gothic construction in Lübeck, Rostock and Danzig.

No less important than the actual masonry was the lime-mortar in which it was set. Lime-burning must have been carried on wherever limestone outcropped in suitable proximity to fuel supplies, for the limekiln was a voracious consumer of wood. Ypres appears to have obtained its lime by cart from kilns built on the chalk of Artois, near Calais and Saint-Omer, but the lime over most of the north European plain and the whole of Scandinavia can have been supplied only by imports by sea. Of this aspect of medieval industry all too little is known.

Salt and other minerals. The medieval alchemist used a variety of salts and minerals, whose true nature he rarely understood. Most entered into commerce on such a minute scale that they are omitted from this discussion. There were, however, a number of minerals which held some value as decorative or colouring material and as the raw material of the glass and ceramic industries. Among these were alabaster and marble which, towards the end of the Middle Ages, began to be used in sculpture. Their occurrence was highly localised, and, unlike ordinary building materials, they were valuable enough to bear the cost of long distance transport.

Pottery clay was more widely distributed, and coarse ware was made in very many areas for local use. A few areas, such as the lower Rhineland which appears to have inherited the industry from the Roman period (see p. 153), carried on a more specialised and more sophisticated manufacture. Pottery, however, did not enter significantly into long-distance trade, and the wine, which in classical times had been transported in amphorae, was usually carried in wooden barrels.

Glass manufacture was more specialised, demanding more elaborate materials and a more sophisticated technique. It was, in consequence, carried on less extensively. Constantinople had remained a centre of glass-making, and much of the inspiration of medieval glass manufacture derives from the Byzantine or Arab example. Glass was, however, made at Venice and Genoa, at Trier and in Lorraine and in Normandy. Though vases, vessels and figures were made, a great deal of the glass-maker's energies went into the production of window glass, the quality and richness of which had developed greatly during

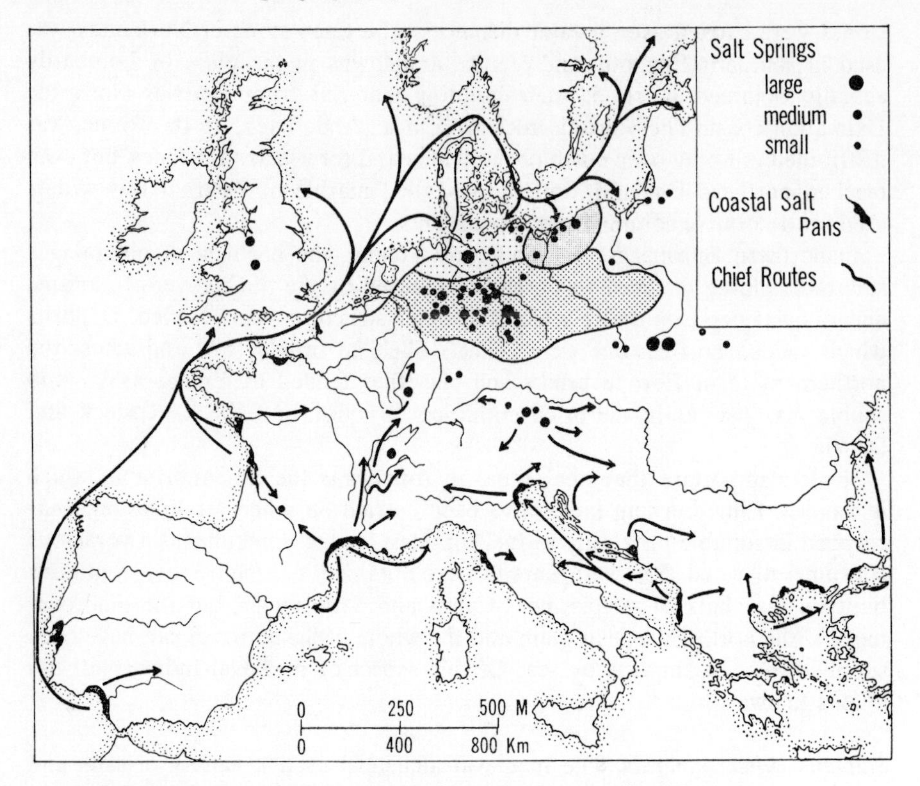

Fig.6.17 Salt production and the salt trade in late medieval Europe

the previous century. Less important, but nevertheless widely used, were mineral pigments, mostly oxides and salts, used in colouring glass, making enamels and preparing paint and tempera.

Most important by far of these mineral substances in the medieval economy was salt. Quite apart from the needs of the human and animal organism, it was used in immense quantities to preserve fish and meat, and on a much smaller scale industrially in the manufacture of ceramics, the preservation of hides, and in numerous medical and alchemical preparations.

Two sources supplied most of the salt of medieval commerce, the salt-pans around the coast and the salt springs of inland regions. To these should be added the salt that was obtained by burning the fenland peat of the Low Countries and removing by solution the salt that remained in the ash.

Coastal salt-pans were the cheapest and most abundant source of salt. They relied upon the natural evaporation of the sea-water, which was admitted gradually into the shallow pans, until enough salt had formed to be scraped up. It was impure, and was sometimes further refined near the place of consumption by being dissolved and again evaporated. This, however, usually required a furnace and was dependent upon an adequate supply of fuel. Hot, dry summers and a flat coastline with firm beaches were desirable. Though salt-pans were found even in the cool, humid conditions of north-west Europe, most lay along the shores of southern Europe. They were found in Macedonia and

Thrace, along the shores of Greece, and to a greater extent on the coast of Albania and at the mouth of the Neretva river. The early fortunes of Venice were founded on the salt trade, and the lagoons continued through the later Middle Ages to furnish salt. Salt was obtained by evaporation around the coast of much of the Mediterranean Sea. There were, however, three important centres of production: the coast of Provence; south-western Spain and southern Portugal, and the west coast of France.

The first, extending from Toulon into Cataluña, produced large quantities of salt, much of which was sent up the Rhône valley. The flat coast of southern Portugal and the region around the mouth of the Guadalquivir were somewhat less important sources of salt, because of their greater distance from the markets. They were, however, gaining in importance in the later Middle Ages.

Though salt was obtained from coastal salt-pans as far north as Normandy, the northern limit of large-scale production was the Bay of Bourgneuf, 'the Bay' *par excellence.* It lay a few miles to the south of the Loire mouth, on the boundary between Brittany and Poitou. Its relative nearness to the markets in north-west Europe and immunity from military disturbance gave it a considerable advantage so that its market was at this time being extended at the expense of inland sources of salt. Salt was also produced to the south, in Aunis and Saintonge, where physical conditions were, if anything, better.

The market for 'Bay' salt had been increasing in Great Britain since the beginning of the century, and was by 1330 spreading into the Low Countries and Baltic. Coastal salt-pans seem to have been no longer used in north-western Europe, and the domestic supply was derived by burning the rapidly diminishing reserves of fenland peat and by evaporating the brine from salt springs. The latter lay in three groups. The first was near the coast of the North and Baltic Seas, and included the salt springs of Lüneburg, the most important of them all, and of Greifswald and Kołobrzeg (fig. 6.17). The second lay to the south and formed a belt extending from Westphalia, through Saxony into Silesia and southern Poland. It included the salt springs of the Bishop of Paderborn at Salzkotten and of the Bishop of Minden at Salzuflen.

The salt springs of the eastern Alps constituted a third group. The chief centres were near Salzburg, where salt had been worked at least since the early Iron Age, and also near Aussee, some 30 miles to the east. The latter was well placed to distribute salt to Styria and eastern Carinthia, and was at this time growing steadily in importance.

In addition, the salt springs of Lorraine and of the Jura, particularly of Château Salins, remained of local importance, but throughout western Europe river transport was bringing marine salt, which seems in general to have been both cheaper and purer, to all places that were accessible to it. Liège, for instance, received its salt from the coast by the river Meuse, and the Loire and Rhône bore salt upstream from the salt-pans near their mouths.

Fuel. Wood was universally used as fuel in medieval Europe, and for this reason alone the woodlands were essential to the medieval economy. The wood was reduced to charcoal not only for use as a metallurgical fuel but also for ease of transport. It is likely that a significant part of the fuel burned in the baking ovens and by the craftsmen in the larger towns was in this form. In Paris at the end of the thirteenth century there were, for example, no less than 23 charcoal merchants, and they must also have been numerous in all towns except the smallest.

Charcoal and wood were supplemented by peat and turf. Peat-cutting was already important in the Low Countries, and satisfied a considerable part of the fuel needs of the towns. But turbaries were also used in other parts of northern Europe. At Amiens the citizens were in 1341 authorised to cut turf in a lowland area – presumably in the Somme valley – that could no longer be used for grazing (see above, p. 13). Rights to peat, and also to the rough wood found along the damp valley bottoms, were jealously preserved.

The growing scarcity and high price of timber, at least in the more urbanised parts of Europe, was in the fourteenth century directing people's attention to coal. Coal outcropped in many places, and could be recognised and dug as easily as peat, at least until the accumulation of water forced the abandonment of the pits. No doubt the peasants scratched the surface for coal, but for a coal trade there is little evidence except in the populous regions of north-western Europe. Coal was being mined in the early fourteenth century in the Meuse valley, near Liège, where the abbey of Val-Bénoit reserved for itself all mining rights on its lands, and in the vicinity of Mons, where the abbey of St Ghislain controlled the coal from its own estates.

There were probably other centres of coal-mining between these two extremities of the exposed Belgian field, and coal was also obtained, doubtless in very small quantities, from the small coalfields of the Lyonnais, Franche Comté and the south of France. It is nevertheless surprising to find that there were 25 coal merchants – greater than the number of charcoal dealers – in Paris about 1300.[106] The known coalfields of France were not at this time accessible by boat from Paris, and we may assume that coal, at least in quantities sufficient to employ so many merchants, was not sent overland to Paris. It is probable, therefore, that it came from the Liège area, and was taken down the Meuse, along the coast and up the Seine. This is confirmed by the appearance of coal in the tables of tolls levied at Mantes, between Paris and Rouen, on boats using the Seine.[107]

TRANSPORT AND TRADE

Despite the prevailing self-sufficiency of the medieval economy, there was a considerable though indeterminate amount of trade, both local and long-distance. To this every village community contributed a little. The rents and dues which it paid in kind went to supply the monastery or castle and the surplus was fed into the market. The transition to money rents, which was taking place in western and central Europe in the early fourteenth century, led to an increased volume of trade. The peasants sold directly to the market, and used part of the money they received to discharge the obligations to their lords, who were turning increasingly to the market to satisfy their needs for foodstuffs and other goods.

Means of transport were still poorly developed, and, as a general rule, only the lighter and more valuable goods entered into long-distance trade. Bulk goods, like grain and lumber, charcoal and building stone, made only short journeys from producer to consumer. There were, however, exceptions. River navigation had developed greatly, and there was an important coast-wise trade around most of Europe's coastline. This greatly facilitated the handling of bulky and low-valued goods, so that many of the larger urban centres were supplied with grain in this way. The increasing facility with which such specialised products as wine and salt could be obtained from distant sources had already led to the abandonment of many of the marginal and less profitable centres of production.

Local trade

Local trade lay at the foundation of the medieval commercial structure. It operated through local markets; collecting the agricultural surpluses of the village communities, and supplying such goods as salt, cloth and iron, which the peasants could not provide for themselves. Nothing is more indicative of economic growth in the later Middle Ages than the rapid increase in the number of markets. They constituted the dominant economic function of most of the dwarf towns which were founded in Germany, and in France and England they were often established at village sites.

The establishment of markets was a response to an economic need; it was also a source of revenue to the seigneur, who thus obtained market tolls. A market was usually authorised by a formal act, occasionally after some kind of inquiry into local conditions. In England this most often took the form of a charge to the sheriff to discover whether a proposed market would be to the detriment of existing markets. It became the practice here to assume that markets should not be closer than $6\frac{2}{3}$ miles to one another. This rule, in effect, defined the maximum density of markets. It did not, however, prevent many from succumbing to the competition of their neighbours, nor did it ensure that the density of markets even approached this level in all areas.

The density of markets was governed by the normal length of a peasant's journey with his cart. That this optimum density was not realised in much of Europe is apparent, and it must be assumed that the fewer the markets, the greater was the degree of local self-sufficiency and less the dependence on a money economy. At the local markets the peasants sold to one another, but commerce was dominated by the petty traders who bought up local surpluses of grain and animal products, and made available the few specialised goods needed by the peasants.

The peasants fed their products into the market in two ways, directly and by way of the rents and dues which they paid in kind to their lay or ecclesiastical seigneurs. The latter sometimes passed part on to the market, and, conversely, were sometimes compelled to turn to the market to supplement their income in kind. In the early fourteenth century an increasing proportion of the product of the rural community was, at least in western Europe, tending to move directly to the market, while feudal obligations were discharged by cash payments.

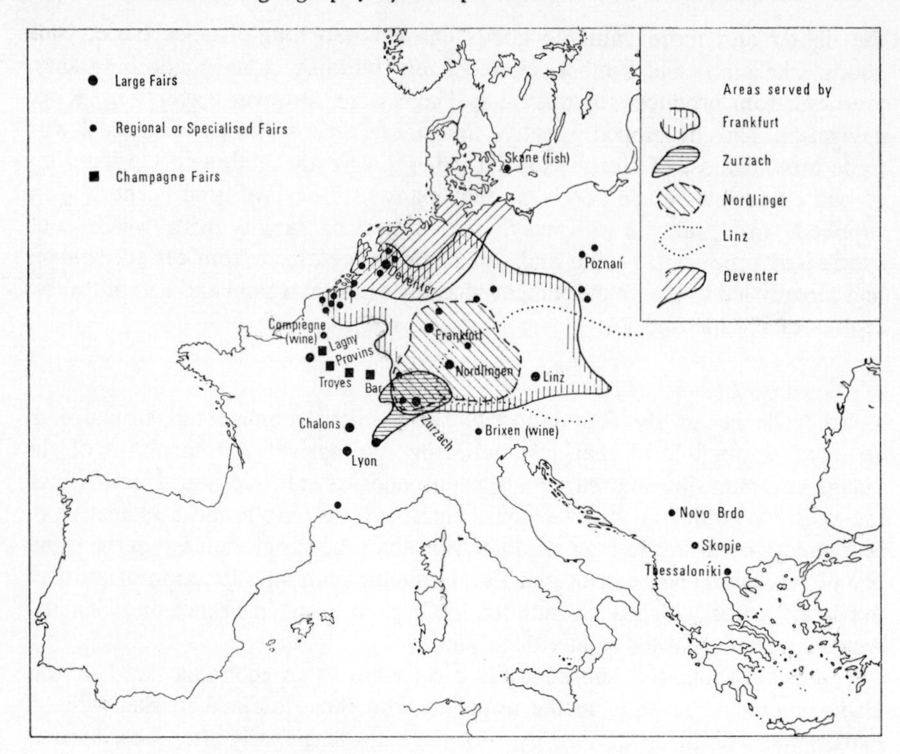

Fig.6.18 European fairs in the fourteenth and fifteenth centuries, and the areas which they served (The Champagne fairs were of very slight importance by the early fourteenth century.)

Long-distance trade

Goods handled in the local market were, as a general rule, bulky and of low value, and they never travelled very far from producer to consumer. It has often been assumed, on the other hand, that long-distance trade was mainly in goods of small bulk and high value. 'The glitter and glamour of trade in luxuries ought not to overshadow the heavy trade in cheap bulky goods that went on at the same time.'[108] The local market regularly fed goods into the regional and inter-regional markets, and the greater part of this commerce consisted of primary products, like grain, wool, lumber, wine and salt.

Fairs. It is sometimes assumed that fairs played a role in long-distance trade analogous to that of markets in local trade. They were, in fact, only one of several means of trade, and their role was changing in the early fourteenth century and their importance on the whole declining. Fairs were annual, or at least periodic gatherings of merchants for a short period of intense commercial activity. The periodicity of some types of economic activity, such as the wine-harvest, continued to favour annual fairs, such as that of Compiègne, at which the wines of the Paris basin were sold to merchants of the Low Countries, and the herring fair of Scania. But fairs were now proving to be less useful as a medium of exchange for goods in regular and continuing demand.

This was particularly conspicuous in the changing role of the fairs of Champagne. These fairs, which were held in turn at the towns of Lagny, Provins, Troyes and Bar-sur-Aube, had served for a good deal more than a century as the chief meeting place for Italian merchants with those of Flanders and northern France. Much of the cloth exported by the Low Countries had been handled in the Champagne fairs. But several changes had taken place in the late thirteenth and early fourteenth century. The opening of more easterly alpine passes had led much of this commerce to take a more direct route by way of the Rhine valley and St Gotthard Pass; Champagne had ceased to be an island of security and peace in a disturbed Europe. Above all, both the scale and the mechanics of Italian trade were changing. The richer and more important merchants now sat at their desks in Florence or Milan, and employed agents or factors in the northern cities, instead of themselves facing the confusion of the fairs and the dangers of travel.

But the age of fairs was far from over. That of Lendit, near Paris, continued to prosper, perhaps because it had become an appendage to the city of Paris, whose commercial facilities it augmented at fair time. The role of the Champagne fairs was in some measure taken over by those which were growing up in the Rhône valley and the Rhineland. This eastward shift in the activity of the fairs is in line with the change in the geographical pattern of trade. It reflects also the new structure of industry, with the emergence of a large number of small manufacturing centres, especially of cloth, in the Rhineland and south Germany.

None of the newer fairs, however, could rival those of Champagne in the volume of their trade or the extent of their commercial ramifications. Each served a very much smaller area, and its chief function was to link a restricted area of small but mostly urban manufacturing centres with the European market.

The largest and most important of these fairs lay in south Germany, where the need was greatest for the kind of service which they provided. The Frankfurt fairs served the largest area. The navigable rivers which converged on the Frankfurt region gave them relatively easy connections with much of western and central Europe, so that Frankfurt served to collect merchandise from as far away as Zeeland and Bohemia.

The chief fair in south Germany was at Nördlingen, situated between Nürnberg and Ulm. It served a more restricted area than that of Frankfurt, consisting in the main of south-western Germany. Towards the south its area overlapped that of the important Swiss fair of Zurzach, and the sphere of the latter was being cut into by the Geneva fair, which at this time was growing in importance and competing also with that of Chalon-sur-Saône.

Several smaller fairs existed in south Germany, either supplementary to the trade of the towns where they grew up, or subordinate to the larger fairs of the region. Among them were the fairs of Friedberg, to the north of Frankfurt; of Nürnberg. Strasbourg and Basel. The fair of Linz, on the Danube, was a meeting place for merchants who did business in the middle Danube valley, and the Bozen fair in the southern Tirol was primarily a fair for Italian wine which was sold in Austria and Germany.

The chief single item of commerce at the south German fairs was cloth, which was collected here for export either to Italy or northern Europe. At the northern end of the Rhine commercial axis there developed another, though smaller group of fairs. Foremost amongst them was that of Deventer, situated on the navigable river IJssel, which linked the Rhine with the Zuider Zee and thus with the shipping routes of northern Europe. The older fair at Utrecht was of diminishing importance, but small fairs were also held in the towns of Zwolle, Zutphen and Arnhem.

Two centuries earlier fairs had been an important means of trade even within Flanders, but trade in woollen cloth had largely passed to the towns. The fairs, however, continued to be held, and others grew up, particularly in Brabant, but they were concerned by and large with farm products, and horses and livestock were prominent amongst the items traded.

Fairs played an essential role in regions where the volume of production and the amount of trade were insufficient to justify permanent institutions. For this reason they were declining in western Europe, just as they were increasing in importance in central and eastern Europe and in the Balkans, where surplus production was small, and was best handled at periodic gatherings of merchants. Fairs had thus grown up at Poznań and Gniezno in Poland, and in the Balkans they were perhaps the chief means of conducting trade. Not only were there important fairs at Thessaloníki, to which goods were brought from Macedonia, but also at a number of inland centres, including Prizren, Skopje and the mining centre of Novo Brdo.

Commercial towns. The role which the fairs had formerly played was in the fourteenth century passing gradually to the towns. Long-distance commerce was becoming a continuing, not an intermittent process, and periodic fairs had long ceased to be adequate. Petty merchants in the small towns sold to merchants who operated on a larger scale in towns of intermediate size, while in the largest towns were to be found the agents of the Italian commercial and banking companies.

All towns were obliged to purchase at least a part of their food supply. In the case of the smaller this could normally be obtained through the medium of the local market from neighbouring villages. Most of the larger towns were in some degree dependent upon long-distance trade for the supply of the necessities of life. In the case of the largest, the purchase and transmission of grain and other foodstuffs not only employed a sizeable body of merchants, but was also a matter of continual concern to the city government (see below, p. 426).

The supply of industrial raw materials was second in importance only to the procurement of food. Sources of such goods were often fairly restricted, and the supply of some was conditioned by political factors. The diplomatic correspondence of the time is filled with promises to respect or protect traders and their merchandise. Towns, lastly, purchased a very wide range of wares, which, if not all belonging to the categories of luxury goods and *objets d'art*, nevertheless had only a narrow market among the rich and the privileged.

Towns lived by the sale of their manufactured goods and by the services which their citizens performed in banking and finance and in wholesale and retail trade. In most cities there was also an inflow, in cash and in kind, in the form of rents from rural land-ownership. This was, relatively at least, most important in the Italian cities; least so in those of eastern and northern Europe. It was, however, the major source of income of urban religious institutions in most parts of Europe. Only in the rarest cases is it possible, on the basis of surviving accounts, to assess the relative importance of manufacturing, services and rents in the support of a town. Those in which manufacturing was the basic occupation probably constituted a minority. In many the service function appears to have been uppermost. The small cluster of towns along the IJssel river appear, for example, to have been heavily engaged in handling the commerce between the northern towns and the Rhineland, and a significant part of its population must have been engaged in loading and unloading seagoing vessels and in operating river craft. The total employment in legal, if unnecessary operations, such as the exercise of the *Stapelrecht* in river towns, is incalculable, but must have been large.

Long-distance trade was overwhelmingly between towns and was carried on by urban merchants. One is constantly amazed, in reading the records of their transactions, at the immense distances which they covered, and at the complex pattern of trade which resulted. Certain themes in fourteenth-century commerce stand out; the movement of English wool to neighbouring parts of the continent; the trade in the cloth of the Low Countries to the Mediterranean and Baltic regions; the distributions of certain specialised products, from silks and spices to woad and alum. But beneath this dominant pattern of trade, which changed only slowly, was a confusing and fluctuating movement of goods of all kinds.

This secondary trade can be illustrated only by a few examples. Dollinger has published a study of the commercial relations of Strasbourg and Fribourg in Switzerland.[109] Merchants of the former shipped chiefly wool and cloth, but also salted fish and dyestuffs to the latter, and bought there hides and skins, in which Fribourg had established something more than a local reputation, and also cloth. The extent to which Strasbourg cloth was sold in Fribourg, a town with a quite considerable cloth industry of its own, would be truly surprising if it could not be paralleled in other places.

In fact, different towns prided themselves on the peculiarities of colour, size and texture in their products, and jealously protected them from imitation elsewhere. Inter-urban trade in those commodities which they all produced in common was not necessary, but it was clearly encouraged by the preference of some consumers for the more distant product and by their ability to pay for it.

The merchants. Between the eleventh and the fourteenth century trade had become big business, in which a small number of merchant houses dominated. Most of the richest were Italian. De Roover has prepared a reference list of about 200 Italian commercial and banking companies.[110] Not all were active about 1330; nor, on the other hand, is the list complete. Some were highly capitalised, and Giovanni Villani wrote in glowing terms of the wealth of the Acciaiuoli of Florence for whom he worked.[111] The wealthier maintained agents or factors in the commercial cities of western Europe. The Peruzzi, also of Florence, in 1336 had no less than 88 such agents distributed over at least 17 cities, and five years later the Acciaiuoli had 53 in as many places. These ranged from Cyprus and Chiarenza in the Peloponnese to London, Paris and Bruges.[112]

In the fourteenth century Paris became an important centre of activity for the Italian merchants and their agents, who had abandoned the Champagne fairs for the greater continuity and doubtless also comfort of operating in a large city. The 'Lombards' were also numerous and active in the Low Countries, where a number of them were summoned in 1309 to discuss a tax-levy with the Emperor Henry VII. In Lorraine, Florentines even leased the salt-works at Moyenvic and Marsal, belonging to the Bishop of Metz, and Italians from Asti and Chieri were active also in banking and money changing. The Italian merchant houses controlled the trade of southern Italy; they were to be found in the Mediterranean colonies of the Genoese and Venetians, whose ships they used, and they played a dominant role in remitting to Avignon the proceeds of papal taxation throughout western Christendom. There is said to have been a colony of no less than 5,000 Italians at Avignon, a large number of whom were merchants or financiers.

Italian merchants dominated trade in the eastern Mediterranean, except along the coast of the Levant, where they were at the mercy of the Arabs. Both Genoese and Venetians had their quarters in Constantinople; they maintained trading stations in the Black Sea; the Genoese monopolised the Phocaean alum mines, and Venetian merchants were established in Thessaloníki, as well as in their colonies in the Peloponnese. Italians controlled the shipment of goods from the Balkan and Black Sea ports, but it is not likely that they played more than a minor role in bringing the goods *to* the ports. The latter were thoroughly cosmopolitan, and among the peoples who travelled the hinterland in search of merchandise to sell to the Italians, the Armenians were the most numerous and important.

The Italians also dominated, where they did not control, the trade of the Dalmatian coast. It was Italian merchants who bought the silver and lead of Bosnia and the wool and hides of Serbia, but they served mainly as middlemen, and the actual commerce of the interior of the Balkans was largely in the hands of the slavicised Dalmatians and the Vlachs, who formed temporary associations for the purpose.

The Italians were the wealthiest and the most conspicuous of the merchants engaged in long-distance trade at this time, but they were by no means alone. The operations of the Bonis brothers of Montauban are known through their account books. And if such a company had its seat at a town of no great size or consequence such as this, there must have been hundreds of others operating out of all except the smallest towns of western and central Europe. Even in Italy the Italian merchants were far from monopolising commerce. A tabulation of merchants active in Genoa shows, among the numerous foreigners, a predominance of Flemings, but French merchants from Troyes, Provins, and Verdun were present, as well as men from Liège and Hollanders, Germans, Swiss and English.

Italians and Germans predominated in the trade which was now beginning to move in increasing quantities across the more easterly of the alpine passes. From early in the thirteenth century the German merchants had possessed a house in Venice, where they stored their merchandise, conducted their business, and lived as long as they were in the city. The number of German merchants frequenting Venice grew, and early in the fourteenth century, the *Fondaco*, a massive, masonry structure beside the Grand Canal, was rebuilt and enlarged. What is known of its activities suggests that the number of German merchants using it was considerable, and the volume of merchandise passing through it large (see below). In the later Middle Ages, the membership of the *Fondaco* included merchants from as far away as Cologne, Saxony, Silesia and Moravia.

In the later fourteenth century, the clothing industry of south Germany (see above, p. 391) became more organised. Previously its merchants had sent their linen and other goods to the Champagne fairs; now they shipped it down the Rhine to the ports and cities of the Low Countries, or across the central and eastern passes of the Alps to Italy. Milan, near the approach to both the Simplon and the St Gotthard passes, became the most important of their southern emporia. German merchants, operating mainly from such cities as Cologne, Frankfurt and Nürnberg, were becoming very active in eastern Europe; they carried on a considerable trade with Poland, and had become the middlemen in the trade between the Czech lands and Venice.

Hanseatic League

The greatest participation of German merchants in European commerce at this date was through the Hanseatic League. Since early in the twelfth century they had dominated the growing trade with the Baltic region. They had never tended to form large, highly capitalised companies, such as were a feature of Italian commercial organisation. Instead, they usually operated individually. This left them in a relatively weak position, but for this they compensated by establishing a dominant position within the cities. The policy of the quasi-independent city-states of north Germany became, in effect, the policy of the merchants who lived there. During the thirteenth century a number of urban leagues had been formed to protect the interests of the towns. Many of the Westphalian, Saxon and Wendish towns were united in this way. These groups clearly had a common purpose – the preservation of their independence and the prosecution of their trade. From the end of the thirteenth century they began to adopt a common policy. Thus the 'Hanse of Cities' came gradually into being, an informal but nevertheless effective organisation dedicated to the pursuit of trade in the Baltic and north European region.

About 1330 the Hanseatic towns were struggling – in the end successfully – to make good their independence of their territorial lords. They were about to face the more formidable opposition of the Kingdom of Denmark, whose territories lay astride the Sound, the most used route between the Baltic and North Seas. The hostility and insecurity with which the merchant in northern Europe was faced in the later Middle Ages ultimately obliged the Hanseatic cities to replace their informal association with a more precise constitution, but this lay forty years in the future. The circumstances which made this necessary

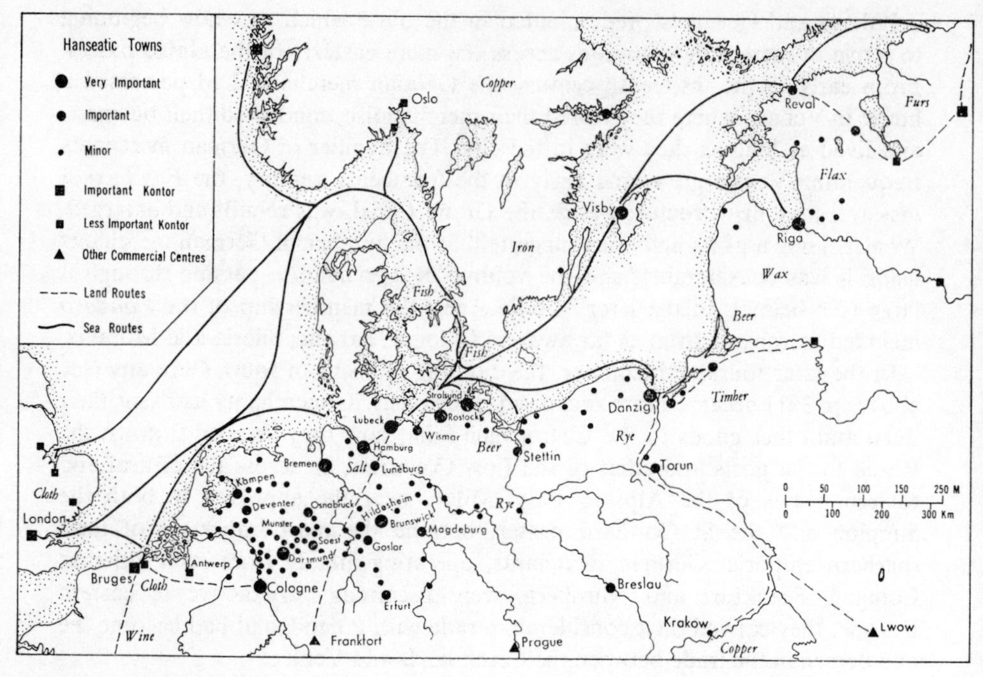

Fig.6.19 The Hanseatic League in the fourteenth century (Its composition changed later.)

were, as Postan has emphasised,[113] an indication of the worsening political and economic conditions during the later Middle Ages.

The German merchants of the Hanse exercised a somewhat insecure control over the trade of the Baltic. They had 'factories' at Novgorod in Russia, and at Polotsk and Kaunas in Lithuania, and most of the coastal and riverine cities from Dorpat and Reval to Lübeck and Kiel were members of the association. The small ports of the lower Rhine and IJssel mostly belonged to the League, but the commercial and industrial cities of Flanders, which provided a significant part of the eastbound freight of the ships of the Hanse, were not. Bruges and its outports; Antwerp; London, and the ports of the east coast of England contained factories at which, as at the *Kontor* in Novgorod and the *Fondaco* in Venice, the German merchants were able to conduct their business. Indeed the Germans constituted the most important group of foreign merchants and outnumbered the Italians in the towns of the Low Countries and England.

English merchants were at this time increasingly important in the ports of the Low Countries, to which they shipped their 'staple' product, wool, and, though on a small scale at this date, 'adventured' their woollen cloth. In the British Isles, the merchants of the Hanse were the most frequent visitors, though small groups of Italians were beginning to make the hazardous sea voyage by way of the Strait of Gibraltar and the Atlantic Ocean to Flanders and Britain. Though English ships and merchants occasionally penetrated to the markets of Scandinavia and Iceland, this was predominantly a preserve for the merchants of the Hanse. Bergen, though not a member of the League, was 'a mart town for al the merchants of Germanie',[114] and Wisby on the island of Gotland, an almost exclusively German town, played a similar role for Sweden and the north Baltic.

Trade routes and travel

It is a commonplace of medieval history that travel was difficult and trade obstructed by tolls and threatened by insecurity and violence. It is a remarkable tribute to the persistence of the merchant class and the pressure of consumer demand that commerce was able to develop as much as it did by the fourteenth century. Goods were moved by road, river and sea. It is impossible to measure with any precision the relative importance of these three means of transport, but rivers were relatively more important than in later centuries, and maritime transport probably less so.

The road system. The medieval road system fluctuated and was poorly maintained. Nevertheless, a 'system' of roads was unquestionably in existence by the fourteenth century. It was known to travellers and recorded in numerous 'theoretical' itineraries[115] and descriptions of travel. It had developed during the previous five or more centuries in response to the needs of travellers. In general it owed little to the system which the Romans had created. Though it made use of Roman roads it did not in general allow itself to be shaped by them. The medieval road was 'essentially where one went'; usage made it a road, and it remained a road only as long as it was used as such. Between neighbouring towns there was usually not one road but several, and travellers chose between them, according to the weather, their means of travel and the state of the ground. In the following pages the word 'road' refers to the path actually traversed, while the term 'route' is used for the complex of alternate roads which linked neighbouring places.

The route was nevertheless reduced not infrequently to a single road, as, for example, where the variety of paths converged to cross a bridge or causeway or to traverse a mountain pass. The multiplication of roads between markets was easy in the heavily forested east of Europe, but in the west this encroachment on settled and cultivated land was likely to be resented and resisted, and it is unlikely that the road system was in fact quite as casual as this suggests. Nevertheless every important road had a number of variants, any of which could be changed or diverted, in the words of the Carolingian *capitularium* on the subject, 'ubi nunc necesse est'.[116]

There is little evidence that much was done to maintain the surface of medieval roads. Where they ran across naturally dry terrain, such as outcrops of limestone and patches of sand and gravel, they could probably be used by vehicles for most of the year. Where soils were of clay or in other ways poorly drained, this was far from the case. In the Franc de Bruges, in the Flanders plain, 'la terre est tele et si mole tenace et parfonde en la saison d'iver, que on n'y puet mener par charroy vivres, denrées, ne marchandises d'une ville à autre'.[117] At this time of year commerce had to rely exclusively on 'cours ou courans d'eaues . . . qui ont plusieurs branches servans aux villes dudit pays vers occident'.

In England the law placed the obligation to repair and maintain the so-called King's Highway fairly and squarely on the landowner across whose lands it passed, and there is evidence that the courts enforced this obligation. In France also the landowners appear to have had the duty of maintaining the roads but in general appear to have done very little about it. Bridges, which were no less important than the roads, were on the other hand the object of many benevolent bequests. Towns not infrequently owed their prosperity in part to a bridge, on which they sometimes lavished immense care and attention. Bridges were, however, relatively few and, though supplemented in some degrees by ferries, were of great importance in determining the course of the major routes.

It may be assumed that the traveller was likely to experience difficulties whenever his path crossed clay lowlands and the alluvial flood plains of the rivers. Nevertheless he often made remarkably good speeds, even though he customarily moved in a group made up of both horsemen and pedestrians, and even carts. Fifteen or sixteen miles a day was not unusual, and distances of thirty miles were sometimes covered. Transalpine travel, of which many records survive, was in general slower, and the traveller was usually willing, when the opportunity offered, to take a boat for part of his journey.

Ordinary travellers walked or rode on horses or mules. Merchants used pack animals or wagons, usually four-wheeled vehicles resembling the peasant's cart of eastern Europe today. The former were the principal means of carrying merchandise across the Alps, where the roads were often too steep, narrow and twisting to be negotiated by any kind of wheeled vehicle.

The European route system in the early fourteenth century can be reconstructed only from the itineraries and route books which have survived. These are numerous, but most give only the names of key places – often quite widely separated – along the route. There often remains considerable ambiguity regarding the actual road that is recommended. Compendia or guides for the assistance of merchants, such as the *Trattura* of Pegolotti,[118] also provide hints on the best routes for transporting merchandise. Of all these guides for merchants and travellers the so called *Itinerary of Bruges* is the most ambitious and complete.[119] This road-book was compiled in the late fourteenth century, and describes the road net radiating from the city of Bruges to all parts of Europe. The itinerary includes the well-known pilgrim routes to Campostello, Rome and Jerusalem, but it adds routes which formed no part of any established pilgrim itinerary and led rather to the sites of the more important fairs and commercial towns. The work appears to be a compilation derived from several sources and intended to serve a variety of purposes.

Its chief feature is the clustering of routes which ran from north-western Europe, between the mouth of the Seine and that of the Rhine, south-south-eastwards to northern Italy, thus linking the two major centres of industrial and commercial development at this time. Together they constituted the principal commercial axis of the later Middle Ages. They were continued through Rome into southern Italy; branches ran south-westwards encircling the Pyrenees and reaching the extremities of the Iberian peninsula, and eastwards to Prussia, Livonia, Bohemia, Hungary and the Byzantine Empire. The

Itinerary, for all its detail, presents an incomplete picture of the road system of late medieval Europe. Not only was there an immense number of lesser routes, but many highly significant routes were omitted.

The groups of routes between northern Italy and the Low Countries consisted essentially of the following:

1. Routes from the Paris basin to the Mediterranean coast by way of both the Massif Central and the Saône and Rhône valleys.

2. Routes from eastern France which used the cols of the Jura and the passes of the western Alps, primarily the Great and Little St Bernard passes, to reach the plain of Lombardy.

3. Routes which followed the valley of the Rhine and its tributaries, and used the central passes – particularly the St Gotthard and San Bernardino – or the more easterly passes of the Grisons and Tyrol, to reach Milan or Venice.

To these highly generalised routes must be added those which crossed the French Alps from the plain of Lombardy to the Rhône valley. The most important of these were the Little St Bernard Pass, between the Isère Valley and the Val d'Aosta, and the Mont Cenis Pass, which offered the more direct route from the Arc valley to the Dora Riparia and Turin. In the first half of the fourteenth century, however, the establishment of the Papacy at Avignon was attracting a heavier traffic across the passes of the French Alps which converged on the Durance valley, particularly the Col du Génèvre and the Col de Larche. The coastal road through Nice and also the route across the Col di Tenda to Cuneo were also used.

To the medieval traveller the alpine crossing was the most formidable part of his journey. The mountains were viewed with a peculiar horror. The choice of an alpine route was determined not only by its length, but also by the roughness of the track and the dangers to be expected from snow, ice and avalanches; by tolls and political restraints along the approach roads, and the availability of hostels for the overnight reception of travellers. The last had been in existence at least since Carolingian times, but in the later Middle Ages their number was increased, thus adding considerably to the range of choice available to travellers in deciding their route.

Anything like a traffic count is impossible for any of the passes, but the available records do suggest a changing pattern of use. The Great St Bernard and the Mount Cenis had previously been the most used, but the Simplon, a relatively low and easy pass with a southern approach of great difficulty, and the St Gotthard came into use in the thirteenth century. The latter, opened up by the construction of a road through the Reuss gorge on its northern approach, permitted the whole mountain range to be crossed by one relatively simple ascent and descent. Not only did it have a profound effect on the political development of the Swiss, but it was the shortest and very quickly became the most used of the alpine routes.

Of the numerous possible routes between Italy and the upper Rhine valley, the Lukmanier, San Bernardino, Splügen, Septimer and Julier were all used. The Septimer, possibly the most frequently used of this group, had recently been opened up to wheeled vehicles. Both the Splügen and the San Bernardino suffered from the difficulties of their approach road, the *Via Mala*, along the

Hinter Rhein. All, however, had hospices near their summits. It is interesting to note that the *Itinerary of Bruges* contains no mention of any of this group of passes. East of the St Gotthard, only two routes across the eastern Alps are described in the *Itinerary*: first, that which used the Arlberg Pass and the low and relatively easy Reschen–Scheideck Pass from the Unter Engadin at Finstermünz to the upper Adige, and, second, the Brenner route from Innsbrück to Bolzano.

The St Gotthard route and the roads which used the passes of the Grisons converged on Milan. The Reschen–Scheideck and Brenner routes no less clearly focussed on Verona, and offered the easiest routes from north-west Europe to Venice. During the fourteenth century the eastern passes were increasing in both relative and absolute importance. The Bruges *Itinerary* contains no less than three routes which converged on the Reschen–Scheideck. Furthermore the routes which linked the Danubian plain with the head of the Adriatic were coming into use. The *Itinerary* gives in very summary fashion a route which ran by way of the Semmering and, in all probability, the Pontebba passes from Vienna to Venice. Almost certainly the Predil Pass and the routes from Gorizia and Trieste to the Sava valley were also used.

River navigation. No one travelled by road if it was possible to use the rivers. Water transport alone offered the means of cheap transport for bulky commodities, and no town could reach even moderate size without this means of bringing in food, building materials and other merchandise. Small sailing boats were used on the larger rivers, such as the Rhine and the interlacing waterways of its estuary. Rowboats were much used, but on all the smaller rivers the greatest reliance was placed on haulage by man or beast. But no river was really navigable without a towpath (*via navigii*) and on some of the more important of the navigable rivers, there were towpaths along both banks.

It was nowhere easy to keep a river open for navigation, and even on the most easily managed waterways traffic tended to be seasonal. If ice presented no great problem, floods could sometimes be disastrous, not only sinking boats but destroying the towpaths and washing away the sluices. The smaller rivers made considerable use of sluices to separate different levels along the courses of the rivers and also, in the Low Countries, to exclude the tide. Boats were commonly moved from one level to another either by means of locks with gates which moved vertically in masonry grooves, or by dragging them with the help of a capstan over wooden ramps.

Bridges, mills, fish-weirs and low water in summer presented yet other obstacles. On some rivers the speed of the current made downstream traffic hazardous and upstream impossible. Such streams were often used only for floating lumber from the forests which bordered their mountain tracts. The Black Forest was thus a prolific source of lumber for the Rhineland.

There can be no doubt that on some of the rivers of western Europe there was a quite intense traffic. A very large part of it was local. On the Rhine the *Marktschiff* served the same purpose as the peasant's cart in bringing foodstuffs to the urban markets, and a large part of the food supply of such towns as Frankfurt was transported in this way. The Ypres accounts recorded the passage within a period of 122 days in 1297 of 3,250 *escuttes* (small sailing boats) and 87 *marctsceipen* (market boats) at one of the *overdraghes* or sluices near the city, an average of 27 ships per day. The Flanders plain was, of course, usually dependent upon waterborne transport. The evidence for the river Seine shows the passage at this time of at least one boat a day at Meulan, 40 kilometres below Paris. This would almost certainly have been a larger ship than those which were dragged across the sluices near Ypres, and if we can add a similar volume of traffic directed towards Paris along the Oise, Marne and upper Seine, this probably represents a very considerable dependence upon waterborne traffic.

The Garonne was used from Toulouse to the sea, and the English kings had taken steps to improve navigation on both the mainstream and its tributaries, such as the Lot and Tarn, which flowed through important wine-growing areas. The Rhône was used, particularly by downstream traffic, from Lake Geneva to the sea, with the exception only of the defiles above Seyssel, around which it was usual to transport goods by pack animals. The Loire was one of the most used of Europe's rivers, especially for the seasonal movement of wine and salt. The Saône, Somme and the tributaries of the Scheldt were all much used. Much of the traffic on the Somme was in wine, though the locally produced woad (*guède*), wool, cloth and salt herring were also transported. The movement of wine from Burgundy to Amiens and Abbeville will serve to illustrate the extent of the dependence on rivers. It was carried overland from Beaune to Cravant, at the limit of navigation on the Yonne. Thence it was taken by boat to Paris, where, joined by wines brought by boat from Auxerre and overland from Orléans, it was shipped up the Oise to Compiègne, and thence by wagon to Péronne, where it was again loaded on boats for the voyage down the Somme. Some was even transshipped at St Valéry for the voyage to Flanders.

The tributaries of the Scheldt were used in particular for the shipment of wheat and other bread-crops from Artois and Picardy to the large Flemish towns. These goods were taken by wagon to river ports such as Lens, Douai, Valenciennes and Mons, which lay near the limit of navigation, and were shipped downstream. There was at this date also a small, but growing upstream movement of grain which had been imported through the coastal ports (see below, p. 427). The Meuse formed a navigable highway far into eastern France, and had attracted a not inconsiderable amount of industry to its banks. Its tributary, the Sambre, was used upstream almost as far as Maubeuge.

The importance of the Rhine itself has already been mentioned. It was used regularly and intensively, despite its many natural hazards at this time, from the falls of Schaffhausen to the sea. Its navigable tributaries gave access to the Swiss plateau, to Franconia, Swabia and Lorraine. The branching waterways of its delta linked the Rhineland, in the one direction, with the Meuse (Maas) and the waterways of Flanders, and, in the other, with the Zuider Zee and the maritime route to the Baltic.

Any comparison of the volumes of traffic using Europe's rivers at this time is impossible, but the fragmentary data suggest that the Rhine was by far the most important of them. It formed part of the most frequently used route between northern Italy and the Low Countries; its tributaries linked it with the basins of the Danube and the Rhône, and it carried a more varied long-distance traffic than any. A humorous poem of the Renaissance told how the citizens of Zürich undertook to cook the most noted of their local dishes and to deliver it, still hot, to the citizens of Strasbourg. It describes the voyage down the swift Limmat and Aare and through the rapids of the upper Rhine. These ended at Rheinfelden, and:

> der Rein fangt an
> Zurinnen reyn und still davon,
> Das er sicht wie ein eben feld,
> Und unbetrübt sich forthin stellt.[120]

Nothing shows more clearly than this poem how natural it had become in the later Middle Ages to take to the river and to use it on every possible occasion. This is apparent in the activities of the *Marktschiffe* at every important city along the Rhine.

There was less traffic on the rivers to the east of the Rhine, but there can be no doubt that all waterways capable of floating the small river-craft of the time were in fact used, if only intermittently. The Weser and also the lower courses of its principal headwaters, the Fulda and Werra, were used to ship the local agricultural surpluses down to the seaport of Bremen. It appears, however, to have been difficult to keep the river clear of fish-weirs and other obstructions, as long as the volume of shipping remained small. The Elbe was used more regularly, and, with its downstream shipments of wheat, rye, wax, honey and timber, contributed to the seaborne commerce of Hamburg. The Oder was less valuable. Its course in Silesia was so encumbered with fish-weirs and mills that it ceased, in fact, to be navigable; Frankfurt was able to interrupt the passage of all boats coming northwards from the upper river, and traffic was significant only between Frankfurt and the seaport of Stettin (Szczecin). The lower Vistula was regularly used; much of the goods exported from Danzig had been brought downriver, and the river towns of Chełmno and Torún were, in fact, members of the Hanse.

The Danube was very much less important than more westerly rivers, largely because there was little trade with the regions towards which it flowed. The upper river was used to send Austrian wine westwards to Bavaria, but even here the volume of traffic was small. Below Vienna trade was probably quite negligible.

The Mediterranean rivers, with steep gradients and seasonal flow, were with few exceptions of little use for navigation. There seems to have been a fairly intense local traffic on the Guadalquivir in southern Spain, and it is probable that the lower courses of the Tagus and Ebro were also used. But the most important Mediterranean river was without question the Po. It was regularly navigated between Piacenza and the sea and was used upstream to Padua, and the *Itinerary of Bruges* actually prescribed a boat for this segment of the journey from Bruges to Venice. Many of its tributaries, as well as rivers like the Brenta and Adige, which shared a common delta with the Po, were used.

The lack of navigable river was felt so acutely by Milan and Bologna, that they even cut canals to connect themselves with the system of navigable waterways. But, except in northern Italy and Flanders, no real attempts were made to improve on nature and to create navigable waterways where none had existed before.

The river boat was a slow and uncertain means of transport, and only the most bulky and least valuable goods were ever trusted to it. Cargoes were at the mercy of the weather, and there was always a serious risk of a boat foundering or letting in water. The surviving schedules of river tolls show that the cargoes were of the simplest. Wine was one of the most important of them. Sealed in wooden barrels or tuns, it was immune to most of the hazards of a river voyage, and furthermore profited from the smoother journey by water. Salt also figured prominently, especially on the Loire, where 989 boats out of 1,397 passing the toll station at Champtoceaux, above Nantes, in one year were loaded entirely with salt.[121] Grain was also important on many rivers, especially those which flowed down to the Flanders plain and those which carried the rye and wheat exports of eastern Germany and Poland down to the Baltic coast. Salt meat, salt fish and other non-perishable foodstuffs, such as olive oil, together with wool and leather, lumber, millstones, ironware and coal made up most of the cargoes. Building stone was sometimes a prominent item, as, for example, limestone was on the Oise for the construction work of Paris.

Tolls. All medieval travel and transport was heavily burdened by tolls. In a very few instances these were intended to cover the cost of improvements along the river. Most often they were exactions demanded by a seigneur from all traffic that could not avoid crossing his lands. His right to do this seems never to have been questioned; only occasionally would he excuse from payment the goods owned by the Church or by a particularly influential layman.

Tolls were easiest to collect from river traffic. Road traffic was sometimes able, by a judicious choice of roads, to evade some of these impositions, but the course of the boat was fixed by nature. It was possible for any landowner with a water frontage to exact his tolls. On the Garonne between Toulouse and Bordeaux were no less than 31 such toll stations. There were 18 on the river Seine, of which five were owned by the abbey of Saint-Denis. In 1273 the Parlement de Paris had forbidden the creation of more toll stations, but illicit tolls continued to be exacted until they were suppressed by the king late in the fourteenth century. The number of toll stations increased similarly along the Rhine, where there was no royal authority capable of terminating the abuse. By the end of the Middle Ages a cargo travelling the whole length of the navigable river paid toll about 35 times, while there are said to have been 30 toll stations on the Weser and at least 35 on the Elbe.

No less obstructive were the staple rights claimed by the larger riverine towns. This consisted of the right of the municipal authorities to require that certain commodities be unloaded and offered for sale. They were allowed to continue their journey only if there was no local demand for them. Dordrecht, for example, was in 1304 made the staple for all oats, wine and timber carried on the Lech and Merwede rivers, and only the burgesses of the other Zeeland towns of Zierikzee, Middelburg and Platten were relieved of the obligation to leave their merchandise in Dordrecht for fourteen days. Five years later the Archbishop of Magdeburg ruled that the Magdeburg *Altstadt* should be the staple for grain transported on the Elbe. Paris, the larger Rhineland towns, Bremen, Münden on the Weser, and Frankfurt-on-Oder all possessed the right of staple with the risk which it posed of delay and interruption of trade.

Tolls were scarcely less obstructive on the most used river of southern Europe, the Po. Almost every feudal lord through whose territory the river flowed, exacted his tribute, and the citizens of Cremona even cut a canal to evade the impositions of their neighbour, Mantua.

On the Rhine there was the further obligation at Cologne and Mainz to reload the goods into a *different* boat. This *Umschlagsrecht* derived from the different physical conditions of the lower, middle and upper river, which, it was held, necessitated boats of different size and draught. It had, however, become an abuse, from which the shipping of the whole Rhineland suffered, and only the boatmen's guilds and the privileged towns could benefit.

Any particular road was less likely to be encumbered by tolls, because the stations could be bypassed with greater ease. Toll stations were nonetheless numerous along the more important of them, and tolls were particularly burdensome at bridges and near the boundaries of political jurisdiction. Trade between Flanders and northern France paid toll at Bapaume, Péronne, Roye, Compiègne or Crespy, according to the road used, and the officials of the Péronne and Nesle toll stations were in 1302 forbidden by the Parlement to discriminate in their charges against merchants from Italy. Much of the traffic tended to pass through Bapaume, whose toll records show how great was its variety.

Maritime trade. The seaborne trade of Europe in the fourteenth century was centred in two groups of ports, with but a tenuous link beween them. The first was the Mediterranean basin and the second, the North and Baltic Seas, and the infrequent voyages by way of the Atlantic sea route provided their only connection by sea.

The Mediterranean trade was made up of both the long-distance voyages of the Venetian and Genoese galleys, and of short-distance trips made by small vessels. The latter sailed from one small port to the next, collecting local surpluses and delivering them to the larger ports which lay on the routes of the galleys.

The Italian ports lay at the centre of trade at both these levels. Along the shores of Cataluña, Provence, Italy and Dalmatia lay a large number of small ports. The Italian industrial and commercial cities were obliged to import part of their food supply and much of their raw materials. Wheat from Apulia, the coastal lowlands of eastern Italy and the plain of Albania was regularly collected by small craft and delivered to the port cities of northern Italy. Lumber from Corsica, Sardinia, the mountainous south of Italy, and the Dalmatian coast went to supply the building and naval construction industries of Venice and Genoa. Salt was collected from Languedoc and the Adriatic coast; iron-ore was shipped from Elba, wool from southern Italy and Spain, and English wool was brought overland to the ports of southern France, and taken along the coast to Tuscany; olive oil and fruits from southern Italy and Sicily, and wine from Crete and the Peloponnese were landed at the Italian ports. All this amounted to a very substantial volume of trade, and it probably accounted, in volume at least, for the greater part of Italy's seaborne commerce. It was requited by the export of cloth, some of it linen and much of poor quality; of metal goods, including armour and weapons, of leather goods, and of a few luxuries.

Narbonne was typical of the less important ports at this time. It exported wine, oil and, when the harvests permitted, wheat. Its closest ties were with Genoa and the small ports of the Riviera, such as Nice, Savona and Ventimiglia, and, in the opposite direction, with Cataluña. Much of its commerce was evidently a small-scale coasting traffic, but Narbonne had a few larger ships, and a few well-placed merchants, who at this date did business as far afield as Sicily and North Africa.

Ragusa, or Dubrovnik, showed a similar pattern of local and long-distance trade, with greater importance, however, attaching to the latter. Incapable of providing its own food supply, it imported grain, oil and fruits from other parts of the Dalmatian and Epirote coast, as well as from the opposite coast of Italy. It was a gathering point for salt, obtained from coastal pans from Valona to the Neretva mouth, and a shipping point for metals, skins and hides from the interior. Its merchants also participated, under the auspices of Venice, in long-distance trade between Italy, the Aegean and the Levant.

The long-distance maritime trade of Venice and Genoa must be measured, wrote Luzzatto, 'by the quality much more than the quantity of merchandise brought home'. The total annual import of eastern wares at Venice could at this time have been brought in by three convoys of no more than two to four ships each and the dead weight of the largest ship is unlikely to have exceeded 500 tons. All the Venetian fleets 'can have transported only some 2,000 tons of merchandise from the East' in a year,[122] and those of Genoa could have brought no more. Add the commerce of the lesser ports, and the total volume of oriental goods brought into European ports cannot have greatly exceeded 5,000 tons annually.

Yet it was a high value cargo. With the exception of wine from southern Greece and Crete, there were few bulky commodities. Raw cotton, raw silk and alum from Anatolian Phocaea constituted the chief industrial raw materials. Most of the cargoes were, for the rest, made up of spices, particularly pepper, of sugar, decorative woods, dyestuffs, precious and ornamental stones, and slaves.

These imports were paid for by the export of the manufactured products of the west. These included not only silks and the quality products of the *Arte di Calimala* but also humbler fustians and linens. Cargoes from Italy also included metal goods and weapons, soap and glass from Venice, Flemish and German cloth, copper, lead and tin.

Both Venice and Genoa traded principally with their own Aegean and Black Sea dependencies, though both had 'factories' in Constantinople and made less frequent voyages to the coast of Asia Minor, the Levant and Egypt. The most important Venetian bases were Clarenza and Modon, in the Peloponnese; the Negropont, and Crete, but they also had a small colony at Thessaloníki and visited the Black Sea for slaves. The Genoese maintained their bases on the opposite coast of the Aegean, particularly on the islands of Lesbos, Chios and Samos, which gave them control of Phocaean alum deposits. They also controlled a number of commercial centres around the shores of the Black Sea. Foremost among these were the towns of Kaffa and Soldaia, in the Crimea, but Bratianu[123] lists at least ten other places where the Genoese had some kind of a foothold. From here they shipped the silks, brought overland from the east, as well as less valuable goods of local origin: skins, hides and wheat. It is doubtful whether much of the grain of the Russian steppe reached Italy; it is more likely that most of it went to the provisioning of Constantinople.

Other commercial cities of the western Mediterranean played a very much smaller role in the Aegean region, and it was the policy of both the Venetians and the Genoese to exclude them whenever possible. Nevertheless, a Catalan or Majorcan ship paid an occasional visit to Greece, where there was still a small Catalan colony, and ships from Ragusa and the ports of southern Italy undoubtedly traded with the Byzantine Empire.

Ships of Languedoc, Cataluña and southern Spain visited most parts of the western basin of the Mediterranean, and shipped cloth and occasionally grain to the Berber ports of North Africa. Spain at this time also maintained close ties not only with the Balearic Islands, but also with Sardinia and Sicily.

The Italians were slow to extend their trading voyages westwards to the Atlantic and north-western Europe. This was in part because their ships were ill-suited to the rougher waters of the ocean; in part through a natural unwillingness to interrupt their established pattern of trade. The former difficulty was removed by the adoption of a vessel more like the cog of the northern seas. At the same time, the Champagne fairs were declining; Italian merchants were tending more and more to take up their residence in the northern commercial cities, and the older pattern of overland trade became fluid.

It was at this juncture that the first Genoese ships sailed to Flanders, preceded in all probability by the Majorcans. The first documented voyage from Genoa was in 1277–8. Subsequent voyages were highly irregular and the fleets were generally small. There was frequently only one vessel a year in the period preceding 1330, and in some years none at all came. It was not until 1312 that Venice began to take part in the Atlantic trade, but from 1325 a fairly large fleet began to sail each spring from the Venetian lagoons for the Thames, Zwin and Scheldt, stopping in all probability at a number of ports in Dalmatia, southern Italy and Spain on the way.

It is generally assumed that the cargoes carried by the Venetian and Genoese ships to northern ports were made up of pepper, cloves, nutmegs and cinnamon, apothecaries' wares, saffron and gum arabic, silks, velvets and oriental carpets. In fact, it was alum which predominated in the early voyages. The return freight from north-western Europe was made up of cloth from Flanders and Brabant and wool from England, but lead, ironware, copper, and Cornish tin were among the goods sent back to the Mediterranean.

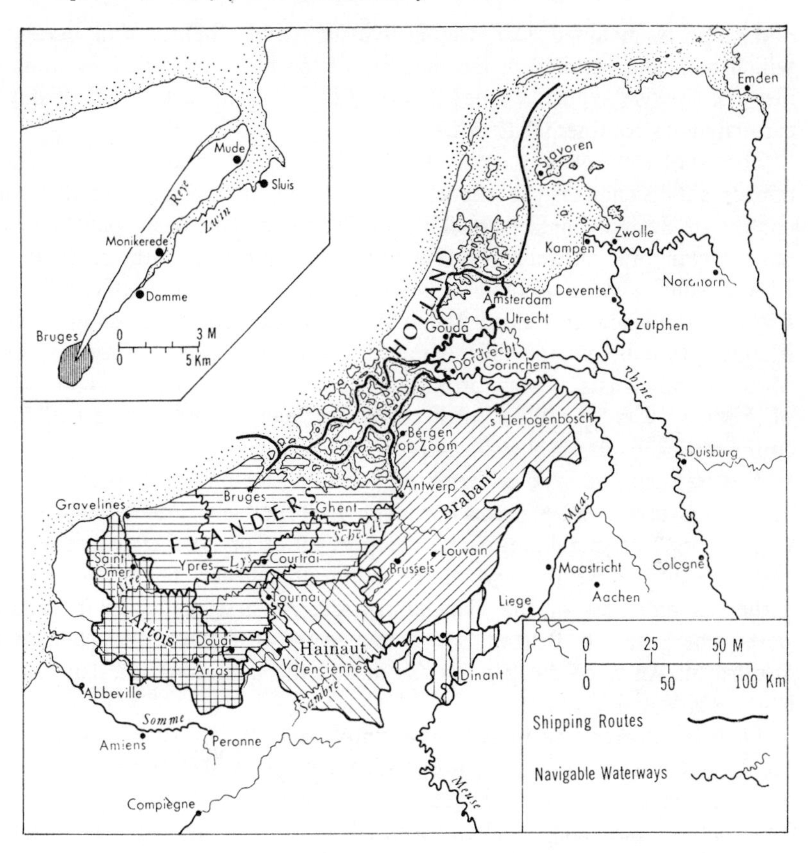

Fig.6.20 The Low Countries in the mid-fourteenth century

The northern trade. The focus of the northern trade was the Low Countries:

> For the little land of Flanders is
> But a staple to other lands ywis.[124]

By the early fourteenth century four separate, though interconnected streams of maritime traffic converged on the ports and estuaries which lay between Calais and the IJssel river. First was that which originated in the Bay of Biscay. It was concerned with two commodities: wine from Anjou and Poitou and more recently from Gascony, and salt. The latter originated mainly in the Bay of Bourgneuf (see p. 401), though there were other sources along the Biscay coast. There was also a coastwise shipping between the ports of northern France and a smaller and less regular trade in cereals, vegetable dyestuffs and iron from the Basque territories.

The second stream of commerce was between the Low Countries and the British Isles, and was similarly specialised. It was dominated by the import of English wool, without which the manufacture of quality cloth would have been impossible in the Low Countries, and to a lesser degree by the export of cloth to England. Lead, tin and cloth supplemented wool among English exports, and imports included salt, alum, dyestuffs and oriental and luxury goods which had been retailed in the markets of the Low Countries. In some degree also the English export of wool was used to defray the costs of English military campaigns in continental Europe.

The wool trade was not only taxed but also controlled by the king. It was convenient to channel the export to the continent through a small number of centres or even a single port. This wool 'staple' had been located in turn at Saint-Omer and Antwerp. Then a number of ports in both England and the Low Countries were used in succession or even simultaneously, before it was finally established at Calais. Much of the wool went to Brabant rather than Flanders (see below, p. 430); Brabant merchants, led by those of Louvain, Malines and Brussels, were prominent in the English wool trade, and the port of Antwerp was by far the most convenient port of entry, and had for many commodities displaced the silting ports of Bruges.

The Baltic trade was in terms of volume, and probably also of value, the most important of the streams converging on the Low Countries. It was largely in the hands of merchants from the port cities of northern Germany, which had already, before the end of the previous century, begun to associate together as members of the German Hanse. At first they had frequented the ports which served Bruges (see below, p. 424); latterly they tended to use Dordrecht, Antwerp, Bergen op Zoom and the numerous smaller ports of the lower Rhine and IJssel.

Their commerce was in general a simple one. It consisted in the import into the Low Countries of grain, predominantly rye, and of forest products such as skins, wax and honey. These were supplemented by smaller quantities of low-quality linen and woollen cloth, flax, animal fats, herring, lumber and wood ash. These commodities were paid for by the export of cloth of all kinds, salt, and wine.

The Hanseatic merchants operated individually or in small associations, and their home ports might be Riga, Lübeck or others of the major cities of the League, but might as readily be Harderwick or Zierikzee or some other of the small ports of the Netherlands or Mecklenburg coast. The normal practice of the Hansards was to lay up their vessels in their home port in winter for a period which, in ports like Reval in the northern Baltic, might be as much as five months. In spring they sailed for Flanders, laden with grain from the previous autumn's harvest and the products of the winter hunting. Some even proceeded to the Biscay coast for salt; others took on their cargoes of salt, along with cloth and wine, in the Low Countries.

Some ships made the voyage between the Low Countries and the eastern Baltic by way of the Danish Sound, where they could pick up a cargo of salted herring from Skåne in southern Sweden. Some – possibly those less suited to sail the stormy North Sea – unloaded in Lübeck, whence goods were either reshipped or transported overland to Hamburg. No less than twenty-five ports lay along the Baltic coast from Kiel to Reval or on the rivers of the hinterland. Most were small, and their trade probably consisted largely of sending cargoes of grain to the larger entrepots, like Stettin and Danzig for transshipment and export to the west.

The exports of the Baltic ports were gathered from a very extensive hinterland. Rivers were much used, prominent among them the Oder, Vistula and Dwina. It was no accident that some of the most important of the Hanseatic cities lay near the mouths of these rivers. Local rulers, like the Duke of Pomerania and the Grand-Master of the Teutonic Order, attempted to canalise this trade through a few ports, but cities like Riga were not slow to establish staple rights of their own. Much of the overland traffic in the northern Baltic region took place in winter, when movement was easier over the frozen ground. Commercial forays into the Baltic hinterland were sometimes long and usually dangerous. In 1323 the Prince of Novgorod had assured the German merchants of protection and freedom of travel up the Neva to Novgorod itself, free from interference by the Swedes and the townsmen of Viborg (Viipuri). The Hanseatic records contain numerous instances of requests for or promises of protection to merchants journeying deep into Sweden, Russia, Lithuania and Ruthenia. That such promises, once granted, were not always honoured is apparent from the retaliatory measures taken by the Hansards when the Polish king excluded them from Ruthenia. The merchants themselves were usually described as German; whether they were so in fact is not always clear. The richer and more important were almost certainly from immigrant German families, but it seems probable that they had assimilated a number of the local peoples, whether Poles, Lithuanians or Latvians.

Though the Baltic trade of the Hanseatic merchants is often described in very simple terms: 'in mit solt unde ut mit roggen'[125] – there were in fact a number of local variations in this pattern. Though salt dominated the Baltic imports in volume, if not also in value, grain – chiefly rye with some wheat – was far more important from Pomerania and Prussia than it was from the east Baltic region. Here the export of rye, supplemented by oats was far from regular, and it was probably forest products that filled the holds of the Hansard's cogs as they sailed each spring for the west. In the fourteenth century the export of metals, especially copper and iron, was becoming important from the Swedish ports, particularly Stockholm and Kalmar.

The fourth of the streams of trade which converged on the Low Countries was by far the least important. It derived from Norway and the North Atlantic islands. It brought fish from Bergen and other Norwegian ports, and from Iceland, and it took in exchange salt to cure the fish, as well as grain and cloth.

Within the Low Countries the pattern of trade and the location of the chief commercial centres had been changing in recent decades. Bruges had ceased to be an international market of the first rank, and its decline cannot be attributed wholly to the progressive silting of its ports. It remained a centre of manufacture of high-quality cloth and a market for the village industries of west Flanders, but the seaborne and riverborne trade of the Low Countries had shifted towards the east.

The new ports lay where the ships of the Hansards and the Italians met the river boats of the Dutch and the Germans. Foremost was Antwerp itself, not yet a world market, and still the victim of the political hostility of the cities it had displaced, but nevertheless expanding its population, its manufactures, and its commerce rapidly at this time. The rise of Antwerp was accompanied by that of the ports of Breda and Bergen op Zoom, with which it shared the commerce of the lower Maas, and by Middelburg, in Zeeland. Even more striking was the rise of ports along the lower Rhine and its distributaries, the Lek, Merwede, Oude Rijn and the IJssel. They owed their importance to the economic rise of the Rhineland and the eastward shift of the commercial axis of Europe from Flanders, Champagne, Lorraine and the Rhône valley to a line from Brabant and the Netherlands, up the Rhine and across the central alpine passes to Italy.

The multiplicity of ports in Brabant and the Netherlands reflected in part the diversity of rivers and the number of possible routes from the coast to the Rhine, but they also owed something to the political fragmentation of this region. The chief port on the lower Rhine itself was Dordrecht, successor to Tiel and Dorestad. Utrecht, lying on the rapidly silting Oude or Kromme Rijn, was having difficulty in maintaining its connections both with the Zuider Zee and with the Rhine (Lek), and had probably passed the peak of its medieval commercial importance.

A route from the Zuider Zee, by way of the small but growing port of Amsterdam, to Gouda and the Waal was also used, but the IJssel river itself offered the shortest and easiest route between the North Sea and the lower Rhine. It diverged from the Lek above Arnhem and entered the Zuider Zee below Kampen. Along both the river and the shore, particularly the eastern, of the Zuider Zee there emerged a succession of small ports: Doesburg, Zutphen, Deventer, Zwolle and Kampen on the one; Harderwijk, Elburg, Stavoren on the other. The Vecht, a tributary of the IJssel was used to gain access to the Münster region of north-west Germany. Along the lower Rhine were numerous small ports, whose merchants played a modest role in the maritime trade of northern Europe, and below its division into Lek and Waal lay the two important ports of Arnhem and Dordrecht.

While traffic was becoming more intense in the Rhine delta and the number of its small ports was growing, the estuary of the Zwin was slowly silting. This small river, at most 15 miles long, had been formed as recently as the late eleventh and early twelfth centuries by the small rise in sea level. It consisted, in fact, of a double channel, the Reie and the Oude Zwin, which flowed north-eastwards from Bruges to enter the sea near Kadzand. In the twelfth century sea-going vessels may have been able to reach Bruges, but silting was rapid. In 1180 a dam was built across the Reie and the small port which resulted, Damme, was for a time the limit for most sea-going vessels, though a small lock actually allowed some to reach Bruges. Damme maintained its position as the chief port for Bruges into the fourteenth century, though gradually supplemented and in the end replaced by a succession of smaller ports nearer the sea: Monikerede, Hoeke and Mude. The most important of these, however, was Sluis, which lay at the mouth of the Reie and opposite the island of Kadzand, and seems to have been first used about 1300. It grew quickly in importance and by 1330 had come almost to monopolise the seaborne trade of the Zwin. The fifteenth-century *Libel of English Policie* described how merchandise from all quarters

Is into Flanders shipped full craftily,
Unto Bruges as to her staple fayre:
The Heaven of Scluse hir Haven for her repayre
Which is cleped Swyn tho shippes giding:
Where many vessels and fayre are abiding.[126]

The silting of the Zwin is commonly blamed for the decay of Bruges. It should be pointed out, however, that sea-going vessels had ceased to sail up the river to the city long before it had reached its apogee, and throughout Bruges' period of high prosperity the river bed was being steadily transformed into water-meadows. It is more realistic to link the decay of Bruges' commerce to the growing importance of the Rhineland in Europe's commercial pattern and to the structural changes, which have already been discussed, in the clothing industry itself. The citizens of Bruges were in the early fourteenth century playing a steadily diminishing role as merchants, and their city had in fact become a 'concentration of alien merchants',[127] amongst whom the Hansards were foremost and the Italians and English were numerous.

The commodities of trade

The previous pages have been concerned primarily with where the merchants went, and the nature of the wares carried has been only incidental to the discussion. The concluding section of this chapter examines the range of goods which made up this commerce. Commodities entering into trade – and it is primarily long-distance trade that is being considered – can be grouped under the three headings: foodstuffs, raw materials, and manufactured goods. Under the first came grain, wine and salt, which together made up a large part of the total volume. Raw materials were dominated by wool, and manufactured goods by cloth. These commodities must have made up perhaps 80 to 90 per cent of the total volume of trade. The remaining 10 to 20 per cent was made up of leather, metals, and a few other foodstuffs and raw materials, as well as the vast range of spices, dyestuffs and luxury wares.

The grain trade. The growth of towns and the development of manufacturing were necessarily accompanied by the evolution of a grain market. At no period in medieval history, in all probability, was the self-sufficing village community universal, or even widespread. Gras has taught us to think rather in terms of groups of manors or villas, between which there was an active exchange.[128] In England a class of corn-mongers or dealers arose as early as the twelfth century, and bought up manorial surpluses for sale wherever there was a market. The irregularity of grain yields – and all the evidence confirms that in the Middle Ages cropping was very uneven – together with local variations in soil and climate, encouraged trade in agricultural surpluses, but the chief reason for the emergence of a grain market was unquestionably the growth of the towns.

Yet grain was a bulky commodity of fairly low value; its transport, at least over long distances, was difficult unless it could be moved by boat. The records of prices that survive for Spain, France and England show extreme fluctuations in any one area as bad seasons succeeded good. The market mechanism worked slowly, and was incapable of bringing grain prices to an approximately even level over any considerable area. To this extent there was a multiplicity of markets, rather than a single market.

The regular provision of grain was in many countries an object of direct government concern. Frequently one finds the prince ordering the purchase of grain, or requesting another ruler to facilitate its transport. Nowhere was a grain policy more necessary, or more complex than in the Low Countries and in parts of northern Italy.

In the early fourteenth century, we find in Europe certain regions which could be expected, on any reasonable calculation of probabilities, to have a grain surplus. These would include Apulia, Sicily and Prussia, and also large areas of the Paris basin and Artois. By contrast, there were regions which with an at least equal degree of probability would be food-deficit areas. Among them would certainly have been Flanders and Tuscany, and to these we should add the regions surrounding every large group of towns. The rest of Europe might be said, on average, to have broken even, which means in fact that in most years its component regions either had a small deficit or a small surplus.

The grain trade between the surplus and the deficit regions was highly organised. There could, however, have been no regular grain trade in the intermediate regions, other than that which took place between the towns and their surrounding villages. Here one must assume a regional self-sufficiency, tempered by occasional surplus and scarcity. Normandy was such a region. The visitations of Archbishop Eudes of Rouen revealed conditions of real scarcity in parts of his diocese, while a ship returning from the Bay of Biscay with a cargo of salt or wine would call at a port on the Norman coast and pick up a small surplus.

The Low Countries probably showed the greatest deficiency of any part of Europe at this time. Not only were the towns large and their functions specialised, but much of the cropland was itself of poor quality. Even Holland, one of the less urbanised areas at this time, was obliged to import grain from the lower Rhineland. Flanders, which is unlikely to have been able to produce more than a half of the bread-crops needed, obtained a large part of the remainder from Artois, Picardy and other parts of northern France, where it was produced on estates like that of Thierry d' Hireçon (see above, p. 372). This was, in the main, a wheat–oats region, and it was wheat which as a general rule was taken to the shipping points on the Lys, Scarpe and Scheldt to be sent downstream to the Flanders towns. That this vital movement of bread-crops was liable to interruption is all too apparent, and fierce were the struggles between the riverine cities to secure their share. The variations in grain supply reflected the harvest, but were increased by the machinations of the merchants who knew all too well how to manipulate the supply and increase prices. This bore harshly on the Flemings, many of whom were employed at a fixed wage.

The most important shipping point was probably Douai, on the Scarpe. Its citizens owned land deep in Artois, and lived by sending their wheat down the river to Ghent. But Douai faced competition from Valenciennes on the Scheldt,

Lens on the Deule, and Aire on the Lys. Not only were these river ports competing for the downstream market, particularly that of Ghent, but the Flemish towns were also competing among themselves. Ghent itself established a staple in grain, so that Termonde and Antwerp could receive only what Ghent did not want. Upstream towns, particularly Courtrai and Tournai, were able to get to the grain boats before Ghent, and Tournai, somewhat later, actually created a staple of its own. Ypres, which did not lie on the Scheldt system, attempted to tap into one of these lines of supply by building a road from the Ieperlee to the Lys at Warneton, about 5 miles away. Ghent protested and in the end succeeded in preventing Ypres from poaching in this way on its own grain preserve. Ypres was thus obliged to rely on imports brought up the Ieper from Nieuport on the coast.

Brussels, Malines and Louvain similarly derived grain from the good agricultural lands which lay to the south of them. Fortunately, however, the urbanised region of the Low Countries was not obliged to rely only on grain from Artois, Hainaut and southern Brabant. Even within the urbanised region of north-western Europe there were areas with a regular grain surplus. Amongst these were the small County of Alost and the more extensive districts of Guelders, Julich and Kleve, which regularly sent grain down the Rhine to Holland, Zeeland and probably also Antwerp and parts of Flanders. Wheat, probably in fairly small quantities, was brought to Flanders, Zeeland and Holland from Normandy and other parts of northern France by the ships that had taken cloth and fish to Gascony and Spain. But in the late thirteenth century the German merchants opened up a source of grain capable of satisfying most of the needs of the Low Countries. This was the Baltic region, which was to continue into the nineteenth century to be a major source of grain for western Europe. In the early fourteenth century, however, this trade was still growing; a century later Flemish peasants were abandoning grain cultivation in the face of cheap Baltic rye.

Other areas of north-western Europe also suffered from a shortage of grain, and were obliged to rely on long-distance trade to maintain their supply. Among them was the Paris region. Grain produced in the immediate environs of the city was supplemented from much of the Paris basin. Melun, Corbeil and St Denis were important collecting points, but grain also came by way of the Seine and its tributaries from Burgundy, Beauce, Champagne and even Artois and Vermandois. Normandy, it should be noted, appears not to have contributed to the bread supply of Paris. This was due, in part at least, to the stranglehold which Rouen maintained over the traffic of the lower Seine. The effect of this was to turn the available grain towards the coast and thus to the Low Countries.

The Rhineland, though highly urbanised, was probably not, taken as a whole, a deficit area, and the river system was able to distribute grain from surplus areas within the Rhineland, such as Kleve, Jülich and Alsace, to the cities which lay along its banks. The urbanised coast of Cataluña also had a deficit which it made good with imports from Sicily.

Northern Italy, the second region of high urban density and developed manufactures, was also an important food importer. Tuscany, it is said, could produce enough grain for only five months in the year. Pisa had a small export, but the great part of the import both for Florence and Genoa came from southern Italy and Sicily. The Lombardy plain was capable of producing more heavily, but the local grain production was supplemented, especially for Venice, which at this date held very little territory within the plain itself, from Romagna, Apulia, Sicily and Albania. In most other areas serious deficits appear to have been only temporary. Lisbon for a time had recourse to North African grain, and Venetian settlements in Greece appear to have derived most of their cereals from southern Italy and Sicily. Constantinople, lastly, relied very heavily on overseas grain, but its coastal position between two seas gave it unusual advantages. Some came from Thrace and Asia Minor, but the most important source appears to have been the fertile lands along the lower Danube, particularly Moldavia, which shipped large quantities of grain to Constantinople in return mainly for cloth.

In northern Europe there were two areas which generally had a grain surplus. One was a broad belt of territory covering much of northern France and extending eastwards through Hainaut and southern Brabant to the lower Rhine plain, the Börde of Hanover, and Saxony. The other was a rather indeterminate area which enclosed the Baltic region on the south and east. East of the Rhine the principal export grain was rye with, from some areas, small quantities of wheat. Grain was sent down the Weser and its tributaries from the loess region of the Börde. Hamburg collected the grain from Holstein, the Altmark and parts of Saxony. The Wendish cities, Wismar, Rostock, Stralsund and Greifswald, shipped the export of Mecklenburg; Stettin handled the surplus from the Oder valley, particularly the Neumark, and Danzig that of the Vistula. As far to the north-east as Riga rye was the dominant export, and even in Reval its shipment was occasionally important.

In southern Europe areas with a regular grain surplus were fewer and very much less extensive. They included tracts of lowland around the coast of peninsular Italy, including the Pisan Maremma and the Romagna, but the chief sources were Apulia and Sicily. To these should be added the small plain of Albania, parts of Thrace and Macedonia, and the Black Sea coastlands from which Constantinople continued to draw much of its food supply.

The wine trade. If the grain trade resulted from the growing concentration of people in cities and towns, the wine trade stemmed from the limited physical conditions under which the grape-vine yields the greatest amount of quality wine (see above, p. 379). Wine was produced over much of southern Europe; most was consumed locally, and little entered into long-distance trade. The important 'export' vineyards lay of necessity close to ports and navigable rivers, and foremost amongst them were those which had been developed under English rule in Gascony. The rivers Dordogne, Lot and Garonne were used to convey the barrels down to Bordeaux, Libourne, and the small ports of the Gironde. In the early years of the century as much as 90,000 to 100,000 tons had been cleared for north-western Europe.[129] Much of this export went to England, but the remainder was taken to the French Channel ports and to Bruges, Antwerp and the ports of the Netherlands. From the former it travelled inland, generally by river boat, to Rouen, Paris, Amiens. From the latter it was reshipped for distribution by water to the towns of the Low Countries and Rhineland and by the ships of the Hansards to the ports of Scandinavia and the Baltic.

The wine market of northern Europe was also supplied, though on a smaller scale, by the vineyards of Burgundy, the Rhineland and Tirol. Wine from these areas moved northwards and westwards by navigable rivers: the Seine system, the Meuse, Moselle, Rhine, and Danube. Along these waterways it was met by Gascon wines brought up-river from the coast. Burgundian wine also travelled down the Rhône to supply the Papal court at Avignon.

Within the Mediterranean region a surplus of wine was produced in Provence, and shipped along the coast to the urban centres of Cataluña, Liguria and Tuscany, as well as to Rome and the Mediterranean islands. Wine from Romagna and southern Italy was sent to Venice and the Adriatic ports; the Venetian galleys brought back the wines of Greece, of which the Malmsey from the Peloponnese (cf. Monemvasia) was for a time one of the most favoured wines in north-west Europe.

The salt trade. The production of salt was more localised than that of wine, and its use even more widespread (see above, p. 400). It was thus a bulky and important element in both long- and short-distance trade. Each salt-spring supplied a surrounding area, whose size and extent depended in the main on political circumstances and conditions of transport. The map (fig. 6.17) attempts to show the areas tributary to the principal groups of salt-springs. Long-distance movement of a commodity of such low intrinsic value was clearly dependent on cheap transport. The rivers Po, Rhône and Loire formed important channels for the movement of salt into the interior of the continent. The greatest volume, however, was taken by sea from the 'Bay' to the ports of north-western and northern Europe. From here it was distributed, also mainly by river. It is doubtful whether by 1330 it had reached the Baltic on anything more than a small scale, and the herring were probably cured very largely with salt from Lüneburg.

The pattern of trade in salt bore a certain resemblance to that in wine. The heaviest production of both was in the more southerly parts of the continent, and the northward movement, by river and overland routes, competed near the northern and north-western coasts with seaborne traffic from the Mediterranean and Bay of Biscay.

Industrial raw materials. The raw materials which supplied the limited range of medieval manufacturing rarely entered into long-distance trade. As a general rule their low value and high bulk restricted movement. The location of smelting was determined almost entirely by the availability of ore and charcoal; tanning and leather dressing were carried on where the hides were produced. The exception, however, was the materials of the textile industry. High quality wool was by far the most important industrial raw material in medieval trade, and the other textile materials: silk, cotton and flax, were also important in some parts of Europe. To these should be added the subsidiary materials of the textile industry: alum, fuller's earth, dyestuffs and materials (see p. 393), such as gold and silver thread, used in embroidery.

The wool trade. Sheep-rearing was as widespread as the cloth industry itself. No area was without sheep, and in some, such as the Meseta of Spain, the limestone uplands of England and the downs of Artois, they were reared in numbers that far exceeded the total of all other animals. The village and small town cloth industries used the wool from the local flocks for the weaving of a poor quality fabric for local sale. The areas of more intensive sheep-rearing produced a wool of higher quality, which entered into long-distance trade.

The quantity of wool varied greatly, and with it the value of the cloth that could be produced. Its value showed an immense range per sack from the best English wool from the Welsh Border counties to that commodity from the south-western peninsula which the Parliament Roll dismissed as 'Cornish hair'. Though the more highly priced wools dominated the long-distance market, there are many instances of the poorer wools – from Scotland and northern England, for example – being shipped to the Low Countries. Reasons for this lie not only in the growing markets for middle quality textiles and the practice of blending it with the better English wool but also in the shortages created by the government's manipulation of the wool trade in the interests of policy.

The most important areas of wool production for export were, first and foremost England, where wool formed the staple *par excellence*; the Spanish Meseta, and certain parts of France, including Languedoc and Provence. The English wool went very largely to the Low Countries, where it was considered indispensable for the manufacture of the older draperies and was also used in that of the 'new'. Next in importance was the Italian, particularly the Tuscan, market. Pegolotti clearly regarded English wool as essential for the *Arte della Lana* of his native town. Not only did he list the English monasteries that produced the best wool, but indicated also the route by which it could most easily be shipped to Italy. In general it was taken by sea from the English ports to the Gironde, unloaded at Libourne on the Dordogne, conveyed across to the Mediterranean coast and shipped again from Aigues Mortes or Montpellier.

French wool from the plains of northern France moved to Flanders, where it was in the early days of the cloth industry a major source of supply, and continued to be important throughout the Middle Ages. There were important centres of cloth manufacture in France, each of which must have drawn its raw material from a considerable area. Wool figured in the lists of goods which paid toll at the stations along the Seine, and the large flocks of Languedoc and Provence provided wool, though not probably of a particularly high quality, for export to Italy.

Manufactured goods. The manufactured goods entering into long-distance commerce were dominated by cloth, with metalwork of a secondary importance. It was in clothing that the rich could give rein to their vanity, and the demand among the few for fine and expensive fabrics knew no limit. The special cloths of each of the towns of the Low Countries were very familiar to the Cardinals in Avignon and the princes of Italy, and the merchants knew just which fabrics would tempt their wealthy patrons. Yet a change was taking place in the fourteenth century. A long-distance market was beginning to develop in the cheaper textiles, more suited to both the taste and the pocket of the growing middle classes. This was reflected, not only in the rise of new centres of production, but also in the development of new routes and directions of trade.

Expensive fabrics probably constituted only a small part of the cloth now made in the Low Countries. The 'new draperies' and fabrics incorporating some flax were being marketed in France and the Rhineland and were now breaking into the growing Baltic market, where the luxury products had never achieved more than a trifling success. It was not only French Flanders and Brabant which began to sell medium quality fabrics. The 'grey' cloth of the north German towns, the woollens of central Poland, the linen and fustian of south Germany, and the fabrics of numerous French and Catalan towns entered the market in increasing quantities. Some moved to Italy for export to the Balkans, Levant and North Africa. Much was merely sold in other towns, where it was preferred to the local drapery, not because it was superior in quality, but rather because it was different in colour or texture. If any clear pattern of trade had emerged, one cannot easily recognise it in the confused mass of data on cloth purchases and shipments at this time.

Silk, despite its high reputation, was of only minor importance. It was too expensive to command more than a very small market. It was made in Lucca and Venice – silk-weaving centres beyond the Alps had not yet developed – and was dispatched across the mountains to northern centres of wealth and luxury.

Amongst other manufactured or processed goods the metals were probably the most important. Iron, smelted in the forests, was carried for fabrication to the towns. Most towns for which evidence of occupational structure has survived, had a guild – sometimes surprisingly large – of smiths, and many towns were noted for their manufacture of knives or other small pieces of hardware. These must have been distributed widely, if only in very small quantities. Milan, particularly noted for its ironwork, was making armour for the nobility of much of Europe.

The non-ferrous metals, more localised in their occurrence than iron, formed a small but important item in long-distance trade. Lead was carried from its producing areas to construction sites over all of central and western Europe. England sent tin to continental Europe, and in the fourteenth century the Flanders towns had become the emporium of the tin trade, despatching tin to Venice and the Levant. The anonymous author of the fifteenth-century *Libel of English Policie* equated tin with cloth and wool among the exports of England to the continent:

> Thus these galeys for this licking ware,
> And eating ware, bare hence our best chaffare:
> Cloth, woll, and tinne[130]

Copper was commercially more significant than tin, and its importance was greatly enhanced with the coming of bronze or 'brass' cannon. There was, as with tin, a movement of copper ingots from the mining and smelting areas to the urban workshops. Add to these articles of everyday use the costly goods of the luxury trade: the products of the gold- and silversmiths' art, the stained and painted glass, pewter ware for the table, manuscripts and religious books, cut gemstones and sculptures in alabaster or marble, which were transported in an almost random fashion, and we have a pattern of trade of immense variety and complexity. A quantitative analysis of this trade is possible only for a very few commodities like Gascon wine and English wool. For the rest, the picture of European trade about 1330 is entirely qualitative, and estimates of the volume and significance of different articles of commerce are in the main subjective.

CONCLUSION

The Europe of 1330 was familiar with pestilence and famine, but plague had not, at least for many centuries, been among its afflictions. Overpopulation, malnutrition, and bad harvests, which may have been recurring with increasing frequency, appear to have slowed the rate of economic growth. The crude indicators that are available even suggest a stagnation in the economy of much of Europe early in the fourteenth century. An economy as self-sufficing as that of medieval Europe, and as little dependent on the mechanism of the market, could not experience the kind of cyclical economic expansion and contraction familiar to modern societies. Economic growth consisted in the creation of a surplus of labour and goods which could be invested or consumed. In the thirteenth century this surplus had in part been consumed by the feudal classes; in part invested in buildings, land reclamation and development. There must have been an increasing *per capita* national income through these years, but there can be little doubt that most of this increase was used in ways from which the peasant classes benefited not at all.[131]

At some time in the late thirteenth or early fourteenth centuries this growth was slowed or even brought to a halt. In the absence of other indicators, some would put the turning point during that run of bad harvests which began in 1315. It probably was a combination of plague, famine and war which brought on the recession of the later fourteenth century.

The Black Death was brought to Europe in the winter of 1347–8 almost certainly from Caffa, on the Black Sea coast. On its way it infected Constantinople, where John Cantacuzene described its course in words strongly reminiscent of Thucydides on the plague of Athens. From here it spread to Sicily; then to Genoa and from port to port along the shore of the Mediterranean. Boccaccio described it in Florence. Mortality in Avignon was severe, and the Pope withdrew to his apartment, cutting himself off from all communication with the town during the height of the plague. So it spread across Europe, reaching Ireland in 1350, and the Baltic in 1351.

Estimates of mortality varied greatly. There were communities, secular and monastic, which were almost wiped out by the plague; there were others which distance and isolation preserved from infection. The overall mortality has been put as high as a half; it is unlikely to have been less than a quarter, and may well have been a third of the total population. The result was an economic crisis of proportions which Europe had not hitherto witnessed. There was a shortage of labour, and the numerous recurrences of the plague did nothing to improve the situation. For a time there was an acute shortage of foodstuffs, and prices rose sharply as harvests were allowed to rot in the fields for lack of harvesters. But eventually prices began to drop as demand contracted. Agriculture was restructured in many parts of Europe, and the constraints exercised by the medieval community, already weakening before the plague, were in many areas swept away entirely.

Along with the plague came war. Feudal warfare was far from unknown before the plague, but it now became more frequent and even more violent and destructive. The loosening of the bonds of society brought into the field bands of footloose men, mercenaries in wartime, bandits in time of peace. They provided the sinews of almost continuous war, and, where they passed they ravaged like the plague. The hearth lists of Burgundy record for village after village that all had been 'destruiz et brulez par les Ecorcheurs'.[132]

Yet there were breaks in the heavy overcast of the later Middle Ages. The blood-letting of the great pestilence was, in effect, a Malthusian check on too great a population. In the long run the level of welfare of the peasantry may have been improved in consequence. When in the later fifteenth and sixteenth centuries population again began to increase, it was in a technically different age, better equipped to handle it. Nor did warfare strike everywhere; for many areas the fifteenth century was one of quiet prosperity. There was again capital to invest in major building operations, though few of these were now churches or monastic foundations. The cloth industry – no longer a luxury industry for a restricted market – was developing new ranges of production; mining was developing, and from the tin mines of Cornwall to the silver of Bohemia there was expansion and growth. The difficulty in sensing the direction of change lies largely in the fact that high promise co-existed with the blackest depression. The latter never disappeared from Europe, but the areas of light and hope were in the fifteenth century widening into the Renaissance.

Notes

CHAPTER 1

1 Joseph W. Shaw, 'Shallow-water Excavation at Kencheai', *Am. Jnl. Arch.*, 71 (1967), 223–31.
2 Richard J. Russell, 'Geomorphology of the Rhône Delta', *Ann. A.A.G.*, 32 (1942), 149–254.
3 Stanisław Kurnatowski, 'Uwagi o kształtowaniu się stref zasiedlenia dorzecza Obry w czasie od środkowego okresu epoki brązu do późnego średniowiecza', *Arch. Pol.*, 8 (1963), 181–221; Maurice Braure, 'Etude économique sur les Châtellenies de Lille, Donai et Orchies', *Rev. Nord.*, 14 (1928), 85–116, 165–200.
4 Joseph Garnier, *La Recherche des Feux en Bourgogne* (Dijon, 1876).
5 Franz Firbas, *Spät- und nacheiszeitliche Waldgeschichte Mitteleuropas nördlich der Alpen* (Jena, 1949–52), I, 354–63; Alice Garnett, 'The Loess Regions of Central Europe in Prehistoric Times', *Geog. Jnl.*, 106 (1945), 132–41.
6 The final version of this map, which has passed through several editions, is Blatt 10 of *Atlas Östliches Mitteleuropa* (Bielefeld, 1959).
7 Notably by H. Jaeger, 'Zur Erforschung der mittelalterlichen Landesnatur', and Charles Higounet, 'Les forêts de l'Europe occidentale du Ve au XIe siècle', *Agricoltura e Mondo Rurale in Occidente nell Alto Medioevo, Sett. Spol.*, 13 (1966), 343–98.
8 P. George, 'Les sols et les forêts en région méditerranéenne', *Ann. Géog.*, 42 (1933), 194–9.
9 Columella, I, *praefatio*, 1–2.
10 Printed in *Walter of Henley's Husbandry*, trans. Elizabeth Lamond (London, 1890), pp. 60–81; see especially p. 71.
11 J-M. Richard, 'Thierry d'Hireçon: agriculteur artésien', *Bibl. Ec. Chartes*, 53 (1892), 383–416, 571–604.
12 Columella, II. 9. 1.
13 Columella, III. 3. 4.
14 Paul Perceveaux, 'Structures et relations économiques en Dombes à la fin du Moyen Age', *Cah. Hist.*, 12 (1967), 339–57.
15 André Déléage, *La Vie Rurale en Bourgogne jusqu'au début du onzième siècle* (Mâcon, 1941), p. 520, claims that fifteen canons and their servants could be supported by a hundred farms.
16 David Hume, 'On the Populousness of Ancient Nations', *The Philosophical Works of David Hulme*, (Boston, 1854), III, 410–93; see p. 454.

CHAPTER 2

1 Pseudo-Aristotle, *Oeconomica* I. 2.
2 Plato considered that the state should 'be allowed to grow so long as growth is compatible with unity'. *Republic*, IV. 523.
3 Aristotle, *Politics*, I. 1. 8.
4 Pausanias, VIII, 37. 2–5; see also N. J. G. Pounds, 'The First Megalopolis', *P.G.*, 17 (1965), No. 5, 1–5.
5 Pausanias, X. 4. 1.
6 Thucydides, I. 10; II. 16–17.
7 V. Ehrenberg, *The Greek State* (Oxford, 1960), p. 29.
8 'What . . . is Greece . . . how do you define Greece? For most of the Aetolians themselves are not Greeks. No! the countries of the Agraae, the Apodotae, and the Amphilochians are not Greece.' Polybius, XVIII. 5. 7–9. See also A. J. Toynbee, *Hannibal's Legacy: the Hannibalic War's Effects on Roman Life* (Oxford, 1965), I, 72.

9 Herodotus, VIII. 73.
10 Pausanias, VII. 27. 2–5.
11 Oscar Broneer, 'The Corinthian Isthmus and the Ishmian Sanctuary', *Ant.*, 32 (1958), 80–8.
12 A. J. Toynbee, *A Study of History* (Oxford, 1935), II, 43.
13 A. Jardé, *The Formation of the Greek People* (New York, 1926), pp. 91–6.
14 A. Jardé, *The Formation of the Greek People*, p. 86; also H. D. Westlake, *Thessaly in the Fourth Century B.C.* (London, 1935), pp. 1–2; A. Andrewes, *The Greek Tyrants* (London, 1956), p. 143.
15 Stanley Casson, 'A Greek Settlement in Thrace', *Ant.*, VII (1933), 324–8.
16 There is a considerable literature on the political unity of Chalcidice: Allen B. West, *The History of the Chalcidie League*, Bull. of the Univ. of Wisconsin, History Series, vol. IV, No. 2; E. Harrison, 'Chalkidike', *Class. Quart.*, 6 (1921), 93–103, 165–78; Franz Hampl, 'Olynth und der Chalkidische Staat', *Herm.*, 70 (1935), 177–96; Ulrich Kahrstedt, 'Chalcidic Studies', *Am. Jnl. Phil.*, 57 (1936), 416–44.
17 William A. McDonald, *The Political Meeting Places of the Greeks*, 124; see also J. A. O. Larsen, *Greek Federal States* (Oxford, 1968), pp. 3–11.
18 B. D. Meritt, H. T. Wade-Gery and M. F. McGregor, *The Athenian Tribute Lists* (Cambridge, Mass., 1950), III, 239.
19 N. J. G. Pounds, 'The Urbanization of the Classical World', *Ann. A.A.G.*, 59 (1969), 135–57.
20 See the reply of Gelo of Syracuse to the envoys sent from Greece; Herodotus, VII. 157–8. It should be remembered that the Sicilian Greeks were at this time being threatened by the Carthaginians.
21 *Odyssey*, E. V. Rieu's translation (Harmondsworth, 1946), IX. 130–41.
22 The periplus of Skylax gives an incomplete list: *Geographi Graeci Minores*, ed. Carl Muller, vol. I.
23 The claim that the Adriatic settlements were commercial, put forward by R. L. Beaumont, 'Greek Influence in the Adriatic Sea before the Fourth Century B.C.', *Jnl. Hell. St.*, 56 (1936), 159–204, has been effectively refuted by M. I. Finley, 'Trade and Politics in the Ancient World: Classical Greece', 2me *Conf. Int. Hist. Econ.*, 1962, vol. I, 11–35.
24 D. Randall-MacIver, *Greek Cities in Italy and Sicily* (Oxford, 1931), p. 2.
25 Otto-Wilhelm von Vacano, *The Etruscans in the Ancient World* (London, 1960), pp. 32–3.
26 M. Pallottino, *The Etruscans* (Harmondsworth, 1955), p. 69.
27 R. C. Carrington, 'The Etruscans and Pompeii', *Ant.*, 6 (1932), 5–23.
28 Raymond Bloch, *The Origins of Rome* (London, 1960), p. 84.
29 F. E. Adcock, 'The Conquest of Central Italy', *Camb. Anc. Hist.*, VII (1928), p. 583.
30 R. S. Conway, 'The Venetic Inscriptions', in *The Prae-Italic Dialects of Italy*, I, 5.
31 D. Randall-McIver, *The Iron Age in Italy* (Oxford, 1924), p. 61.
32 J. Whatmough, in *Prae-Italic Dialects*, II, 147–8, 154–6.
33 The most ambitious attempt to reconstruct language maps for prehistoric Europe is K. Jażdżewski, *Atlas do Pradziejów Slowiań* (Lódź, 1949). Many articles have appeared, especially in *Slavia Antiqua*, but all show perhaps too great a readiness to identify linguistic with culture-areas.
34 J. Whatmough, *Prae-Italic Dialects*, II, 6, 148, 153.
35 V. Gordon Childe, 'Prehistory', *The European Inheritance* (Oxford, 1954), I, 146.
36 V. Gordon Childe, *Prehistoric Migrations in Europe* (Oslo, 1950), p. 200.
37 T. G. E. Powell, *The Celts* (London, 1958), p. 76.
38 Antonio Arribas, *The Iberians* (London, 1964), pp. 41, 56.
39 J. M. de Navarro, 'The Coming of the Celts', *Camb. Anc. Hist.*, VII, pp. 41–74.
40 V. Gordon Childe, in *Encycl. Brit.*, sub 'Scythia'.
41 Ellis H. Minns, *Scythians and Greeks* (Cambridge, 1913), p. 35.
42 Herodotus, IV, 47.
43 A. L. Mongait, *Archaeology in the U.S.S.R.* (Harmondsworth, 1961), pp. 158–63.
44 Herodotus, IV. 106.
45 Stanley Casson, *Macedonia, Thrace and Illyria* (Oxford, 1926), pp. 325–7.

46 *Encycl. Brit.*, (1962), sub 'Thrace'.
47 Herodotus, V. 3.
48 Aristotle, *The Nicomachean Ethics*, IX. 10.3: 'Ten people would not make a city, and with a hundred thousand, it is a city no longer.'
49 Plato, *Laws*, V. 737.
50 A. H. M. Jones, *Athenian Democracy* (Oxford, 1957), p. 179.
51 T. R. Glover, *The Challenge of the Greek* (Cambridge, 1943), p. 88.
52 G. B. Grundy, *Thucydides and the History of His Age* (Oxford, 1961), I, 89—90; A. French, *The Growth of the Athenian Economy* (London, 1964), gives at least 250,000. For a summary statement of earlier estimates see A. B. Wolfe, 'The Economics of Population in Ancient Greece', *Facts and Factors in Economic History* (Cambridge, Mass., 1932), pp. 18–39.
53 *Inscr. Graec.* II^2, 1672 (The Eleusis tables).
54 Demosthenes, XX. 31–2; for an analysis of Demosthenes' figures see A. H. W. Jones, 'The Social Structure of Athens in the Fourth Century B.C.', *Econ. Hist. Rev.*, 8 (1955), 141–55.
55 Julius Beloch, *Die Bevölkerung der Griechisch-Römischen Welt* (Leipzig, 1888).
56 M. Rostovtzeff, *Iranians and Greeks in South Russia* (Oxford, 1922), p. 66, states: 'The cities of the Athenian confederation tried to import as much foodstuffs as possible: the other cities did the same.'
57 Jacques Heurgon, *La vie quotidienne chez les Etrusques* (Paris, 1961), p. 180.
58 In *Ciba Foundation Symposium on Medical Biology and Etruscan Origins* (Boston, 1959), pp. 80–1.
59 Pseudo-Dicaearchus; this is J. G. Frazer's translation in *Pausanias' Description of Greece*, I, xliii–xlvii. In Muller's text, *Geographi Graeci Minores*, I, 101.
60 Erik J. Holmberg, *The Swedish Excavations at Asea in Arcadia*, Skrifter Utgivna av Svanska Institutet i Rom, vol. XI (Lund, 1944).
61 *Geographi Graeci Minores*, ed. C. Muller, I, 99. The translation is that of Frazer, *Pausanias*, I, xliii–xlvii.
62 Rodney S. Young, 'An Industrial District of Ancient Athens', *Hesp.* 20 (1951), 139.
63 Demosthenes, *Olynthiac*, III. 25.
64 *Geographi Graeci Minores*, Dicaearchus, I. 104.
65 Old Smyrna is alleged to have had a rectilinear plan in the seventh century B.C.; see Ekrem Akurgal, 'The Early Period and the Golden Age of Iona', *Am. Jnl. Arch.*, 66 (1962), 369–74; J. M. Cook, 'Old Smyrna', *Ann. Brit. Sch. Ath.*, 53–4 (1958–9), 1–34.
66 Aristotle, *Politics*, II. 5. 1.
67 Diodorus Siculus, XII. 10. 7.
68 Thucydides, I. 89.
69 *Ibid.*, I. 90.
70 Gosta Saflund, 'The Dating of Ancient Fortifications in Southern Italy and Greece', *Opuscula Archaeologica*, 1 (Lund, 1935), 87–119; it is repeated in Rhys Carpenter's discussion of the city walls of Corinth, *Corinth.* vol. III, pt. II, 'The Defences of Acrocorinth and the Lower Town' (Cambridge, Mass., 1936).
71 R. E. Wycherley, *How the Greeks Built Cities* (London, 1962), p. 39.
72 A. W. Gomme, *A Historical Commentary on Thucydides* (Oxford, 1956), II, 63–5.
73 A. W. Gomme, *ibid.*, II, 39.
74 Census of 1961; figures taken for the *nomos* of Khalkidhiki and part of that of Thessaloníki. This area includes the semi-urbanised country around the city of Thessaloníki.
75 J. Beloch, *Die Bevölkerung der Griechisch-Römischen Welt*, p. 506; Allen B. West, *The History of the Chalcidic League*, Bull. of the University of Wisconsin, History Series, vol. IV, p. 755.
76 N. J. G. Pounds, 'The Urbanization of the Classical World', *Ann. A.A.G.*, 59 (1969), 135–57.
77 David M. Robinson and J. Walter Graham, *Excavations at Olynthus*, pt. VIII, The Johns Hopkins Studies in Archaeology, No. 25 (Baltimore, 1938), pp. 43–4. This confirms the conclusions reached by A. B. West, *The History of the Chalcidic League*, 154–6.

78 W. A. Eden, 'The Plan of Mesta Chios'. See also H. A. Ormerod, *Piracy in the Ancient World* (London, 1924), pp. 38–58.
79 This appears to be what Plato recommended for his ideal city, *Laws*, VI. 778E–779B.
80 F. Haverfield, *Ancient Town-Planning* (Oxford, 1913), pp. 57–62, seems to have doubted whether the Etruscans could have built Marzabotto, but to Paul MacKendrich, 'Roman Town Planning', *Arch.*, 9 (1956), 126–33, 'Etruscans were great city planners'. See also Guido A. Mansuelli, 'La cité étrusque de Marzabotto et les problèmes de l'Etrurie Padane', *Acad. Inscr. C. R.* (1960), 65–84; Nereo Alfieri and Paulo Enrico Arias, *Spina* (Munich, 1958).
81 Gosta Saflund, 'Ancient Latin Cities of the Hills and Plains', *Opuscula Archaeologica* (Lund, 1935), I, 64–86.
82 Ddzisław Rajewski, *Biskupin-Wykopaliska* (Warsaw, 1960); also *Historia Polski*, vol. I, pt. 1 (Warsaw, 1958), pp. 46–8.
83 A. W. Gomme, *The Population of Athens in the Fifth and Fourth Centuries B.C.* (Oxford, 1933), pp. 26–47.
84 Xenophon, *Oeconomicus*, IX; XII.
85 Gutorm Gjessing, 'Circumpolar Stone Age', *Acta Arc.* (Dk.) Fasc. II (1964), 46–7.
86 Thucydides, I. 2.
87 Gustave Glotz, *Ancient Greece at Work* (London, 1926), p. 256; the present province of Attica has an area of 1,457 square miles.
88 Menander, *Arbitrants*, lines 25–30.
89 Euripides, *Bacchae*, 677–9.
90 Euripides, *Electra*, lines 410–12.
91 Plato, *Laws*, VI. 761; VIII. 844. Dionysius of Halcarnassus, I. 61. 2, describes how cultivation in the valleys of Arcadia was prevented by the poor drainage; the passage relates to a mythical event, but the physical circumstances were doubtless regarded as possible.
92 Theophrastus, IV. 4. 10.
93 Figures for the late 1930s have been used, in order to exclude recent improvements; see *Statistique annuelle agricole de l'élevage de la Grèce, 1937* (Athens, 1938); the crop yields are for 1948–53.
94 Aristophanes, *Acharnians*, lines 33–6. See also V. Ehrenberg, *The People of Aristophanes* (Oxford, 1951), pp. 73–94.
95 Moses I. Finley, *Studies in Land and Credit in Ancient Athens, 500–200 B.C.*, (New Brunswick, New Jersey, 1951), pp. 58–9.
96 Pseudo-Demosthenes, XLII; see also G. E. M. de Ste Croix, 'The Estate of Phaenippus', *Ancient Society and Institutions* (Oxford, 1966), pp. 109–14.
97 *Oeconomicus*, IX; XII.
98 Aristotle, *The Oeconomica*, I. 2–6; this work may in part have been derived from Xenophon.
99 Gustave Glotz, *Ancient Greece at Work* (London, 1926), p. 247.
100 Aristophanes, *Acharnians*, lines 994–8.
101 George Thomson, 'On Greek Land Tenure', in *Studies Presented to David Moore Robinson*, ed. G. E. Mylonas and D. Raymond (St. Louis, Mo., 1951), II, 840–50.
102 Hesiod, *Works and Days;* see A. F. S. Gow, 'The Ancient Plough', *Jnl. Hell. St.*, 34 (1914), 249–75.
103 Horses or mules were also used; see Sophocles, *Antigone*, lines 338–40, and comment by W. E. Heitland, *Agricola* (Cambridge, 1921), p. 33.
104 Isaeus, VIII. 42.
105 Aristophanes, *Clouds*, lines 71–2; see also V. Ehrenberg, *The People of Aristophanes* (Oxford, 1951), pp. 73–94.
106 Aristophanes, *Knights*, line 480.
107 'The Thebans in their distress sent to the Thessalian port of Pagasae for grain': Xenophon, *Hellenica*, V. iv. 56.
108 A. Jardé, *The Formation of the Greek People* (New York, 1926), p. 130.
109 *Odyssey*, IX. 144; IV. 79.
110 Aristophanes, *Wasps*, lines 838, 897; *Peace*, line 250.
111 Livy, II. 9. 6; II. 24. 3–7; IX. 36. 12.
112 Livy, IX, 13. 7; X. 2. 6–8.

113 Gudmund Hatt, 'Prehistoric Fields in Jyland', *Acta Arch.*, 2 (1931), 117–58.
114 J. Brønsted, *Danmarks Oldtid* (Copenhagen, 1938); see also J. G. D. Clark, *Prehistoric Europe*, p. 99, and for a more recent view of the Danish fields, Ole Klindt-Jensen, *Denmark before the Vikings* (London, 1957), pp. 94–5.
115 M. Stenberger and O. Klindt-Jensen, *Vallhagar, a Migration Period Settlement* (Copenhagen, 1955), vol. I, *passim*.
116 E. Cecil Curwen, 'Furrows in Prehistoric Fields in Denmark', *Ant.*, 20 (1946), 38–9; also *Vallhagar*, I, 263; E. Cecil Curwen, 'Evidence of Early Ploughing from Holland', *Ant.*, 20 (1946), 158.
117 H. Helbaek, 'The Botany of the Vallhagar Iron Age', *Vallhagar*, II, 653–99.
118 Shimon Applebaum, 'The Agriculture of the British Early Iron Age as exemplified at Figheldean Down, Wiltshire', *Proc. Prehist, Soc.* (1954), 103–14.
119 E. H. Minns, *Scythians and Greeks* (Cambridge, 1913), pp. 159–61.
120 Friedrich Oerted, 'Zur Frage der attischen Grossindustrie', *Rhein. Mus. Philol.*, N.F., 79 (1930), 233–52. The literature is reviewed by E. Will, 'Trois quarts de siècle de recherches sur l'économie grecque antique', *Ann. ESC.* 9 (1954), 7–22.
121 Lysias, 12. 19; 24. 6.
122 Marcus N. Tod. 'The Economic Background of the Fifth Century', *Camb. Anc. Hist.*, 5 (1953), 15.
123 Thucydides, VII. 27.
124 Richard H. Randall, 'The Erechtheum Workmen', *Am. Jnl. Arch.*, 57 (1953), 199–210. The deme affiliation indicates their 'home' rather than their place of residence. Clearly they lived nearer Athens than some of these affiliations would suggest.
125 On technical aspects of this and other branches of Greek manufacturing see Hugo Blumner, *Technologie und Terminologie der Gewerbe und Kunste bei Griechen und Romern*, 4 vols. (Leipzig, 1875–86). Vol. I was revised and republished in 1912. Also Charles Singer, E. J. Holmyard and A. R. Hall, *A History of Technology*. vol. I (Oxford, 1954).
126 Pollux, I. 149. The arts and crafts of Sparta were probably pursued by the *Perioeci*; see R. M. Cook, 'Spartan History and Archaeology', *Class. Quart.*, 56 (1962), 156–8.
127 Lysias 12. 19; Demosthenes 'Against Aphobus', I. 9–11, *Private Orations*, 27.
128 According to Xenophon; H. Bolkestein, *Economic Life in Greece's Golden Age* (Leiden, 1958), p. 52, considers this to be exaggerated, but not without substance.
129 Livy, XXVIII. 45.
130 Raymond Bloch, *The Etruscans* (London, 1958), p. 121.
131 Christopher and Jacquetta Hawkes, *Prehistoric Britain* (Harmondsworth, 1949), pp. 117–8.
132 S. Lauffer, *Die Bergwerkssclawen von Laureion*, Akad. Wiss Mainz, Abh. Geistes- u. Sozialwiss. 1955, No. 12; 1956, No. 11.
133 R. Pittioni, 'Prehistoric Copper Mining in Austria', *Annual Report, Institute of Archaeology*, London University, 7 (1951), 16–43; *ibid.*, 'Recent Researches on Ancient Copper Mining in Austria, *Man*, 48 (1948), No. 133, 120–2.
134 *Ora Maritima*, ed. A. Schulten, *Fontes Hispaniae Antiquae*, I (Barcelona, 1955).
135 Diodorus Siculus, V. 22. 1–4.
136 Plato, *Republic*, II. 370. e.
137 V. Ehrenberg, *The People of Aristophanes*, pp. 113–46.
138 J. Beloch, 'Die Handelsbewekung im Altertum', *Jahrb. Nat. Stat.* (1899), 626–31; *Griechische Geschichte* (Strasbourg, 1893-1904), III, 324ff.
139 K. Bücher, *Die Entstehung der Volkswirtschaft*, 6th edn. (Tübingen, 1908), p. 361; Johannes Hasebroek, *Trade and Politics in Ancient Greece* (London, 1933), pp. 71–96; for a judicious though brief summing up, see A. French, *The Growth of the Athenian Eonomy* (London, 1964), pp. 107–29, and Fritz Heichelheim, 'Sitos', *Paul. Wiss.* Suppl. Br. 6 (Stuttgart, 1935), cols. 833–44.
140 A. Koster, *Das antike Seewesen* (Berlin, 1923).
141 Augustus Boeckh, *The Public Economy of Athens*, trans. G. C. Lewis (London, 1847), p. 54.
142 H. Michell, *The Economics of Ancient Greece* (Cambridge, 1940), p. 277; Athenaeus, XIII. 586; 596.
143 Xenophon, *Hellenica*, V. 4. 56.

144 Xenophon, *Hellenica*, IV. 7. 1.
145 Xenophon, *Hellenica*, VI. 1. 11.
146 J. M. F. May, *Ainos: its History and Coinage* (Oxford, 1950). 3–4. 80ff.
147 Herodotus, VII. 158.
148 L. Gernet, *L'Approvisionnement d'Athènes en Blé au Ve et au IVe Siècles* (Paris, 1909), p. 314.
149 Thucydides, VIII. 96. Michell raises the question, p. 261, that grain from Euboea may in fact have been grown farther afield and shipped to a Euboean port, such as Chalchis or Eretria, to be sent on by land to Athens. Economic considerations would seem to be against this suggestion.
150 H. Michell, *The Economics of Ancient Greece*, p. 259.
151 Xenophon, *Oeconomicus*, XX. 27.
152 Thucydides, VI. 90.
153 Euripides, *Cyclops*, 395.
154 Pseudo-Dicaearchus, *Geographi Graeci Minores*, I, 104; the translation is that of Frazer.
155 H. Michell, *The Economics of Ancient Greece*, p. 287.
156 Athenaeus, I. 4; I. 27.
157 Aristophanes, *Knights*, 22. 311–12.
158 Athenaeus, III. 116.
159 Xenophon, *On the Revenues*, I. 6–7. See also Isocrates, *Panegyricus*.
160 Xenophon, *Hellenica*, VI. 2. 9.
161 M. Cary, 'The Greeks and Ancient Trade with the Atlantic', *Jnl. Hell. St.*, 44 (1924), 166–79; see also Carl Roebuck, *Ionian Trade and Colonization*, American Archaeological Institute (New York, 1959), pp. 94–101.
162 Jean J. Hatt, 'Les problèmes des influences Helleniques dans le domain celtique aux VIe siècles avant notre ère', *Actas de la IV Sesion, Madrid, 1954, Int. Congr. of Prehist. and Protohist. Sci.* (Zaragoza, 1956), pp. 823–8.
163 J. M. De Navarro, 'Massilia and Early Celtic Culture', *Ant.*, 2 (1928), 423–42.
164 Roger Dion, 'Le problème des Cassiterides', *Lat.* 11 (1952), 306–14.
165 H. O'N. Hencken, 'The Ancient Cornish Tin Trade', *Proc. First Int. Congr. of Prehist. and Protohist. Sci., London, 1932* (Oxford, 1934), pp. 283–4; also *Archaeology of Cornwall and Scilly* (London, 1932), pp. 158–87.
166 Diodorus Siculus, V. 22. 1–4.
167 J. M. de Navarro, 'Prehistoric Routes between Northern Europe and Italy defined by the Amber Trade', *Geog. Jnl.* (1925), 66 vol., 481–507.
168 See Stephen Foltiny, 'Athens and the East Hallstatt Region', *Am. Jnl. Arch.*, 65 (1961), 283–97; A. M. Snodgrass, 'Iron Age Greece and Central Europe', *ibid.*, 66 (1962), 408–10.

CHAPTER 3

1 This tradition has been generally accepted, at least since the time of Gibbon. 'That tradition,' wrote A. H. M. Jones, 'is not a very good one.' He presents arguments to show that the age was neither as prosperous nor as contented as it has been represented; see also F. W. Walbank, *The Decline of the Roman Empire in the West* (London, 1946), p. 7.
2 Aelius Aristeides, *To Rome*, 70; see also James H. Oliver, 'The Ruling Power: A study of the Roman Empire in the Second Century after Christ through the Roman Oration of Aelius Aristedes', *Trans. Am. Phil. Soc.*, 43 (1953), 869–1003.
3 Tertullian, *De anima*, 30. 3.
4 A. N. Sherwin-White, 'Roman Imperialism', in *The Romans*, ed. J. P. V. D. Balsdon (London, 1965), pp. 76–100.
5 *The Letters of Cassiodorus*, ed. Thomas Hodgkin (London, 1886), Book 8, No. 31, p. 378.
6 Tacitus, *Annals*, I. 11.
7 'The Roman Frontier System', in *Essays by Henry Francis Pelman*, ed. F. Haverfield (Oxford, 1911), pp. 164–78.
8 Olwen Brogan, 'An Introduction to the Roman Land Frontier in Germany' *Gr. and R.*, 3 (1933–4), 23–30.

9 This is reflected in the conflicting lines shown in different, and even in the same historical atlases. See also Vasile Christescu, 'Le Limes romain de la Valachie', *Résumés des Communications présentées au Congrès, VII*e *Congr. Int. Sci. Hist.* (Warsaw, 1933), I, 72–3.
10 F. Millar, 'The Emperor, the Senate and the Provinces', *Jnl. Rom. St.*, 56 (1966), 156–66. See also W. T. Arnold, *The Roman System of Provincial Administration to the accession of Constantine the Great* (Oxford, edn. of 1906), pp. 100–68; Mason Hammond, *The Augustan Principate* (New York, 1968), pp. 54–64.
11 *Notitia Galliarum, M.G.H., Chronica Minora*, 9 (1892), 584–612. This source belongs to the late fourth or early fifth century.
12 M. Rostovtzeff, *The Social and Economic History of the Roman Empire* (Oxford, 1926–57), I, 135.
13 Pliny, *Natural History*, III. 18.
14 G. H. Stevenson, *Roman Provincial Administration*, p. 160, suggested that many had in the meantime disappeared or been merged with others.
15 Rudi Thomsen, *The Italic Regions*, Classica et Medievalia: Dissertationes IV (Copenhagen, 1947), vol. II.
16 Juvenal, *Satires*, III. 58–125.
17 Lucian, *Negrinus*, esp. pp. 112–13, 116–17 and *passim* in the Loeb edition of Lucian, vol. I.
18 Strabo, III. 2. 15.
19 Ramsay MacMullen, 'Barbarian Enclaves in the Northern Roman Empire', *Ant. Class*, 32 (1963), 552–61; also *id.*, *Enemies of the Roman Order* (Cambridge, Mass., 1966).
20 Strabo, V. 1. 4; see also Pliny's '15 peoples of Lucus, unimportant and bearing outlandish names', *Natural History*, III. iii. 28.
21 R. G. Collingwood and J. N. L. Myres, *Roman Britain and the English Settlements* (Oxford, 1937), pp. 208–25.
22 The evidence is summarised by Georges Dottin, *Manuel pour servir à l'étude de l'Antiquité Classique* (Paris, 1915), pp. 134–9; Camille Jullian, *Histoire de la Gaule* (Paris, 1920), VI 110–15.
23 Ramsay MacMullen, 'The Celtic Renaissance', *Hist.*, 14 (1965), 93–104.
24 Gaius, *Institutes*, III. 93 (*Font. Jur. Rom. Antejust.* II, 120–1).
25 St. Jerome, *Pat. Lat.*, XXXVI, col. 382.
26 Horace, *Epistolae*, II. 1, lines 156–7.
27 Strabo, VI. 1. 2.
28 Seneca, *Ad Helviam Matrem de Consolatione, Dialogues*, XII. 6, 2–3.
29 Juvenal, *Satires*, III. 62.
30 Tenney Frank, 'Race Mixture in the Roman Empire', *Am. Hist. Rev.*, 21 (1915–16), 689–708.
31 Ramsay MacMullen, *Soldier and Civilian in the Later Roman Empire* (Cambridge, Mass., 1963), p. 96.
32 Ramsay MacMullen, 'Barbarian Enclaves in the Northern Roman Empire', *Ant. Class.*, 32 (1963), 552–61.
33 J. P. Gilliam, 'Roman and Native A.D. 122–197', *Roman and Native in North Britain*, ed. I. A. Richmond (London, 1958), p. 75; A. H. A. Hogg, 'The Votadini', *Aspects of Archaeology*, ed. W. F. Grimes (London, 1951), p. 211; see also George Jobey, 'Homesteads and Settlements of the Frontier Area', *Rural Settlement in Roman Britain*, pp. 1–14.
34 On the reliability of Tacitus see *Cornelii Taciti de Origine et Situ Germanorium*, ed. J. G. C. Anderson (Oxford, 1938), XIX–XXXVII.
35 E. H. Bunbury, *A History of Ancient Geography* (London, 1879–1959), II, 495; see also J. Oliver Thompson, *History of Ancient Geography* (Cambridge, 1948), pp. 242–4.
36 Konrad Jażdżewski, *Atlas do Pradziejów Słowiań*, plate 9 of atlas; pp. 63–7 of text; *Cornelii Taciti de Origine et Situ Germanorum*, ed. J. G. C. Anderson, pp. 216–17.
37 Marija Gimbutas, *The Slavs*, p. 62; Tacitus, *Germania*, 46; Pliny, *Natural History*, IV. 97; Ptolemy, III. 5, 7, 8, 56; see Ferdinand Lot, *Les Invasions Barbares* (Paris, 1942), pp. 206–7.
38 J. G. C. Anderson's edition of *Germania*, 198–9; L. Schmidt, *Histoire de Vandales* (Paris, 1953), pp. 7–12.

39 Tacitus, *Germania*, 45; see J. G. C. Anderson's comment, p. 209.
40 Ptolemy, III. 5, 9. The medieval Galinda and Suduva lived to the north of the Narew, in East Prussia; see Marija Gimbutas, *The Balts* (London, 1963), pp. 22–4.
41 Pliny, *Natural History*, IV. 97; XXXVII. 11–12; Tacitus, *Germania*, 45.
42 Tacitus, *Germania*, 46, 3–4.
43 T. H. Hollingsworth, 'The Importance of the Quality of the Data in Historical Demography', *Daed.*, 97 (1968), 415–32.
44 Arthur E. R. Boak, *Manpower Shortage and the Fall of the Roman Empire in the West* (Ann Arbor, Mich., 1955), pp. 15–16.
45 Summarised by Tenney Frank, 'Roman Census Statistics from 225 to 28 B.C.', *Class. Philol.*, 19 (1924), 329–41.
46 *Augusti Res Gestae*, II. 8. 2–4; Tacitus, *Annals*, XI. 25.8.
47 Tenney Frank, *An Economic Survey of Ancient Rome*, V, 'Rome and Italy of the Empire', p. 141.
48 Harold Mattingly, s.v. 'Population (Roman World)', pp. 718–19.
49 This is also the conclusion of the most recent writer on the subject, P. A. Brunt, *Italian Manpower 225 B.C. – A.D. 14* (Oxford, 1971), 121–30, 198–203.
50 Russell Meiggs, *Roman Ostia* (Oxford, 1960), pp. 532–3. Tenney Frank's estimate of 'about 100,000 inhabitants' is clearly too large: 'The People of Ostia', *Class. Jnl.*, 29 (1934), 481–93.
51 V. M. Scramuzza, 'Roman Sicily', in *An Economic Survey of Ancient Rome*, III, 334.
52 G. J. Beloch, 'La popolazione antica della Sicilia'. *Arch. Stor. Sic.*, 14 (1889), 3–83, esp. 64, but see also M. I. Finley, *Ancient Sicily* (London, 1968), pp. 122–136, esp. p. 133.
53 Julius Beloch, 'Die Bevölkerung Galliens zur Zeit Caesars', *Rhein. Mus. Philol.*, N.F., 54 (1899), 414–45; Camille Jullian, *Histoire de la Gaule* (Paris, 1921), II, 3–13.
54 Pliny, *Natural History*, III. iii. 28. This, of course, relates to Galicia, Asturias and northern Portugal.
55 Julius Beloch, *Die Bevölkerung der Griechisch–Römischen Welt*, p. 448.
56 U. Kahrstedt, 'Geschichte der Bevolkerungsbewung', p. 668.
57 Pliny, *Natural History*, III, 1. 7; Rolf Nierhaus, 'Zum wirtschaftlichen Aufschwung der Baetica zur Zeit Trajans und Hadrians', *Les Empereurs Romains d'Espagne* (Paris, 1965), pp. 181–94.
58 J. J. van Nostrand, 'Roman Spain', in Tenney Frank, *An Economic Survey of Ancient Rome*, III, 148.
59 R. G. Collingwood and J. N. L. Myres, *Roman Britain and the English Settlements*, p. 180.
60 R. E. M. Wheeler, 'Mr Collingwood and Mr Randall', *Ant.*, 4 (1930), 91–5.
61 S. S. Frere, *Britannia* (London, 1967), pp. 310–11; this total includes 320,000 persons 'who had to be fed by the labours of the agricultural population'.
62 Andreas Mócsy, *Die Bevölkerung von Pannonien bis zu den Markomannenkriegen*, (Budapest, 1959), p. 124; 'Pannonia', *Paul. Wiss.*, Suppl. Bd. IX, cols. 667–81; on urban development, cols. 596–610.
63 G. W. Bowersock, *Augustus and the Greek World*, (Oxford, 1965) p. 91.
64 H. Furneaux, ed., *De Germania* (Oxford, 1894), pp. 14–16.
65 J. B. Bury, *The Invasion of Europe by the Barbarians* (London, 1928), p. 42; see also Ferdinand Lot, *Les Invasions Barbares et le Peuplement de l'Europe*, 2 vols. (Paris, 1937–42).
66 In some areas the cultivated area may have been more extensive than it is today; see M. Stenberger, 'Remnants of Iron Age Houses on Öland', *Acta Arch. (Dk)*, 2 (1931), 93–104.
67 A. E. R. Boak, *Manpower Shortage and the Fall of the Roman Empire in the West* (Ann Arbor, 1955), but see critique by M. I. Finley, *Jnl. Rom. St.*, 48 (1968), 156–64, who demonstrates the inadequacy of Boak's assumptions.
68 Tacitus, *Agricola*, 21 (trans. H. Mattingly, Penguin).
69 S. S. Frere, 'The End of Towns in Roman Britain', *The Civitas Capitals of Roman Britain* (Leicester, 1966), pp. 87–100.
70 A. H. M. Jones, *The Later Roman Empire*, II, 712; *id.*, 'The Economic Life of the Towns of the Roman Empire', *Rec. Soc. Jean Bodin*, 7 (1955), Pt. 2, 161–94;

Frank Frost Abbott and Allan Chester Johnson, *Municipal Administration in the Roman Empire* (Princeton, N.J., 1926), pp. 56–68.

71 On the nature of *civitas*, see J. C. Mann, 'Civitas: Another Myth', *Ant.*, 34 (1960), 222–3; Sheppard Frere, 'Civitas – A myth?', *Ant.*, 35 (1961), 29–36. The present writer agrees with Frere in regarding the *civitas* as 'the whole tribal area, its capital (*caput*) normally ranking merely as a *vicus* where the government happened to have its seat'. See also Norman J. DeWitt, *Urbanization and the Franchise in Roman Gaul* (Lancaster, Penn., 1938), pp. 18–19; J. J. Hatt, *Histoire de la Gaule Romaine*, pp. 90–1; A. Grenier, *Manuel d'Archéologie Gallo-Romaine*, I, 143–5; and *The Civitas Capitals of Roman Britain*, ed. J. S. Wacher.

72 Strabo, III. 2. 5 (143c); 4. 13 (163c).

73 Graham A. Webster, 'Fort and Town in Early Britain', *Civitas Capitals of Roman Britain*, pp. 31–45.

74 Strabo, V. 1. 12 (218c).

75 For attempts to do so see Ferdinand Lot, 'Recherche sur la Superficie des cites remontant à la période gallo-romaine', *Bibl. Ec. Ht. Et.*, 287 (1945); 296 (1950); 307 (1953).

76 A. Grenier, *Manuel d'Archéologie Gallo-Romain*, I, 284-420; Adrien Blanchet, *Les Enceintes Romaines de la Gaule* (Paris, 1907), gives city plans for all Gallic cities for which they can be constructed.

77 These figures are from R. M. Butler, 'Late Roman Town Walls in Gaul', *Arch. Jnl.* 116 (1961), 25–50; see also A. Grenier, *Manuel d'Archéologie Gallo-Romain*, I, 420; A. Blanchet gives somewhat different figures.

78 Strabo, IV. 1. 11 (186).

79 Pliny, *Natural History*, III. 1. 7; he uses the term *oppida* normally reserved for pre-Roman fortified settlements.

80 R. Thouvenot, 'Essai sur la Province romaine de Bétique', *Bibl. Ec. franc.*, 149 (1940), 363–78.

81 J. Beloch, *Die Bevölkerung der griechischrömischen Welt*, pp. 427–35.

82 Pliny, *Natural History*, III. XV. 115–xviii.

83 Strabo, V. 1. 4–12.

84 A. L. F. Rivet, 'Summing-up: Some Historical Aspects of the Civitates of Roman Britain', *Civitas Capitals of Roman Britain*, pp. 107–10.

85 Axel Boëthius, 'Roman and Greek Town Architecture', *Göteborgs Högskolas Årsskrift*, 54 (1948), No. 3, 8.

86 Vitruvius, *De Architectura*, II. 3–6.

87 Strabo, V. 3. 7 (235c) and R. C. Carrington, 'The Ancient Italian Town-House', *Ant.*, 7 (1933), 133–52.

88 A. Boëthius, 'Urbanism in Italy', in *The Classical Pattern of Modern Western Civilization* (Copenhagen, 1958), IV, 87–107; also Fiechter, s.v. 'Römisches Haus', *Paul. Wiss.*, 1 (1914), cols. 981–2.

89 Lily Ross Taylor, *The Divinity of the Roman Emperor* (Middletown, Conn., 1931), pp. 181–204, 212.

90 S. S. Frere, *Britannia*, p. 257.

91 A. Boëthius, 'Roman and Greek Town Architecture', *Göteborgs Högskolas Årsskrift*, 54 (1948), No. 3, 6.

92 Axel Boëthius, *The Golden House of Nero: Some Aspects of Roman Architecture* (Ann Arbor, Mich., 1960); the house did not last long; it was replaced first by the baths of Titus and then by those of Trajan.

93 The rich did not necessarily change their place of residence; they acquired several villas, one of which might have been a 'town-house'. See the *Letters* of the Younger Pliny, especially II. 17.

94 Jérôme Carcopino, *Daily Life in Ancient Rome*, ed. Henry T. Rowell (London, 1941), p. 20, says 'at least 1,000'.

95 Juvenal, *Satires*, I. 3, lines 190–8; this satire was probably written soon after A.D. 100 Cicero had been hte owner of such property; see *Ad Atticum*, XIV. 9.1; 'Two of my shops (*tabernae*) have fallen down and the rest are cracking: so not only the tenants, but even the mice have migrated.' See also Z. Yavetz, 'Living Conditions of the Urban Plebs in Republican Rome', *Lat.*, 17 (1958), 500–17.

96 Tacitus, *Annals*, XV. 38–41.

97 Tacitus pointed out (*Annals*, XV. 43) that the 'narrow streets and high houses

had provided protection against the burning sun, whereas now the shadowless open spaces radiated a fiercer heat'.

98 Suetonius, *Nero*, 38. 2.

99 Giantilippo Carettoni, Antonio M. Colini, Lucos Cozza and Guglielmo Gatti, *La Pianta Marmorea di Roma Antica: Forma Urbis Romae*, 2 vols. (Rome, 1960); also Paul Bigot, *Rome Antique au IVe Siècle ap. J. C.* (Paris, 1942).

100 Dio Cassius, XXXIX. 61. 1–2.

101 Tacitus, *Annals*, XV. 18.

102 Helen Jefferson Loane, *Industry and Commercc of the City of Rome*, (Baltimore, 1938), pp. 13–14; on the warehouses (*horrea*), see R. A. Staccioli, 'Tipi di *horrea* nella documentazione della *Forma Urbia*', *Hommages à Albert Grenier*, Collection Latomus 58 (Brussels,.1962), pp. 1430–40, and Geoffrey Rickman, *Roman Granaries and Store Buildings* (Cambridge, 1971).

103 Strabo, V. 3. 8. (235c).

104 Frontinus, *De Aquis*, I. 11, 16.

105 That of 54 B.C.; Dio Cassius, XXXIX. 61. 1–2.

106 Suetonius, *Augustus*, 30. 1. See J. Le Gall, *Le Tibre, fleuve de Rome dans l'antiquité* (Paris, 1953), pp. 113–28.

107 Tacitus, *Annals*, I. 76. 1. Missuse of the river, however, continued. Nero, for instance, had grain that had deteriorated dumped into the river: Tacitus, *Annals*, XV. 18.

108 Dionysius of Halicarnassus, III. 44. 1–4.

109 Russell Meiggs, *Roman Ostia* (Oxford, 1960), p. 61.

110 O. G. S. Crawford, 'Our Debt to Rome?' *Ant.*, 2 (1928), 173–88.

111 A. J. Toynbee, *Hannibal's Legacy* (London, 1965), II, 247–52.

112 See J. Ward-Perkins, 'Etruscan Towns, Roman Road and Medieval Villages: the Historical Geography of Southern Etruria', *Geog. Jnl.*, 128 (1962), 389–405.

113 Guy Duncan, 'Sutri', *Papers Brit. Sch. Rome*, 26 (1958), 63–134; G. D. B. Jones, 'Capena and the Ager Capenas', *Papers Brit. Sch. Rome*, 30 (1962), 116–208; 31 (1963), 100–58; *Ibid.*, 18 (1963), 100–58; R. M. Ogilvie, 'Eretrum', *ibid.*, 33 (1965), 70–112.

114 Horace, *Satires*, I. 5; Silius Italicus, *Punica*, VIII. 379–83.

115 Pliny the Younger, II. 17.

116 Suetonius, *Domitian*, vii. 2. See also Antonio Sirago, *L'Italia agraria sotto Traiano* (Louvain, 1958), p. 193.

117 F. Haverfield, 'Centuriation in Roman Britain', *Eng. Hist. Rev.*, 33 (1918).

118 F. Oelmann, 'Ein gallorömischer Bauernhof bei Mayen', *Bonn. Jahrb.*, 123 (1928), 51–152; R. de Maeyer, *De Romeinsche Villa's in België: een archeologische studie* (Antwerp, 1937); excavation has shown a very similar history for the villa at Lullingstone, Kent; see G. W. Meates, *Lullingstone Roman Villa* (London, 1962), pp. 6–8. See also E. Sadee, 'Gutsherrn und Bauern im römischen Rheinland', *Bonn Jahrb.*, 128 (1923), 109–17.

119 Albert Grenier, *Manual d'Archéologie Gallo-Romaine*, II, Pt 2, 795; O. Brogan, *Roman Gaul* (London, 1953), pp. 120–6.

120 This accords with the description given by Tacitus, *Germania*, 16, and Vitruvius, *De Architectura*, II. 1. 4. See A. Grenier, *Manuel d'Archéologie Gallo-Romaine*, II, Pt. 2, 752–63; *id.*, 'Habitations gauloises et Villas latines dans la Cité des Médiomatrices', *Bibl. Ec. Ht. Et.*, 157 (1906), 49.

121 Roger Agache, 'Recherches aériennes de l'habitat rural gallo-romain en Picardie', *Mélanges d'Archéologie ea d'Histoire offerts à André Piganiol*, ed. Raymond Chevallier (Paris, 1966), I, 49–62; *ibid.*, 'Aerial Reconnaissance in Picardie', *Ant.*, 38 (1964), 113–19; compare R. de Maeyer, *De Romeinsche Villa's in België: een archeologische studie* (Antwerp, 1937). See also Franz Oelmann, 'Gallo-Röminsche Strassensiedlungen und Kleinhausbauten', *Bonn. Jahrb.*, 128 (1923), 77–97.

122 Strabo, 3. 2. 15 (151c); 3. 3. 5 (154c).

123 Gudmund Hatt, 'Dwelling-houses in Jutland in the Iron Age', *Ant.*, 11 (1937), 162–73; also Herbert Jahnkuhn, *Vor- und Frühgeschichte*, pp. 129–41.

124 'Fortified Refuge and Village at Borremose, Denmark', *Am. Jnl. Arch.*, 52 (1948), 225–6.

125 Mårten Stenberger, 'Remnants of Iron Age Houses on Öland', *Acta Arch. (Dk)*, 2 (1931), 93–104.

126 Mårten Stenberger and Ole Klindt-Jensen, *Vallhagar: a Migration Period Settlement on Gotland/Sweden*, 2 vols.
127 Tacitus, *Germania*, 46; see also A. W. Brogger, 'The History of the Settlement of Northern Norway', *Proc. First Internat. Congr. Prehist. and Protohist. Sci, 1932* (Oxford, 1934), 292.
128 A. H. M. Jones, *The Later Roman Empire*, (Oxford, 1964), II, 770.
129 The view that soil deterioration was economically significant during the later Empire has been effectively refuted by A. P. Usher, 'Soil Fertility, Soil Exhaustion, and Their Historical Significance', *Quart. Jnl. Econ.*, 37 (1922–3), 385–411, and Norman H. Baynes, 'The Decline of Roman Power in Western Europe: some Modern Explanations', *Jnl. Rom. St.*, 33 (1943), 29–35.
130 Cato, *De Agri Cultura*, III. 5; VI. 1–2.
131 Varro, I. vii. 9–10; XVI. 3.
132 Columella, III. i. 2–ii, 31.
133 Pliny, *Natural History*, xviii, 44. 149.
134 Shimon Applebaum, 'Peasant Economy and Types of Agriculture', *Rural Settlement in Roman Britain*, pp. 99–107.
135 Columella, II. 9. 1–2; II. 10. 4, 22.
136 Strabo, III. 3. 7 (155c).
137 Varro, II. i. 9–10; III. xvii. 9; the text of the *Lex Agraria* is printed in *Font. Jur. Rom. Antejust*, I, 103–21, sect. 26.
138 Silius Italicus, *Punica*, VII. 364–6.
139 The Younger Pliny mentioned sheep, cattle and horses as wintering near his villa on the coast of Latium at Laurentinum. *Letters*, II. 17. 3.
140 *Corp. Inscr. Lat.*, IX, 2438; this inscription of *c.* A.D. 168 was found at Altila, near Monte Matese.
141 *Corp. Inscr. Lat.*, IX, 2826.
142 Pliny the Younger, *Letters*, V. 6; also VIII. 17.
143 Pliny, *Natural History*, XVIII. 7. 35; K. D. Knight, 'Latifundia', *Inst. Class. St.*, Bull. 14 (1967), 62–79.
144 Pliny the Younger, *Letters*, V. 6; II. 17; Richard Duncan-Jones, 'The Finances of the Younger Pliny', *Papers Brit. Sch. Rome*, 33 (1965), 177–88.
145 H. Döhr, *Die italische Gutshöfe nach den Schriften Catos und Varros* (Niederzier, 1965), pp. 42–45.
146 *Corp. Inscr. Lat.*, XI, Pt. 1, No. 1147 (pp. 208–31).
147 *Cicero's Letters to Atticus*, ed. D. R. Shackleton Bailey (Cambridge, 1965), I, 4.
148 John Day. *An Economic History of Athens under Roman Domination* (New York, 1942), p. 235.
149 Dio Chrysostom, VII. 34.
150 Vergil, *Eclogues*, II. 66, and *Georgics*, I. 169–75.
151 Pliny, *Natural History*, XVIII. 48. 171–3.
152 Varro, De Lingua Latina, V. 31. 135.
153 Pliny, *Natural History*, XVIII. 30. 296.
154 Palladius, *Agric.*, VII. 2. 2–4 (Teubner edn., p. 172).
155 Tertullian, *De anima*, xxx. 3.
156 *The Letters of Cassiodorus*, XII. 24, ed. Thomas Hodkin (London, 1884), p. 517.
157 Plutarch, *Life of Caesar*, 58. 4–5.
158 Varro, I. 7. 8.
159 Tacitus, *Germania*, 5. 26.
160 Tacitus, *Germania*, 5.
161 Ramsay MacMullen, 'The Celtic Renaissance', *Hist.*, 14 (1965), 93–104.
162 *Font. Iur. Rom. Antejust.*, I, No. 104, pp. 498–502.
163 Pliny, *Natural History*, III. 30.
164 Cassiodorus, III. 25.
165 Witold Hensel and Aleksander Gieysztor, *Archaeological Research in Poland* (Warsaw, 1958), p. 43.
166 Tacitus, *Annals*, XIII. 53.
167 Cassiodorus, XII. 24.
168 The text, with cartographic reconstruction is given in Konrad Miller, *Itineraria Romana*. (Stuttgart, 1916), and in O. Cuntz, *Itineraria Romana*, vol. I (Leipzig-Berlin, 1929).

169 Denis van Berchem, *Les Distributions de Blé et d'Argent à la Plèbe romaine sous l'Empire* (Geneva, 1939), p. 80.
170 Juvenal, VII. 22.
171 Martial, I. 82; VI. 11. 7.
172 E. Espérandieu, *Recueil Général des Bas-Reliefs de la Gaule Romaine*, vol. VI (Paris, 1915), No. 5268. The fulling process is illustrated in Luxembourg, *ibid.*, V (Paris, 1913), Nos. 4125, 5136.
173 Louis C. West, *Imperial Roman Spain*, p. 3.
174 Glanville Downey, 'Libanius' Oration in Praise of Antioch (Oration XI)', *Proc. Am. Phil. Soc.*, 103 (1959), 652–86.

CHAPTER 4

1 *The Russian Primary Chronicle*, ed. S. H. Cross, p. 56.
2 Einhard, XVII, XXVI.
3 *Mon. Germ. Hist. Capitularia Regum Francorum*, ed. A. Boretius, I, 81.
4 H. W. Garrod and R. B. Mowat, *Einhard's Life of Charlemagne* (Oxford, 1925), p. xlvii.
5 Einhard, VII.
6 Einhard, XV.
7 Einhard, XIII; *Russian Primary Chronicle*, ed. S. H. Cross, p. 56.
8 Adam of Bremen, *History of the Archbishops of Hamburg-Bremen*, trans. and ed. Francis J. Tschan (New York, 1959), pp. 64–5.
9 *The Russian Primary Chronicle*, ed. S. H. Cross, pp. 52–3.
10 *King Alfred's Orosius*, ed. Henry Sweet (1883), p. 20. E.E.T.S.
11 Vita Anscharii, *Scriptores Aerum Suevicarum*, II (Uppsala, 1828), 232, states of Courland: 'regnum vero ipsum quinque habebat civitates'.
12 Jesse D. Clarkson, *A History of Russia* (New York, 1962), p. 25.
13 *De Administrando Imperio*, vol. II, Commentary, ed. R. J. H. Jenkins (London, 1962), p. 23; George Vernadsky, *The Origins of Russia* (Oxford, 1959), pp. 198–213.
14 *The Gothic History of Jordanes*, trans. and ed. C. C. Mierow (Princetown, N.J., 1915), III. 23 (p. 56).
15 *Constantine Porphyrogenitus de administrando Imperio*, X, in the edition G. Moravcsik (Budapest, 1949).
16 Constantine Porphyrogenitus, XXXVII. There are grave difficulties in this dating. The westward movement of the Pechenegs is ascribed by Reginon of Prüm to 889. In any case it took place considerably later than the period which we are considering, but is typical of the movements which occurred in the steppe.
17 Ibn Rusta, as quoted in *Camb. Med. Hist.*, IV, 197.
18 Constantine Porphyrogenitus, L; see also *De Administrando Imperio*, vol. II, Commentary, ed. R. J. H. Jenkins, p. 186.
19 Peter Charanis, 'The Slavic Element in Byzantine Asia Minor', *Byz.*, 18 (1948), 69–83.
20 J. C. Russell, 'Late Ancient and Medieval Population', *Trans. Am. Phil. Soc.*, 48 Pt. 3 (1958), 88.
21 E. Levasseur, *La Population Française* (Paris, 1889), I, 137–9.
22 J. C. Russell, *op. cit.*, p. 93.
23 Procopius, VII. 14. 29–30.
24 J. C. Russell, *op. cit.*, p. 148.
25 David Talbot Rice, *The Byzantines* (London, 1962), p. 23.
26 G. M. Morant, *The Races of Central Europe* (London, 1939), p. 315.
27 G. Des Marez, *Le Problème de la Colonisation Franque et du Régime agraire* (1926); see also Gordon East, *An Historical Geography of Europe* (London, 1935), pp. 66–70.
28 Ch. Dubois, 'L'influence des chaussées romaines sur la frontière linguistique de l'Est', *Rev. Belge Phil. Hist.*, 9 (1930), 455–94.
29 Franz Petri, *Germanische Volkserbe in Wallonien und Nordfrankreich* (Bonn, 1937).

30 St Jerome, Migne, *Pat. Lat.*, XXVI, col. 382; see also P.-F. Fournier, 'La persistance du Gaulois au VIe siècle d'apres Gregoire de Tours', *Recueil de Travaux offert à Clovis Brunel* (Paris, 1955), I, 448–53.
31 Peter Charanis, 'The Chronicle of Monemvasia and the Question of the Slavonic Settlements in Greece', *Dumb. Oaks Papers*, No. 5 (1950), 139–66.
32 Gregory of Tours, *The History of the Franks*, ed. O. M. Dalton (Oxford, 1927), I, 175.
33 Vita S. Willibaldi, *Acta Sanctorum*, July, II, 504; see also 'The Hodœporicon of Saint Willibald', *Pal. Pilgr. Text. Soc.*, 3 (1895).
34 Quoted by Alfred Rambaud, *L'Empire Grec au dixième Siècle* (Paris, 1870) p. 228.
35 Constantine Porphyrogenitus, 29; eight cities are named, including Ragusa. Salona is stated to have been taken by the Slavs.
36 Constantine Porphyrogenitus, 30, 87–8.
37 Albert S. Cook and Chauncey B. Tinker, *Select Translations from Old English Poetry* (Boston, 1926), pp. 56–7.
38 'Where life might be more secure', Isidore of Seville, *Etymologiarum*, Lib XV, c.b., in *Pat. Lat.*, 82, col. 537. See also R. R. Darlington, 'The Early History of English Towns', *Hist.*, 23 (1938–9), 141–50.
39 As argued by J. N. L. Myres, 'Some Thoughts on the Topography of Saxon London', *Ant.*, 8 (1934), 437–42.
40 R. E. M. Wheeler, 'Mr Myers on Saxon London: A Reply', *Ant.*, 8 (1934), 443–7. See also Xavier de Planhol, *The World of Islam* (Ithaca, N.Y., 1959), pp. 14–23, on the distortion of the street pattern of Moslem cities under a weak city government.
41 For the detailed study of an example, see Sir Cyril Fox, 'The Siting of the Monastery of St Mary and St Peter in Exeter', *Dark-Age Britain*, ed. D. B. Harden (London, 1956), pp. 202–17; Gabriel le Bras, 'L'invasion de l'église dans la cité', *Urbanisme et Architecture* (Paris, 1954), pp. 187–98.
42 Steven Runciman, *Christian Constantinople* (London, 1952), 59, 67.
43 J. G. Russell assumes that at the technical level represented by the Byzantine Empire about 1.5 per cent of the population would live in its capital city. A city of a million would thus require a state of 67 million, which is excessive: *Late Ancient and Medieval Population*, pp. 68–71, 93.
44 Constantine Porphyrogenitus, 30, 143. See also Louis Bréhier, *Le Monde Byzantin: Les Institutions de l'Empire Byzantin*, (Paris, 1949), p. 204.
45 Procopius, VI. 1. 17–18.
46 Procopius, IV. 3. See also N. J. G. Pounds, 'The Urbanization of East-central and South-east Europe: An Historical Perspective', in *Eastern Europe*, ed. George W. Hoffman (London, 1970), pp. 45–78.
47 Text and translation in Peter Charanis, 'The Chronicle of Monemvasia and the Question of the Slavic Settlements in Greece', *Dumb. Oaks Papers*, No. 5 (1950), 148.
48 Richard Koebner, 'The Settlement and Colonisation of Europe', *Camb. Econ. Hist.*, I (1941), 1–88.
49 *Ibid.*, 33. For a case study see Shimon Applebaum, 'The Late Gallo-Roman Rural Pattern in the Light of the Carolingian Cartularies', *Lat.*, 23 (1964), 774–87.
50 Albert Grenier, 'La conquête du sol française', *Ann. Hist. Econ. Soc.*, 2 (1930), 26–47, gives a moderate statement of this view.
51 Shimon Applebaum, 'The Late Gallo-Roman Rural Pattern in the Light of the Carolingian Cartularies'.
52 B. H. Slicher van Bath, *The Agrarian History of Western Europe A.D. 500–1850* (London, 1963), p. 41.
53 For an English example of continuity see H. P. R. Finberg, *Roman and Saxon Withington*, University of Leicester Occasional Papers, No. 8 (1955).
54 Auguste Longnon, *Les Noms de Lieux de la France* (Paris, 1920–9).
55 Marc Bloch, 'Réflections d'un historien sur quelques travaux de toponymie', *Ann. Hist. Econ Soc.*, 6 (1934), 252–60.
56 Ferdinand Lot, 'De l'origine et de la signification historique et linguistique des noms de lieux en *-ville* et en *-court*', *Rom.*, 59 (1933), 199–246.
57 Charles Rostaing, *Les Noms de Lieux* (Paris, 1961), pp. 71–7.

58 Compare the English suffix *-ing*; see Allen Mawer, *The Chief Elements used in English Place-Names, E.P.N.S.*, I, Pt. 2 (Cambridge, 1924), 41–2.
59 Marc Bloch, in *Ann. Hist. Econ. Soc.*, 6 (1934), 257.
60 For example, the Dark Age village excavated at Gladbach in the Lower Rhineland: *Germ.*, 22 (1938).
61 For a notable example of this method see S. W. Wooldridge and D. L. Linton, 'Some Aspects of the Saxon Settlement in South-East England Considered in Relation to the Geographical Background', *Geog.*, 20 (1935), 161–75.
62 Charles Higounet, L'occupation du sol du pays entre Tarn et Garonne au Moyen Age', *Ann. Midi*, 65 (1953), 301–30.
63 Jacques Boussard, 'Essai sur le peuplement de la Touraine du Ier au VIIIe siècle', *Moy. Age*, 60 (1954), 261–91.
64 R. Gradmann, 'Die ländlichen Siedlungsformen Württembergs', *Pet. Mitt.*, 56 (1910), 183–6, 246–9.
65 André Déléage, *La Vie rurale en Bourgogne jusqu'au début onzième siècle* (Mâcon, 1941), pp. 101–3.
66 James W. Thompson, 'East German Colonization in the Middle Ages', *Ann. Rept. Am. Hist. Assoc.*, (1915), 125–50.
67 *Relatio Ibrahim Ibn Jakub de Itinere Slavico*, ed. T. Kowalski, *Mon. Pol. Hist.*, New Series I (Kraków, 1946), pp. 2–3.
68 Procopius, VII. xiv. 24.
69 Quoted in Magnus Olsen, *Farms and Fanes of Ancient Norway*, Institutet for Sammenlignende Kulturforskning (Oslo, 1928), p. 34.
70 Magnus Olsen, *ibid.*, p. 148.
71 Alfons Dopsch, *Die Wirtschaftsentwicklung der Karolingerzeit*, 3rd edn., (Cologne, 1962), I, 306–9; Adriaan Verhulst, 'La genèse du régime domanial classique en France au haut moyen age', *Agricoltura e Mondo Rurale in Occidente nell' Alto Medioevo, Sett. Spol.*, 13 (1966), 135–60.
72 Seminantur ad hibernaticum buaria X, de modiis XL, et ad tremissem bunaria X, de modiis LX. Bunaria X interjacent. The text of this very short polyptyque was published by B. Guérard, *Polyptyque de l' Abbé Irminon* (Paris, 1844), *Prolégomènes*, Pt. 2, 925–6.
73 *Registrum Prumiense*, cap. 99.
74 *Liber Traditionum Sancti Blandiniensis*, ed. Arnold Fayem in *Cartulaire de la Ville de Gand*, 2nd series (Ghent, 1906), I, 10–52, cap. 3.
75 *Polyptyque de l'Abbaye de Saint-Rémi de Reims*, ed. B. Guérard (Paris, 1953); on the inequality of winter and spring grains see E. Perrin, 'De la condition des terres dites *Ansingae*', *Mélanges d'Histoire du Moyen Age offerts à M. Ferdinand Lot* (Paris, 1925), pp. 619–40.
76 J. Warichez, 'Une *Descriptio villarum* de l'abbaye de Lobbes à l'époque carolingienne', *Bull. Com. Roy. Hist.*, 78 (1909), 245–67. The spelt was given in part in *corbes*, which have been converted at the rate of 12 *modii* to the *corbis*.
77 *Registrum Prumiense*, cap. 97.
78 *Capitulare de Villis, Mon. Germ. Hist., Legum Sectio* (Hannover, 1883), I, 82–91.
79 Reginon of Prüm, *Pat. Lat.*, 132, 77–8.
80 'They go to the vineyards with two carts at the festival of Saint-Rémy.' Lobbes polyptyque, cap. 10.
81 *Brevium exempla ad describendas res ecclesiasticas et fiscales, Mon. Germ. Hist., Legum Sectio*, vol. II, 250–56, cap. 27. See also Joseph Halkin, 'Etude sur la culture de la vigne en Belgique', *Bull. Soc. Ant. Hist. Liège*, 9 (1895), 1–146, esp. 73–7.
82 *Liber tranditionum Sancti Petri Blandiniensis*, ed. A. Fayem, in *Cartulaire de la Ville de Gand* (Ghent, 1906), I, 10–52, cap. 2.
83 *Liber traditionum Sancti Petri Blandiniensis*, cap. 7.
84 *Brevium exempla*, cap. 25–35.
85 Lobbes is an exception; is possessed a small number of dairy cattle (*vaccae*).
86 K. Lamprecht, *Deutsches Wirtschaftsleben im Mittelalter* (Leipzig, 1886).
87 *Brevium exempla*, cap. 2–16.
88 See especially *Eyrbyggja Saga*, trans. Paul Schach and Lee M. Hollander (Lincoln, Neb., 1959), pp. 61, 107.
89 For example, *Eyrbyggja Saga*, p. 121.

90 It should be remembered, however, that the compilation of the Prüm polyptique followed a raid by the Norsemen into this region; the *Registrum* refers, caps. 49 and 103, to destruction 'by the pagans'.
91 J. W. Thompson, 'Serfdom in the Medieval Campagna', in *Wirtschaft und Kultur: Festschrift zum 70. Geburtstag von Alfons Dopsch* (Leipzig, 1938), pp. 363–81.
92 Léonce Auzias, *L'Aquitaine carolingienne (778-987), Bibl. Merid.*, 2nd series, 28 (Toulouse, 1937), p. 72.
93 J. W. Thompson, 'Serfdom in the Medieval Campagna'.
94 *Codice dipl. del monasterio di S. Colombano di Bobbio*, ed. C. Cipolla, *Inst. Stor. Ital.* 52 (1918), I, 184–217.
95 *Inventarium omnium bonorum eorumque reddituum monasterii sanctimorialium S. Juliae brixiensis, Codex Diplomaticus Langobardiae*, vol. 13 (1873), No. 419, col. 706–27. The polyptyque, which is incomplete, is dated 905–6; see also Gino Luzzatto, 'Mutamenti nell' economia agraria italiana', *I Problemi communi dell' Europa post-carolingia, Cent. It. Alto, Med.* (Spoleto, 1964); an abridged translation in *Early Medieval Society*, ed. Sylvia L. Thrupp (New York, 1967), 206–18.
96 *In curte Griliano sunt . . . furcas ferreas II, et alio ferro libras C.*, col. 706.
97 The Greek text was published by Walter Ashburner, *Jnl. Hell. St.*, 30 (1910), 85ff; an English translation and commentary is published as 'The Farmer's Law II', *ibid.*, 32 (1912), 68–95, and also, in part, in Roy C. Cave and Herbert H. Coulson, *A Source Book for Medieval Economic History* (Milwaukee, Wisc., 1936), pp. 7–13.
98 This is the opinion of Georg Ostrogorsky, 'Agrarian Conditions in the Byzantine Empire in the Middle Ages', *Camb. Econ. Hist.*, I, 194–223. See also Kenneth M. Stetton, 'On the Importance of Land Tenure and Agrarian Taxation, from the Fourth Century to the Fourth Crusade', *Am. Jnl. Phil.*, 74 (1953), 225–59. It must be pointed out that Paul Lemerle, 'Equisse pour une historie agraire de Byance', *Rev. Hist.*, 219 (1958), 32–74, 254–84, does not find so marked a revolution in land tenure, and tends to play down the significance of the Farmer's Law.
99 Kenneth M. Setton, 'On the Importance of Land Tenure', 236.
100 Saint-Germain polyptyque, cap. 13.
101 *Codex Diplomaticus Langobardiae*, vol. 13, No. 419, esp. col. 706.
102 William Stubbs, *Select Charters* (Oxford, 1929), p. 69.
103 Loup de Ferrières, *Correspondence*, ed. L. Levillain, No. 106.
104 *Ibid.*, Nos. 84 and 85.
105 Prüm polyptyque, cap. 41.
106 H. Pirenne, 'Draps de Frise ou Draps de Flandre', *Vj. S. Wg.*, 7 (1909), 308–15.
107 Ermoldus Niger, *Carmina* I, lines 115–30.
108 L. Halphen, *Etudes critiques sur l'Histoire de Charlemagne* (Paris, 1921), pp. 290–2; K. T. von Inama-Sternegg, *Deutsche Wirtschaftsgeschichte*, I (Leipzig, 1909).
109 *Mon. Germ. Hist., Epistolae*, vol. 4 (*Epis. Kar Aevi*), No. 100, pp. 144–5.
110 The text is published in Edwin H. Freshfield, *Roman Law in the Later Roman Empire* (Cambridge, 1938).
111 *Rutilii laudii Namatiani de Reditu Suo*, ed. C. H. Keene (London, 1907).
112 *Patrologia Graeca*, vol. 93, cols. 1615–68, cap. 9.
113 Saint-Germain polyptyque, cap. 13: 'faciunt caropera propter vinum in Andegaro', (Anjou).
114 *Recueil des Chartes de l'Abbaye de Stavelot-Malmèdy*, ed. J. Halkin and C-G. Roland (Brussels, 1909), I, 69.
115 Ermoldus Niger, I, 97–114.
116 F. Vercauteren, 'L'interpretation économique d'une trouvaille de monnaies carolingiennes faite près d'Amiens en 1865', *Rev. Belge Phil. Hist.*, 13 (1934), 750–8.
117 Karl F. Morrison, 'Numismatics and Carolingian Trade: A Critique of the Evidence', *Spec.*, 38 (1963), 403–32, warns against a too ready acceptance of evidence drawn from coin finds.
118 Capit. of Theonville, *Mon. Germ. Hist. Leges*, Sectio I, No. 44, p. 123, cap. 7.
119 Böhmer-Muhlbacher, *Registrum*, I, 329.

120 Saint-Germain polyptyque, cap. 9, 16; See Guérard's edition, *Prolégomènes*, pp. 784–90.
121 *King Alfred's Description of Europe*, ed. Sweet, E.E.T.S. (London, 1906), cap. 20–3.
122 *The Book of the Prefect*, cap. 10.
123 R. S. Lopez, 'East and West in the Early Middle Ages: Economic Relations', *Rel. X Cong. Int.*, 3, 113–63.
124 *Charlemagne: Oeuvre, Rayonnement et Survivances*, Deuxième Exposition sous les Auspices du Conseil de l'Europe (Aachen, 1965), pp. 13–16.

CHAPTER 5

1 'Passim condidam ecclesiarum vestem indueret', Raoul Glaber, ed. Maurice Prou (Paris, 1886), iv, 13 (p. 62).
2 William of Tyre, *A History of Deeds Done beyond the Sea*, trans. and ed. E. A. Babcock and A. C. Krey (New York, 1943), I, 117.
3 N. J. G. Pounds, 'The Origin of the Idea of Natural Frontiers in France', *Ann. A.A.G.*, 41 (1951), 146–57.
4 Otto of Freising, *The Two Cities* (New York, 1953), VII, cap. 5.
5 Quoted in Marcel Pacaut, *Louis VII et Son Royaume* (Paris, 1964), II. Peter did, however, refer several times to nostra Gallia: *The Letters of Peter the Venerable*, ed. Giles Constable (Cambridge, Mass., 1967), 130, 144, 381.
6 Otto of Freising, *Gesta Frederici, Mon. Germ. Hist.*, SS, II, 13–15.
7 Otto of Freising, VII, 29. The translation is that of Charles C. Mierow, *The Two Cities* (New York, 1928), pp. 438–9.
8 Gino Luzzatto, *An Economic History of Italy from the Fall of Rome to the Beginning of the Sixteenth Century* (London, 1961), p. 84.
9 'The Poem of the Cid', trans. W. S. Merwin, in *Medieval Epics* (New York, 1963), pp. 441–590, parts 44 and 23.
10 See Ramón Menendez Pidal, *The Cid and His Spain*, trans. Harold Sunderland (London, 1934), pp. 22–6; this view is criticised by Jaime Vicens Vives, *An Economic History of Spain* (Princeton, N.J., 1969), p. 124.
11 See map of tenth-century Poland in Karol Buczek, *Ziemie Polskie przed tysiącemlat*, Polska Akademia Nauk (Kraków, 1960).
12 See particularly *Vita Ottonis Episcopi Babenbergensis Herbordie* (Lwów, 1872), II, 1 and 10; *Galli Chronicon*, in *Mon. Pol. Hist.*, 1 (1864), I, 25.
13 *Galli Chronicon*, II, 42.
14 Stewart Oakley, *A Short History of Sweden* (New York, 1966), p. 30; see also Carl Hallendorf and Adolf Schück, *History of Sweden* (London, 1929), p. 32.
15 Snorre Sturlason, *Heimskingla* (London, Everyman edition), p. 163.
16 *The First Nine Books of the Danish History of Saxo Grammaticus*, trans. Oliver Elton (London, 1894), p. 9.
17 *Egil's Saga*, trans. E. R. Eddison (Cambridge, 1930), p. 25.
18 Carlo Cipolla, Jean Dhont, M. M. Postan, Philippe Wolff, 'Anthropologie et Démographie', *IXe Congrés International des Sciences Historiques*, I, *Rapports* (Paris, 1950), 56.
19 Gino Luzzatto, *An Economic History of Italy from the Fall of Rome to the Beginning of the Sixteenth Century*, pp. 69–80.
20 George T. Beech, *A Rural Society in Medieval France: The Gâtine of Poitou in the Eleventh Century* (Baltimore, 1964), p. 25.
21 'Valleys and rocks encircle the Ardennes; it has a heavy rainfall; nothing grows in its soil; we think that it has never been cultivated. Not even the needy will settle in this land.' Quoted in Rita Lejeune, 'L'Ardenne dans la Littérature médiévale', *Anc. Pays Ass. Etats*, 28 (1963), 43.
22 *The Chronicle of the Slavs by Helmold*, trans. F. K. Tschan (New York, 1935), I, 89.
23 'a dreadful and vast forest', Vita Ottonis Episcopi Babenbergensis, II. 4; 10, *Mon. Pol. Hist.*, II.
24 'a small but populous island', Vita Ottonis Episcopi Babenbergensis, 73.
25 Karl Lamprecht, *Deutsches Wirtschaftsleben* (Leipzig, 1886), I, 163.

26 See Wilhelm Abel, *Geschichte der deutschen landwirtschaft* (Stuttgart, 1962), pp. 25–6; J. C. Russell, 'Late Ancient and Medieval Population', *Trans. Am. Phil. Soc.*, n.s., 48, Pt. 3 (1958), 95.
27 Julius Beloch, 'Die Bevölkerung Europas im Mittelalter', *Zeitsch. Soz.*, 3 (1900), 405–23.
28 Egon Vielrose, 'Ludność Poleki od X do XVIII wieku', *Kw. Hist. Kult. Mat.*, 5 (1957), 3–49; Irena Gieysztorowa, 'Badania nad historia zaludnienia Polski', *Kw. Hist. Kult. Mat.*, 11 (1963), 523–62; *ibid.*, 'Recherches sur la démographie historique et en particulier rurale en Pologne', *Kw. Hist. Kult. Mat.*, 12 (1964), 509–28.
29 Gy. Györffy, 'Einwohnerzahl und Bevölkerungsdichte in Ungarn bis zum Anfang des XIV Jahrhunderts', *Et. Hist.* (Budapest), 1 (1960), 163–92.
30 Especially *Géographi d'Edrisi*, trans. P. Amédée Jaubert (Paris, 1836–40), II, 12–234; E. Lévi-Provençal, *Séville musulmane au début du XIIe siècle* (Paris, 1947); E. Lévi-Provençal, *La Péninsule Ibérique au Moyen Age d'après ... Ibn 'Abd al-Munim al Himyaṛt* (Leiden, 1938). See also *The Itinerary of Benjamin of Tudela*, trans. and ed. Marcus Nathan Adler (Oxford, 1907), pp. 2–3, and Jaime Vicens Vives, *An Economic History of Spain*, pp. 105–6.
31 Otto of Freising, *The Two Cities*, VII, 5.
32 *Chronicle of the Slavs*, 168, 234, 242, 247.
33 'Chronicon quod conservatur in Monte S. Georgii', ed., Béla Pukánsky, *Script. Rer. Hung.*, 2 (1938), 280.
34 *Anonymi Gesta Hungarorum*, cap. 50.
35 'Legenda Sancti Gerhardi Episcopi', ed. Emericus Madzsar, *Script. Rer. Hung.*, 2 (1938), 461–506, cap. 8.
36 Geoffrey de Vinsauf, I, xx, in *Chronicles of the Crusades* (London, 1903), p. 92.
37 *The Alexiad of the Princess Anna Comnena*, trans. Elizabeth A-S. Dawes (London, 1928), VI. 14; VII. 2; X. 3; XIV. 8. She uses classical names, such as Scythians and Sarmatians, for these peoples who were in fact quite different both racially and linguistically.
38 *Ibid.*, XIV. 8; but see also J. M. Hussey, 'The Later Macedonians, the Comneni and the Angeli', *Camb. Med. Hist.*, IV, Pt. 1 (1966), 213.
39 *Gesta Francorum et aliorum Hierosolimitanorum*, I. 4; see edition by Rosalind Hill (London, Nelson's Medieval Texts, 1962), p. 8.
40 *Alexiad*, II. 9.
41 *Alexiad*, IV. 8, and William of Tyre, I. 18.
42 *Alexiad*, VI. 7.
43 *Alexiad*, V. 5.
44 *Alexiad*, VIII. 3.
45 *Alexiad*, X. 3; the whereabóuts of the 'Zygum' have been exhaustively, but inconclusively, discussed.
46 William of Tyre, II. 17.
47 *Anonymi Gesta Hungarorum*, cap. 9, 25, 26.
48 M. Dinić, 'The Balkans, 1018–1499', *Camb. Med. Hist.*, 4, Pt. 1 (1966), 560–1.
49 Gy Györffy, 'Formation d'états au IXe siècle suivant les *Gesta Hungarorum* du Notaire Anonyme', *Nouv. Et. Hist.* (Budapest), I (1965), 27–51.
50 Irving A. Agus, *Urban Civilization in Pre-Crusade Europe* (New York, 1965), No. 42.
51 Benjamin of Tudela, 14.
52 William of Tyre, I. 29; see also Otto of Freising, *Gesta Frederici*, I. 38.
53 A. Andréadès, 'The Jews in the Byzantine Empire', *Econ. Hist.*, 3 (1934–7), 1–23.
54 See E. A. Kosminsky, *Studies in the Agrarian History of England in the Thirteenth Century* (Oxford, 1956), who estimates, p. 322, that in 1086 only 5 per cent of the population of England lived in towns.
55 *Slavic Chronicle*, I. 12.
56 Albrecht Meitzen, *Siedlung und Agrarwesen der Westgermanen und Ostgermanen, der Kelter, Römer, Finnen und Slaven* (Berlin, 1895).
57 *Vita Sancti Anselmi*, ed. R. W. Southern (London, 1962), II, 29; 31 (pp. 106–8).
58 Robert Latouche, 'Agrarzustände im westlichen Franreich während des Hochmittelalters', *Vj. S. Wg.*, 36 (1929), 105–13; *id.*, 'Un aspect de la vie rurale dans le Maine au XIe et au XIIe siècle; l'éstablissement des bourgs', *Moy. Age*, 3rd

series, 13 (1937), 44–64; *id*, 'L'économie agraire et le peuplement des pays bocagers', *Rev. Syn.*, 59 (1939), 45–51.

59 'a certain enclosure lying at the end of the marsh, where Gausbert, Grand's villein, lives, together with an orchard and ploughland next to the enclosure', George T. Beech, *A Rural Society in Medieval France: The Gâtine of Poitou in the Eleventh Century*, p. 29.

60 Procopius, VII. 14. 24. Also Jovan Cvijić, *La Péninsule Balkanique* (Paris, 1918), pp. 216–20.

61 Helmold, I. XXXIV.

62 *Ibid.*, I. LXXV.

63 Philip Jones, 'Medieval Agrarian Society in its Prime: Italy', *Camb. Econ. Hist.*, I (2nd Edition, 1966), 340–431, especially 393–5.

64 *Egil's Saga*, Book XIV, 25.

65 *Ibid.*, Book XXIX, 57; the passage relates to Iceland, but conditions in Norway were closely similar.

66 *Heimskringla*, 319; see also 162.

67 Based on Ferdinand Lot, 'L'état des paroisses et des feux de 1328', *Bibl. Ec. Chartes*, 90 (1929), 51–107, 256–315.

68 *Heimskringla*, 162, 179.

69 Helmold, *Slavic Chronicle*, I. 88–9.

70 *Heimskringla*, 261.

71 In the impact of the Cistercians see the numerous papers by R. A. Donkin. These are in some degree summarised and most of them listed in 'The Cistercian Order in Medieval England: Some Conclusions', *Inst. Brit. Geog.* (1963), 181–98, and 'The Growth and Distribution of the Cistercian Order in Medieval Europe', *Stud. Mon.*, 9 (1967), 275–86.

72 See J. Puig i Cadafalch, *La Géographie et les Origines du Premier Art Roman* (Paris, 1935), pp. 55–77.

73 Lambert of Hersfeld, *Mon. Germ. Hist. SS Rerum Germanicarum in Usum Scholarum*, 38 (Hannover, 1874), p. 105 (*sub.* 1073).

74 The Bayeaux Tapestry illustrates such motte-and-bailey castles in Normandy and Brittany and also shows the construction of that at Hastings.

75 See *The Letters of Gerbert* (of Aurillac), trans. Harriet Pratt Lattin (New York, 1961), No. 63.

76 Gino Luzzatto, *An Economic History of Italy from the Fall of Rome to the Beginning of the Sixteenth Century* p. 48; *The Letters of St Bernard of Clairvaux* (London, 1953), No. 143. See also F. Gregorovius, *History of the City of Rome in the Middle Ages*, III, 516–62; IV, 248–54.

77 P. Feuchère, 'Les origines urbaines de Lens-en Artois', *Rev. Belge Phil. Hist.*, 30 (1952), 91–108.

78 The castle mound is visible on the Van Deventer map of 1560; it has since disappeared.

79 *Annales Sancti Bertini*, quoted in R. Latouche, *The Birth of the Western Economy* (London, 1961), p. 248.

80 See J. Lestocquoy, *Les Villes de Flandre et d'Italie sous le Gouvernement des Patriciens* (Paris, 1952), p. 47; Edouard Perroy, 'Les origines urbaines en Flandres d'après un ouvrage récent', *Rev. Nord*, 29 (1947), 49–63.

81 J. Massiet du Biest, 'Les origines de la population et du patricial urbain à Amiens', *Rev. Nord*, 30 (1948), 113–32. See also Marie-Thérèse Morlet, 'Les noms de personne à Beauvais au XIVe siècle', *Bull. Phil. Hist.* (1955–6), 295–309, and 'L'origine des habitants de Provins aux XIIIe et XIVe siècles', *ibid.*, 161, 95–114.

82 Carl Haase, *Die Entstehung der Westfälischen Städte* (Münster, 1960).

83 Notably by Carl Haase, *Die Entstehung der Westfälischen Städte*, p. 32.

84 'Relatio Ibrāhim Ibn Ja'kub de Itinere Slavico', ed. Tadeusz Kowalski, *Mon. Pol. Hist.*, 1, new series, (1946), 2. They were almost certainly the same as the *civitates* of the Bavarian geographer: Geograf Bawarski, *Mon. Pol. Hist.*, 1 (Lwów, 1864), 10–11.

85 Anna Niesiołowska-Hoffman, 'Ze studiów nad budownictwem plemion kultury Luzyckiej', *Kw. Hist. Kult. Mat.*, 10 (1963), 25–130.

86 Eckhard Müller-Mertens, 'Zur Geschichte der mittelalterlichen brandenburgischen Städte: ein Forschungsbericht', *Kw. Hist. Kult. Mat.*, 10 (1962), 536–45,

87 František Kavka, 'Der Stand der Forschungen über den Anfang der Städte in der Tschechoslowakei', *Kw. Hist. Kult. Mat.*, 10 (1962), 546–9; see also Evžen and Jiří Neustaupný, *Czechoslovakia*, (London, 1961), p. 185.
88 'Relation Ibrāhīm Ibn Ja'kub de Itinere Slavico', ed. Tadeusz Kowalski, *Mon. Pol. Hist.*, n.s. I (Kraków, 1946), 3. See also Ivan Borkovský, 'Der altböhmische Přemysliden-Fürtensitz Praha', *Hist.* (Prague), 3 (1961), 57–72.
89 William of Tyre, I. 18, 20 and 30.
90 *Géographie d'Edrisi*, trans. P. Amadee Jaubert (Paris, 1836–40), II, 288–90; see also William of Tyre, II. 13–15; and *Alexiad*, V. 4–5.
91 David Jacoby, 'La Population de Constantinople à l'époque byzantine: un problême de démographie urbaine', *Byz.* 31 (1961), 81–109.
92 'De Legatione Constantinopolitana', in *The Works of Liudprand of Cremona*, ed. F. A. Wright (London, 1930), pp. 233–77.
93 *The Itinerary of Benjamin of Tudela*, trans. Marcus Nathan Adler (Oxford, 1907), pp. 12–14.
94 William of Tyre, II. 9.
95 Horatio F. Brown, 'The Venetians and the Venetian Quarter in Constantinople to the Close of the Twelfth Century', *Jnl. Hell. Stud.*, 40 (1920), 68–88.
96 A translation is given in E. Lévi-Provençal, *Séville Musulmane au début du XIIe siècle* (Paris, 1947); see also *Géographie d'Edrisi*. II, 19ff.
97 E. Lévi-Provençal, *Histoire de l'Espagne Musulmane* (Paris, 1953), III, 331.
98 L. Genicot, 'Donations de *villae* ou défrichements: les origines du temporel de l'abbaye de Lobbes', *Miscellanea Historica in honorem Alberti de Meyer* (Louvain, 1946), I, 286–96.
99 Paul Ourliac, 'Les villages de la région toulousaine au XIIe siècle', *Ann. E.S.C.*, 4 (1949), 268–77.
100 Thérèse Sclafert, 'A propos due déboisement des Alpes du Sud', *Ann. Geog.*, 42 (1933), 266–77, 350–60.
101 Karl Lamprecht, 'Beiträge zur Geschichte des französischen Wirtschaftsleben im elften Jahrhundert', *Staats- und socialwissenschaftliche Forschungen . . . von Gustav Schmoller*, I, Heft 3 (1878), pp. 1–38.
102 Suger, 'De rebus in administratione sua gestis', cap. X, in *Oeuvres complètes de Suger*, ed. A. Lecoy de la Marche (Paris, 1867), pp. 164–5; the translation is that of *The European Inheritance*, ed. Ernest Barker (Oxford, 1954), I, 499–500.
103 George T. Beech, *A Rural Society in Medieval France: The Gâtine of Poitou in the Eleventh Century*, p. 29.
104 'The peasant and the villein. The one from the bocage; the other from the plain.' Robert Wace, *Roman de Rou*, ed. Hugo Andresen, II, Pt. 3 (Heilbronn, 1879), lines 819–20. On the other hand Mr de Boüard 'Paysage agraire et problèmes de vocabulaire: le bocage et le plaine dans la Normandie médiévale, *Rev. Hist. Droit Franc. Etr.*, 31 (1953), 327–8, considers that the phrase is an adaptation of the formula, 'in bosco et in plano' and in no way implies a perception of contrasted landscapes.
105 *Suger, Vie de Louis VI le Gros*, XIX, *Classiques de l'Histoire de France au Moyen Age*, XI, 135.
106 Lambert of Ardres, *Rec. Hist. Gaules France*, II (1876), cap. 100 (p. 300).
107 *Actes des Comtes de Flandre 1071–1128*, Com. Roy. Hist., Brussels, 1938, No. 37: 'novam terram . . . que per jactum maris jam crevit et que in posterum accrescet'.
108 *Ibid.*, No. 42: 'novam terram que usque ad hanc diem de palude facta est.'
109 'Terra illa et potest decrascsere [*sic.*] per fractum et potest crescere per jactum ejusdem maris'; quoted by M. Mollat. 'Les hôtes de l'abbaye de Bourbourg', *Mélanges d'Histoire du Moyen Age dédiés à la mémoire de Louis Halphen* (Paris, 1951), n. 384.
110 Helmut Jäger, 'Die Entwicklung der Kulturlandschaft im Kreise Hofgeismar,' *Göttinger Geographische Abhandlungen*, Heft 8 (1951).
111 A. E. Verhulst, 'Probleme der mittelalterlichen Agrarlandschaft in Flandern', *Zeitschr. Agrargesch.*, 9 (1961), 13–19.
112 Bertrand Gille, 'Recherches sur les instruments du labour au moyen age', *Bibl. Ec. Chartes*, 120 (1963), 5–38; for a stimulating debate on this subject between Joan Thirsk and J. Z. Titow, see *P & P.*, No. 29 (1964); 32 (1965); 33 (1966).

113 The censier of St Emmeram is published in P. Dollinger, *l'Evolûtion des Classes rurales en Bavarie* (Paris, 1949), pp. 504–12.
114 Galbert of Bruges, *The Murder of Charles the Good*, cap. III, ed. J. B. Ross (New York, 1960), p. 88.
115 M. Gysseling and A. Verhulst, *Het Oudste Goederenregister wan de Sint-Baafsabdij te Gent* (Ghent, 1964).
116 'Censier de St Emmeram', in P. Dollinger, *L'Evolûtion des Classes rurales en Bavarie*, pp. 504–12.
117 Guillaume de Poitiers, ed. Raymonde Voreville, *Les Classiques de l'Histoire de France au Moyen Age*, Vol. 23 (Paris, 1952), cap. 37–8, 87, 91.
118 *Géographie d'Edrisi*, II, 357–63.
119 *Actes des Comtes de Flandres, 1071–1128, Com. Roy. Hist.*, (1938) No. 95, pp. 213–16.
120 *Guillaume de Poitiers, Les Classiques de l'Histoire de France au Moyen Age*, vol. 23 (Paris, 1952), cap. 44, pp. 109–11.
121 Snorre Sturlason, *Heimskringla* (Everyman edn.), p. 335. *The Laxdoela Saga*, trans. and ed. A. Margaret Arent (Seattle, Wash., 1964), cap. 35, describes transhumance in Iceland.
122 *Egil's Saga*, trans. and ed. E. R. Eddison (Cambridge, 1930), Book 29, p. 57.
123 *Ibid.*, Book 30, p. 59.
124 *Recueil des Chartes de l'Abbaye de Stavelot-Malmédy*, ed. J. Halkin and C. G. Roland (Brussels, 1909), No. 144, pp. 292–3.
125 Benjamin of Tudela, pp. 10, 12–14.
126 *Geographie d'Edrisi*, pp. 43, 50–1; this figure would appear to be excessive.
127 R. Lopez, 'Les Influences Orientales et l'Eveil Economique de l'Occident', *Cah. Hist. Mond.*, 1 (1953–4), 594–622.
128 See Irving A. Agus, *Urban Civilization in Pre-Crusade Europe* (New York, 1965), for the texts of the *Responsa* of the Talmudic scholars in these cases; the author, however, surely exaggerates the role played at this time by Jewish merchants.
129 *Libellus de vita et miraculis S. Godrici . . . de Finchale*, ed. J. Stevenson, Surtees Soc., 1945; see also Walther Vogel, 'Ein seefahrender Kaufmann um 1100', *Hans. Geschbl.*, 18 (1912), 239–48. On the origin of merchants, see A. B. Hibbert, 'Origins of the Medieval Town Patriciate', *P. & P.*, No. 3 (1953), 15–27.
130 André Joris, 'Itinéraires routiers entre Rhenanie et pays mosans à la fin du XIIème siècle', *Beiträge zur Wirtschafts- und Stadtgeschichte* (Wiesbaden, 1965), pp. 253–69.
131 *The Letters of St Bernard of Clairvaux*, trans. B. S. James (London, 1953), No. 188, p. 262; *The Letters of Peter the Venerable*, I, 144, 168. See also Lambert of Hersfeld's description of Henry IV's journey to Canosa, in 1077, cap. 285–6.
132 *Vita Sancti Anselmi*, ed. R. W. Southern (London, 1962), I. 4; II, 34.
133 Lambert of Hersfeld, cap. 285.
134 *Imperial Lives and Letters of the Eleventh Century*, ed. Theodor E. Mommsen and Karl F. Morrison (New York, 1962), p. 135.
135 E. Lévi-Provençal, *Seville Musulmane au début du XIIe siècle* (Paris, 1947); *Géographie d'Edrisi*, II, 43–8.
136 *The Itinerary of Benjamin of Tudela*, p. 9.
137 'Gesta Roberti Wiscardi', *Mon. Germ. Hist.*, S.S., 9, 239–98, lines 476–85.
138 *The Itinerary of Benjamin of Tudela*, p. 11.
139 Benjamin of Tudela, 12–13.
140 *Géographie d'Edrisi*, II, 288–90.
141 Carl Haase, *Die Entstehung der westfälischen Stadte* (Münster, 1965).

CHAPTER 6

1 Daniel Waley, *Medieval Orvieto* (Cambridge, 1952), xiv–xviii.
2 K. M. Setton, *Catalan Domination of Athens 1311–1388* (Cambridge, Mass., 1948).
3 M. M. Postan, 'Some Economic Evidence of Declining Population in the Later Middle Ages', *Econ. Hist. Rev.*, 2 (1950), 221–46.

4 Elisabeth Charpentier, 'Autour de la peste noire: Famines et épidémies', *Ann. ESC*, 17 (1962), 1062–92.

5 Ferdinand Lot, used a multiplier of 5. J. C. Russell, *British Medieval Population* (Albuquerque, New Mexico, 1948), has argued for a small household in England of less than 4. This argument does not seem conclusive, and the author prefers the conclusions of H. E. Hallam, 'Some Thirteenth Century Censuses', *Econ. Hist. Rev.*, 10 (1958), 340–61, and J. Krause, 'The Medieval Household: Large or Small?', *Econ. Hist. Rev.*, 9 (1957), 420–32, who both prefer the larger household. François Maillard and Robert-Henri Bautier found an average household of 5.1 at two villages in southern France in 1306: 'Un dénombrement des feux, des individus et des fortunes dans deux villages du Fenouillèdes en 1306', *Bull. Phil. Hist.* (1965), 309–28.

6 Philippe Wolff estimated that in fourteenth-century Toulouse the clerical population made up 5 per cent of the total: *Commerce et Marchands de Toulouse* (Paris, 1954), p. 71.

7 J. C. Russell, 'Late Medieval Population Patterns', *Spec.*, 20 (1945), 157–71.

8 Ferdinand Lot, 'L'état des paroisses et des feux de 1328', *Bibl. Ec. Chartes*, 90 (1929), 51–107, 256–315.

9 Joseph R. Strayer, 'Economic Conditions in the County of Beaumont-le-Roger, 1261–1313', *Spec.*, 26 (1951), 277–87.

10 P. Gras, 'Le registre paroissial de Givry (1334–1357)', *Bibl. Ec. Chartes*, 100 (1939), 295–308.

11 Guy Fourquin, 'La population de la région parisienne aux environs de 1328', *Moy. Age*, Series 4, 11 (1956), 63–91.

12 A. Molinier, 'La Sénéchaussée de Rouergue en 1341', *Bibl. Ec. Chartes*, 44 (1883), 425–88.

13 Roger Mols, *Introduction à la Démographie Historique des Villes d'Europe* (Louvain, 1956), III, 91.

14 P. Dollinger, 'Le chiffre de population de Paris au XIVe siècle: 210,000 ou 80,000 habitants?', *Rev. Hist.*, 216 (1956), 35–44.

15 Raymond Cazelles, 'La population de Paris avant la Peste Noire', *Acad. Inscr. C.R.* (1966), 539–50; Bronislaw Geremek, 'Paris la plus grande ville de l'occident médiéval?', *Acta. Pol. Hist.*, 18 (1968), 18–37.

16 Édouard Baratier, 'La Démographie Provençale du XIIIe au XVIe Siècle', Démographie et Société, 5, *Ec. Prat. Ht. Et.* (Paris, 1961).

17 'Polyptychum Rotomagensis Diocesis', in *Recueil des Historiens des Gaules et de la France*, (Paris, 1894), XXIII, 228–329.

18 *Documents relatifs au Comté de Champagne et de Brie*, ed. Auguste Longnon (Paris, 1891), Nos. 7304–404.

19 Hektor Ammann, 'Bevölkerung der Westschweiz', *Festschrift F. E. Welti* (Aargau, 1937), pp. 390–447.

20 W. Bickel, *Bevölkerungsgeschichte und Bevölkerungspolitik des Schweiz seit dem Ausgang des Mittlealters* (Zurich, 1947), p. 41.

21 Karlheinz Blaschke, 'Vevölkerungsgang und Wüstungen in Sachsen Während des späten Mittelalters', *Jahrb. Nat. Stat.* 174 (1962), 414–29; see also *ibid.*, *Bevölkerungsgeschichte von Sachsen bis zur Industriellen Revolution* (Weimar, 1967), pp. 67–79.

22 G. W. von Raumer, *Die Neumark Brandenburg in Jahre 1337* (Berlin, 1837).

23 *Das Landregister der Herrschaft von Sorau von 1381*, ed. Johannes Schultze (Berlin, 1936), xi–xiv, xxvii–xxxiii.

24 T. Ladenberger, *Zaludnienie Polskina początku panowania Kazimierza Weilkiego* (Lwów, 1930); J. Mitkowski, 'Uwagi o zaludnieniu Polski na początku panowania Kazimierza Wielkiego', *Rocz. Dziej. Społ. Gosp.*, 10 (1948), 121–30; W. Kula, 'Stan i potrzeby badań nad demografią historyczną dawniej Polski', *ibid.*, 13 (1951), 23–110; Egon Vielrose, 'Ludność Polski od X do XVIII wieku', *Kw. Hist. Kult. Mat.*, 5 (1957), 3–49.

25 Irena Gieysztorowa, 'Research into the Demographic History of Poland: A Provisional Summing-up', *Acta Pol. Hist.*, 18 (1968), 5–17; see also Tadeusz Ladogórski 'Etudes sur le peuplement en Pologne au XIVe siècle', *Kw. Hist. Kult. Mat.*, 12 (1964), 529–34.

26 *Historia Polski*, Tom I, *Mapy*, No. 5, Polska Akademia Nauk (Warsaw, 1960).

27 Peter Charanis, 'A Note on the Population and Cities of the Byzantine Empire

in the Thirteenth Century', *The Joshua Starr Memorial Volume, Jewish Social Studies*, Publication 5 (New York, 1953), pp. 135–48; see also D. Jacoby, 'Phénomènes de démographic rurale a Byzance sur XIIIe, XIVe et XVe siècles', *Et. Rur.*, 5–6 (1962), 161–86.

28 Györffy, 'Einwohnerzahl und Bevölkerungsdichte in Ungarn bis zum Anfang des XIV Jahrhunderts', *Et. Hist.*, 1 (1960), 163–92.

29 Joseph Kovacsics, 'An account of reasearch work in historical demography in Hungary', *Actes Coll. Int. Dém. Hist.*, 249–72.

30 Stefan Pascu, 'Les sources et les recherches démographiques en Roumanie', *Actes Coll. Int. Dém. Hist.*, 283–303.

31 Karl Julius Beloch, *Bevölkerungsgeschichte Italiens*, 3 vols. (Berlin, 1937–9).

32 David Herlihy, 'Population, Plague and Social Change in Rural Pistoia, 1201–1430', *Econ. Hist. Rev.*, 2nd series, 18 (1965), 225–44; *ibid., Medieval and Renaissance Pistoia* (New Haven, 1967), pp. 55–116.

33 *Croniche di Giovanni, Matteo e Filippo Villani* (Trieste, 1857), Book 11, cap. 94.

34 D. Herlihy, *Medieval and Renaissance Pistoia*, p. 113; Beloch suggests (II, 179) 55,000 in the city and (II, 191) 120,000 for the *contado*.

35 Calculated from figures given by K. J. Beloch. *Bevölkerungsgeschichte Italiens* (Berlin, 1961), III, 65–8, 342.

36 Calculated from E. Carpentier's figures, *Une Ville devant la Peste: Orvieto et la Peste Noire de 1348* (Paris, 1962), pp. 30–2.

37 'Descriptio provincial Romandiole facta anno 1371', in Theiner, *Codex Diplomaticus Dominii temporalis Sancti Sedis*, II, 490ff.

38 *Cartes de los antiguos reinos de Aragon*, quoted by J. C. Russell, 'Late Ancient and Medieval Population', *Trans. Am. Phil. Soc.*, n.s., 48 Pt. 3 (1958), 116. J. V. Vives suggests 450,000 for Cataluna and 200,000 for Aragon and Valencia: *An Economic History of Spain*, p. 176.

39 Notably by D. Herlihy, 'Population, Plague and Social Change in Rural Pistoia, 1201–1430', *Econ. Hist. Rev.*, 18 (1965), 225–44.

40 Constantin Jireček, *Die Romanen in den Städten Dalmatiens während des Mittelalters* (Vienna, 1902–4).

41 See Cecil Roth, *The Jews of Medieval Oxford* (Oxford, 1951), p. 83ff.

42 *Grundbuch des Kölner Judenviertels 1135–1425*, ed. Adolf Kober (Bonn, 1920), pp. 9–15.

43 *Nürnberg Chronicle*, quoted in Gerald Strauss, *Nuremberg in the Sixteenth Century* (New York, 1966), p. 118.

44 Gerald Strauss, *op. cit.*, p. 118; Arnd Müller, *Geschichte der Juden in Nürnberg 1146–1945* (Nürnberg, 1968).

45 'Itinéraires des Pélerinages du Chevalier Arnold von Harff (1496–1499)', in Gilles le Bouvier, *Le Livre de la Description des Pays*, ed. E.-T. Hamy, pp. 217–239.

46 J. Starr, *The Jews in the Byzantine Empire*, 641–1204 (Athens, 1939), p. 38; Andréadès put the total at 15,000.

47 Heinrich Bechtel, *Wirtschaftsgeschite Deutschlands* (Munich, 1951), I, 255–6.

48 Perhaps this is true of most south European towns; see Philippe Wolff, 'Villes et Campagnes dan le Midi Français médiéval', *France Méridionale et Pays Ibériques: Mélanges Géographiques offerts ... à Daniel Faucher*, 2 (1949), 677–85.

49 Philippe Wolff, 'Les bouchers de Toulouse de XIIe siècle', *Ann. Midi*, 65 (1953), 375–93.

50 Karl Bücher, *Die Bevölkerung von Frankfurt am Main im XIV and XV Jahrhundert* (Tübingen, 1886), pp. 103–11.

51 Chronicle of Giovanni Villani, Book XI ch. 94 (1336–8), quoted in Robert S. Lopez and Irving W. Raymond, *Medieval Trade in the Mediterranean World* (Oxford, 1955).

52 *Réglements sur les Arts et Métiers de Paris*, ed. G.-B. Depping (Paris, 1837).

53 *Les Livres de Comptes des Frères Bonis*, ed. Edouard Forestié, *Arch. Gasc.*, 20 (1890), xix–cxliii.

54 *Comptes Royaux* (1285–1314), ed. Robert Fawtier, vol. 1, *Comptes Généraux* (Paris, 1953), esp Nos. 184, 188, 192, 4525, 4644, 5191, 5274.

55 Maurice Beresford, *The Lost Villages of England*, London, 1954; Wilhelm Abel, *Die Wüstungen des ausgehenden Mittelalters* (Stuttgart, 1955); Karlheinz

Blaschke, 'Bevölkerungsgang und Wüstungen in Sachsen während des späten Mittelalters', *Jahrb. Nat. Stat.*, 174 (1962), 414–29.

56 Archibald R. Lewis, 'The Closing of the Medieval Frontier', *Spec.*, *33* (1958), 475–83.

57 Edouard Perroy, 'Les Crises du XIV siècle', *Ann. ESC*, 4 (1949), 167–82.

58 Wilhelm Segin, 'Von der Kleinsiedlung zur Grossiedlung im oberen Almegebiet', *Festgabe für Alois Fuchs*, ed. W. Tack (Paderborn, 1950), pp. 437–62. Hans Mortensen, 'Die mittelalterliche deutsche Kulturlandschaft und ihr Verhältnis zur Gegenwart', *Vj. S. Wg.*, 45 (1958), 17–36, postulated a migration from small settlements to larger.

59 Friedrich Stuhr, 'Die Bevölkerung Mecklenburgs am Ausgang des Mittelalters', *Jahrb. Ver. meckl. Gesch.*, 58 (1893), 232–78.

60 *The Chronicle of Jean de Venette*, trans. Jean Birdsall (New York, 1953), sub. 1358, p. 85; also sub. 1360, p. 99.

61 *Annales Gandenses*, ed. Hilda Johnstone (London, 1951), pp. 5, 55.

62 T. A. M. Bishop, 'The Rotation of Crops at Westerham, 1297–1350', *Econ. Hist. Rev.*, 9 (1938–9), 38–44.

63 Quoted by Charles Higounet, 'L'assolement triennal dans la plaine de France au XIIIe siècle', *Acad. Inscr. C.R.* (1956), 507–12: 'Sciendum est quod totum territorium Vallis Laurencii dividitur in tres aristas. Prima arista segetis continet XVIII.XX et V arpennos et dimidium et VI perches. Secunda arista que est in iasqueaa continet XVI.XX et .III arpennos et IX perches. Tercia arista que est en marcesche continet .XVI.XX et XIII arpennos et X perches.'

64 Charles Higounet, 'La Grange de Vaulerent', *Ec. Prat. Ht. Et.*, VIe Section, 10 (Paris, 1965), 42–3.

65 Odile Martin-Lorber, 'L'exploitation d'un grange cistercienne à la fin du XIVe et au début du XVe siècle', *Ann. Bourg.*, 29 (1957), 161–80.

66 Leo Verriest, 'Polyptyque du Chapitre de Sainte-Waudru de Mons, 1278–1279', *Analectes pour servir à l'Histoire Ecclesiastique de la Belgique*, 38 (1912), 48–64, 146–80, 245–27, 333–64; 39 (1913), 5–20, 133–48.

67 A. van Zuylen van Nyevelt, 'Les grandes fermes en Flandre vers 1300', *Ann. de la Soc. d'Emulation de Bruges*, 74 (1931), 40–52.

68 *Le Livre de l'Abbé Guillaume de Ryckel* (*1249–1272*): *Polyptyque et Comptes de l'Abbaye de Saint-Trond au milieu du XIIIe siècle*, ed. Henri Pirenne (Gent, 1896), p. 232.

69 *Le Polyptyque de l'Abbaye de Villiers* (*1272*), p. 146n.

70 Leon Lahaye, 'Fragments d'un polytyque de la collègiable Saint Jean Evangéliste à Liège, de l'an 1250', *Bull. Comm. Roy. Hist.*, 107 (1942), 199–292.

71 David Herlihy, *Medieval and Renaissance Pistoia* (New Haven, Conn., 1967), pp. 143–4.

72 Etienne Juillard, 'L'assolement biennal dans l'agriculture septentrionale: le cas particulier de la Basse-Alsace', *Ann. Géog.*, 61 (1952), 34–45.

73 *Walter of Henley's Husbandry*, trans. Elizabeth Lamond (London, 1890); see also Eileen Power, 'On the Need for a New Edition of Walter of Henley', *Trans. Roy. Hist. Soc.*, 4th series, 17 (1934), 107–16.

74 'which are given to the lord abbot to feed his horses and entertain casual visitors.' *Die Urbare von S. Pantaleon in Köln*, ed. Benno Helliger, Rheinische Urbare, I, 20 (Bonn, 1902), B, cap. 88 (p. 256); the date was 1322–4.

75 J. A. Mertens and A. E. Verhulst, 'Yield-Ratios in Flanders in the Fourteenth Century', *Econ. Hist. Rev.*, 19 (1966), 175–82.

76 J.-M. Richard, 'Thierry d'Hireçon: agriculteur artesien', *Bibl. Ec. Chartes*, 53 (1892), 399.

77 W. H. Beveridge, 'The Yield and Price of Corn in the Middle Ages', *Econ. Hist.*, 1 (1926–9), 155–67; see also M. Wretts-Smith, 'Organization of Farming at Croyland Abbey', *Jnl. Bus. Econ. Hist.*, 4 (1931–2), 168–92.

78 B. H. Slicher van Bath, 'De oogstopbrengsten van verschillende geswassen, voornamelijk granen, in verhouding tot het zaaizaad, ca 810–1820', *A. A. G. Bijd.*, 9 (1963), 29–125; *ibid.*, 'Yield Ratios, 810–1820', *A. A. G. Bijd.*, 10 (1963) (whole issue).

79 *The Register of Eudes of Rouen*, trans. Sydney M. Brown (New York, 1964).

80 For a review and critique of the evidence see R. Lennard, 'The alleged Exhaustion of the Soil in Medieval England'. *Econ. Jnl.*, 32 (1922), 12–27.

81 David Herlihy, *Medieval and Renaissance Pistoia*, p. 144.
82 N. G. Svoronos, 'Sur quelques formes de la vie rurale à Byzance: Petite et grande exploitation', *Ann. ESC*, 11 (1956), 325–35.
83 *Le Compte Général de 1187, connu sous le nom de 'Gros Brief' et les institutions financières du comté de Flandre au XIIe siècle*, ed. A. Verhulst and M. Gysseling (Brussels, 1962). See discussion of this document in Bruce Lyon and A. E. Verhulst, *Medieval Finance: A Comparison of Financial Institutions in Northwestern Europe* (Providence, R. I., 1967), pp. 12ff.
84 *Urzář zboží rožmberského z roku 1379*, ed. Josef Truhlář Abhandlungen der Königl. Böhmischen Gesellsch. d. Wissenschaften vom Jahre 1879 und 1880 (Prague, 1880).
85 'Censier de St Emmeram de Ratisbonne', in P. Dollinger, *L'Evolution des Classes Rurales en Baviere* (Paris, 1949), pp. 504–12.
86 'Monumenta San-Vlricana – Codex Traditionum', *Monumenta Boica*, 22 (1814), 131–60.
87 *Cartulaire de l'Abbaye de Redon*, ed. Aurélien de Courson, Colln. de Doc. Inéd. (Paris, 1863), No. 262.
88 *De expugnatione Lyxabonensi*, ed. Charles W. David (New York, 1936), p. 179.
89 Walter Kuhn, 'Die Beseidlung des Reichthaler Haltes in Niederschlesien', *Festschrift Hermann Aubin zum 80. Geburtstag* (Wiesbaden), I, 95–125.
90. Th. Sclafert, 'A propos du déboisement des Alpes du sud: le rôle des troupeaux', *Ann. Géog.*, 43 (1934), 126–45.
91 Ph. Dollinger, 'Les transformations du régime dominal en Bavière au XIIIe siècle d'après deux censiers de l'abbaye de Baumburg', *Moy. Age*, 56 (1950), 279–306.
92 *Documents relatifs au Comté de Champagne et de Brie 1172–1361*, ed. Auguste Longnon (Paris, 1904), pp. 185–97.
93 *L'Administration et les Finances du Comté de Namur du XIIIe au XVe Siècle*, vol. 1, *Cens et Rentes du Comté de Namur au XIIIe Siècle*, ed. D. D. Brouwers (Namur, 1910).
94 *Chartes de l'Abbaye de Saint-Hubert en Ardenne*, ed. G. Kurth, No. 409, pp. 556–7.
95 *Die Landesfürstlichen Urbare Nieder– und Oberösterreichs aus dem 13. und 14. Jahrhundert*, ed. A. Dopsch, CLXXVIII.
96 E. Coornaert, *La Draperie-Sayetterie d'Hondschoote (XIVe–XVIIIe siècles)* (Rennes, 1930), p. 4.
97 *Documents relatifs à l'Histoire de l'Industrie et du Commerce en France*, ed. Gustave Fagniez (1900), II, 173–9.
98 Giovanni Villani, Book XI. ch. 94, as quoted in Robert S. Lopez and Irving W. Raymond, *Medieval Trade in the Mediterranean World* (Oxford, 1955), p. 72.
99 'Les Livres de Comptes des Frères Bonis', ed. Edouard Forestié, *Arch. Gasc.*, 20 (1890), 52–4.
100 Felix Bourquelot, *Etudes sur les Foires de la Champagne aux XIIe, XIIIe et XIVe siècles* (Paris, 1865).
101 A. Varron, 'Silks of Lyons', *Ciba Rev.*, 6 (1938).
102 'Trade Routes and Dye Markets in the Middle Ages', *Ciba Rev.*, 10 (1938).
103 These place-names indicate respectively 'lead', 'silver' and 'mine'.
104 Vannochio Biringuccio, *De la Pirotechnia* (1540), edn. of C. S. Smith and M. T. Gnudi (New York, 1943), p. 61.
105 Gino Luzzatto, *An Economic History of Italy* (London, 1961), pp. 107–8.
106 *Paris sous Philippe-le-Bel d'après des documents originaux*, ed. H. Gévand (Paris, 1937), p. 494.
107 Gustave Guilmoto, *Étude sur les Droits de Navigation de la Seine de Paris à Roche-Guyon du XIe au XVIIIe siècle* (Paris, 1889), pp. 61, 94.
108 *Medieval Trade in the Mediterranean World*, ed. Robert S. Lopez and Irving W. Raymond (Oxford, 1955), p. 116.
109 Philippe Dollinger, 'Commerce et Marchands strasbourgeois à Fribourg en Suisse au Moyen Age', *Beiträge zur Wirtschafts– und Stadtgeschichte: Festschrift Hektor Ammann* (Wiesbaden, 1965), pp. 124–43.
110 R. de Roover, 'The Organization of Trade', *Camb. Econ. Hist.*, III, 42–118 (pp 75–6).

111 Ferdinand Schevill, *History of Florence* (New York, 1936), pp. 301ff.; A. Sapori, *Le Marchand Italien au Moyen Age*, (Paris, 1952).
112 R. de Roover, 'The Organization of Trade', p. 86, n. 517.
113 M. M. Postan, 'The Trade of Medieval Europe: the North', *Camb. Econ. Hist.*, II (1952), esp. 223–32.
114 From Albertus Krantzius, *History of Norway*, in Richard Hakluyt, *The Principal Navigations, Voyages, Traffiques and Discoveries of the English Nation* (Glasgow, 1903), II, 68.
115 The phrase is that of Yves Renouard, 'Routes, étapes et vitesses de Marche de France à Rome au XIIIe et au XIVe siècles d'après les itinéraires d'Eudes Rigaud (1254) et de Barthélémy Bonis (1350)', *Studi in Onore di Arnintore Fanfani* (Milan, 1962), II, 403–28.
116 'Wherever it might be necessary.' *Mon. Germ. Hist., Capitularia*, I, 301, No. 148.
117 *Recueil des Ordinnances de Pays-Bas*, ed. Charles Laurent, 2e série, I, 444.
118 Francesco Balducci Pegolotti, *La Practica della Mercatura*, ed. Allen Evans (Cambridge, Mass., 1936).
119 'Itinerarium de Brugis', published as Appendix 4 Gilles le Bouvier, *Le Livre de* la *Description des Pays*, ed, E. -T. Hamy (Paris, 1908), pp. 157–216. It was first published by J. Lelewel, *Géographie du Moyen Age, Epilogue* (Brussels, 1857).
120 Johannes R. Frischart, *Das Glückhafte Schiff von Zurich*, 1577, lines 449–52.
121 Charles Higounet, 'Un mémoire sur les péages de la Garonne au début du XIVe siècle', *Ann. Midi*, 61 (1948–9), 320–4.
122 Gino Luzzatto, *op. cit.*, p. 87.
123 G. I. Bratianu, *Recherches sur le Commerce Génois dans la Mer Noire au XIIIe siècle* (Paris, 1929), pp. 205–49.
124 *Libel of English Policie*, in Richard Hakluyt, *The Principal Navigations, Voyages, Traffiques and Discoveries of the English Nation* (Glasgow, 1903), II, 117.
125 'In with salt and out with rye', W. Stieda, *Revaler Zollbeicher*, No. 1381.
126 *Libel of English Policie*, pp. 115–6.
127 J. A. van Houtte, 'The Rise and Decline of the Market of Bruges', *Econ. Hist. Rev.*, 19 (1966), 29–47.
128 N. S. B. Gras, *The Evolution of the English Corn Market from the Twelfth to the Eighteenth Century* (Cambridge, Mass., 1915), p. 17.
129 Margery K. James, 'The Fluctuations of the Anglo-Gascon Wine Trade during the Fourteenth Century', *Econ. Hist. Rev.*, 4 (1951), 170–96; *ibid.*, 'Les activités commerciales des négociants en vins gascons en Angleterre durant la fin du moyen âge', *Ann. Midi*, 65 (1953), 35–48.
130 *Libel of English Policie*, p. 124.
131 For the situation in England, see Edward Miller, 'The English Economy in the Thirteenth Century: Implications of Recent Research', *P. & P.*, No. 28 (1964), 21–40.
132 J. Garnier, *La Recherche des Feux en Bougogne aux XIVe et XVe Siècles* (Dijon, 1876); Yves Renouard, 'La peste noire de 1348–1350', *Etudes d'Histoire Médiévale* (Paris, 1968), pp. 143–55; *ibid.*, 'Conséquences et intérêt démographiques de la peste noire de 1348', *Etudes d'Histoire Médiévale*, pp. 158–64.

Index